Physical Aspects of Polymer Self-Assembly

Physical Aspects of Polymer Self-Assembly

P. R. Sundararajan

WILEY

Published by John Wiley & Sons, Inc., Hoboken, New Jersey
Published simultaneously in Canada

For general information on our other products and services or for technical support, please contact our Customer Care Department within the United States at (800) 762-2974, outside the United States at (317) 572-3993 or fax (317) 572-4002.

Wiley also publishes its books in a variety of electronic formats. Some content that appears in print may not be available in electronic formats. For more information about Wiley products, visit our web site at www.wiley.com.

Library of Congress Cataloging-in-Publication data has been applied for:

Set in 10/12pt Warnock by SPi Global, Pondicherry, India

10 9 8 7 6 5 4 3 2 1

To my wife Kumuda

Contents

Preface

Studies on molecular self-assembly has gained enormous momentum in the past two decades, driven by concurrent interest and developments in the areas of nanotechnology, molecular switches, shape memory devices, etc. Close to 130 000 publications have so far appeared in the literature on the topic of self-assembly and over 200 000 on the concept of self-assembly. Various disciplines of chemistry, biology, physics, and engineering have contributed to its emergence as a key research area. Novel synthetic methods, advances in the ability to manipulate molecules even at the atomic level, and device fabrication have been instrumental in its advancement. Amongst these, polymers have played a significant role and the self-assembly studies have explored various types of polymers, from everyday commodity plastics to high-performance polymers.

Several books and reviews have been published on the self-assembly of polymeric materials. Most of them have focussed on select themes such as supramolecular polymers. In this book, I have given an overview of various physical aspects of polymer self-assembly. Leaving the synthetic methods to the experts in that field, physical chemistry such as the morphology, etc., as it relates to polymer self-assembly is highlighted. This is an elaboration of a half-term graduate/fourth-year undergraduate course, "Macromolecular Nanotechnology," that I have been teaching for the past few years on the topic of polymer self-assembly as it relates to nanotechnology.

A brief introduction is given in Chapter 1 on the basic properties of polymers that distinguish them from small molecules as well as the basic definition of self-assembly. A look back at the literature on the emergence of the field of self-assembly, in terms of the number of publications from 1960 to 2015 is included.

As we emphasize noncovalent interactions as the trigger for self-assembly, I have given on overview of the force fields ("potential functions" as they used to be called) that were developed over the past decades to calculate the energies of interactions for van der Waals, hydrogen bonds, torsional contributions, and Coulombic interactions in Chapter 2. I myself have published several papers using these formulations, and was associated with the groups of G. N. Ramachandran (who devised the (Φ, Ψ) map) and Paul Flory during the decade that saw the development of these force fields. In addition to those, the halogen interactions and C—H···π interactions also have been discussed, as these have been investigated in detail in recent years through crystallographic and nuclear magnetic resonance (NMR) studies.

Chapter 3 deals with the features that have been observed during the self-assembly of molecules and these include self-sorting; polymer-guest molecule assembly; the moieties that are used to functionalize polymers to induce self-assembly; sergeant–soldier phenomenon; chain folding; foldamers; and single-polymer, single-crystals, and single-polymer nanoparticles. Using these concepts, supramolecular polymers are discussed in Chapter 4. Since the International Union of Pure and Applied Chemistry (IUPAC) definition distinguishes between the terms polymers and macromolecules, examples of supramolecular assemblies that lead to polymers as well as macromolecules are presented in the chapter.

Block copolymer morphology is discussed in Chapter 5. This field is enormous with publications dating back to several decades, and several books and reviews have been published. I have tried to give a summary of the di- and triblock copolymer morphologies as well as those with organic/inorganic blocks. The gyroid morphology, grapho-epitaxy, and applications of block copolymer morphologies to nanoporous materials, metamaterials, and nano-reactors are discussed. The supramolecular assembly between linear polymers and cyclics that leads to rotaxanes is described in Chapter 6. A brief description of how these could be used for molecular switches, machines, and muscles is given. Although we consider liquid crystal as the fourth state of matter, one cannot ignore the gel state which has several applications. For the organo- or polymer gels, this state is a result of hydrophobic or solvo-phobic interactions. The gel morphologies of various synthetic polymers are discussed in Chapter 7.

Chapter 8 treats the self-assembly of small molecules in polymer matrices. Polymers such as polycarbonate are used as matrices to hold functional small molecules that are applicable to opto-electronic devices. Even the plasticizer technology uses small molecules in polymers. Hence, the interaction between the small molecules in a polymer matrix, whether they self-assemble or not, becomes important for device applications. In the chapter, I discuss the morphology of such composites when the small molecule possesses self-assembly attributes, such as spontaneous association driven by hydrogen bonds or π-interactions. The utilization of this attribute in techniques such as polymerization-induced phase separation and reaction-induced phase separation is illustrated. This then carries on to the application of these techniques for the fabrication of liquid crystal displays. Studies on polymer–surfactant interactions could not be ignored.

Although I became an academic late in my career, I benefited immensely from the contributions of my students and postdoctoral associates, and I thank them sincerely for their dedication. I thank Carleton University for providing me with a sabbatical break during which I completed parts of the book. On a personal note, I appreciate the support and encouragement from my wife Kumuda throughout our lives together. I am indebted to my brother Chellappa for supporting me during my university years. I thank my daughter Purnima, another nonmedical doctor in my family, for encouraging me to accept the invitation from Wiley to write the book.

P. R. Sundararajan
Carleton University
Ottawa, Canada
June 2016

1

Introduction

Polymers are ubiquitous in our daily lives. Natural polymers such as the DNA, RNA, and proteins have played a major role in the evolution of life itself. Cellulose, hemicelluloses, starch (amylose), and other naturally occurring polymers have been studied thoroughly, modified for useful applications and have been the key components of industrial advancement. Familiar to many is the chronological transition of the belief of these substances as colloids to the concept of "polymers." The "Rise of the Macromolecular Hypothesis" was discussed by Flory [1]. Supporting the Staudinger school of thought were, among others, the X-ray diffraction studies of cellulose by Meyer and Mark [2, 3]. Cellulose is perhaps the first polymer for which oriented X-ray fiber diffraction was recorded [2]. Synthetic polymers have also entered the scene around that time, with the synthesis of polystyrene [4] in 1839.

According to the International Union of Pure and Applied Chemistry (IUPAC) nomenclature [5], the terms polymer and macromolecule do not mean the same thing. A polymer is a substance composed of macromolecules. All macromolecules are not polymers.

1.1 Polymer Tacticity

As a polymer (Figure 1.1) is built up of many small molecular monomer units, the dimer (diad), trimer (triad), tetramer (tetrad), pentamer (pentad), and hexamer refer to two, three, four, five, and six monomers, respectively, linked together. Longer sequences of up to about 50-mers are called "oligomers." However, it is not uncommon in the studies on self-assembly to call such short chains polymers. Although natural or biopolymers such as poly(nucleic acids) (e.g., DNA), proteins and polysaccharides are also long chain molecules, the commonplace notion is that a polymer refers to the synthetic variety. Of these, vinyl polymers such as polyethylene (PE) and polystyrene are popularly known as plastics. The simplest of polymers is PE, with just a sequence of (CH_2) units (Figure 1.2). Polymerization of ethylene, vinyl fluoride, vinylidene

Physical Aspects of Polymer Self-Assembly, First Edition. P. R. Sundararajan.

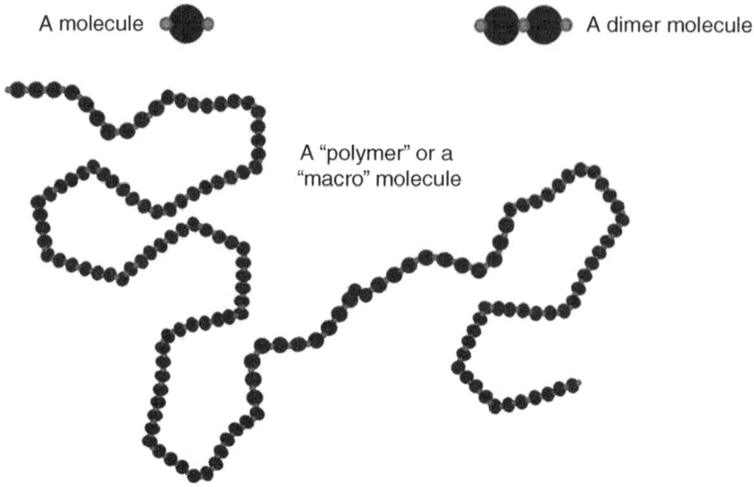

Figure 1.1 Schematics of a monomer, dimer, and a polymer.

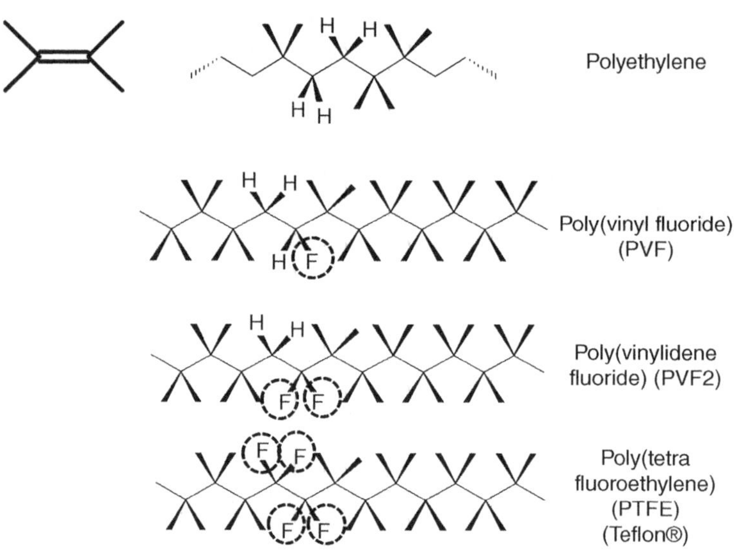

Figure 1.2 Chemical structures of polyethylene and analogous fluorinated polymers. The dashed circles highlight the fluorine atoms.

fluoride, and tetrafluoroethylene lead to PE, poly(vinyl fluoride) (PVF), poly(vinylidene fluoride) (PVF2), and poly(tetrafluoroethylene), respectively. The latter is the well-known Teflon®.

With bulkier substituents, polymers such as polystyrene, poly(methyl methacrylate), and poly(propylene) are obtained (Figure 1.3). Note that in the

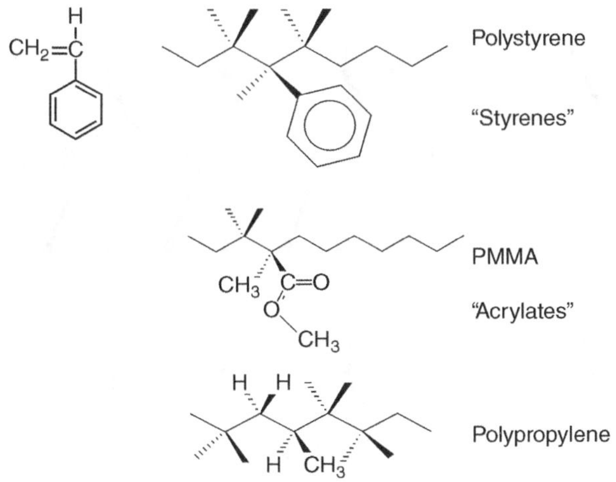

Figure 1.3 Chemical structures of polystyrene, poly(methyl methacrylate), and polypropylene.

case of PE, PVF2, and PTFE, the substitution is symmetric, whereas with PVF, polystyrene, PMMA, and polypropylene, it is asymmetric. In the schematics shown in Figures 1.2 and 1.3, the bulkier substituent (R) is shown above the plane of the page. A general rendition is shown in Figure 1.4. In the figure [6], with the skeletal bonds in the all-*trans* conformation, the configuration with R above the page at the asymmetric carbon is designated as *d* and as *l* if it is below. If the substituents on two successive asymmetric carbons are in the same *dd* configuration, it is known as a "*meso*" (*m*) diad.

Perpetuation of such *meso* sequence would lead to an isotactic chain (Figure 1.4a). If the R groups are up and down the page, that is, *dl*, it is the racemic (*r*) diad (Figure 1.4b). The repetition of the racemic diad (*dl* sequence) would result in a syndiotactic polymer. Random occurrence of *m* and *r* results in an atactic polymer. The designation of *d* and *l* for the configuration at the asymmetric carbon is arbitrary. It follows the convention developed by Flory [7], which then led to the formulation of statistical weight matrices to calculate the statistical chain conformations. If the chain is rotated through 180° about a vertical axis such that the ends of the chain are reversed, all the *d* would become *l* and vice versa. But the chain configuration would remain the same as long as the chain ends are indistinguishable. The *ll* would define the meso diad. With the diad configurations *m* and *r* defined earlier, a triad could have sequences of *mm, rr,* and *mr* (or *rm*). A tetrad could have sequences *mmm, mmr, rmr, mrm, rrm,* and *rrr,* of which only two of them would lead to stereoregular isotactic or syndiotactic chains. A pentad could have 10 such sequences, 8 of which will be heterotactic. Determining the distribution of such sequences in vinyl polymer chains that were prepared by various synthetic procedures was an active field, concurrent with advances in the NMR

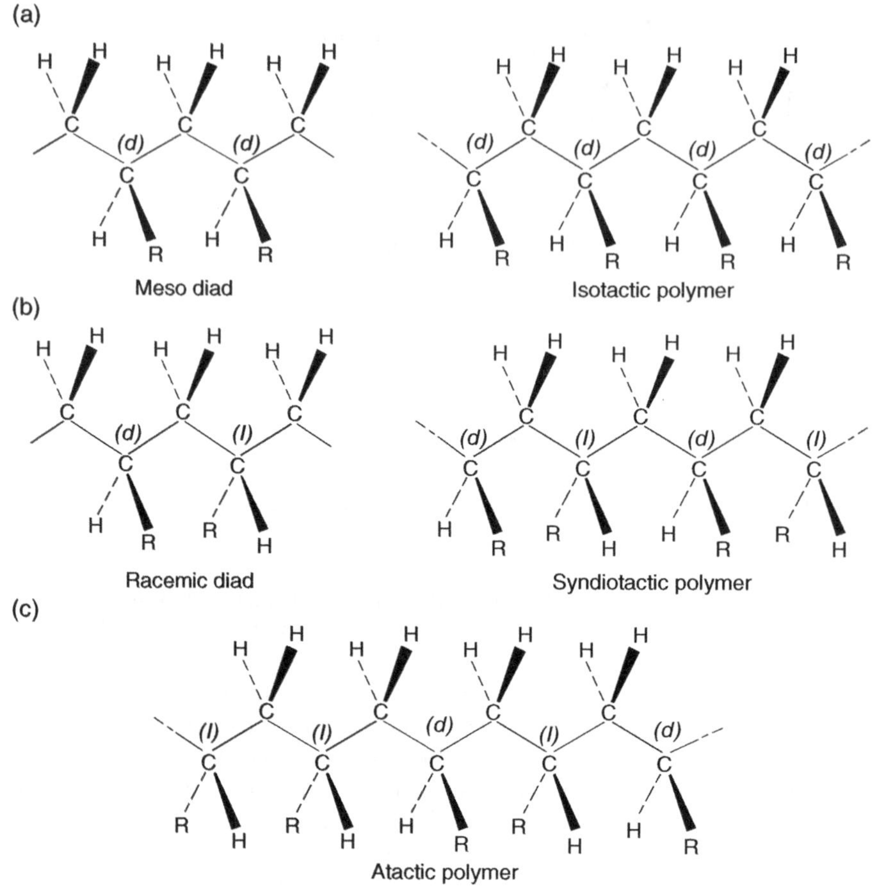

Figure 1.4 (a)–(c) Schematic of the definition of tacticity for asymmetric chains. (Source: Sundararajan [6]. Reproduced with permission of Springer.)

techniques [8–11]. Although synthetic methods for stereoregular polymers have been developed, some of them such as isotactic polystyrene are of academic interest rather than of commercial use. Atactic polystyrene, in the form of, for example, Styrofoam, finds widespread applications. Likewise, PMMA in its atactic form, is used as a substitution for glass in the form of Plexiglass, as well as in microelectronic chips, etc. Studies on highly isotactic or syndiotactic polymers led to the understanding of aspects such as polymer crystallization and chain folding. In the studies of self-assembly of polymers, tacticity was seldom taken into consideration.

Some of the other polymers commonly used in studies of polymer self-assembly are polyesters, polyamides, polyurethanes, and polysiloxanes, the constituent units of which are shown in Figure 1.5. A summary of the synthetic procedures

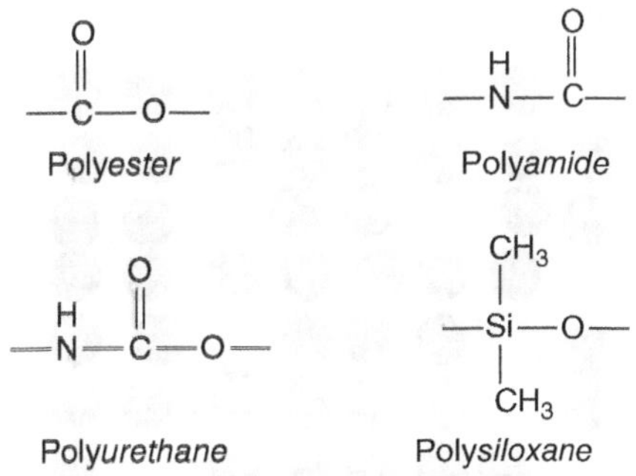

Figure 1.5 Schematics of the segments of polyester, polyamide, polyurethane, and polysiloxane.

adopted for various types of polymers and their primary characterization properties was given by Sundararajan [6].

1.2 Big versus Small

As the small molecule "mono" mer units are joined together to build a "poly" mer, the properties change as the polymer increases in length. For example, ethane has boiling temperature of −89° and melts at −183°C. After growing by a few units, hexadecane melts at 18.5°C. With a further increase in chain length, triacontane [CH_3—$(CH_2)_{28}$—CH_3] melts at 65°C, and does not boil. High-molecular-weight PE melts at 138°C. While the small molecules could melt and vaporize, polymer molecules melt and degrade rather than vaporizing. Most polymers show a glass–rubber transition (T_g), whereas not all small molecules show a glass transition. Both could crystallize, but polymers should have a regular sequence of monomers (e.g., isotactic or syndiotactic) to be able to crystallize. Those with random sequence remain amorphous. The T_g is an important property of a polymer since it dictates the processing conditions for industrial applications. Small molecules such as D-glucose have a finite number of conformations, for example, the C1 chair form. However, amylose, which is a high-molecular-weight polymer of D-glucose, could have a large number of conformational sequences between contiguous D-glucose units due to rotations about the interunit C_1—O—C_4 single bonds. The number of accessible conformations without steric overlap determines the "flexibility" of a polymer.

1.3 Entanglement

When a small molecule is dissolved in a solvent medium (lattice model) as shown pictorially in Figure 1.6a, both molecules can move and exchange lattice points just by diffusion and thermal motion. However, when a polymer is placed in a solvent (Figure 1.6b), coordinated movement of the monomer units

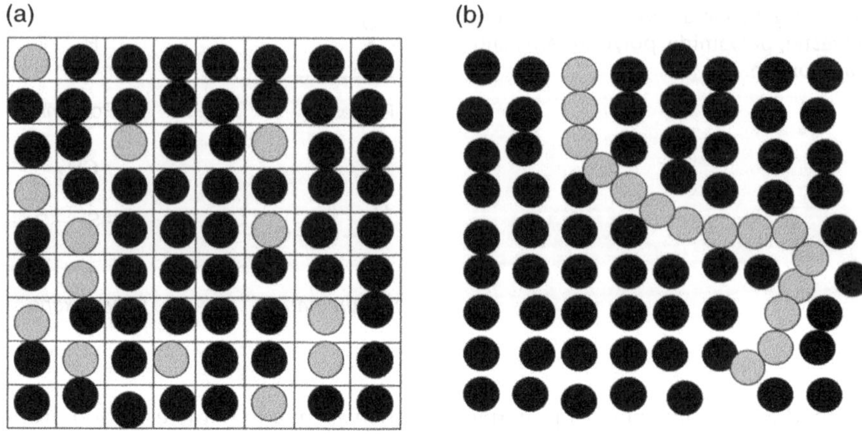

Figure 1.6 Schematic of (a) a small molecule and (b) a polymer in a solvent medium. The black circles denote the solvent molecules.

is necessary for the chain to diffuse. This results in loss of entropy of mixing. This would increase the viscosity of the solution compared to the small molecule/solvent mixture. One of the most important properties of a polymer is the "entanglement" between its segments. A small molecule cannot entangle. As the polymer chain grows in length, due to coiling in solution or the melt, entanglement would set in. Most text books would mention a bowl of spaghetti as an example of entangled polymer chains. The polymer must be of a certain minimum length for entanglement to occur. For a bowl of cooked spaghetti to be an entangled stock, the length of the dry noodle should be about one foot. The average length of commercially sold dry spaghetti is 12–14 in., which assures the entanglement upon cooking. The critical molecular weight M_c for entanglement for polystyrene is about 37 000, while it is only about 5000 g/mol for polycarbonate [12]. Conformational analysis showed that polycarbonate chain can adopt flat helical as well as extended chain shapes with equal probability [13]. Thus, a low molecular weight is sufficient for entanglements to occur in the case of polycarbonate. M_c varies depending on the polymer structure and conformation. Another parameter is the entanglement molecular weight, M_e, which corresponds to the average molecular weight between entanglement points. The value of M_e is usually calculated from the plateau modulus G_N^0 using oscillatory shear experiments. The entanglement plays a significant role in the mechanical properties. Figure 1.7 shows the log–log plot of melt viscosity and molecular weight. It is seen that up to a certain molecular weight, the curve is linear, the region in which the viscosity scales with the first power of molecular weight. Beyond a certain molecular weight, the viscosity increases more sharply, with a slope of 3.4. The point at which the slope

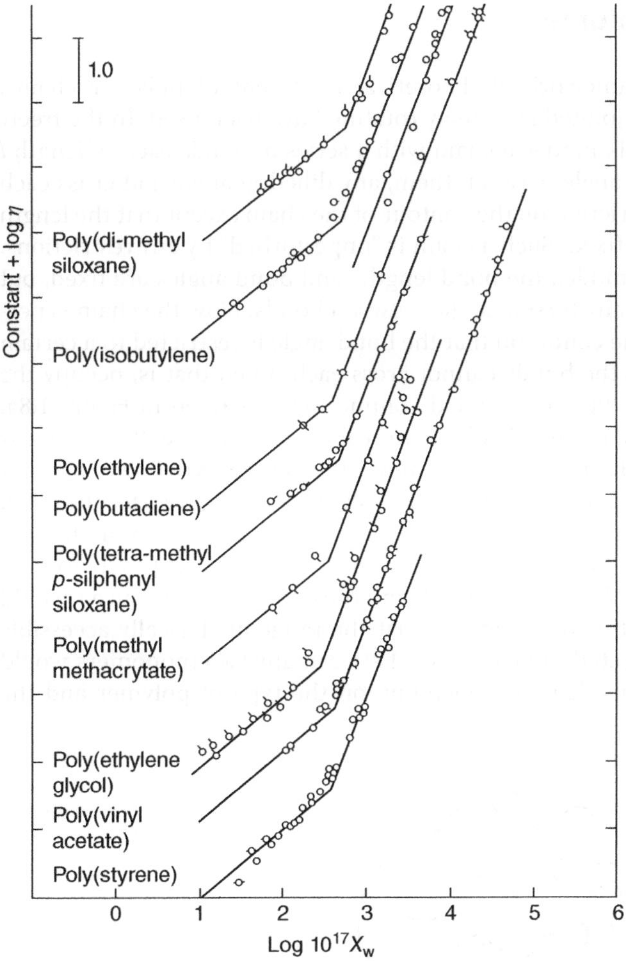

Figure 1.7 Log–log plot of zero shear viscosities and molecular weight for a number of polymers. For clarity, the curves have been arbitrarily shifted along the ordinate. (Source: Graessley [14]. Reproduced with permission of Springer.)

changes from 1 to 3.4 is the one corresponding to M_c. The following relationship generally applies to most linear polymers:

$$\eta = KM \quad M < M_c \tag{1.1}$$

$$\eta = K'M^{3.4} \quad M > M_c \tag{1.2}$$

Here, K and K' are constants.

1.4 Excluded Volume

In the conceptual framework of theoretical treatment of polymer chains, models such as freely jointed and freely rotating have been used. In the freely jointed or random walk model, a chain with a series of bonds each of length l can adopt any (bond) angle between them, any dihedral angle and cross each other. There is no restriction on the contour of the chain except that the length of each step (bond) is fixed. Such a chain is "unperturbed" by any restrictions. In the freely rotating model, the bond lengths and bond angles are fixed, but the chain can assume any torsion angle between bonds. Now the chain is perturbed by imposing the condition that the bond angle is restricted to a certain value. In a real chain, the bonds cannot cross each other, that is, occupy the same space, which results in "excluded volume," as illustrated in Figure 1.8a. We find the concept of excluded volume in real life everywhere. If someone is sitting on a park bench or in a seat in a movie theater, that volume of space is "excluded" for others. In a hypothetical chain shown as in Figure 1.8a, the chain segments (bonds) are allowed to cross each other (indicated by *), as in the case of freely jointed model. Since the excluded volume would prevent such overlap, the chain would expand as in Figure 1.8b and the end-to-end distance r would increase. Another factor to consider is that the range of sterically accessible dihedral angles, illustrated in Figure 1.8c, between any two monomers would be restricted to various degrees depending on the type of polymer and the

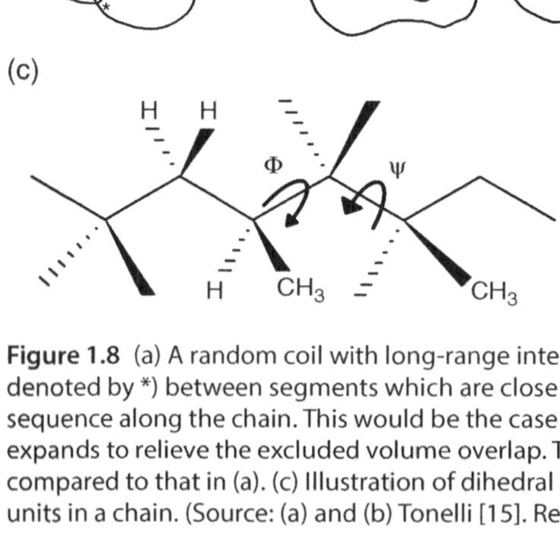

Figure 1.8 (a) A random coil with long-range interaction (excluded volume overlap, denoted by *) between segments which are close in space but with large separation in the sequence along the chain. This would be the case in a freely jointed chain. (b) The chain expands to relieve the excluded volume overlap. The end-to-end distance r increases compared to that in (a). (c) Illustration of dihedral angles Φ and Ψ between contiguous units in a chain. (Source: (a) and (b) Tonelli [15]. Reproduced with permission of Wiley.)

interactions between contiguous monomers. These are short-range interactions which occur between contiguous neighbors, and are determined by the bond angles, torsion angles, and nonbonded forces such as van der Waals, hydrogen bonding, etc. Thus, the random coil of a real chain would be more expanded than the hypothetical free jointed or freely rotating models.

The role of excluded volume in the solution conformations of polymers has been discussed in detail by Flory [1]. As mentioned earlier, the long-range interaction gives rise to the excluded volume effect and causes chain expansion. The "long range" here does not mean in terms of the physical distance between the segments but the relative positions of the segments in the sequence along the chain, for example, the 5th and the 100th monomer unit, when they come close to each other. The average conformation of a randomly coiled chain in solution is governed by the short- and long-range interactions as well the polymer–solvent interaction. In a good solvent, the polymer–solvent interaction would be stronger than the mean of the energies between the polymer–polymer and solvent–solvent pairs. The chain would tend to expand to maximize the contacts between the polymer segments and the solvent molecules. In a poor solvent, the polymer would tend to adopt compact conformations to increase the number of contacts between the intrapolymer segments (subject to excluded volume effects) and to minimize the polymer–solvent interaction. In a particular solvent, at a certain temperature, the expansion due to excluded volume may be compensated exactly by the contraction due to the poor solvent. At this temperature, a balance occurs between the excluded volume (mutual volume exclusion of the segments), which expands the chain and the positive energy of mixing in the solvent which promotes first-neighbor contacts between the polymer segments resulting in a compact conformation. This temperature is known as the Flory Θ temperature. The chain would be unperturbed by long-range interactions and its conformation would be governed by short-range interactions between contiguous neighbors. By taking the rotations Φ and Ψ marked in Figure 1.8c to be discrete conformational states such as *trans, gauche,* and *gauche minus* (*t-, g-, and g-rotational isomers,* respectively), Flory [7] introduced the rotational isomeric-state treatment of the statistical chain conformations in the unperturbed state (i.e., Θ condition). The Θ temperature depends on the solvent–polymer pair. Measurements and calculations of the Θ temperatures and chain dimensions for various polymers were a significant activity during the 1960s, extending to even the 1990s. The characteristic ratio of the unperturbed end-to-end distance of a chain with infinite molecular weight, defined as

$$C_\infty = \frac{\langle r^2 \rangle_0}{nl^2} \tag{1.3}$$

was related to the experimental measurements at the Θ temperature [7]. It was also during this period that experimental (neutron scattering, deuterated polymers, etc.) and theoretical work confirmed that the chain conformation in the amorphous, glassy state corresponded to that in the Θ condition. Properties such as the entanglement molecular weight (M_c), distance between entanglements (M_e), and viscoelastic properties have been correlated to the characteristic ratio [16–20]. A collection of the experimental results on the theta temperatures for various solvent/polymer pairs and the chain conformations was published [21] in 2007.

1.5 Free Volume

The excluded volume concept discussed earlier relates to the chain conformation and the physical properties associated with it. Another important factor is the "free volume," which is related to the glass transition temperature of a polymer. This is similar to the void space between spheres that are packed in a container. As the polymer coils pack themselves in space, intermolecular steric repulsions as well as thermal motions create free volume in a polymer matrix. The free volume would then depend on the temperature as well as the ease with which the chain can change its shape (conformation). For example, the rotational states of contiguous units of a poly(dimethyl siloxane) (PDMS) chain encounter very little energy barrier and the chain is hence highly flexible. As the temperature is lowered, the chain motion would cease at a particular temperature, which is the T_g. Due to its high conformational flexibility, the molecular notion in the case of PDMS does not freeze until a very low temperature of −125°C. For atactic polystyrene, the T_g is about 100°C; for poly(arylene sulfone), it is 250°C. Thus, the free volume above T_g depends on the conformational flexibility, and interchain interactions. Main chain aromatic polymers in which π–π interaction is possible and those with interchain hydrogen bonds (e.g., polyamides) exhibit high glass transition temperatures. Other features that influence the T_g are long side chains, tacticity, etc. The molecular weight distribution would also affect the free volume. If the bulk sample has a large number of short chains in its midst, the number of chain ends would be significant. Since there must be a certain steric space between these chain ends, the free volume would increase, and the T_g would decrease. The low-molecular-weight fractions would act as internal plasticizers. The same would occur if the polymer is exposed to a solvent vapor or if a small-molecule plasticizer is added. These would have a lubricating effect enabling chain movement and increase the free volume. The effect of free volume on the self-assembly of small molecules in polymer matrices is discussed in Chapter 8.

1.6 Self-Assembly

The research related to "self-assembly" spans a wide range of areas. It is believed to have originated with the studies on biological systems such as the DNA and RNA that led to an understanding that mutual recognition between molecular segments enables them to associate with one another. It is interesting to note the ever-increasing interest in this area, as shown in Table 1.1. These numbers were obtained from a search of the SciFinder Scholar (Chemical Abstracts Service of the American Chemical Society).

The search term was "self-assembly," with no other filter except the range of years. Thus these numbers include publications in all languages as well as patents, on not only molecular but also colloidal and other component self-assembly. While there were just a few articles in the 30-year period from 1960 to 1990, the number increased dramatically since then. It might be that it was during the late 1990s and the 2000s that the idea of using self-assembled systems for nanotechnology was gaining momentum.

Pelesko [22] quoted definitions of several authors as to the meaning of self-assembly. It is now generally defined as a process by which molecules or parts of molecules ***spontaneously*** form organized structures driven by specific ***non-covalent*** interactions between themselves, without an external stimulus. Further, the static self-assembly is a situation where the ordered state leads to equilibrium structures such as crystals. Although organized structures such as crystallization could occur in a large number of systems, the self-assembly is characterized by ***spontaneous*** and ***reversible*** organization of the molecular components. When the components are molecules, it is "molecular self-assembly." Whitesides [23, 24] presented scenarios of making useful devices at various length scales by self-assembly. Pelesko [22] discussed the self-organization of colloidal and other particle systems by the application of electrostatic force.

Table 1.1 Number of publications from SciFinder Scholar on "self-assembly" as entered and the "concept of self-assembly."

Publication years	Self-assembly	Concept of self-assembly
1960–1964	0	5
1965–1969	43	65
1970–1974	76	111
1975–1979	204	280
1980–1985	402	591
1986–1990	491	838
1991–1995	1612	3441
1996–2000	6167	12960
2001–2005	19319	34351
2006–2010	39639	63411
2011–2015	60011	87474
Total	**127964**	**203527**

The self-assembly can thus be "directed" by specific external fields. Grzelczak et al. [25] reviewed the process and applications of directed self-assembly of nanoparticles. The term self-assembly has been used even in assemble-it-yourself furniture systems, where the assembly is modular [26, 27].

Although the majority of studies of self-assembly focus on functionalizing molecules with codes for hydrogen bonding, π-interaction or Coulombic interactions, other factors such as hydrophobic or solvo-phobic interactions also play a role in self-assembly. For example, phthalocyanines and porphyrins are insoluble in most organic solvents, but they could be made soluble in common solvents such as chloroform by peripheral substitution. Addition of a nonsolvent to such solutions in a controlled manner would trigger phase separation and the self-assembly via π-interaction, leading to nanowires and similar morphologies [28]. Even when a molecule is coded with, for example, hydrogen bonding, mutual steric and geometric registry must be favorable for self-assembly to occur.

1.7 Polymer Self-Assembly

As for the self-assembly of polymers, those with hydrogen bonding groups, such as polyamides, with aromatic interactions, for example, polyaramids, have been studied extensively in the past. As shown in Figure 1.5, the amide segment is similar to the amino acid moiety. Similar to the trend shown in Table 1.1, the research activity in the area of polymer self-assembly became intense only in the late 1990s. Table 1.2 shows the results of a search of the SciFinder Scholar. Although studies on hydrogen bonded polymers and similar systems were active before the 1990s, they do not appear significantly in the table, perhaps because the words "self-assembly" did not gain fame at that time in the area of synthetic polymers. In attempts to understand the molecular recognition between the nucleic acid base pairs, synthetic analogues were synthesized a few decades ago [29–39]. To this end, vinyl polymers of the nucleic acid bases, poly(vinyl adenine), poly(vinyl cytosine), poly(vinyl uracil), and poly(vinyl thymine), were synthesized and their interactions were studied since the 1970s.

Most articles on molecular self-assembly would point to the molecular recognition in biological molecules and suggest that Nature should be imitated when designing novel materials. To quote Ronald Breslow [40], "We learned the principle of wings from the birds but not the details of how to use them and power the flight. We take the principles of Nature, not the blueprint. The jumbo jet is not a scaled up version of the pigeon."

In considering self-assembly as it relates to polymers, the aspects are the self-assembly of polymers themselves and the self-assembly of individual units to create a polymer. In the former case, the polymer self-assembly could be

Table 1.2 Number of publications from SciFinder Scholar on "polymer self-assembly" as entered and the "concept of polymer self-assembly"

Publication years	Polymer self-assembly	Concept of polymer self-assembly
1960–1964		
1965–1969		
1970–1974		3
1975–1979		8
1980–1985		18
1986–1990		39
1991–1995	7	275
1996–2000	25	1204
2001–2005	93	3370
2006–2010	182	6250
2011–2015	215	9191
	522	**20358**

driven by association elements such as hydrogen bonds. The self-association in the case of block copolymers would be driven by the incompatibility of the segments and to some extent by "self-recognition," "self-sorting." Using the principles of Nature, polymers functionalized with specific recognition units have been designed [41–48], and these led to various potential applications in areas such as nanotechnology. Using such methodologies to create folded structures, intramolecular and intermolecular association of chromophores with oligomeric or polymeric linkages by chain folding have created opportunities for optical applications [49–52]. We will discuss some of these in the following chapters.

References

1 P. J. Flory, *Principles of Polymer Chemistry*, **1969**, Cornell University Press.
2 K. H. Meyer and H. Mark, *Ber. Dtsch. Chem. Ges. A/B*, **1928**, *61*, 593.
3 K. H. Meyer and H. Mark, *Ber. Dtsch. Chem. Ges. A/B*, **1928**, *61*, 1939.
4 E. Simon, *Ann. Pharm.*, **1839**, *31*, 265.
5 R. C. Hiorns, R. J. Boucher, R. Duhlev, K.-H. Hellwich, P. Hodge, A. D. Jenkins, R. G. Jones, J. Kahovec, G. Moad, C. K. Ober, D. W. Smith, R. F. T. Stepto, J.-P. Vairon, and J. Vohlídal, *Pure Appl. Chem.*, **2012**, *84*, 2167–2169.

6 P. R. Sundararajan, in *Physical Properties of Polymers Handbook*, 2nd Ed., **2007**, J. E. Mark, Ed., Chapter 1, Springer.
7 P. J. Flory, *Statistical Mechanics of Chain Molecules*, **1969**, John Wiley & Sons, Inc., Chapter VI (Republished (**1989**) by Oxford University Press).
8 F. A. Bovey, *Polymer Conformation and Configuration*, **1969**, Academic Press.
9 F. A. Bovey, P. Mirau, and H. S. Gutowsky, *Nuclear Magnetic Resonance Spectroscopy*, 2nd Ed., **1988**, Academic Press.
10 A. E. Tonelli, *NMR Spectroscopy and Polymer Microstructure: The Conformational Connection*, **1989**, Wiley-Interscience.
11 A. E. Tonelli and J. L. White, in *Physical Properties of Polymers Handbook*, 2nd Ed., J. E. Mark, Ed., **2007**, Chapter 20, Springer.
12 Z. Dobkowski, *Rheol. Acta*, **1995**, *34*, 578–585.
13 P. R. Sundararajan, *Macromolecules*, **1987**, *20*, 1534–1539.
14 W. W. Graessley, *Adv. Polym. Sci.*, **1974**, *16*, 1.
15 A. E. Tonelli, *Polymers from the Inside Out*, **2001**, John Wiley & Sons, Inc., New York.
16 S. M. Aharoni, *Macromolecules*, **1983**, *16*, 1722.
17 T. A. Kavassalis and J. Noolandi, *Macromolecules*, **1988**, *21*, 2869.
18 T. A. Kavassalis and J. Noolandi, *Macromolecules*, **1989**, *22*, 2709.
19 R. P. Wool, *Macromolecules*, **1993**, *26*, 1564.
20 L. J. Fetters, D. J. Lohse, D. Richter, T. A. Witten, and A. Zirkel, *Macromolecules*, **1994**, *27*, 4639.
21 P. R. Sundararajan, in *Physical Properties of Polymers Handbook*, **2007**, J. E. Mark, Ed., Chapter 15, Springer.
22 J. A. Pelesko, *Self-Assembly*, **2007**, Chapman & Hall.
23 G. Whitesides, *Science*, **2002**, *295*, 2418.
24 G. Whitesides, *Mater. Res. Soc. Bull.*, **2005**, *30*, 736.
25 M. Grzelczak, J. Vermant, E. M. Furst, and L. M. Liz-Marzán, *ACS Nano*, **2010**, *4*, 3591–3605.
26 M. Hackwood, G. Karmon, J. Adorian, Y. Tal, and M. Leczycki, Self-Assembly Furniture System, US Patent 5,244,271, **1993**.
27 C. Niziol, Self-Assembly Furniture, US Patent 4,421,366, **1983**.
28 J. Gao, C. Cheng, X. Zhou, Y. Li, X. Xu, X. Du, and H. Zhang, *J. Colloid Interface Sci.*, **2010**, *342*, 225–228.
29 E. Yashima, T. Tajima, N. Miyauchi, and M. Akashi, *Biopolymers*, **1992**, *32*, 811–817.
30 L. Maggiora, S. Boguslawski, and M. P. Mertes, *J. Med. Chem.*, **1977**, *20*, 1283–1287.
31 J. Pitha, *Cancer Res.*, **1976**, *36*, 1273–1277.
32 F. H. Reynolds Jr., P. M. Pitha, R. Chuang, T.-C. Cheng, H. H. Kazazian, Jr., and D. Grunberger, *Mol. Pharmacol.*, **1975**, *11*, 708–715.

33 H. Kaye and H. J. Chou, *J. Polym. Sci. Polym. Phys. Ed.*, **1975**, *13*, 477–492.

34 J. Pitha, P. M. Pitha, and E. Stuart, *Biochemistry*, **1971**, *10*, 4595–4602.

35 P. M. Pitha and J. Pitha, *Biopolymers*, **1970**, *9*, 965–977.

36 H. Kaye, *J. Polym. Sci. Polym. Lett. Ed.*, **1969**, *7*, 1–5.

37 L. Noronha-Blob and J. Pitha, *Biochim. Biophys. Acta*, **1978**, *519*, 285–290.

38 F. Reynolds, D. Grunberger, J. Pitha, and P. M. Pitha, *Biochemistry*, **1972**, *11*, 3261–3266.

39 P. M. Pitha and A. M. Michelson, *Biochim. Biophys. Acta*, **1970**, *204*, 381–388.

40 R. Breslow, in *Artificial Enzymes*, **2005**, R. Breslow, Ed., Wiley-VCH Verlag GmbH, Weinheim, p. 31.

41 J.-M. Lehn, *Supramolecular Chemistry: Concepts and Perspectives*, **1995**, Wiley-VCH Verlag GmbH, Weinheim.

42 S. I. Stupp, S. Son, L. S. Li, H. C. Lin, and M. Keser, *J. Am. Chem. Soc.*, **1995**, *117*, 5212 (*Science*, **1997**, *276*, 384).

43 A. Ciferri, Ed., *Supramolecular Polymers*, **2000**, Marcel Dekker, New York.

44 O. Ikkala and G. ten Brinke, *Science*, **2002**, *295*, 2407.

45 Y. Guan, S.-H. Yu, M. Antonietti, C. Böttcher, and C. F. J. Faul, *Chem. Eur. J.*, **2005**, *11*, 1305.

46 K. V. Gothelf and R. S. Brown, *Chem. Eur. J.*, **2005**, *11*, 1062.

47 G. B. W. L. Ligthart, H. Ohkawa, R. P. Sijbesma, and E. W. Meijer, *J. Am. Chem. Soc.*, **2005**, *127*, 810.

48 S. Hecht, *Mater. Today (Elsevier)*, March **2005**, pp. 48.

49 A. D. Q. Li, W. Wang, and L.-Q. Wang, *Chem. Eur. J.*, **2003**, *9*, 4594.

50 S. H. Gellman, *Acc. Chem. Res.*, **1998**, *31*, 173.

51 P. D. Yarborough and D. Y. Sogah, *Polym. Mater. Sci. Eng.*, **1997**, *76*, 395 (**1997**, *77*, 601).

52 C. Kübel, M. J. Mio, J. S. Moore, and D. C. Martin, *J. Am. Chem. Soc.*, **2002**, *124*, 8605.

2

Molecular Forces

The mutual recognition that brings any two molecules to self-assemble would depend on the noncovalent interaction segments either in the main chain or as a side group. Molecules could be coded for self-assembly using groups that lead to one or more of these interactions: aromatic π–π, hydrogen bond, van der Waals, electrostatic (including charge transfer), hydrophilic, or hydrophobic. Conjugated polymers and those with porphyrins, phthalocyanines, perylenes, etc., in the main chain or as a side group necessarily involve the aromatic interactions. Hydrogen bonding is perhaps the most used code in designing polymers (or even oligomers and small molecules) for the study of self-assembly. The van der Waals interactions are ubiquitous and commonplace. This chapter gives a brief description of these noncovalent forces.

Developing methods to estimate the contributions of the various forces was an active area of research in the 1950s and the 1960s. The efforts were aimed at analyzing and predicting the conformations of proteins, polypeptides, polynucleotides, and other biological systems. Shneior Lifson, Arieh Warshel, Michael Levitt, Martin Karplus, Herald Scheraga, Paul Flory, and G. N. Ramachandran were some of the early investigators. The latter developed the so-called (φ, ψ) map or the Ramachandran map to compare the relative energies of the conformations of polypeptides and proteins. Starting from what we now call "molecular mechanics," the principles of quantum mechanics were applied to model systems [1]. The 2013 Nobel Prize in chemistry was awarded to Martin Karplus, Michael Levitt, and Arieh Warshel for their pioneering contributions. Figure 2.1 shows the basic equations that were used in the early days.

2.1 Van der Waals Interaction

According to the International Union of Pure and Applied Chemistry (IUPAC) definition [2], van der Waals forces are "The attractive or repulsive forces between molecular entities (or between groups within the same molecular

Physical Aspects of Polymer Self-Assembly, First Edition. P. R. Sundararajan.

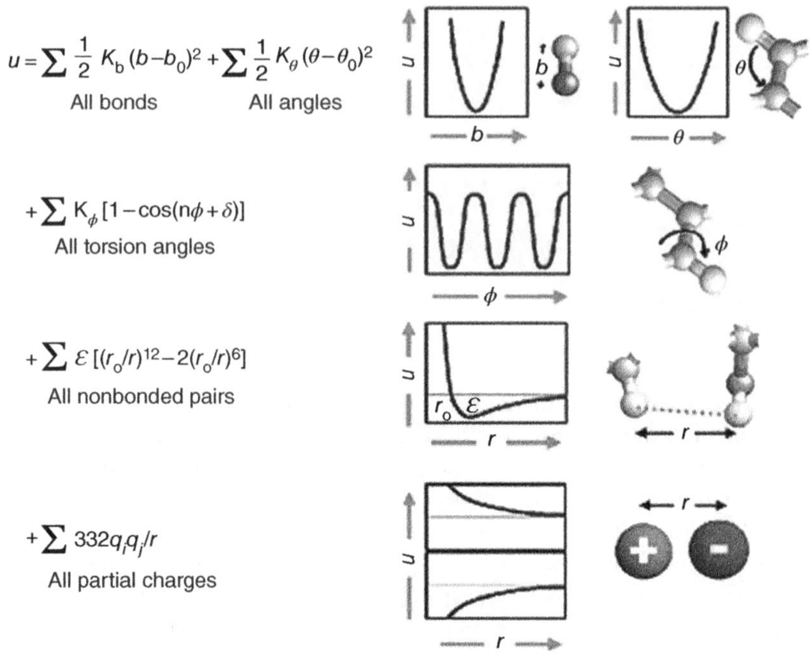

Figure 2.1 Empirical equations used for calculating the conformational energies, as summarized by Michael Levitt [1]. (Source: Levitt [1]. Reproduced with permission of Nature Publishing Group.)

entity) other than those due to bond formation or to the electrostatic interaction of ions or of ionic groups with one another or with neutral molecules. The term includes dipole–dipole, dipole-induced dipole, and London (instantaneous induced dipole-induced dipole) forces. The term is sometimes used loosely for the totality of nonspecific attractive or repulsive intermolecular forces". The general form of this force (also known as potential) is shown in Figure 2.2 and arises from two effects: a strong nucleus–nucleus coulombic repulsion at short distances and the overlap of the electron clouds between the two atoms. In the figure, U is the interatomic potential, r is the distance between two atoms, and r_m is the distance (usually taken as the sum of the van der Waals radii) at which the potential is a minimum. In the simulation of polymers, usually empirical equations are used to calculate the potential between any two nonbonded atoms, which take the form

$$E_{vdW} = -\frac{A_{ij}}{r_{ij}^6} + \frac{B_{ij}e^{Dr}}{r_{ij}^C} \tag{2.1}$$

Figure 2.2 Variation of van der Waals force U with interatomic distance r.

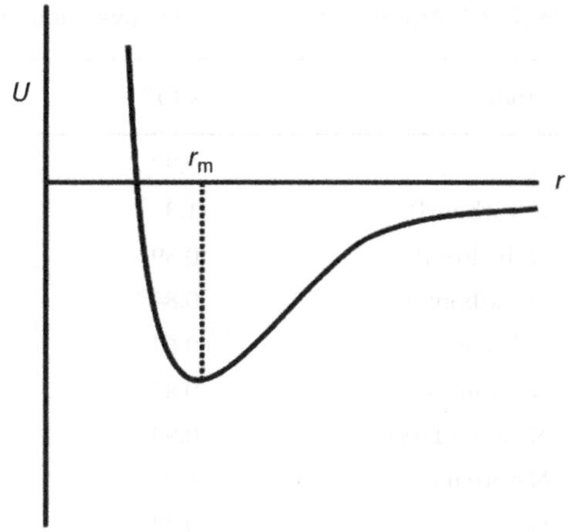

where r_{ij} is the distance between atoms i and j and A and B are constants. In this equation, if $D = -4.6$ and $C = 0$, it is called the Buckingham function, of the form

$$E_{\mathrm{vdW}} = -\frac{A_{ij}}{r_{ij}^{6}} + B_{ij}^{-4.6r} \tag{2.2}$$

If $D = 0$ and $C = 12$, it is the more popular Lennard-Jones function, which is

$$E_{\mathrm{vdW}} = -\frac{A_{ij}}{r_{ij}^{6}} + \frac{B_{ij}}{r_{ij}^{12}} \tag{2.3}$$

The parameters A_{ij}, characterizing the attractive part, are calculated from the effective number of outer shell electrons N_{eff} and the atomic polarizabilities α by the application of the Slater–Kirkwood equation.

$$A_{ij} = \frac{(3/2)e\left(\hbar / m^{1/2}\right)\alpha_i \alpha_j}{(\alpha_i / N_i)^{1/2} + (\alpha_j / N_j)^{1/2}} \tag{2.4}$$

Here, e and m are the charge and mass of an electron, α_i and α_j are the atomic polarizabilities of the interacting atoms i and j, and N_i and N_j are the effective number of polarizable electrons on these atoms. The values of α_i and α_j are usually taken from Ketelaar [3] and these parameters for the atoms C, O, N, and H are listed in Table 2.1.

More recent polarizability values based on experimental and theoretical work has been discussed, for example, by Teachout and Pack [4] and by Anslyn

Table 2.1 Atomic polarizability, effective number of outer shell electrons.

Atom	$\alpha \times 10^{24}$	N_{eff}	R (Å)
C	0.93	5.2	1.7
C (carbonyl)	1.3	5.2	1.7
O (hydroxyl)	0.59	7.0	1.5
O (carbonyl)	0.84	7.0	1.5
O (ether)	0.64	7.0	1.5
N (primary)	0.87	6.1	1.5
N (secondary)	0.93	6.1	1.5
N (tertiary)	1.03	6.1	1.5
H	0.42	0.9	1.2
CH_2	1.77	7.0	1.8
C double bond	0.58		
C triple bond	0.86		

Source: In part from Ketelaar [3].

and Dougherty [5]. The values of B_{ij} in Eqs. 2.2 and 2.3 are calculated by requiring that the energy E_{vdW} be minimum at a distance R_{ij} (r_m in Figure 2.2). This is usually the sum of the van der Waals radii of the pair of atoms. With this condition, using Eq. 2.3,

$$\left(\frac{dE}{dr}\right)_{r=R_{ij}} = \frac{6A_{ij}}{R_{ij}^{7}} - \frac{12B_{ij}}{R_{ij}^{13}} = 0 \tag{2.5}$$

$$\text{Then, } B_{ij} = \frac{1}{2} A_{ij} R_{ij}^{6} \tag{2.6}$$

Thus, both A_{ij} and B_{ij} in Eqs. 2.2 and 2.3 can be calculated as described earlier. Variations of Eq. 2.3 have been used in some of the commercial software (see Tables 2.2 and 2.3).

Polyethylene is a unique example of a polymer, the crystalline structure of which is stabilized by only the van der Waals interactions. The chain prefers all-trans conformation of the C—C bonds. Infrared (IR) spectroscopy is a common technique that can be used to get an estimate of the strength of the van der Waals interactions between these CH_2 groups. Coleman et al. [6] and Suzuki et al. [7] related the shift of the CH_2 stretching frequencies and the half-widths of the peaks to the strength of the van der Waals interactions. For example, in the case of polyurethane model compounds with alkyl side chains,

Table 2.2 Equations and parameters used in the DRIEDING force field [15].

Interaction		Equation
Valence terms	Bond energy	$E_b = 1/2K_b(R - R_0)^2$
	Angle energy	$E_\theta = 1/2K_\theta(R - R_\theta)^2$
	Torsion energy	$E_\phi = \sum_{n=1}^{2} 1/2K_\phi n[1 \pm \cos(n\phi)]$
	Out-of-plane angle	$E_{inv} = K_{inv}(\cos\chi - \cos\chi_0)^2$
Nonbond terms	Coulombic interaction	$E_{Coul} = 332.0637 q_i q_j / \varepsilon r_{ij} \mathrm{erfc}(r_{ij}/\beta)$
	Hydrogen bonding	$E_{hb} = \varepsilon_0\left[5(R_0/R)^{12} - 6(R_0/R)^{10}\right]\cos^4\varphi$
	Van der Waals energy	$E_{vdW} = \varepsilon_0\left[(\sigma_0/R)^{12} - 2(\sigma_0/R)^6\right]$

Source: Li et al. [56]. Reproduced with permission of Elsevier.

$K_b = 2.93 \times 10^5$ kJ/mol/nm^2, $R_0 = 0.15$nm, $K_\theta = 418$kJ/mol/rad^2, $R_\theta = 1.91$rad, $K_\phi = 20.9$kJ/mol, $K_{inv} = 20.9$kJ/mol, $\chi_0 = 2.09$rad, $\beta = 0.234$nm, $\varepsilon = 1$.

Hydrogen bonding: $\varepsilon_0 = 16.72$kJ/mol; $R_0 = 0.275$nm.
Van der Waals energy:

H atom: $\varepsilon_0 = 0.00042$kJ/mol, $\sigma_0 = 0.3195$nm
O atom: $\varepsilon_0 = 0.4$kJ/mol, $\sigma_0 = 0.3405$nm
C atom: $\varepsilon_0 = 0.6985$kJ / mol , $\sigma_0 = 0.3923$nm
Off-diagonal interactions: $\varepsilon_0 = 1.2607$kJ/mol, $\sigma_0 = 0.4237$nm.

Khanna et al. [8] found that the asymmetric and symmetric CH_2 stretching frequencies were observed at 2935 and 2863 cm^{-1}, respectively, for a C_4 side chain, and these frequencies shifted to lower wavenumbers of 2917 and 2849 cm^{-1}, respectively, with a C_{18} side chain. A decrease in the half-width of the peaks was also seen. These are indicative of an increase in the ordering of the alkyl chains via the van der Waals interactions. Similar frequency shifts were observed by Suzuki et al. [7] in the studies on urea-based gels. Using the Lennard-Jones function similar to that in Eq. 2.3, Abe et al. [9] calculated the van der Waals energy for a CH_2–CH_2 pair to be only −0.5 kcal/mol. However, this interaction becomes significant when long sequences of CH_2 groups interact with each other, as in the polyethylene crystalline structure.

2.2 Hydrogen Bond

The van der Waals interaction discussed before occurs between any two atoms. However, hydrogen bond is directional, since it involves three atoms. The understanding of the hydrogen bond owes its evolution to the crystal structure

Table 2.3 Functional forms used in the COMPASS force field [57].

Interaction		**Equation**
Valence terms	Bond energy	$E_b = \sum_b \left[k_2 (b-b_0)^2 + k_3 (b-b)^3 + k_4 (b-b)^4 \right]$
	Angle energy	$E_\theta = \sum_\theta \left[k_2 (\theta-\theta_0)^2 + k_3 (\theta-\theta_0)^3 + k_4 (\theta-\theta_0)^4 \right]$
	Torsion energy	$E_\phi = \sum_\phi \left[k_1 (1-\cos\phi) + k_2 (1-\cos 2\phi) + k_3 (1-\cos 3\phi) \right]$
	Out-of-plane angle	$E_\chi = \sum_\chi k_2 \chi^2$
	Cross-coupling terms	$E_{\text{cross}} = \sum_{b,b'} k(b-b_0)(b'-b'_0) + \sum_{b,\theta} k(b-b_0)(\theta-\theta_0) + \sum_{b,\phi} (b-b_0)[k_1 \cos\phi + k_2 \cos 2\phi + k_3 \cos 3\phi] + \sum_{\theta,\phi} (\theta-\theta_0)[k_1 \cos\phi + k_2 \cos 2\phi + k_3 \cos 3\phi] + \sum_{\theta,\theta'} k(\theta'-\theta'_0)(\theta-\theta_0) + \sum_{\theta,\theta',\phi} k(\theta-\theta_0)(\theta'-\theta'_0)\cos\phi$
Nonbond terms	Coulombic interaction	$E_{\text{Coul}} = \sum_{ij} q_i q_j / r_{ij}$
	Van der Waals energy	$E_{\text{vdW}} = \sum_{ij} \varepsilon_{ij} \left[2\left(r_{ij}^0 / r_{ij} \right)^9 - 3\left(r_{ij}^0 / r_{ij} \right)^6 \right]$

Source: Li et al. [56]. Reproduced with permission of Elsevier.

studies of small molecules. Reviews of theoretical and crystal structure aspects of hydrogen bond have been published by Steiner [10] and Desiraju [11]. The hydrogen bond interaction involves cumulative effects of electrostatic, van der Waals, and covalent contributions to varying degrees [10]. IUPAC issued a document [12] on the definition of hydrogen bond, along with six criteria that could be used as evidence for its presence. These are described later as they appear in the IUPAC document, since hydrogen bond is a key player in the self-assembly of polymers or small molecules. (Note that IUPAC permits republication or reproduction of the aforementioned document without its formal approval, provided proper acknowledgment is provided). As for the definition,

The hydrogen bond is an attractive interaction between a hydrogen atom from a molecule or a molecular fragment X—H in which X is more

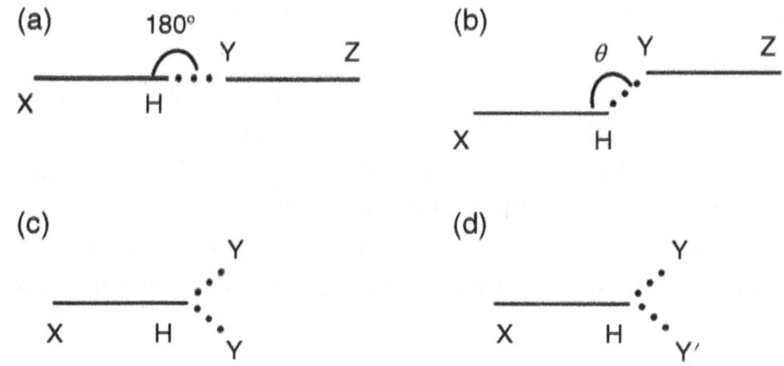

Figure 2.3 Schematic of the hydrogen bond. X—H and Y—Z are the donor and the acceptor, respectively. The hydrogen bond is denoted by three dots. (a) A straight hydrogen bond with 180° between X—H and Y—Z, (b) a bent hydrogen bond with an angle between X—H and Y, and (c) and (d) bifurcated hydrogen bond between X—H and two same atoms of type Y or two types Y and Y′.

electronegative than H and an atom or a group of atoms in the same or a different molecule, in which there is evidence of bond formation. A typical hydrogen bond may be depicted as X—H···Y—Z, where the three dots denote the bond. X—H represents the hydrogen bond donor. The acceptor may be an atom or an anion Y, or a fragment or a molecule Y—Z, where Y is bonded to Z. In some cases, X and Y are the same. In more specific cases, X and Y are the same and X—H and Y—H distances are the same as well leading to symmetric hydrogen bonds. In any event, the acceptor is an electron-rich region such as, but not limited to, a lone pair of Y or π-bonded pair of Y—Z.

The criteria and characteristics set out in the IUPAC document that can be used as evidence for a hydrogen bond X—H···Y—Z include the following:

1) The forces involved in the formation of a hydrogen bond include those of an electrostatic origin, those arising from charge transfer between the donor and acceptor leading to partial covalent bond formation between H and Y, and those originating from dispersion.
2) The atoms X and H are covalently bonded to one another and the X—H bond is polarized; the H···Y bond strength increases with the increase in the electronegativity of X.
3) The X—H···Y angle is usually linear (180°) and the closer the angle is to 180° (as illustrated in Figure 2.3), the stronger is the hydrogen bond and the shorter is the H···Y distance. It is because of this angular dependence that the hydrogen bond is characterized as "directional."
4) The length of the X—H bond usually increases on hydrogen bond formation leading to a red shift in the IR X—H stretching frequency and an increase in the IR absorption cross section for the X—H stretching vibration. The greater the lengthening of the X—H bond in X—H···Y, the stronger the H···Y

bond. Simultaneously, new vibrational modes associated with the formation of the H···Y bond are generated.

5) The X—H···Y—Z hydrogen bond leads to characteristic nuclear magnetic resonance (NMR) signatures that typically include pronounced proton deshielding for H in X—H, through hydrogen bond spin–spin couplings between X and Y and nuclear Overhauser enhancements.
6) The Gibbs energy of formation for the hydrogen bond should be greater than the thermal energy of the system for the hydrogen bond to be detected experimentally.

Similar to the case of van der Waals forces, several types of equations have been developed to calculate the hydrogen bond energy. The Lippincott–Schroeder equation [13] included the dissociation energies of the X—H bond and H···Y. Scheraga et al. [14] used the Lennard-Jones 6–12 function, with different sets of the constants *A* and *B*. The equation in the DREIDING force field is of the following form [15]:

$$E_{hb} = D_{hb}\left[5\left(\frac{R_{hb}}{R_{DA}}\right)^{12} - 6\left(\frac{R_{hb}}{R_{DA}}\right)^{10}\right]\cos^4(\theta_{DHA}) \tag{2.7}$$

Here, E_{hb} is the energy due to the hydrogen bond; D_{hb} and R_{hb} are constants derived from the results on water dimer, with values of 2.75 Å and 9.5 kcal/mol, respectively; R_{DA} is the distance between the donor and acceptor atoms (e.g., X and Y in Figure 2.3); and θ is the angle defined in Figure 2.3b. Typically, the attractive energy contribution of a hydrogen bond is about 3 kcal/mol.

Whereas hydrogen bonding in small molecules can be directly determined from the nonbonded distances in crystal structure analysis, estimating such interaction in polymer solutions, gels, and solid state has to depend on other techniques, such as NMR and IR spectroscopies. In both cases, frequency shifts are associated with hydrogen bonds. Several pioneering studies analyzed the effect of hydrogen bond on the shift in the IR frequencies of X—H bonds. These investigations explored the relationship between the hydrogen bond distances in the small-molecule crystal structures and the IR spectra of the crystals. In one of the earliest papers, Nakamoto et al. [16] analyzed various crystal structures and plotted the X—H IR stretching frequencies and the hydrogen bond distances of the types O—H···O, O—H···N; N—H···O; N—H···N, O—H···Cl; etc. Figure 2.4 shows the plots of some of these from the works of Nakamoto et al. [16] and Pimentel and Sederholm [17]. In these plots, $\Delta\nu$ is the difference between the X—H stretching frequency of the monomer observed in the solid and the frequency in the gas phase or in dilute solutions. These plots show that the slopes are different between the hydrogen types.

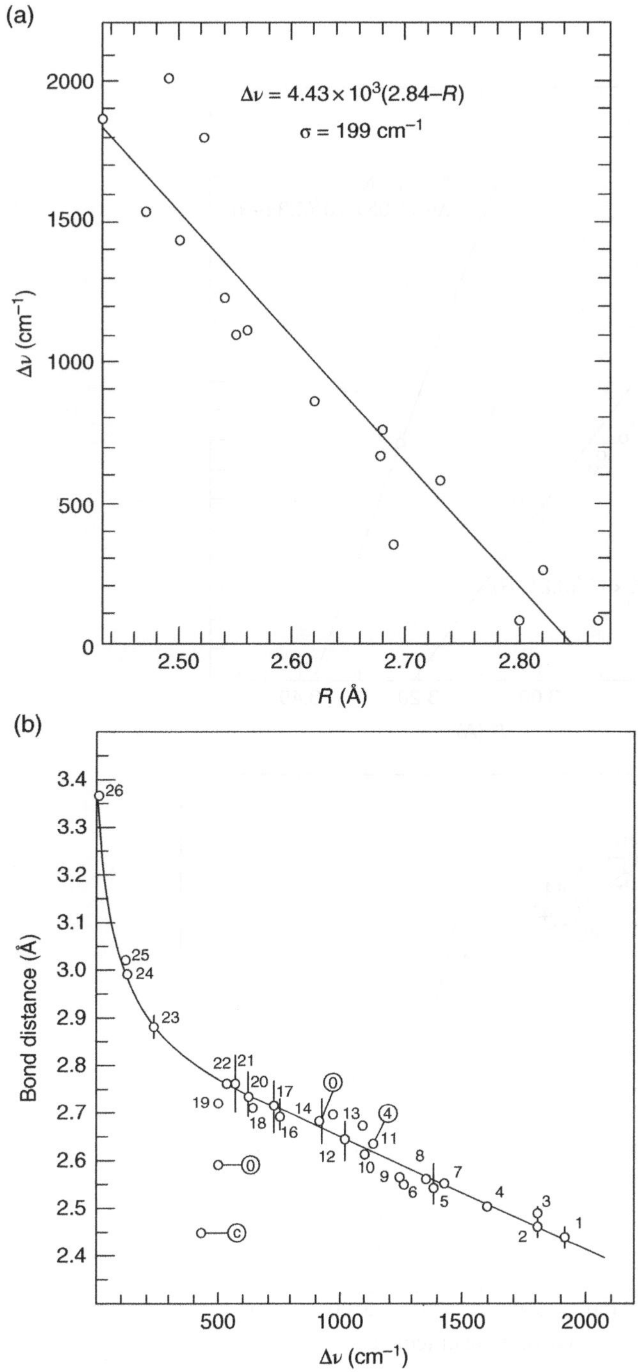

Figure 2.4 Plots of (a) and (b): $\Delta\nu$ vs. distance for O—H···O; (c): N—H···O and N—H···N; and (d): N—H and O—H frequencies vs hydrogen bond distance. (Source: Nakamoto et al. [16]. Reproduced with permission of American Chemical Society.)

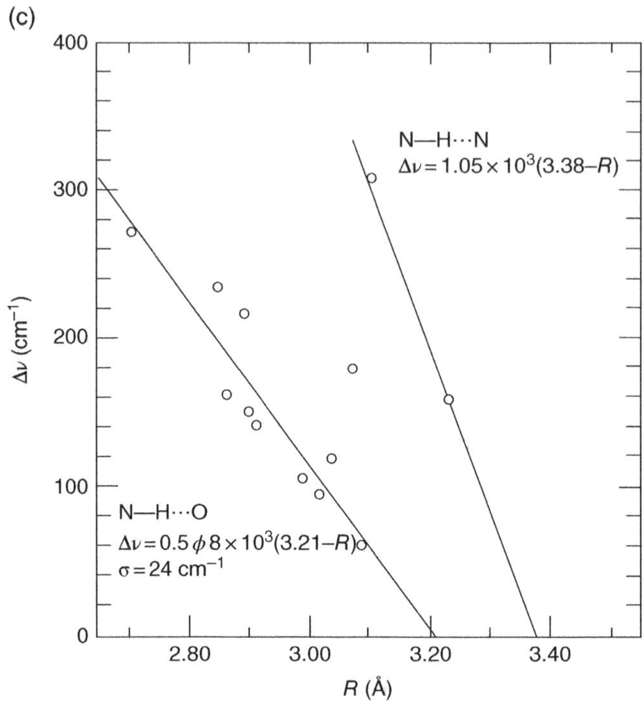

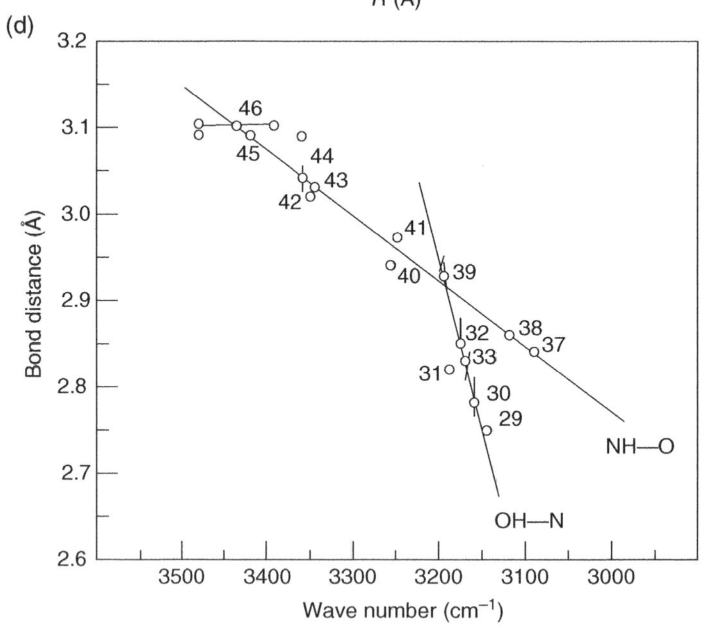

Figure 2.4 (Continued)

Based on these studies, Pimentel and Sederholm [17] proposed a simple relationship between $\Delta\nu$ and the hydrogen bond distance R (Å) as follows:

$$\Delta\nu[\mathrm{O-H\cdots O}](\mathrm{cm^{-1}}) = 4430(2.84 - R) \tag{2.8}$$

$$\Delta\nu[\mathrm{N-H\cdots O}](\mathrm{cm^{-1}}) = 548(3.21 - R) \tag{2.9}$$

$$\Delta\nu[\mathrm{N-H\cdots N}](\mathrm{cm^{-1}}) = 1050(3.38 - R) \tag{2.10}$$

By modeling the hydrogen bond interaction on the Lennard-Jones framework as a sum of attractive and repulsive forces, Bellamy and Owen [18] proposed a relationship between $\Delta\nu$ and the hydrogen bond distance between X and Y as follows:

$$\Delta\nu(\mathrm{cm^{-1}}) = 50\left[\left(\frac{d}{R}\right)^{12} - \left(\frac{d}{R}\right)^{6}\right] \tag{2.11}$$

Here, R is the hydrogen bond distance and d is the sum of the "collision radii" (similar to the van der Waals radii) of the X and Y atoms (d = 3.2, 3.35, 3.4, 3.4, 3.6, 3.85, and 3.9 Å for F—H···F, O—H···O, N—H···F, N—H···O, N—H···N, O—H···Cl, and N—H···Cl hydrogen bonds, respectively). Although these relationships were derived from studies on small molecules, they are of use in estimating the hydrogen bond distances in large and polymer molecules with the aid of IR spectra. Using the given equations, Khan and Sundararajan [19] calculated the N···O hydrogen bond distance in biscarbamates, which are model compounds for polyurethanes. With the observed $\Delta\nu$ of $120\,\mathrm{cm^{-1}}$, the calculated distance was 2.991 Å using Eq. 2.9 and 2.998 Å with Eq. 2.11. The observed N···O distance in the crystal structure of a similar biscarbamate [20] was 2.927 Å. Several papers were published by the team of Coleman and Painter on the IR spectroscopy of hydrogen bonding in conventional polymers. The applications of NMR spectroscopy to study hydrogen bonding and supramolecular assemblies have been discussed in recent reviews [21, 22].

2.3 C—H···π Interaction

The effect of C—H bonds on the NMR spectra of aromatic molecules has been known for over five decades [23]. The conformational preference of an aromatic or alkyl C—H bond with respect to a neighboring phenyl group has been confirmed by a number of crystal structure studies. Theoretical calculations led to the conclusion that additional forces other than dispersion are operative which would include delocalization and Coulombic contributions to varying

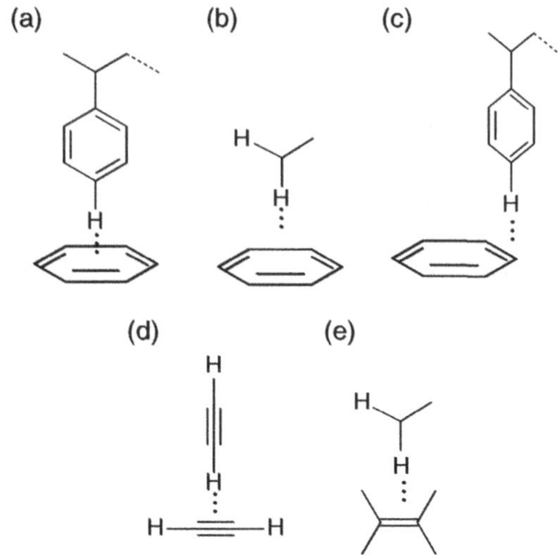

Figure 2.5 Various types of CH/π interactions are illustrated as follows: (a) T-type aromatic–aromatic, (b) alkyl CH-aromatic, (c) L-type, (d) acetylene, and (e) alkyl–ethylene.

degrees [24, 25]. Such an interaction has also been found between acetylenes. Various types of C—H···π dispositions are shown in Figure 2.5. The C—H···π interaction (also denoted as CH/π) exerts a significant influence in the conformation and crystal structures of organic molecules, drug binding, etc. [25]. Its role in chemical and biological recognition has been reviewed by Meyer et al. [26]. It is recognized as a hydrogen bond, as the nonbonded distance between the parent atoms is usually shorter than normal. A database analysis of the crystal structures in the Cambridge Structural Database by Takahashi et al. [27] showed that the C—H···π distances are short in a number of crystal structures of inclusion compounds, such as cyclodextrin complexes, pseudo-rotaxanes, etc. Another study of the aforementioned database by Suezawa et al. [28] focused on the *syn* orientation of the C—H bond in crystal structures of aralkyl compounds of the type $ArCH_2XCH_2Y$ and $ArCHCH_3XCH$ and the energy difference between the *syn* and *anti* conformations were calculated using molecular orbital methods. The CH/π interaction is usually weak. Wilcox et al. [29, 30] designed molecular "torsional balances" of the type shown in Figure 2.6 to determine the strength of the C—H···π hydrogen bond in the edge-to-face conformation, using crystal structure analysis and NMR spectra in solution. It was found that isopropyl esters (Figure 2.6b) exhibited higher folding compared to the phenyl esters and their folding energies were 0.5 ± 0.1 and 0.3 ± 0.1 kcal/mol, respectively. Dispersion forces were considered to be the main contributor. Ab initio calculations showed that C—H···π bonding with hydrocarbons involve −1.82 and −2.06 kcal/mol for ethane and ethylene,

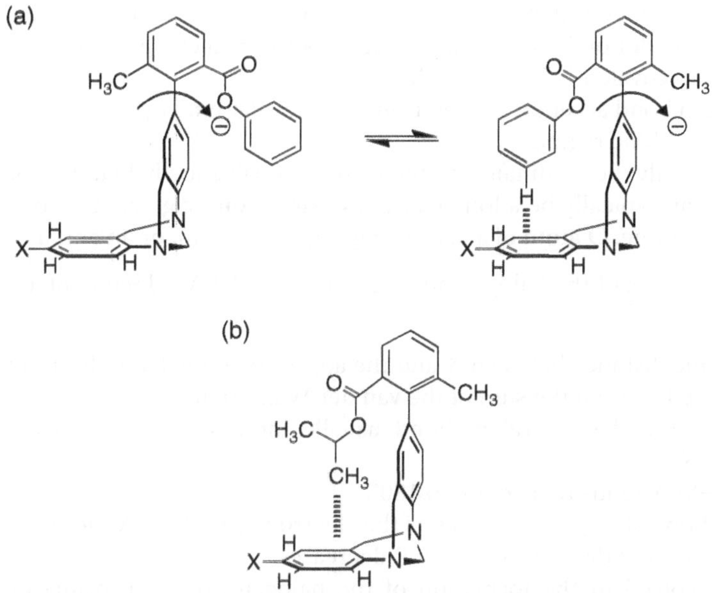

Figure 2.6 (a) Molecular torsional balance described by Wilcox and coworkers [29, 30] to quantify aromatic edge-to-face interactions. T-shaped aromatic interactions are effective only in the folded conformation. (b) Preferred folded state of isopropyl ester. X = NO_2, CN, I, Br, Me, OH, NH_2. (Source: Meyer et al. [26]. Reproduced with permission of Wiley.)

respectively. Similar molecular balances for measuring aliphatic CH/π interactions were reported by Carroll et al. [31]. They used bicyclic N-arylimides as molecular balances to study the C—H···π interactions between the alkyl and arene groups in the folded conformers. These balances were designed with one and two arms (single and two substituents), the latter crystallized in folded conformation. The weak interaction could show up field shifts as small as 0.01 ppm in the ^{1}H NMR spectra of the alkoxy protons.

2.4 Halogen Bond

The halogen bond is gaining recognition in crystal design of small molecules. This is similar to the hydrogen bond but involving a halogen atom. The IUPAC definition [32] states thus:

- A halogen bond occurs when there is evidence of a net attractive interaction between an electrophilic region associated with a halogen atom in a molecular entity and a nucleophilic region in another or the same molecular entity.

- A typical halogen bond is denoted by the three dots in R—X···Y. Here, R—X is the halogen bond donor, X is any halogen atom with an electrophilic region, and R is a group covalently bonded to X.
- Y is the halogen bond acceptor and typically a molecular entity possessing at least one nucleophilic region.
- R—X can typically be a dihalogen molecule, haloalkane, haloarene, or haloimide. Y can typically be a lone pair possessing atom (e.g., N atom of a pyridine or an amine, O atom of a carbonyl group), π-system, or an anion.

Some of the features of the halogen bond listed in the IUPAC document are as follows:

- The interatomic distance between X and the appropriate nucleophilic atom of Y tends to be less than the sum of the van der Waals radii.
- The length of the R—X covalent bond usually increases relative to the unbonded R—X.
- The angle R—X···Y tends to be close to 180°.
- The halogen bond strength decreases as the electronegativity of X increases and the electron-withdrawing ability of R decreases.
- The forces involved in the formation of the halogen bond are primarily electrostatic, but polarization, charge transfer, and dispersion contributions all play an important role. The relative roles of the different forces may vary from one case to the other.

Desiraju et al. [33] summarized the role of halogen bond in crystal engineering and discussed the difference between the hydrogen bond and the halogen bond. Although the two are similar, the hydrogen atom is much smaller than the halogen atom. Several small molecule crystal structures exhibiting halogen bonds are known [33–38]. There has not often been an intentional inclusion of halogen bonding in the studies of self-assembly of polymers. The effect of chlorine and dichloromethane on the crystallization in solution cast films of a polyimide was reported and the polymer solubility in these solvents was attributed to halogen interaction of the solvent with the polymer [39].

2.5 Other Hydrogen Bonds

Other types of hydrogen bonds, which are often weak, have been found in crystal structures and solutions of small molecules and in oligomeric systems. One of these is the hydrogen bond involving the C—H bond [40–48], for example, C—H···O=C, C—H···O—C, or C—H···N. This is also called an "improper hydrogen bond." In a typical hydrogen bonding situation, the corresponding peaks in the Fourier transform IR spectroscopy (FTIR) spectrum would be red-shifted. However, with the C—H···O bond, the absorption frequencies were seen to be blue-shifted [41, 43, 47]. Viswamitra et al. [48] found evidence for O—H···C and N—H···C hydrogen bonds in the crystal structures of alkynes, alkenes, and aromatics. Rozenberg et al. [49] presented crystal structure and spectroscopic evidence for S—H···Ph hydrogen bond, which is similar to the C—H···π interaction discussed earlier.

Thus, while the early studies on biological systems focused on hydrogen bonds involving O—H, N—H, and C=O bonds, various other types of noncovalent, hydrogen bond type of interactions have been discovered, with small-molecule crystal structures and spectroscopic techniques. Still, the conventional hydrogen bond schemes are used as the main elements in the design of self-assembling polymers.

2.6 Coulombic Interaction

Recalling the equations of Levitt in Figure 2.1, the Coulombic force is usually expressed as

$$E_{\mathrm{Coul}} = \frac{332 q_i q_j}{\varepsilon r_{ij}} \tag{2.12}$$

in which 332 is the conversion factor to express the energy in kcal/mol, q_i and q_j are the partial charges on the interacting atoms, r_{ij} is the distance between them, and ε is the "effective" dielectric constant. The value of ε is taken to be 2 for calculations without a solvent and 4 for aqueous media. Experimental evaluation of ε in proteins have been reported [50], by using the equation

$$\varepsilon = \frac{q_1 q_2}{Ur} \tag{2.13}$$

in which U is the electrostatic energy between two charges q_1 and q_2 and r is the distance between them. Rees [50] discussed the calculation of U using the redox potential of proteins. A simple method using bond moments can be used for calculating the partial charges q_i and q_j in Eq. 2.12. For example, the bond moment of the C—O bond is 0.82 D. Taking a bond length of 1.43 Å, the partial charge is given by

$$e = \frac{0.82}{1.43 \times 4.8028} = 0.119$$

For a charge of +1 and −1 separated by 1 Å, the dipole moment is 4.8028 D, which appears in the denominator of the given equation. A value of −0.119 is assigned to O and +0.119 to C. Typical bond moment values used are 0 for

C—C, 0.82 D for C—O, 0.38 D for C—H, 1.53 D for O—H, 2.3 D for C=O, and 0.74 D for a partial CO double bond. While this is a simple method which considers individual bonds, Del Re [51, 52] developed a molecular orbital linear combination of atomic orbital (MO-LCAO) method which takes into account all the bonds that are connected to an atom in a molecule.

It should be mentioned that the calculations of interactions that were discussed earlier are between a pair of atoms in a molecule and are additive. Atoms are treated as point centers. Influence of a neighboring atom on the force field between any pair of atoms is not explicitly taken into account, that is, these are only two-body potentials. Multi-body potentials are not known for organic molecules, although these have been developed for metals [53, 54]. To reduce computational time, the "united atom" approach has also been used for the calculations of conformational energies. In this case, for example, CH_2 or CH_3 groups are treated as a single entity, with the carbon atom as the point center and equivalent van der Waals radii used for the calculations (as in Table 2.1). This is the simplest example of the "coarse grain modeling" approach adopted for computational efficiency. The united atom description has also been used in the assignment of statistical weights for the different conformations of chains for the calculation of the properties of random chain configurations. Flory [55] treated the various conformations of the chains as rotational isomers and the assignment of statistical weights based on the relative energies of the conformers (rotational isomers) was used in the rotational isomeric state (RIS) modeling of polymer chains.

In addition to the force fields that were discussed earlier, contributions due to deviation of bond lengths and bond angles from their "normal" values are also included in the calculation using empirical methods. These are the first two terms on the right-hand side of the Levitt equation in Figure 2.1. Yet another term is the "torsional contribution," which arises due to the deviation of a torsion angle involving three contiguous bonds from staggered conformation and the barrier associated with going from staggered to eclipsed conformation of a bond. This was added to the force field since the early calculations could not reproduce the (gas phase) barrier to internal rotation from nonbonded interactions alone. The general form of this function is shown in the Levitt equation in Figure 2.1, but equations with different terms have been used by various authors and the commercial software vendors. The equations given before for the various interactions are the generic ones and have been modified to different extents in commercial software for molecular simulations. Different forms of these equations applied to the same problem could sometimes lead to differences in the results, as discussed by Li et al. [56] for the case of chain folding of polyethylene. The forms of the equations for the different terms in the DRIEDING (Mayo et al. [15]) and COMPASS (Sun [57]) force fields were discussed by Li et al. [56] and are given in Tables 2.2 and 2.3.

In the studies of molecular self-assembly leading to crystallization, gelation, etc., it is not unusual to combine experimental observations with molecular simulations to model the molecular association. The experimental techniques for measuring the interactions and force fields discussed earlier have been developed and continually refined during the past decades.

References

1 M. Levitt, *Nat. Struct. Mol. Biol.*, **2001**, *8*, 392–393.
2 P. Muller, *Pure Appl. Chem.*, **1994**, *66*, 1077, DOI:10.1351/goldbook.V06597.
3 J. Ketelaar, *Chemical Constitution*, **1953**, Elsevier, p. 91.
4 R. R. Teachout and R. T. Pack, *At. Data*, **1971**, *3*, 195–214. (This publication continues as *Atomic Data and Nuclear Data Tables* (Elsevier)).
5 E. V. Anslyn and D. A. Dougherty, *Modern Physical Organic Chemistry*, **2006**, University Science Books, Sausalito, CA.
6 M. M. Coleman, K. H. Lee, D. J. Skrovanek, and P. C. Painter, *Macromolecules*, **1986**, *19*, 2149–2157.
7 M. Suzuki, Y. Nakajima, M. Yumoto, M. Kimura, H. Shirai, and K. Hanabusa, *Langmuir*, **2003**, *19*, 8622–8624.
8 S. Khanna, M. Moniruzzaman, and P. R. Sundararajan, *J. Phys. Chem. B*, **2006**, *110*, 15251–15260.
9 A. Abe, R. L. Jernigan, and P. J. Flory, *J. Am. Chem. Soc.*, **1966**, *88*, 631.
10 T. Steiner, *Angew. Chem. Int. Ed.*, **2002**, *41*, 48–76.
11 G. R. Desiraju, *Angew. Chem. Int. Ed.*, **2011**, *50*, 52–59.
12 E. Arunan, G. R. Desiraju, R. A. Klein, J. Sadlej, S. Scheiner, I. Alkorta, D. C. Clary, R. H. Crabtree, J. J. Dannenberg, P. Hobza, H. G. Kjaergaard, A. C. Legon, B. Mennucci, and D. J. Nesbitt, *Pure Appl. Chem.*, **2011**, *83*, 1637–1641.
13 E. R. Lippincott and R. Schroeder, *J. Chem. Phys.*, **1955**, *23*, 1099–1106.
14 F. A. Momany, R. F. McGuire, A. W. Burgess, H. A. Scheraga, *J. Phys. Chem.*, **1975**, *79*, 2361–2381.
15 S. L. Mayo, B. D. Olafson, W. A. Goddard III, *J. Phys. Chem.*, **1990**, *94*, 8897–8909.
16 K. Nakamoto, M. Margoshes, and R. E. Rundle, *J. Am. Chem. Soc.*, **1955**, *77*, 6480–6486.
17 G. C. Pimentel and C. H. Sederholm, *J. Chem. Phys.*, **1956**, *24*, 639–641.
18 L. J. Bellamy and A. J. Owen, *Spectrochim. Acta A: Mol. Spectrosc.*, **1969**, *25*, 329–333.
19 M. K. Khan and P. R. Sundararajan, *J. Phys. Chem. B*, **2013**, *117*, 5705–5717.
20 L. L. Alimova, E. G. Atovmyan, and O. S. Filipenko, *Kristallografiya*, **1987**, *32*, 97–101.
21 S. P. Brown, *Macromol. Rapid Commun.*, **2009**, *30*, 688–716.

22 S. P. Brown, *Solid State Nucl. Magn. Reson.*, **2012**, *41*, 1–27.
23 J. S. Waugh and R. W. Fessenden, *J. Am. Chem. Soc.*, **1957**, *79*, 846–849.
24 T. Takagi, A. Tanaka, S. Matsuo, H. Maezaki, M. Tani, H. Fujiwara, and Y. Sasaki, *J. Chem. Soc.*, Perkin Trans. *2*, **1987**, 1015.
25 M. Nishio, Y. Umezawa, M. Hirota, and Y. Takeuchi, *Tetrahedron*, **1995**, *51*, 8665–8701.
26 E. A. Meyer, R. K. Castellano, and F. Diederich, *Angew. Chem. Int. Ed.*, **2003**, *42*, 1210–1250.
27 H. Takahashi, S. Tsuboyama, Y. Umezawa, K. Honda, and M. Nishio, *Tetrahedron*, **2000**, *56*, 6185.
28 H. Suezawa, S. Ishihara, O. Takahashi, K. Saito, Y. Kohno, and M. Nishio, *New J. Chem.*, **2003**, *27*, 1609.
29 S. Paliwal, S. Geib, and C. S. Wilcox, *J. Am. Chem. Soc.*, **1994**, *116*, 4497–4498.
30 E.-I. Kim, S. Paliwal, and C. S. Wilcox, *J. Am. Chem. Soc.*, **1998**, *120*, 11192–11193.
31 W. R. Carroll, C. Zhao, M. D. Smith, P. J. Pellechia, and K. D. Shimizu, *Org. Lett.*, **2011**, *13*, 4320–4323.
32 G. R. Desiraju, P. Shing Ho, L. Kloo, A. C. Legon, R. Marquardt, P. Metrangolo, P. Politzer, G. Resnati, and K. Rissanen, *Pure Appl. Chem.*, **2013**, *85*, 1711–1713.
33 A. Mukherjee, S. Tothadi, and G. R. Desiraju, *Acc. Chem. Res.*, **2014**, *47*, 2514–2524.
34 A. Mukherjee and G. R. Desiraju, *IUCrJ*, **2014**, *1*, 49–60.
35 J. Lieffrig, O. Jeannin, A. Vacher, D. Lorcy, P. Auban-Senzierb, and M. Fourmigue, *Acta Crystallogr.*, **2014**, *B70*, 141–148.
36 M. Saccone, G. Terraneo, T. Pilati, G. Cavallo, A. Priimagi, P. Metrangoloa, and G. Resnatia, *Acta Crystallogr.*, **2014**, *B70*, 149–156.
37 K. Durka, S. Luliński, K. N. Jarzembska, J. Smętek, J. Serwatowskia, and K. Woźniak, *Acta Crystallogr.*, **2014**, *B70*, 157–171.
38 P. Politzer, J. S. Murray, and M. C. Concha, *J. Mol. Model.*, **2007**, *13*, 643–650
39 N. Song, D. Yao, Z. Y. Wang, and P. R. Sundararajan, *Polymer*, **2005**, *46*, 3831–3837.
40 J. Joseph and E. D. Jemmis, *J. Am. Chem. Soc.*, **2007**, *129*, 4620–4622.
41 Y. Tatamitani, B. Liu, J. Shimada, T. Ogata, P. Ottaviani, A. Maris, W. Caminati, and J. L. Alonso, *J. Am. Chem. Soc.*, **2002**, *124*, 2739–2743.
42 V. K. Sharma and R. D. Rogers, *Cryst. Eng.*, **1998**, *1*, 139–145.
43 P. Hobza and Z. Havlas, *Chem. Rev.*, **2000**, *100*, 4253–4264.
44 S. N. Delanoye, W. A. Herrebout, and B. J. van der Veken, *J. Am. Chem. Soc.*, **2002**, *124*, 11854–11855.
45 E. Y. Cheung, E. E. McCabe, K. D. M. Harris, R. L. Johnston, E. Tedesco, K. M. P. Raja, and P. Balaram, *Angew. Chem. Int. Ed.*, **2002**, *41*, 494–496.
46 P. R. Shirhatti, D. K. Maity, S. Bhattacharyya, and S. Wategaonkar, *ChemPhysChem*, **2014**, *15*, 109–117.

47 E. Dahan and P. R. Sundararajan, *Soft Matter*, **2014**, *10*, 5337–5349.

48 M. A. Viswamitra, R. Radhakrishnan, J. Bandekar, and G. R. Desiraju, *J. Am. Chem. Soc.*, **1993**, *115*, 4868–4869.

49 M. S. Rozenberg, T. Nishio, and T. Steiner, *New J. Chem.*, **1999**, *23*, 585–586.

50 C. Rees, *J. Mol. Biol.*, **1980**, *141*, 323–326.

51 G. Del Re, *J. Chem. Soc.*, **1958**, 4031–4040.

52 G. Del Re, *Biochim. Biophys. Acta*, **1963**, *75*, 153.

53 S. B. Sinnott and D. W. Brenner, *MRS Bull.*, **2012**, *37*, 469–473.

54 S. J. Plimpton and A. P. Thompson, *MRS Bull.*, **2012**, *37*, 513–521.

55 P. J. Flory, *Statistical Mechanics of Chain Molecules*, **1969**, John Wiley & Sons, Inc., New York (Reprinted by Hanser Publishers 1989).

56 C. Li, P. Choi, and P. R. Sundararajan, *Polymer*, **2010**, *51*, 2803–2808.

57 H. Sun, *J. Phys. Chem. B*, **1998**, *102*, 7338.

3

Features of Self-Assembly

3.1 Self-Sorting—Small Molecules

As discussed in the previous chapters, self-assembly requires molecular recognition and noncovalent interactions. An additional condition is that there should be geometric registration, that is, would the conformation of the molecule(s) allow for the maximum number of noncovalent interactions. Although the structure of the two molecules might have groups that could lead to hydrogen bonding or any such interactions, the actual formation of such noncovalent bonds would depend on the mutual conformational registry. For example, if two hydrogen bonding groups could not approach each other due to steric restrictions, self-assembly would not occur. These aspects play a major role in mixtures of two or more types of molecules. Each type of molecule might associate with its own kind (self-recognition) and ignore the presence of others (discrimination of the nonself) in the mixture. Or, one type of molecule might prefer to associate exclusively with another type bearing complimentary recognition sites as in the case of receptor–ligand interaction. This is the phenomenon of self-recognition or self-sorting, which was discussed by Lehn et al. [1], in mixtures of complexes of oligobipyridine ligands and metal ions. Figure 3.1 shows the case of oligobipyridines that bind Cu^+ ions. Double helicates formed upon metal–ligand complexation, and these were quite specific, with a mixture of 2, 3, 4, and 5 units. The association occurred between ligands with the same number of units in the oligomer. A similar case of specific recognition is shown in Figure 3.2. The ligand **1** shown in the figure self-assembles with Ni(II) ions as a triple helicate structure with octahedral coordination and the ligand **3a** forms a double helicate with Cu(I) ions with tetrahedral coordination [1, 2]. When a mixture of 3 : 2 equivalent of **1** and **3a** and three equivalents each of Cu(I) and Ni(II) was treated as described in Kramer et al., quantitative yields of the respective double and triple helicates were obtained, with no complexation between **1** and **3a**. The specific association in this case was driven by the tetrahedral and octahedral binding with Cu and Ni ions.

Physical Aspects of Polymer Self-Assembly, First Edition. P. R. Sundararajan.

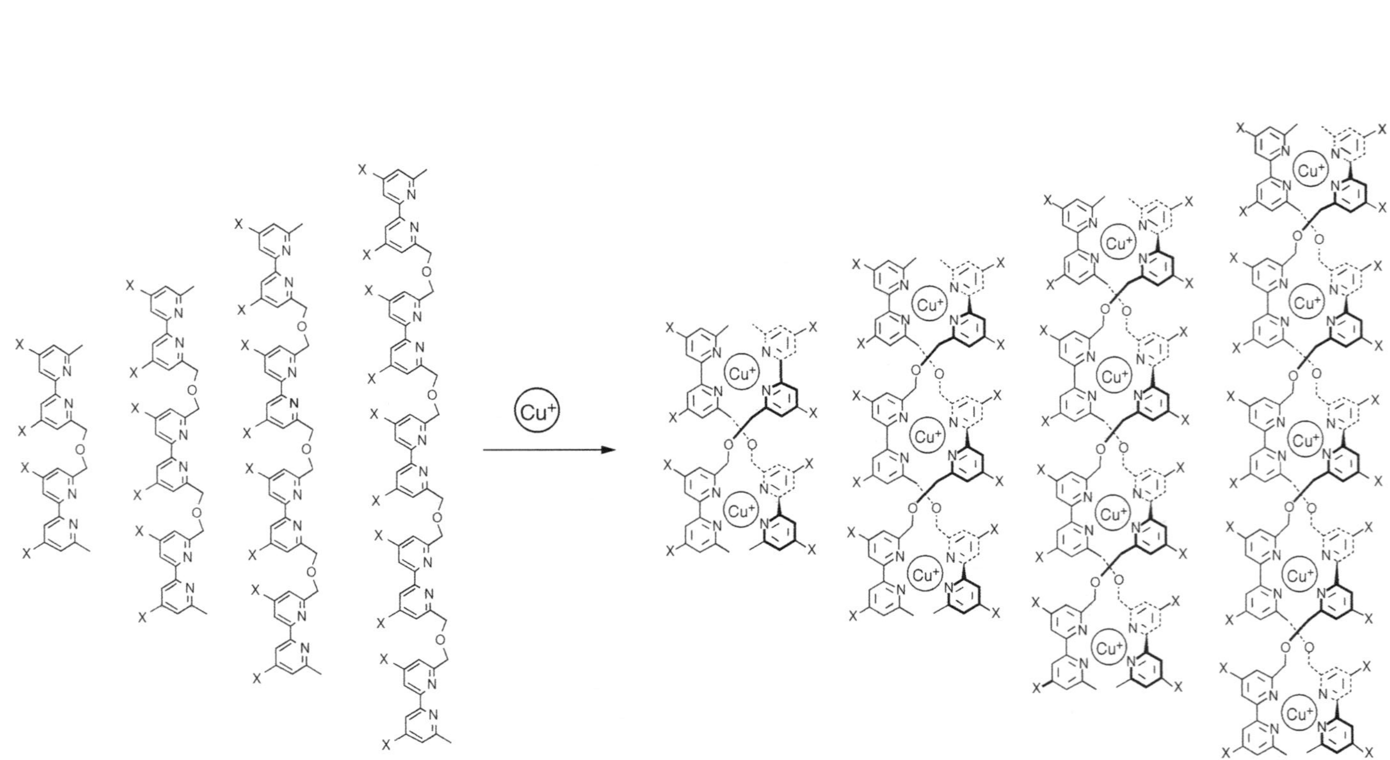

2 3 4 5 ${}^{2}\mathscr{H}(2)$ ${}^{2}\mathscr{H}(3)$ ${}^{2}\mathscr{H}(4)$ ${}^{2}\mathscr{H}(5)$

a X = H
b X = $CONEt_2$
c X = COOEt

Figure 3.1 Self-assembly of oligobipyridine units of the same length upon complexation with the metal ion. (Source: Krämer et al. [1]. Reproduced with permission of National Academy of Sciences.)

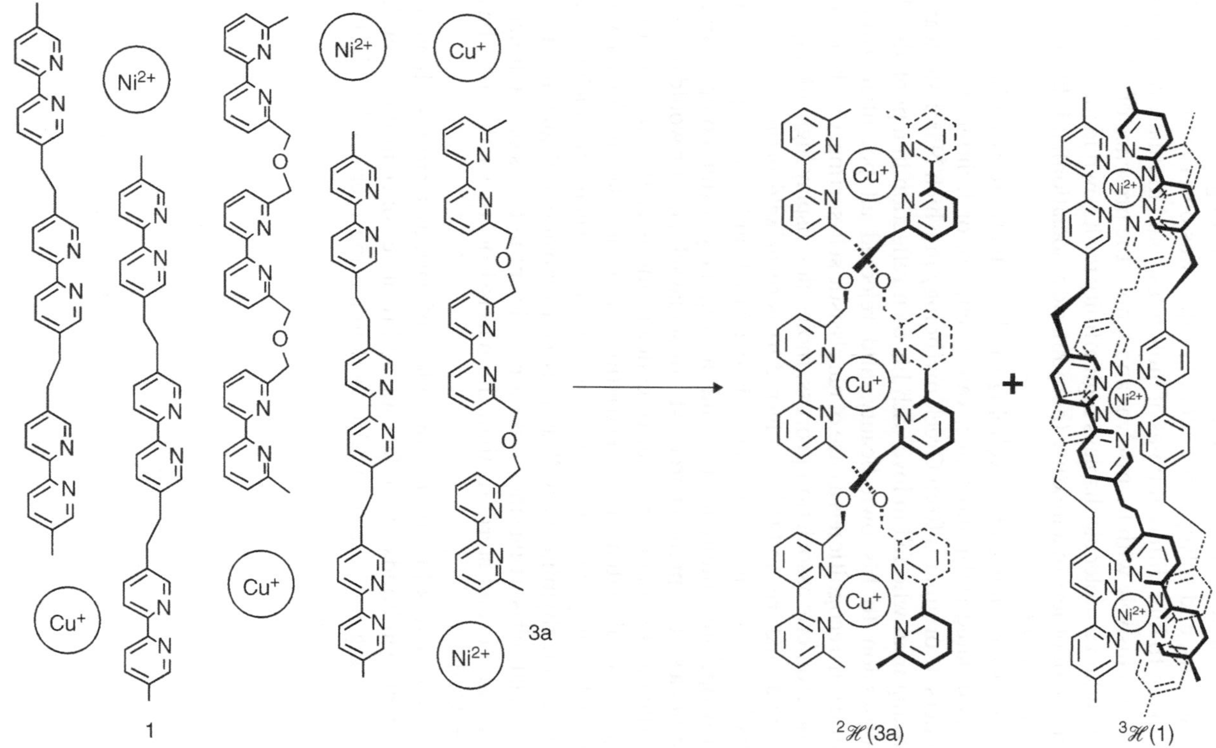

Figure 3.2 Self-recognition in ligand/metal complexes [2]. Parallel formation of a double helicate and a triple helicate by self-selection with self-recognition from a mixture of two different suitably instructed ligands and two different types of metal ions. (Source: Krämer et al. [1]. Reproduced with permission of National Academy of Sciences.)

Lehn termed this phenomenon as "self-recognition," while later authors used the term "self-sorting." This refers to the high-fidelity recognition of self, from the nonself, facilitated by noncovalent codes such as hydrogen bonds, metal–ligand interactions, and solvophobic effect [3–5]. It was observed that closely related sets of up to four different types of molecules could mutually recognize and associate exclusively with themselves and ignore the presence of others in the mixture. When a molecule displays high affinity for its own identical kind, to the exclusion of others in a mixture (homomeric association), it is known as *narcissistic self-sorting.*

"Narcissism" is a term used in psychology and psychiatry to describe conditions such as inflated self-importance, vanity, elitism, or indifference to the plight of others. It originates from Greek mythology in which Narcissus was a beautiful young man who fell in love with his own reflection in a pool, developed an obsession to his own beauty, and rejected all potential lovers. Enantiomeric self-recognition is a case of narcissistic self-sorting. On the other hand, in a heteromeric association process, molecules display high affinity for other types in a mixture, and this is referred to as *social self-sorting.* Such systems display high levels of discrimination between self and nonself species and consequently operate simultaneously and orthogonally within complex mixtures. Whether a system prefers narcissistic or social self-sorting would depend on factors such as temperature, relative concentrations of the species in the mixture, geometrical match, steric restrictions, and the relative efficiency of the hydrogen bonding patterns. The association between the nucleic acid bases adenine and thymine as well as between cytosine and guanine is an example of social self-sorting since it is more efficient in terms of maximizing the hydrogen bonds and steric fit than the narcissistic association of a base with its own kind.

Isaacs and coworkers [4] studied the effect of various factors such as the number of hydrogen bonds, steric factors, and relative concentrations on the mode of self-sorting and hetero-chiral association, using a series of "molecular clips" and "tweezers." These are covalently connected aromatic systems as illustrated in Scheme 3.1, with hydrogen bond groups as side chains.

Figure 3.3 shows four molecular clips studied by Wu et al. [4, 5]. Molecules **1a** and **1b** are achiral meso forms with C_s symmetry, and (±) **2a** and (±) **2b** are chiral C_2-symmetric molecules. Nuclear magnetic resonance (NMR) spectra of these compounds in $CDCl_3$ and X-ray crystal structure analysis showed that they formed stable dimeric aggregates. Apart from two C=O···N—H hydrogen bonds, the aromatic side walls of one clip penetrates deep into the cleft of the pairing clip, resulting in strong π-interaction (as illustrated in Scheme 3.1b). It was also found that with (±) **2b**, only the hetero-chiral aggregate ((+)-**2b**•(−)-**2b**) was formed and enantiomeric self-recognition was not observed. In the latter, only one hydrogen bond is possible, whereas with hetero-chiral association, two hydrogen bonds occur.

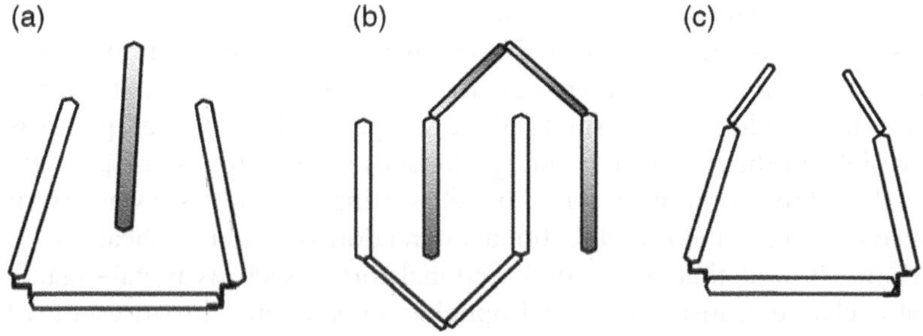

Scheme 3.1 Illustration of (a) and (b) a molecular clip and tweezer (c). ⬭, Aromatic group.

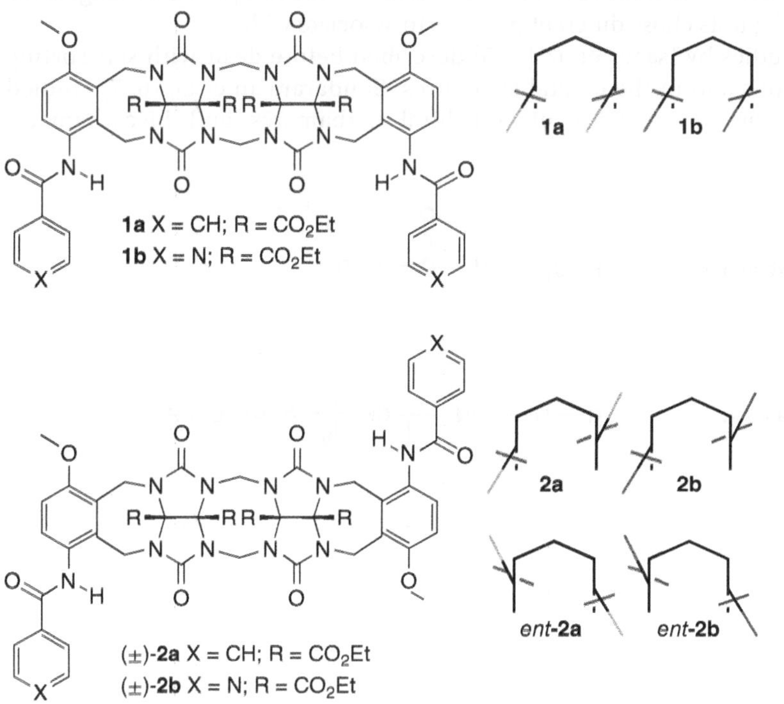

Figure 3.3 Structures of **1a**, **1b**, (±)-**2a**, and (±)-**2b**, and schematic representations of their three-dimensional shapes. Refer to [4] for the type of groups in the schematics.

The Isaacs group also studied mixers of eight types of molecules [3] and using experimental results and simulations concluded that with the concentrations of all the components being the same (i) the components exhibit high degree of thermodynamic self-sorting and (ii) narcissistic self-sorting was observed when the difference in the equilibrium constants (K_{AA}, K_{BB}, K_{AB}) was more than 10-fold. Self-sorting was found to be inefficient when the concentration of one component was significantly larger than those of others in the mixture, the difference in the equilibrium constants of the self-pair and the nonself-pair was less than 10-fold, or the hydrogen bonding efficiency of the heteromeric pair was more than that of the homomeric pair. The self-sorting in these cases was driven by hydrogen bonding, which is a directional interaction. In a later publication, [5] these authors showed that with nondirectional forces such as metal–ligand, electrostatic charge transfer, and hydrophobic interactions, mixtures of 12 components in water socially self-sort. In contrast to the eight-component mixture in which narcissistic self-sorting was observed, the social self-sorting of the 12-component mixture in water did not depend on the dilution or the relative stoichiometry. When the relative concentration of the components changed, the hosts and guests chose different partners to associate with.

The studies by Isaacs et al. [3–5] described before dealt with self-sorting in solutions and in the crystal structures. Sundararajan et al. [6, 7] studied the self-sorting behavior of nonchiral carbamates and biscarbamates

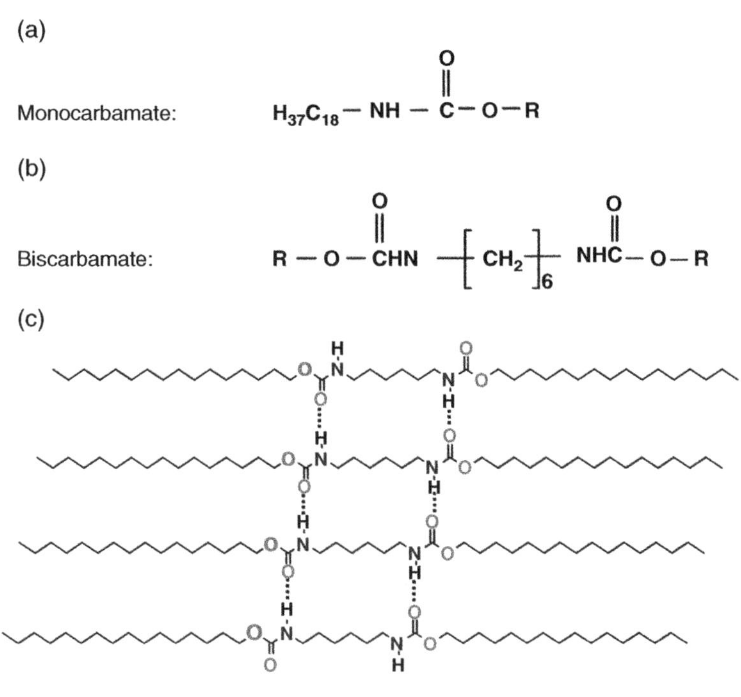

Figure 3.4 (a) and (b) Molecular structures of mono- and biscarbamates, respectively. (c) The hydrogen bond scheme in the latter is shown, where R is C_{12}. The monocarbamate is of the formula C_{18}—NHCOO—R_2, where R_2 varied from C_4 to C_{18}. The biscarbamates are R—OOCNH—C_6—NHCOO—R; R = C_4 to C_{18}.

(Figure 3.4) during crystallization from the melt. These are model compounds for polyurethanes and are cousins of ureas. Their melting temperature ranges from about 70 to 100°C, and the melt viscosity is only about 10 cps, which makes them suitable materials as vehicles for phase change inks in inkjet printing [8, 9]. The monocarbamates have a single hydrogen bonding group, a C_{18} side chain on one side and an alkyl chain on the other, varying in length from C_4 to C_{18}. The biscarbamates have two hydrogen bonding groups separated by a C_6 spacer and alkyl side chains on both sides, symmetrically varying from C_4 to C_{18}. The C=O···NH hydrogen bonds (Figure 3.4) between the biscarbamate molecules are similar to those in peptides. When these are used in inkjet printers, they get ejected in the melt form, and they solidify almost instantly upon hitting a substrate such as paper. The size of the spherulites upon crystallization has to be controlled to be of the order of the wavelength of visible light to avoid haze, etc., in the print. Mixtures of any two monocarbamates or any two biscarbamates were examined with a view to modifying the melting temperature and the spherulite morphology. In a mixture of any two biscarbamate molecules, both of them have the same spacer and hydrogen bond pattern and differ only in the number of CH_2 groups in the side chain. However, it was found that they do not cocrystallize but self-sort, even if the lengths of the side chains differ only by two CH_2 groups. They exhibit two distinct melting endotherms in the differential scanning calorimetry. More recently, such self-sorting was also reported for bis-urea systems by the group of Sijbesma [10].

3.2 Polymer Self-Sorting

The studies described earlier were on the self-sorting of small molecules. As for polymer systems, one might argue that the very phase separated morphology of block copolymers is due to self-sorting, and that there is an extensive literature on this topic. However, in the case of block copolymers, there is an interface between the phase separated domains [11–13]. Similar is the case of phase separated polymer blends.

The concept of self-sorting of polymers was studied more than 20 years ago by Coleman et al. in terms of self-association and interassociation in hydrogen bonded polymer blends [14, 15]. It was illustrated using phenol as an example. Figure 3.5a and b shows mutually hydrogen bonded self-association of a dimer and a trimer. Hydrogen bonding between phenol and ethylpropionate is shown in Figure 3.5c. The latter cannot form its own hydrogen bonded assembly, although it can be an acceptor via its carbonyl group, associating with phenol. Figure 3.5d shows self-association between phenols and interassociation with ethylpropionate.

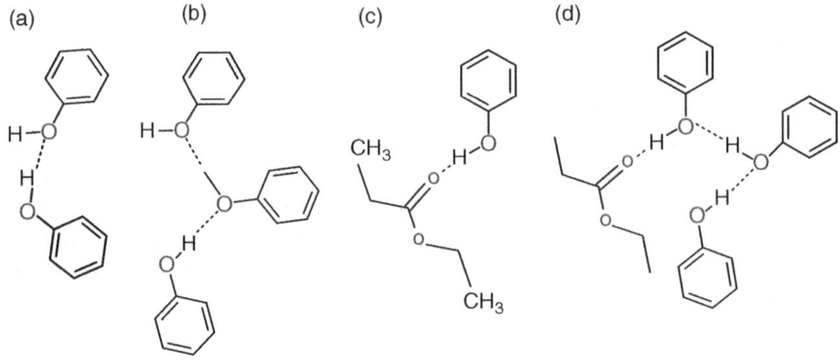

Figure 3.5 Various hydrogen bonding patterns between phenol and ethylpropionate. (a) and (b) Phenol self-association as dimer and trimer, respectively; (c) Inter-association between phenol and ethylpropionate; and (d) Inter- and intraassociations [14].

Miscibility or interassociation between two types of polymers is an exception rather than a rule mainly due to entropic effects. They self-sort. However, if the polymers bear hydrogen bonding groups, they could be miscible, depending on the relative strengths of the intraassociation and interassociation via hydrogen bonds. The free energy of mixing for hydrogen bonded polymer blends is given by

$$\frac{\Delta G_m}{RT} = \left(\frac{\Phi_A}{N_A} \ln \Phi_A + \frac{\Phi_B}{N_B} \ln \Phi_B \right) + \chi \Phi_A \Phi_B + \frac{\Delta G_H}{RT} \quad (3.1)$$

where the first term (in brackets) represents combinatorial entropy which would be small for high-molecular-weight polymers in the blend, the second term refers to the physical forces (e.g., van der Waals repulsions) which are unfavorable and make a positive contribution, and the last term accounts for chemical forces (hydrogen bonds) which would be favorable and make a negative contribution. In this equation, Φ_A and Φ_B are the volume fractions of polymers A and B with degree of polymerization of N_A and N_B, respectively. The interaction parameter χ is given by

$$\chi = \frac{V_r}{RT} (\delta_A - \delta_B)^2 \quad (3.2)$$

where δ_A and δ_B are the solubility parameters which can be calculated from group contribution methods and V_r is a reference volume taken to be 100 cm^3/mol. The first two terms in Eq. 3.1 correspond to Flory–Huggins equation for mixing. It applies to nonpolar chains, randomly mixing with no conformational order, that is, amorphous state. When two molecules of hydrogen bond with each other, a specific interaction precipitates, which also restricts the molecule to a conformation that facilitates the noncovalent bond. The third term in the given equation is to account for the free energy of such interactions.

The difference in the magnitude of self-association versus interassociation determines the extent of miscibility (association or social self-sorting) of two dissimilar polymers. The strength of the ΔG_H/RT term depends on the number of hydrogen bonding sites per unit volume in the blend. An ideal situation would be that hydrogen bonding occurs between all the available pairs throughout the length of the chains in the blend. However, as Coleman pointed out, the system could be diluted depending on the composition of the polymer. To cite their example, poly(vinyl phenol) (PVP) or poly(4-hydroxy styrene) contains a hydroxyl group on every repeat unit in the chain. Copolymerizing PVP with styrene units would reduce the number of hydrogen bonding sites per unit volume. Using such dilution effects on small-molecule analogues, combined with infrared (IR) spectroscopy, Coleman et al. [14, 15] determined self-association equilibrium constants and the strength of the hydrogen bonds. Two dimensionless equilibrium constants K_2 and K_B were defined for dimer and multimer formation, respectively. As an example, to determine the constants K_2 and K_B for 2-propanol, IR spectra were recorded for three different concentrations in cyclohexane (a noninteracting solvent) at 25°C, and for one concentration at various temperatures. The absorption bands at 3630, 3530, and 3350 cm^{-1} were assigned to the free (nonhydrogen bonded) —OH groups, hydrogen bonded dimers, and multimers, respectively. As seen in Figure 3.6, at a low concentration of 0.02 M, the 3630 cm^{-1} peak was predominant. As the concentration was increased to 0.09 M, the intensity of the peaks at 3530 and 3350 cm^{-1} increased comparably indicating the formation of dimers and multimers. With 0.3 M, the peak due to multimers became the most intense. The method for calculating K_2 and K_B has been described in detail by Coleman and Painter. This methodology led to values of 34.8 and 57.6 for K_2 and K_B at 25°C, which decreased to 15.3 and 24.7, respectively, at 70°C. Using the temperature dependence of the constants, the enthalpies of hydrogen bonding were calculated to be −3.6 and −3.9 kcal/mol for dimers and multimers, respectively.

Using an approach similar to that used for low-molecular-weight model compounds, the self-association equilibrium constants were calculated by Coleman and Painter for a few polymers bearing hydroxyl groups. Figure 3.7 shows four polymers (**1**–**4**) and their K_2 and K_B. The effect of steric hindrance on self-association can be clearly seen from these data. The value of K_2 is 21 for PVP or poly(hydroxyl styrene) and reduces to 4.8 for poly(3,5-dimethyl vinyl phenol) (**2**), with two methyl groups on either side of the —OH group. The constant for multimer formation also decreases from 66.8 for **1** to 17.4 for **2**. The mutual access of —OH groups from any two units in the chains is limited by the presence of the bulky methyl groups. A further decrease is seen for the polymer **3**. However, for PVP (**4**), both K_2 and K_B are much larger since the —OH groups have no steric encumbrance in this case.

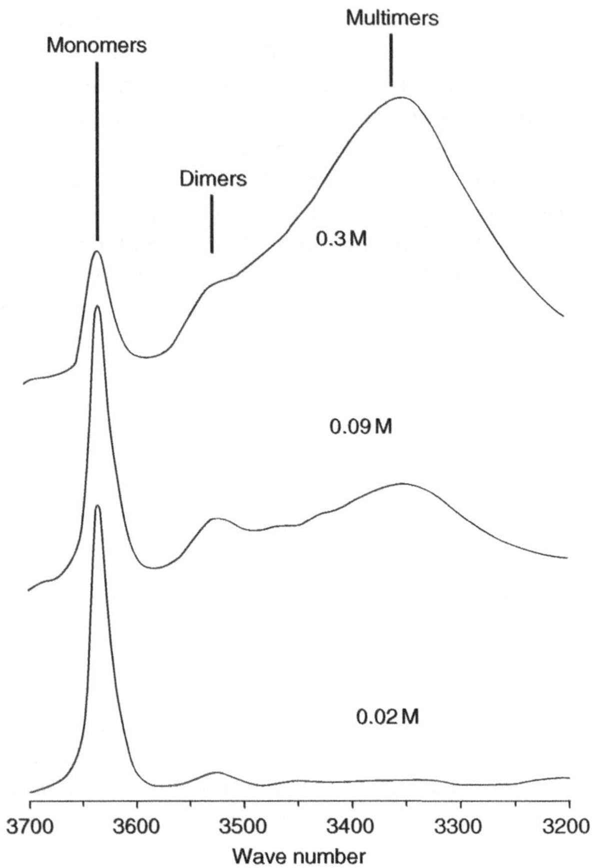

Figure 3.6 IR spectra in the hydroxyl stretching region recorded at 25°C for three different concentrations of 2-propanol in cyclohexane. (Source: Coleman and Painter [14]. Reproduced with permission of Elsevier.)

Extending the analysis to miscible polymer blends, equilibrium constants were determined for self-association (dimer and multimer) and interassociation. These would correspond to the narcissistic self-sorting and social self-sorting, respectively. Table 3.1 lists these values for three hydroxyl-bearing polymers and their association with methacrylate carbonyl or acetoxy carbonyl. The ratio K_A/K_B is indicative of the relative preference that the hydroxyl group exerts between the self- and interassociation.

PVP shows higher preference for self-association than interassociation with methacrylate or acetoxy carbonyl group containing polymers.

Table 3.1 Dimensionless self- and interassociation equilibrium constants determined from miscible polymer blends scaled to a common reference volume (100 cm^3/mol).

	Self-association		Interassociation (K_A)	
Polymer	Dimer formation K_2	Multimer formation K_B	Methacrylate carbonyl 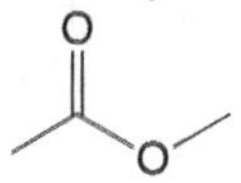	Acetoxy carbonyl 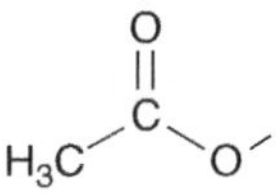
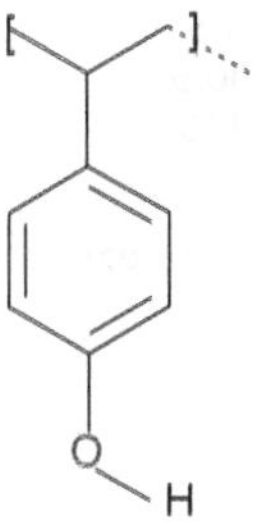	21.0	66.8	37.8 $K_A/K_B = 0.56$	58 $K_A/K_B = 0.87$
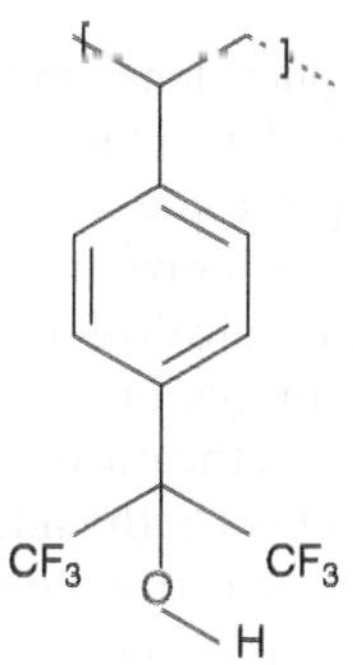	1.3	5.0	21.8 $K_A/K_B = 3.7$	58.7 $K_A/K_B = 10.1$
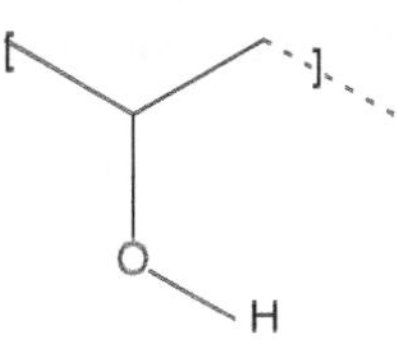	26.7	44.1	—	4.0 $K_A/K_B = 0.09$

Source: Coleman and Painter [14]. Reproduced with permission of Wiley.

Poly(hexafluoro-2-hydroxy-2-propyl styrene) (polymer 3 in Figure 3.7), however, interassociates strongly with these carbonyl group containing polymers. The values of 3.7 and 10.1 obtained for K_A/K_B, respectively, are significantly higher than those for PVP.

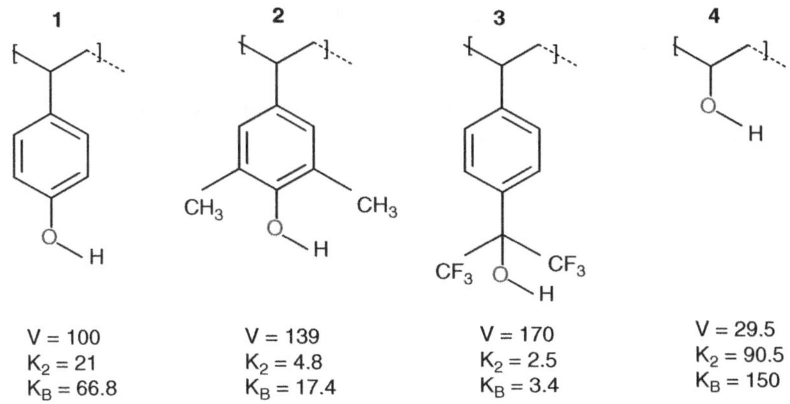

Figure 3.7 Dimensionless self-association equilibrium constants for four hydroxyl-bearing polymers [14]. V, molar volume (cm^3/mol); K_2, dimer; K_B, multimer.

3.3 Concentration-Dependent Association

Mayoral et al. [16] reported a case where the same pair of oligophenylene ethynylene (OPE) molecules showed narcissistic or social self-sorting depending on the relative concentration. One of the OPEs was substituted with polar side chains (OPE-1), and the other, with nonpolar (OPE-2), as shown in Figure 3.8. Individually, each of them formed spherical aggregates of a few nanometers in tetrahydrofuran (THF)/water mixtures. Scanning electron microscopy (SEM) images of spin-coated films also confirmed the spherical morphology. In each case, the micelles were stabilized by the aromatic interactions and the hydrophobic effect of the side chains shielding the molecule from the aqueous environment. Spectroscopic studies upon heating and cooling the solutions showed the association to be thermally reversible, with a single binding constant, which lead to the conclusion that the assembly was isodesmic. Absorption spectra of mixtures of OPE-1 and OPE-2 at very low concentrations (1×10^{-5}–1×10^{-4} M) in THF/water showed simply an overlap of the individual spectra, indicating narcissistic self-sorting. The two types form their own spherical aggregates, as illustrated in Figure 3.9c.

However, upon increasing the concentration to 1 mM, with comparable ratios of OPE-1 and OPE-2, significant changes in the spectra were observed, confirming co-assembly. Upon increasing the temperature, partial dissociation of the aggregates to monomeric forms was observed, and the co-assembly was achieved again upon cooling. Using a plot of the temperature dependence of the fraction of aggregates, it was concluded that the association was cooperative, as described by the nucleation–elongation model. The nucleation involved about six molecules of OPE-1 and/or OPE-2, which, with an increase in local concentration of the molecules, activated the co-assembly. Such social self-sorting led to ribbon-like

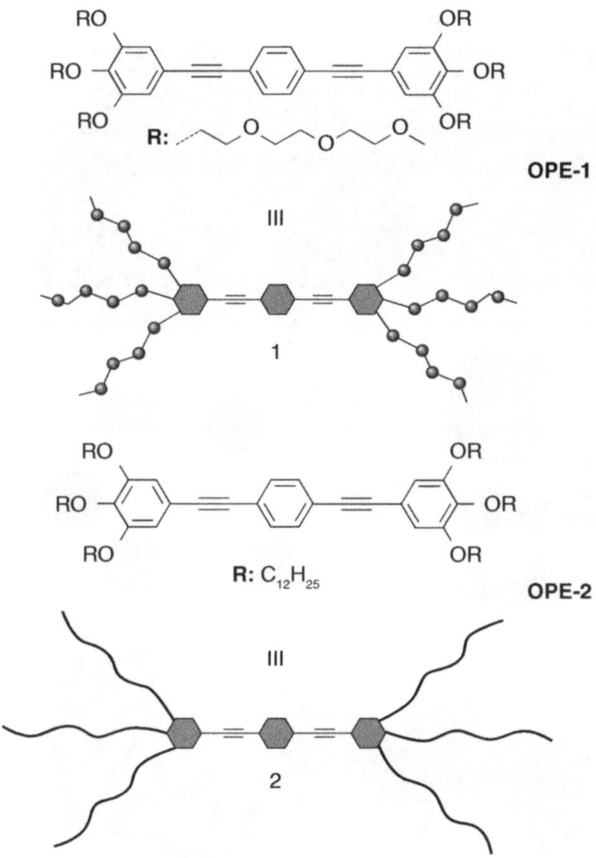

Figure 3.8 Chemical structures and cartoon representation [16] of OPE-1 and OPE-2.

morphology, as in Figure 3.9. Thus, in the case of OPE-1 and OPE-2, which have polar and nonpolar side groups, respectively, narcissistic self-sorting is seen at low concentrations (0.1–0.01 mM) in THF/water mixtures, which transforms to social self-sorting with higher concentrations (1 mM).

3.4 Polymer–Guest Molecule Recognition

Studies on the polymer self- and interassociation discussed earlier were aimed at understanding the miscibility enabled by the presence of, for example, hydrogen bonding groups in the chain and fabricating blends of commodity polymers for specific applications. Later research on self-sorting in polymers was aimed at using recognition elements and their targeted association.

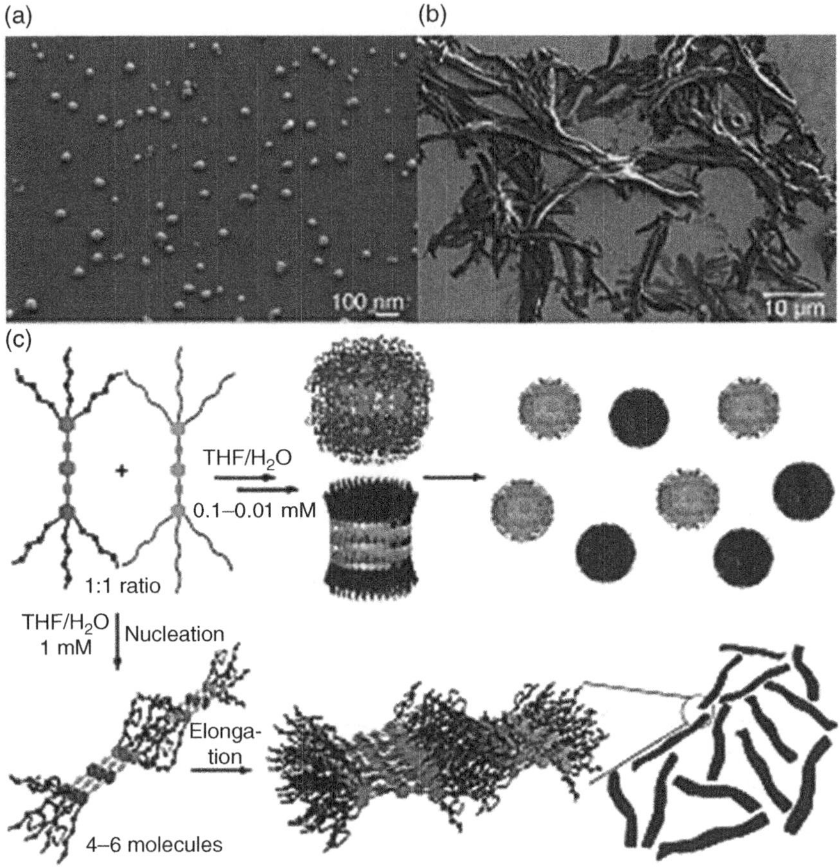

Figure 3.9 SEM images obtained by spin-coating a 1 : 1 mixture of OPE-1 and OPE-2 in THF/water at (a) 0.1 mM and (b) 1 mM on a silicon wafer. (c) Cartoon representation of the concentration-dependent self-sorting behavior of a 1 : 1 mixture of OPE-1 and OPE-2 in THF/water (1 : 1). Narcissistic self-sorting at concentrations between 0.01 and 0.1 mM (top) and social self-sorting at 1 mM (bottom). (Source: Mayoral et al. [16]. Reproduced with permission of Wiley.)

Supramolecular polymers are part of this effort (Chapter 4). In designing such polymers, analogues of nucleic acid bases were often used to create multiple sequential hydrogen bonding sites. For example, the patterns in Scheme 3.2 have a sequence of hydrogen bond donor–acceptor–donor (D–A–D) entities in the fragment shown on the top and the corresponding acceptor–donor–acceptor (A–D–A) sites at the bottom, which resembles the base thymine or uracil. The polymer receptor is then designed with one of the hydrogen bonding patterns as part of the repeat unit or as a side group. The commonly used templates are shown in Figure 3.10. Self-sorting then refers to the targeted

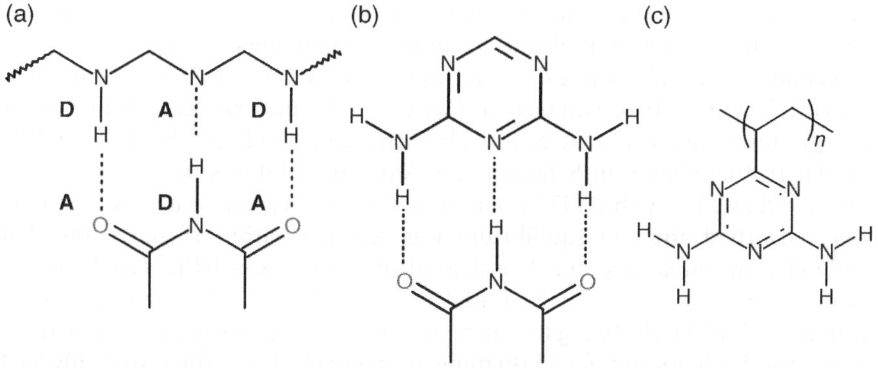

Scheme 3.2 Illustrations of (a) and (b): pairs with donor–acceptor–donor: acceptor/donor/acceptor hydrogen bonding scheme. Top, triazine segment; bottom, part of thymine or uracil. (c) Triazine-based vinyl polymer (PVDAT; see Ref. [17]).

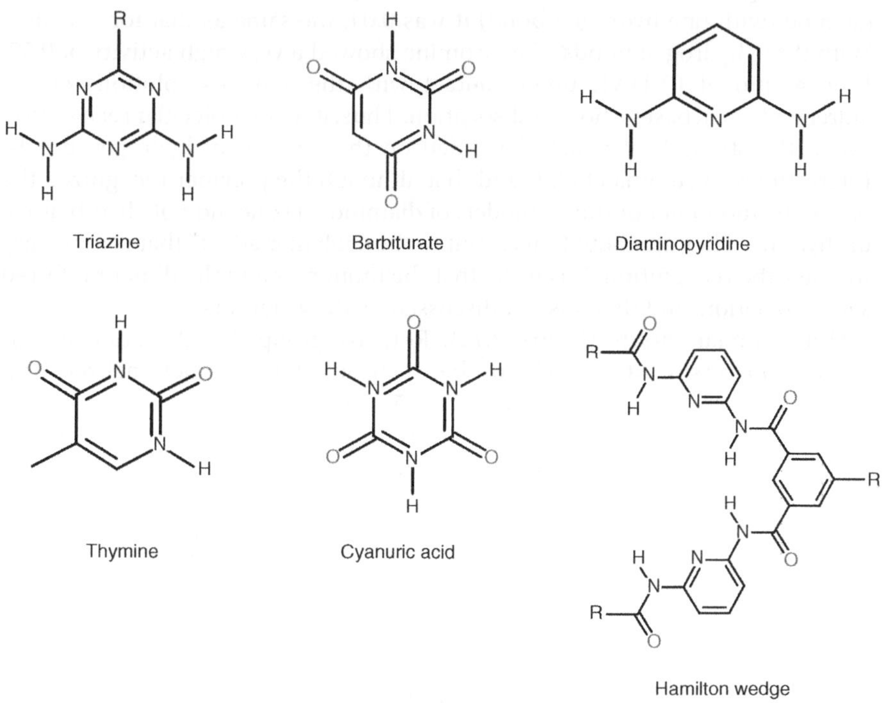

Figure 3.10 Commonly used donor–acceptor templates.

binding of a small-molecule ligand by the receptor-coded polymer. This is similar to the studies on molecularly imprinted polymers [18, 19].

Asanuma et al. [17] synthesized the polymer poly(2-vinyl-4,6-diamino-1,3,5-triazine) (PVDAT) based on triazine (Scheme 3.2c) and examined its molecular recognition of nucleic acid bases. The hydrogen bonding scheme would be similar to that shown in Scheme 3.2b. Aqueous solutions of the bases were incubated at 25°C with PVDAT, and it was found that the uracil and thymine were absorbed and the equilibrium was reached within 15 min. Note that PVDAT is insoluble in water. The absorption activity was defined as the molar ratio of absorbed guest to its initial concentration (1 mM) in water, upon contact with PVDAT (10 mg in 1 ml aqueous solution of the guest). The activity was 0.2 and 0.26 for uracil and thymine, respectively. However, it was only 0.04 with cytosine. The activity which is related to recognition depends on the number of hydrogen bonds between PVDAT and the guest. Both uracil and thymine could form three hydrogen bonds with the diamino triazine side group (as shown in Scheme 3.2b), but cytosine is limited to one hydrogen bond. Similar was the case with purine bases. For adenine, which could form two hydrogen bonds with PVDAT, the absorption activity was 0.13, and for guanine (with one hydrogen bond) it was 0.04, the same as that for cytosine. With three hydrogen bonds, theobromine showed a very high activity of 0.52. Even a film of PVDAT, upon contact with the aqueous solutions of the aforementioned bases, showed absorption. Thus, it was a molecular recognition event, the strength of which depended on the number of hydrogen bonds. Interestingly, Asanuma et al. found that although the polymer recognized the bases, the monomer or dimer models of diamino triazine did not absorb uracil or thymine. They attributed this result to the "polymer effect" that is necessary to cause the recognition. It is likely that the monomer and the dimer preferred self-association, and this was not discussed by these authors.

Using similar motifs (Figure 3.11), Rotello's group [20, 21] studied self-association versus interassociation between polystyrene functionalized with diaminotriazine (polymer **1**) and flavin (**5**). The latter is derived from flavin mononucleotide (**6** in Figure 3.11d). Although Asanuma et al. [17] also used diaminotriazine as the recognition unit, they polymerized vinyl triazine which would connect it directly to the carbon chain backbone. Rotello et al. [20, 21] attached triazine to the phenyl side group of polystyrene, which would give it additional conformational flexibility.

Polymer **1** in Figure 3.11a is substituted with diaminotriazine. It was found that polymer **1** coiled and formed intramolecular hydrogen bonds and its radius of gyration in $CHCl_3$ was 5.6-fold less than that of the parent polymer which had a (nonhydrogen bonding) cyano substitution instead of the triazine. Due to the preference of polymer **1** for intramolecular hydrogen bonds, binding of flavin 5 was very poor, compared to the association (Figure 3.11c) between the triazine monomer (**2**) and flavin. At 298 K, the binding constant K_a was

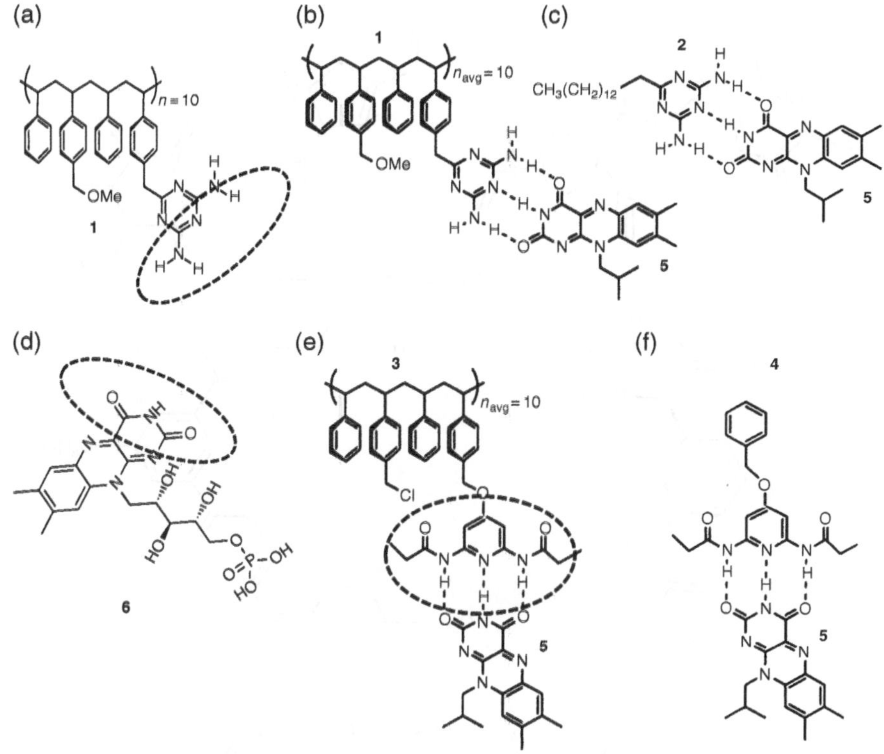

Figure 3.11 (a) Polystyrene functionalized with diaminotriazine (polymer **1**); (b), (c), (e), and (f), respectively, show the hydrogen bonding pattern between flavin (**5**) and polymer **1**, monomer **2**, polymer **3**, and monomer **4**. The molecular structure of flavin mononucleotide (**6**) is shown in (d). The hydrogen bonding fragments are indicated by dashed ellipses. (Source: Ilhan et al. [20, 21]. Reproduced with permission of American Chemical Society & Royal Society of Chemistry.)

$480\,M^{-1}$ between monomer **2** and flavin 5, and was only $36\,M^{-1}$ with polymer **1**. Hence, this is a case of preference of polymer **1** for intraassociation.

Replacing the diaminotriazine side group with diacyldiaminopyridine (polymer **3** in Figure 3.11e) reduced the intramolecular hydrogen bonding, and the K_a for binding between **3** and **5** (Figure 3.11e) was only slightly less than the binding between the corresponding monomers **4** and **5** (as in Figure 3.11f). The association constants were calculated using the NMR chemical shifts. Other authors have used IR frequency shifts for this purpose.

Molecular recognition and self-sorting between pairs in mixtures of polymer host and guest molecules was demonstrated by Burd and Weck [22]. They designed two norbornene-based polymers **1** and **2**, as shown in Figure 3.12, one with a thymine derivative as the recognition unit, attached to the backbone

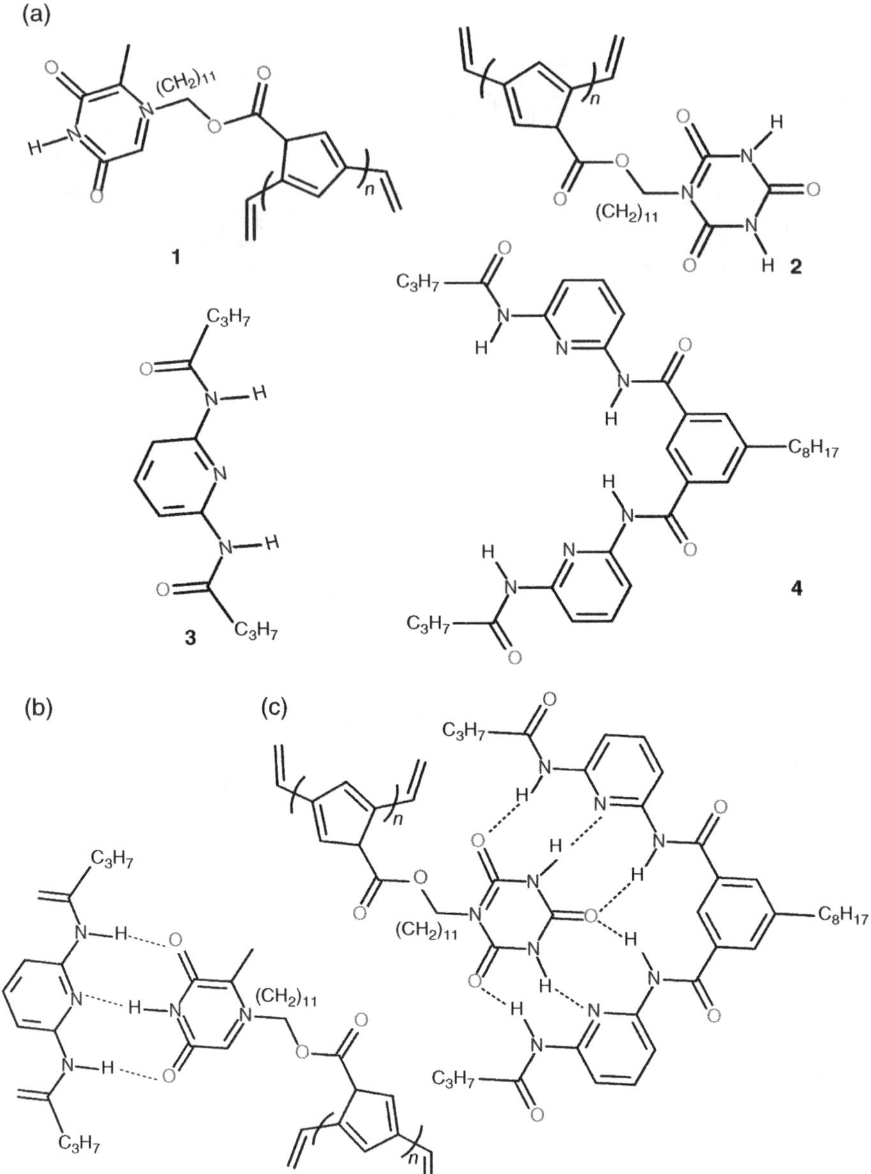

Figure 3.12 (a) Molecular structures of norbornene-based polymers **1** and **2** used by Burd and Weck [22] and the recognition units: dibutylamido pyridine (**3**) and isophthalic wedge-type (**4**) receptors. (b) and (c) Hydrogen-bonded assemblies of **1** with **3** and **2** with **4**, respectively.

via $(CH_2)_{11}$ spacer, and the other with a cyanuric acid side group with the $(CH_2)_{11}$ spacer. The purpose of the long spacer was to decouple the polymer backbones from the recognition units as well as to enhance the solubility. Block copolymers of **1** and **2** (50 : 50) as well as 50 : 50 random copolymers were also synthesized. Dibutylamidopyridine (**3**) and an isophthalic Hamilton wedge (**4**) were used as guests. The thymine derivative attached to polymer **1** would form three hydrogen bonds with **3** as shown in Figure 3.12b. The cyanuric acid attached to polymer **2** would form six hydrogen bonds with **4**, as in Figure 3.12c. For the self-assembly experiments, a solution of 2.8 equivalent of the guest (**3** or **4**) in dioxane:chloroform (15 : 85) was added dropwise to a 0.04 M solution of the polymer in (dioxane:chloroform (15 : 85)). In the initial experiments, only one homopolymer at a time, with its corresponding guest was used, and the association constants K_a for **3** : **1** and **4** : **2** were determined to be 1×10^2 and $3.0 \times 10^2\,M^{-1}$, respectively. These were the same as those for the association of the monomers of **1** and **2** with the respective guests.

Note that the K_a for 2 : 4 is higher due to the presence of six hydrogen bonds. To demonstrate that synthetic polymers also would exhibit self-sorting when coded with segments to serve as receptors, the following experiments were performed and the extent of association was determined using NMR chemical shifts of the imide protons of the thymine or cyanuric acid: (i) a one-pot mixture was prepared with 50 : 50 block copolymer and random copolymer of **1** and **2** with both the guests **3** and **4**. It was found that self-sorting occurred with binding of **3**–**1** and **4**–**2** with the same fidelity as with the homopolymers. The association constants were also similar within experimental error; (ii) in the next set of experiments, only one ligand **3** or **4** was added to the solution of the 50 : 50 random copolymers. When one receptor unit was left unsatisfied with the corresponding ligand, there is now competition between the two receptors on **1** and **2** to bind the available ligand. The binding of **4** was still strong, with only a slight decrease in K_a, whereas that of **3** was weaker; (iii) the dynamic nature of self-sorting was tested by adding one ligand after another to the solution of 50 : 50 block and random **1** : **2** copolymers: equivalent concentration of **3** followed by **4** and vice versa. Although binding to the nonselective receptor occurred upon the addition of one of the ligands, it vanished upon the addition of the second ligand to pair with the appropriate receptor.

3.5 Sergeant–Soldier Phenomenon

Polyisocyanates shown in Figure 3.13 adopt stiff helical conformations, which could be right (P) or left handed (M). The letters P and M refer to positive (clockwise) and negative (minus or counterclockwise) rotations to generate right-handed and left-handed helices, respectively. With an achiral side chain

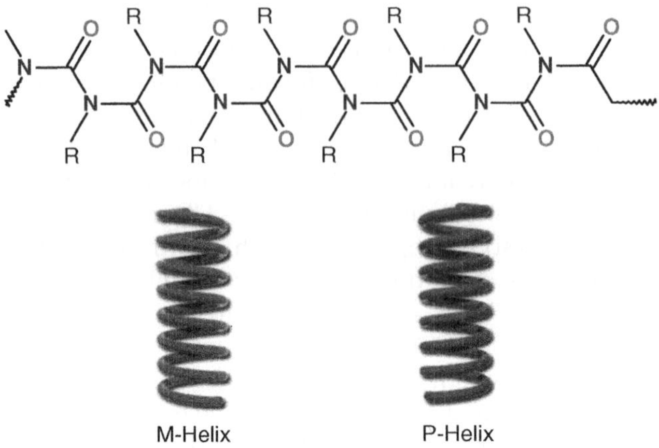

Figure 3.13 Schematic of a poly(isocyanate) chain which adopts a stiff right-handed (P) or left-handed (M) helical conformation.

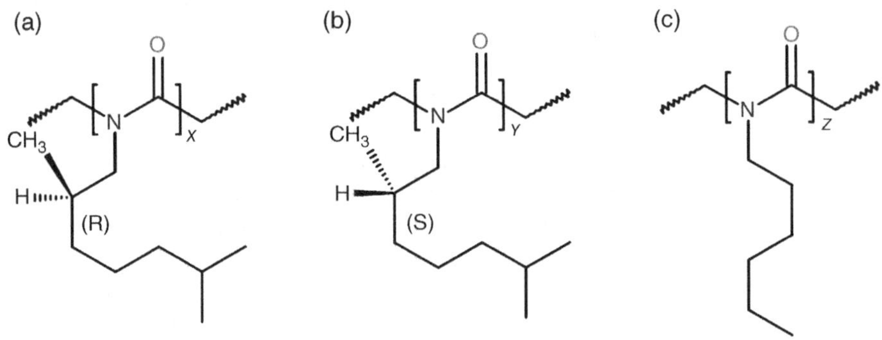

Figure 3.14 Illustrations of (a) and (b) chiral 2,6-dimethylheptyl isocyanate and (c) achiral *n*-hexylisocyanate polymers. *x*, *y*, and *z* denote the degree of polymerization [23, 24].

(Figure 3.14c), both helical senses occur with equal probability along the polymer chain, and hence no optical rotation would be seen. In the studies of Green et al. [23, 24], the chiral poly((*R*)-2,6-dimethylheptyl isocyanate) (Figure 3.14a) showed an optical rotation $[\alpha_D]$ of −514 and −500 at −20 and +20°C, respectively, in chloroform. The achiral poly(hexyl isocyanate) (Figure 3.14c) showed no optical rotation. However, when 15% (mol%) of the chiral monomer (a) was introduced to the achiral polymer ($x = 15$, $z = 85$), the copolymer showed $[\alpha_D]$ of −532 and −480 at −20 and +20°C, respectively, almost the same as that of the polymer containing only the chiral monomer. Even with 0.5% ($x = 0.5$, $z = 99.5$), $[\alpha_D]$ was −140 and −66 at −20 and +20°C, respectively. Figure 3.15 shows the plot of optical rotation of the random

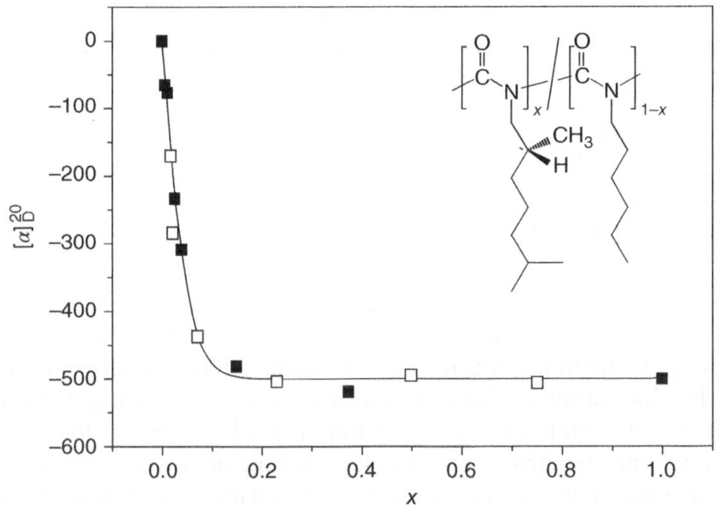

Figure 3.15 Optical rotation of the random copolymer shown in the inset, as a function of the fraction *x* of chiral monomer units. (*Plot made by Mayer and Zentel, from the data of Green et al.* [25]: ■, $c = 0.5$ g/l in $CHCl_3$; □: $c = 1$ g/l in THF. (Source: Mayer and Zentel [26]. Reproduced with permission of Elsevier.)

copolymers of (a) and (c) (Figure 3.14), as a function of the fraction of chiral monomer units of (a).

Thus, the presence of a very small fraction of a chiral unit (a sergeant) directs the rest of the achiral units (soldiers) in the chain to follow the former's sense of the helix to cause significant optical activity. This is the "sergeant–soldier" phenomenon. This is of interest as it provides means to control the sense of the helix precisely. Green et al. [23, 24] also found that chiral solvents ((R)-2-chloroalkanes) led to chiral induction in achiral poly(*n*-hexyl isocyanates). The induced chirality was different among these solvents. The circular dichroism (CD) spectrum in 2-chlorobutane was opposite that of 2-chloroheptane. Thus, apart from the chiral sergeant unit directing the achiral soldiers, even the chiral solvents induce helix sense preference in achiral polymers. To model the energy involved in this process, Lifson et al. [27] developed equations based on the statistical mechanics of chain conformations. For short chains, if $[\alpha]$ is the observed optical rotation and $[\alpha]_m$ is the maximum optical rotation for the helix of one sense,

$$\frac{[\alpha]}{[\alpha]_m} = \tanh\left(\frac{\Delta G_h N}{RT}\right) \tag{3.3}$$

where ΔG_h is the energy difference between the P and M helices and N is the degree of polymerization. For long chains, a parameter L is used which corresponds to the domain lengths between helix reversals and is related to ΔG_r, which is the energy involved in reversing the sense of the helix. L is related to exp (ΔG_h/RT).

$$\frac{[\alpha]}{[\alpha]_m} = \frac{L\Delta G_h / RT}{\left(\left(L\Delta G_h / RT\right)^2 + 1\right)^{1/2}} \tag{3.4}$$

Using this equation for a long chiral deuterated poly(n-hexyl isocyanate), the helix reversal energy was calculated [23] to be about 3.6 kcal/mol. Studies on sergeant–soldier phenomenon continue to be of interest in the literature, even 25 years after the observation of this behavior by Green et al. Recently, Nagata et al. [28] examined the influence of the position of the chiral center along the side chains of various lengths, using random copolymers and sequence-regulated copolymers of poly(quinoxaline-2,3-diyl)s with achiral and chiral side chains. They found that a monomer with (R)-2-octyloxymethyl group stabilized the P helix, while (S)-2-butoxymethyl group stabilized the M-helix. Depending on the sergeant, the helical chirality was also solvent dependent. For example, (R)-2-butoxymethyl side chain showed P-helix induction in chloroform, with a ΔG_h of 0.14 kcal/mol, whereas it showed M-helix in TCE with a ΔG_h of 0.076 kcal/mol. The energies were calculated using the formulations given in the aforementioned equations. The difference in energy between the P and M helices is extremely small. Similar were the conclusions of Green et al. [23, 24] on polyisocyanates. Thus, the energy difference between the P- and M-helices is only a few calories/mol, whereas the energy involved in helix reversal along the chain is about 4 kcal/mol.

Conjugated polymers with chiral side chains exhibit chiroptical properties, which were thought initially to be due to the helicity of the polymer chain. Further studies showed that it is the aggregation of the chains, not a single chain, that caused the circular dichroism, since CD was seen only in poor solvents or in the solid state. As an example, Fiesel and Scherf [29] studied the CD properties of chirally substituted poly(para-phenylene), shown in Figure 3.16. The films of the chiral homopolymer (Figure 3.16a) showed a strong negative Cotton effect, but the chloroform solutions exhibited only a weak CD. Upon using a mixture of chloroform/methanol, to make the solvent medium poor, strong CD spectrum was seen, with the intensity increasing with the nonsolvent methanol concentration. The Cotton effect seen with the film and in the poor solvent medium confirmed that aggregation of individual chains was required for the chiroptic behavior to be seen. Another observation was that the ultraviolet–visible (UV–vis) absorption spectra did not show any significant change with the methanol concentration in the poor solvent mixture. No change in the chain conformation was hence indicated.

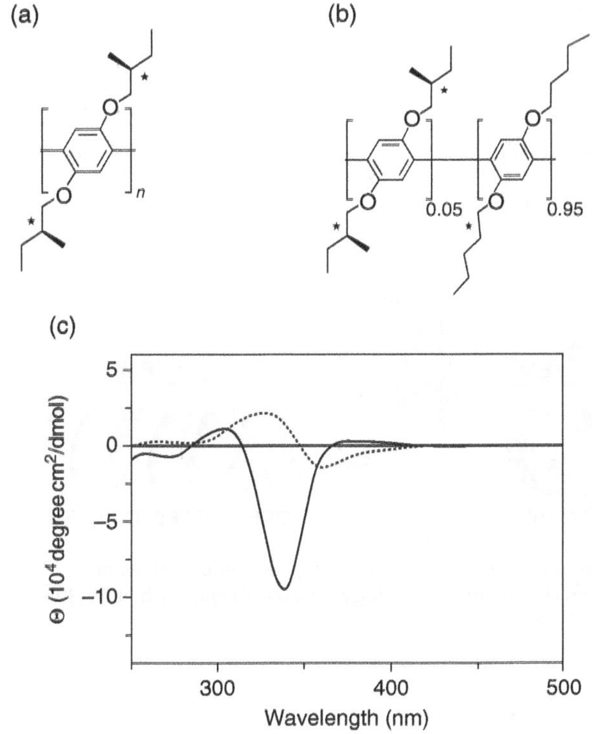

Figure 3.16 (a) Chemical structure of poly[2,5-bis(2-(S)-methylbutoxy)-1,4-phenylene]; (b) poly[2,5-di-*n*-pentoxy-1,4-phenylene-*co*-2,5-bis(2-(S)-methylbutoxy)-1,4-phenylene]; and (c) CD spectrum of (a) (solid line) and the copolymer (b) (dashed line) in chloroform/methanol: 30/70. (Source: Fiesel and Scherf [29]. Reproduced with permission of Wiley.)

Having confirmed that aggregation of the chains is necessary to observe the Cotton effect, Fiesel and Scherf [29] studied the effect of a chiral side chain on the otherwise achiral polymer. The copolymer shown in Figure 3.16b contains only 5% of the chiral monomer, and considering the molecular weight of the copolymer (M_n of about 8200), it was estimated that statistically each chain contains one to two chiral alkoxy group. Figure 3.16c shows that even such a small concentration of the chiral monomer induces significant chiroptical activity, following the sergeant–soldier principle.

The earlier examples dealt with copolymers containing chiral and achiral units, with the former inducing chirality. Even physical mixing of chiral molecules to achiral ones leads to sergeant–soldier effect. Apart from chiral induction, Ajayaghosh et al. [30] described the morphological change caused by sergeant–soldier phenomenon in the case of mixtures of oligo(*p*-phenylene ethynylene)s.

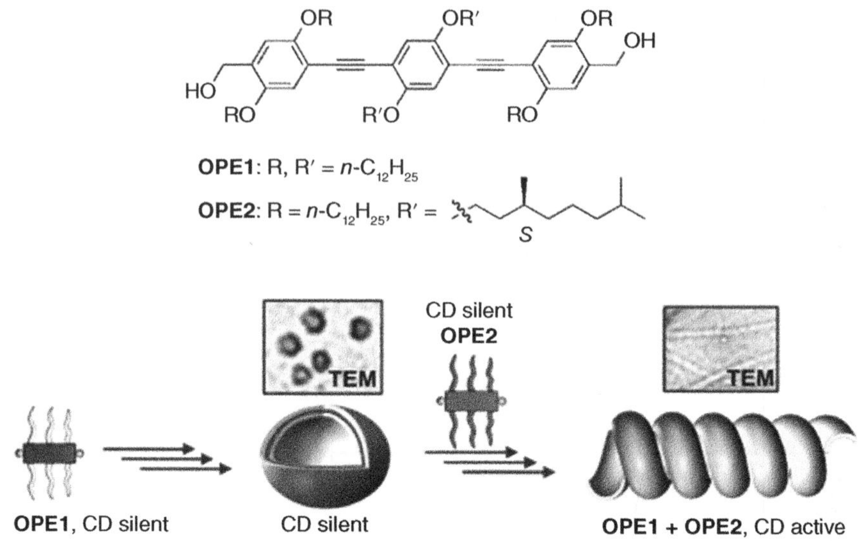

Figure 3.17 Schematics of (top) nonchiral and chiral oligo (*p*-phenylene ethynylene)s; (bottom): sergeant–soldier driven change in morphology. (Source: Ajayaghosh et al. [30]. Reproduced with permission of Wiley.)

The molecule OPE-1 shown in Figure 3.17 is nonchiral. Scattering experiments showed that OPE-1 forms spherical assembly in decane, and the transmission electron microscopy (TEM) analysis of the corresponding films showed them to be vesicles. Although chiral, OPE-2 did not show any CD signal in decane and did not self-assemble. However, upon mixing, OPE-2 induced chirality in OPE-1 and together they became CD active, although individually OPE-2 was not. As shown in Figure 3.17, the chiral induction transformed the vesicular morphology to helical assembly in decane. With 8 mol% of OPE-2 with OPE-1, both nanoparticles and helical structures were seen in the microscope images, and with 25 mol%, only the helical fibers were seen. These fibers were also found to be tubular.

There is at least one example where chirality is not a requirement to observe the sergeant–soldier effect. Organogelation of biscarbamates (Figure 3.4) has been studied by Sundararajan et al. [31]. With short side chains with even number of CH_2 groups, for example, C_8, the gel morphology (with benzonitrile as the solvent) consisted of short fibers as well as some crystals, as in Figure 3.18a. Long gel fibers were seen in the case of side chains with odd number of CH_2 groups. With equimolar concentration of biscarbamates with C_8 and C_9 side chains, only long fibers were seen (Figure 3.18c). The growth of

Figure 3.18 An example of sergeant–soldier effect on the gel fibers of nonchiral biscarbamates with (a) C_8 side chain, (b) C_9 side chain, and (c) C_8/C_9: 50/50. (Source: Khanna et al. [31]. Reproduced with permission of American Chemical Society.)

the gel fibers of the C_9 molecule helped sustain the growth of long fibers of C_8 molecule. Thus, the morphology of the C_8 soldiers was modified by the C_9 sergeant, and neither of them is chiral.

3.6 Majority Rules

In the case of sergeant–soldier behavior, the introduction of a small fraction of a chiral unit to an otherwise achiral polymer results in chiral induction. Green et al. [32] found that in random copolymers of poly(2,6-dimethyl heptyl isocyanate) containing (R) and (S) chiral units, the helix sense was dictated by the majority enantiomer, that is, the majority rules. Figure 3.19 shows the CD spectra of the homo- and copolymers of the (R) and (S)

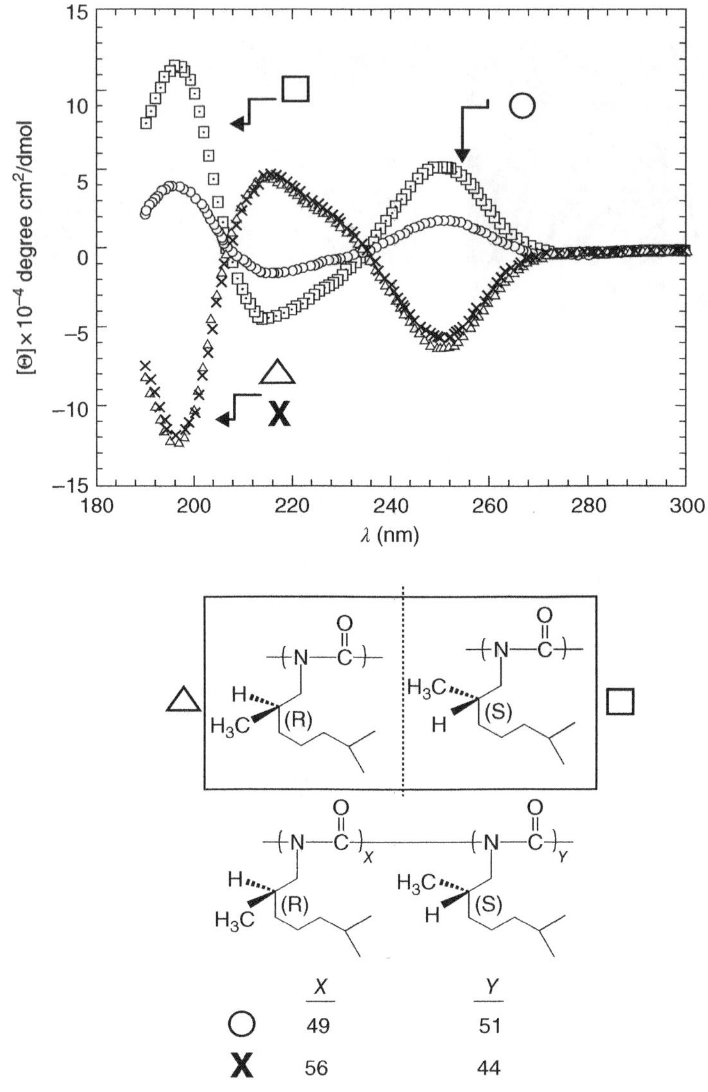

Figure 3.19 CD spectra of poly(2,6-dimethylheptyl isocyanate)s in *n*-hexane at room temperature, with random copolymerization of (R) and (S) enantiomers. The copolymer compositions are shown in the schematic at the bottom. (Source: Green et al. [23]. Reproduced with permission of Wiley.)

enantiomers. The helix corresponding to the (R) homopolymer is left handed and that for the (S) homopolymer is right handed (the curves with the symbols Δ and □, respectively, in Figure 3.19). The random copolymer with 56% (R) and 44% (S) shows the same CD spectrum as the (R) homopolymer

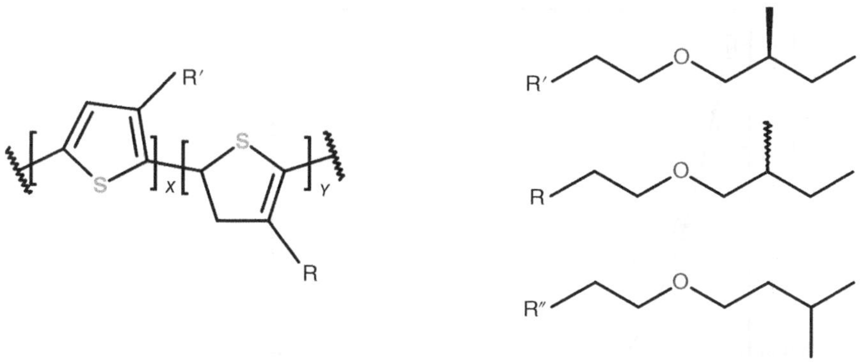

Polymer designation: 1 : x = 0; 2 : x = 0.05; 3 : x = 0.10; 4 : x = 0.20; 5 : x = 0.50; 6 : x = 0.75; 7 : x = 1.0; 8 : x = 1.0; 9 : R = R′ = R″

Scheme 3.3 Chemical structures of 3-substituted polythiophenes [33]. The ee if given by the value of x. Molecular weights of 7 and 8 were 27 and 15.7 kg/mol, respectively. Polymer 9 was achiral.

(symbol X), that is, the majority enantiomer dictates the sense of the helix, and the minority follows the same. It was mentioned earlier that the helix reversal involves an energy of 3.8 kcal/mol. The difference between the left- and right-handed helices is only about 400 cal/mol. Hence, in this case, the monomers of the (S) enantiomer accommodate themselves in the left-handed (R) helix. Similarly, the copolymer with just 2% excess of the (S) enantiomer (49% (R) and 51% (S)) shows a CD spectrum very similar to that of the (S) homopolymer (symbol o in Figure 3.19).

Similar "majority rules" and sergeant–soldier behavior were observed by Meijer et al. [33] in the case of polythiophenes. Scheme 3.3 shows the polymers in their study, with various amounts of enantiomeric excess (ee) of the x-mer. Similar to the case of poly(para phenylene) discussed before, no optical activity was seen in chloroform solutions of polymers 1–7. Aggregation of the chains in 1-decanol, a poor solvent, led to conformational order and an optically active structure, with the CD effect increasing with ee. The bisignate CD spectra confirmed the intermolecular chiral organization. Figure 3.20 shows that the anisotropy factor g at 603 nm strongly, nonlinearly varies with ee. Here, $g = \Delta\varepsilon/\varepsilon$, where ε is the extinction coefficient and $\Delta\varepsilon$ is the difference in the extinction coefficients of the left and right circularly polarized light. The strong dependence of g with ee indicates the cooperative interaction of the chiral side chains leading to chiral superstructures of these polythiophenes and that the principle of majority rules applies.

An interesting behavior was seen with mixtures of 7 or 8 with the achiral polymer 9. When solutions of 8 and 9 in 1-decanol were simply mixed at

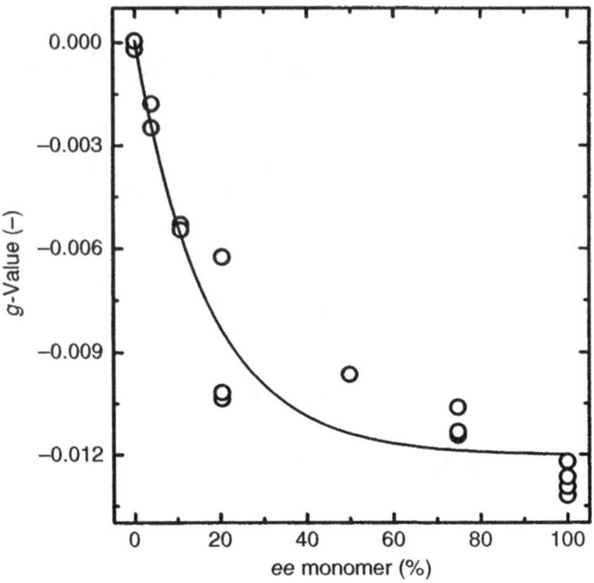

Figure 3.20 Anisotropic factor *g* at 603 nm for polymers 1–7 (see Scheme 3.3), with increasing enantiomeric excess. The multiple data points for the same polymer are for repeat measurements. (Source: Langeveld-Voss et al. [33]. Reproduced with permission of American Chemical Society.)

ambient temperature, the UV–vis spectra were simple superposition of the respective spectra, but the intensity of the CD spectrum increased linearly (symbol o) with the weight fraction of 8, as seen in Figure 3.21. When the mixture was heated to 393 K (to achieve a molecularly dissolved state) and then cooled slowly to ambient temperature, the CD intensity decreased (symbol Δ in Figure 3.21). If the higher molecular weight 7 was mixed with 9, annealed to 393 K, and slow-cooled to 292 K, the CD intensity increased. This behavior was explained by recording the UV spectra at different temperatures and determining the temperature at which disorder to order transition occurred. It was found that the transition temperatures of the polymers were in the order 7 > 9 > 8. The molecular weights are also in this order. When two of these are mixed, "seed aggregates" are formed by the polymer which aggregates at a higher temperature. The seed would be chiral in the case of 7 and nonchiral if 9 forms the seed. The seeds grow, and incorporate the second component within its aggregate. When 8 and 9 were mixed, the nonchiral 9 forms the seed and the activity of 8 is reduced. In the mixture of 7 and 9, chiral seeds of 7 are formed first, and they direct the achiral 9 and the CD increases. Thus, in this scenario of sergeant–soldier behavior, the longer chain controls the chirality of the aggregates.

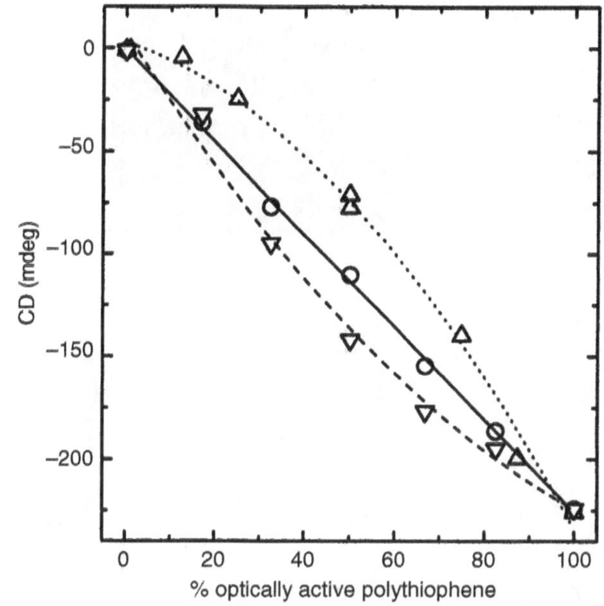

Figure 3.21 CD effects of mixtures of stereoregular 7 or 8 with achiral 9 in 1-decanol. o, Mixing at room temperature; Δ, mixture of 8 and 9 annealed to 393 K and slow-cooled to 293 K; and ∇, mixture of 7 and 9 annealed to 393 K and slow-cooled to 293 K. (Source: Langeveld-Voss et al. [33]. Reproduced with permission of American Chemical Society.)

3.7 Chain Folding

Folding of molecules in biological systems is well known and is a feature that controls their functions. Typically, self-assembly is defined in terms of association between molecules based on recognition facilitated/coded by forces such as hydrogen bond. When different segments of the same chain molecule, whether natural or synthetic, recognize one another and prefer to associate, a change in the conformation of the chain occurs, leading to chain folding. Thus, it is an intramolecular self-assembly process.

Although synthetic polymers with high molecular weight have been synthesized since the 1950s and their crystalline structures studied, it was not until the analysis of single crystals of polyethylene (PE) in the late 1950s that the research on chain folding became intense. In a classic experiment, Keller's group [34a] in Bristol recorded the electron microscope image of a single crystal of PE grown from xylene as well as its diffraction, as shown in Figure 3.22.

The electron diffraction showed that the chain direction corresponded to that shown in Figure 3.22d, that is, the chain direction is along the thickness of the crystal. Each side of the crystal in Figure 3.22a is about 10 μm, but the

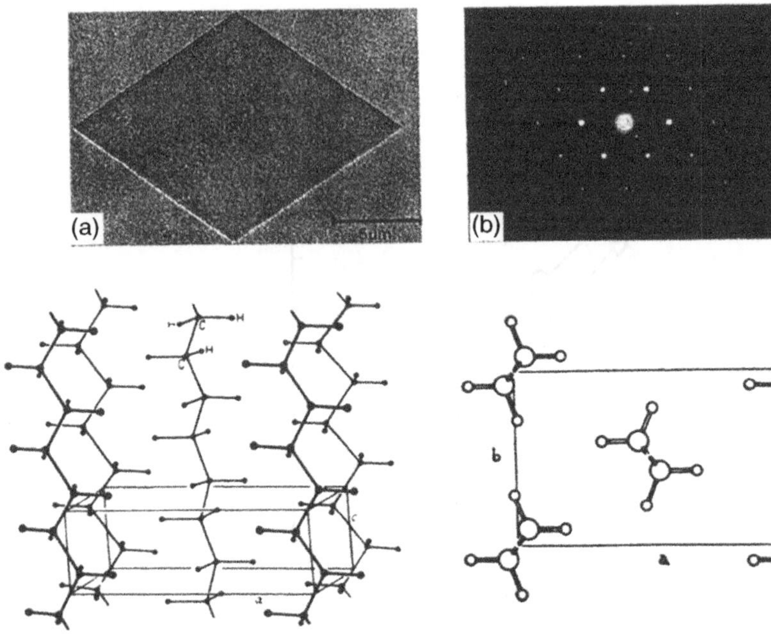

Figure 3.22 (a) An electron microscope image of a PE single crystal from xylene, (b) the electron diffraction of the crystal, (c) a perspective view of the unit cell of PE, and (d) a view down the chain axis. (Source: Keller [34a] and Sperling [34]. Reproduced with permission of Wiley.)

thickness was only about 200 Å. In the extended conformation of PE, the length of each unit is 2.5 Å, which would mean that there were only about 80 units along the thickness of the crystal. Since it was a high-molecular-weight PE, how was the chain accommodated along the thickness of less than 200 Å? This led to the idea of chain folding, that is, the chain has to fold in order to be accommodated in the crystal [34a]. The model proposed by Keller for the folding consisted of regular folds as shown in Figure 3.23a. The chain folded back and forth and it came to be known as "adjacent reentry" model [34–39]. The amorphous parts were relegated to interlamellar regions. Flory [40–43] questioned the validity of the model on kinetic grounds. How can a long chain fold itself regularly so as to reenter adjacent to the previous stem? He proposed the nonadjacent reentry model, also known as "switchboard" model, in which a chain from a crystalline lamella can traverse through the amorphous layer and reenter the same lamella at a distance or become part of another lamella (Figure 3.23b). The top view of a lamella according to this model is shown in Figure 3.23c. This resembles the ancient telephone switchboard through which an operator would connect the callers, and hence the name

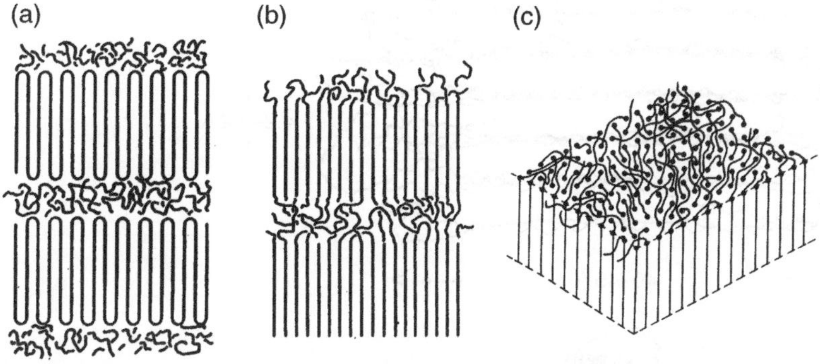

Figure 3.23 (a) Regularly folded adjacent reentry model, (b) nonadjacent (switch-board) reentry model, and (c) top view of a lamella.

"switchboard model." Intense debate on the validity of these models between the "two camps" persisted for more than two decades, until the death of Flory. However, simulations of chain folding emerged, as the necessary computer facilities became available.

With the preference for the trans–trans sequence, the crystalline structure of PE is stabilized by van der Waals interactions between the chains. Chain folding, however, is an intramolecular event during which the trans–trans conformation of a few contiguous bonds has to change to a gauche conformation. Kavassalis and Sundararajan [44] performed molecular dynamics simulations of a 1000-unit single PE chain, in which the chain collapsed into a lamella as shown in Figure 3.24. The stems were about 40 Å in length, and the distance between the stems was about 4.2 Å, similar to that in the crystal structure of PE. The end-on view shows pseudo-hexagonal type of packing of the stems. Note that the stems contain trans–trans sequence of the bonds and are stabilized by van der Waals interactions. The bonds in the folds are in the trans–gauche and gauche–gauche conformations, in order for the chain to fold. The fold length of 40 Å from this simulation is far less than the lamellar dimensions of 100–200 Å observed experimentally. The inconsistency was traced to the soft torsional barrier in the force field used for the simulation, leading to profuse occurrence of gauche conformations and a large number of folds. The torsional barriers (e.g., 2.0–3.0 kcal/mol for the C—C bond) commonly used in these types of simulations were based on gas phase studies. During crystallization from solution or the melt, a chain is not isolated, but would experience frictional force due to the presence of the other chains and/or the presence of the solvent. The results of revised simulations of Sundararajan and Kavassalis [45] for a 500-unit PE, with a tighter torsional barrier of 6 kcal/mol, are shown in Figure 3.25. The chain folds into two stems, with a length of 250 Å, which accords with experimental values. In another set of experiments,

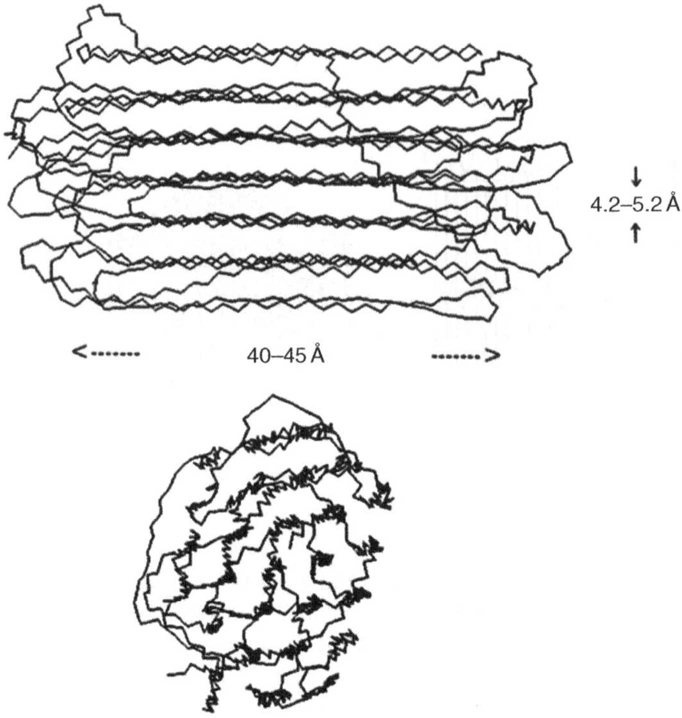

Figure 3.24 Folded chain conformation of a 1000-unit PE during molecular dynamics simulation. The lamellar structure (top) and an end-on view of the lamella are shown. (Source: Kavassalis and Sundararajan [44]. Reproduced with permission of American Chemical Society.)

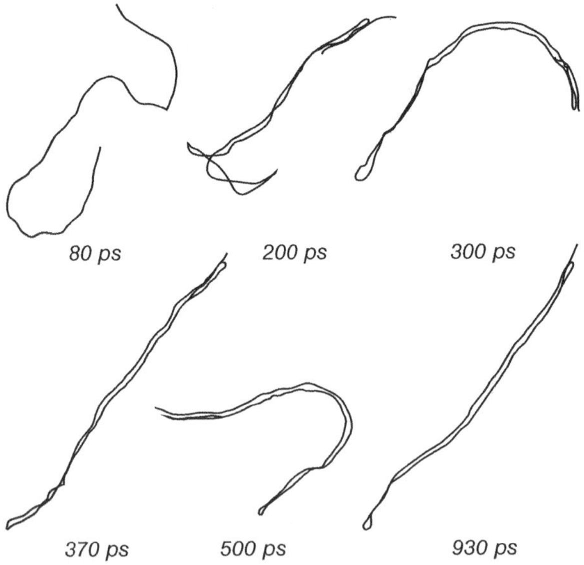

Figure 3.25 Evolution of conformations of a PE chain with 500 CH_2 units during molecular dynamics simulations. A high torsional barrier of 6 kcal/mol was used. The length of the final folded stem is 250 Å. (Source: Sundararajan and Kavassalis [45]. Reproduced with permission of Royal Society of Chemistry.)

Ungar et al. [46, 47] crystallized PE samples of various molecular weights and found that folding occurred only if the chain was at least 150 units in length. Lee and Wegner [48, 49] reached the same conclusion. The experiments of Leung et al. [50] showed a lamellar spacing of 99–117 Å, which corresponds to near-extended conformation of the chain.

With a view to rationalize the aforementioned chain length dependence from energy considerations, Sundararajan and Kavassalis [45] performed molecular dynamics simulations for chains of lengths ranging from 30 to 500 units. With the time evolution of the conformation of a chain of 30 units, there was folding and unfolding, as shown in Figure 3.26a. The fold was not stable. However, with a chain of 150 units, folding occurred within 190 ps of simulation, and it remained stable (Figure 3.26b). In the case of biopolymers, folding is facilitated or dictated by hydrogen bonds, disulfide bridges, etc. However, with PE, it is the only van der Waals force that keeps the stems together, and this should compensate for the higher energy due to the gauche conformations in the folds. Figure 3.27 shows the variation of the various contributions to the conformational energy of PE chains of various lengths ranging from 30 to 250 CH_2 units, after 1000 ps of molecular dynamics (MD) simulation [45]. It is seen that the van der Waals energy decreases with chain length, but the torsion energy and that due to bond angle deformation increase. The total energy becomes negative (i.e., the van der Waals interaction between the straight stems compensates for the higher energy at the folds) only with a chain length of 130 CH_2 units. Hence, shorter chains crystallize with extended conformation in the single crystals.

Although linear chains require a minimum of 150 CH_2 units for stable folding to occur, cyclic alkanes fold with fewer number of CH_2 groups. Of course, cyclic forms of hexane, octane, and even cyclohexadecane do not uniformly fold due to geometric constraints but adopt puckered conformations. Figure 3.28a shows the conformation of $(CH_2)_{36}$ in its crystal structure [51]. The packing of four molecules in the unit cell is shown in Figure 3.28b in which two molecules are seen along the *c*-axis which has a length of 46.3 Å.

The fold region in the crystal structure has the sequence of g g t g g conformations of the bonds (Figure 3.28a), and the distance between the stems is about 4 Å. The stem lengths calculated from MD simulations [52] for cyclic alkanes with 24, 36, 60, and 72 CH_2 units were 13.6, 20.4, 35, and 44 Å, respectively. These are roughly half of the *c*-axis unit cell dimensions in the respective crystal structures, the van der Waals separation of the edges of two molecules along this axis accounting for the difference. Figure 3.29 shows the plot of the calculated fold lengths, the *c*-axis dimensions, and the fold length derived by Lee and Wegner [48, 49] from longitudinal acoustic mode (LAM) frequencies. The fold lengths from simulations and the experimental LAM frequencies agree well. With both linear and cyclic PE, Ungar et al. observed that when folded chain crystallization occurs, the fold length is an integral

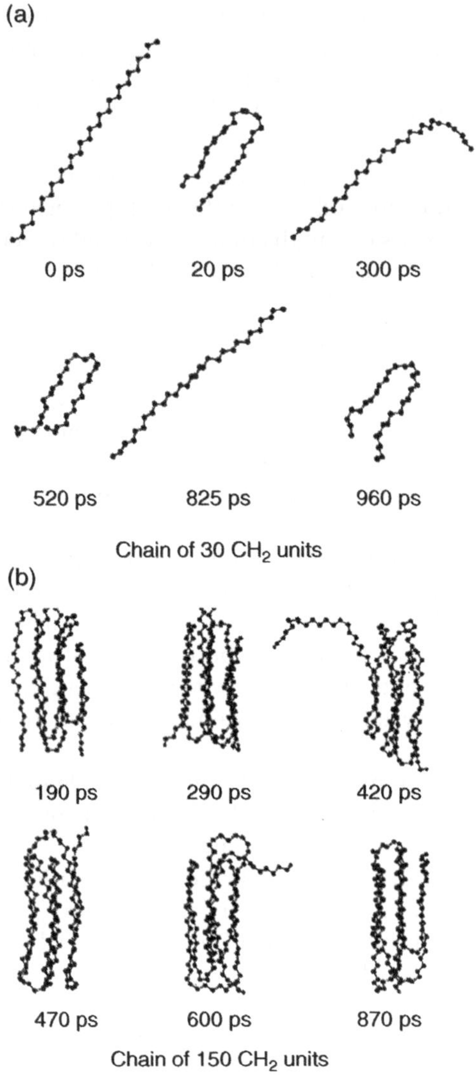

Figure 3.26 Time evolution of conformations during molecular dynamics simulation of PE chains with (a) 30 CH_2 units and (b) 150 CH_2 units. (Source: Sundararajan and Kavassalis [45]. Reproduced with permission of Royal Society of Chemistry.)

reciprocal of the fully extended (chain) length [46, 47, 50]. Nonintegral folds were not observed. Manley also discussed such "quantization" of fold lengths. In a recent publication, Takeshita et al. [53] studied folding in cyclic, linear, and star poly(tetrahydrofuran) (PTHF). Based on X-ray diffraction, they concluded that the crystalline structures were the same for all three types. Quantization of fold lengths was also found in their study. Several papers have been published [54–67], notably by the groups of Muthukumar, Müller-Plathe, Takeuchi,

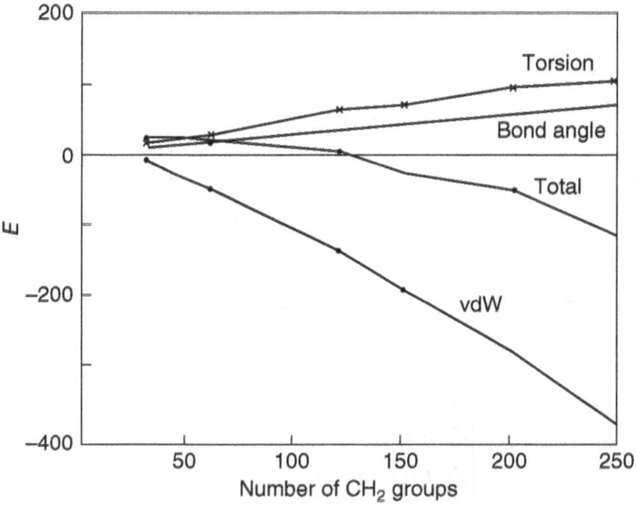

Figure 3.27 Variation of van der Waals, torsion, bond angle deformation, and total energies with chain length. (Source: Sundararajan and Kavassalis [45]. Reproduced with permission of Royal Society of Chemistry.)

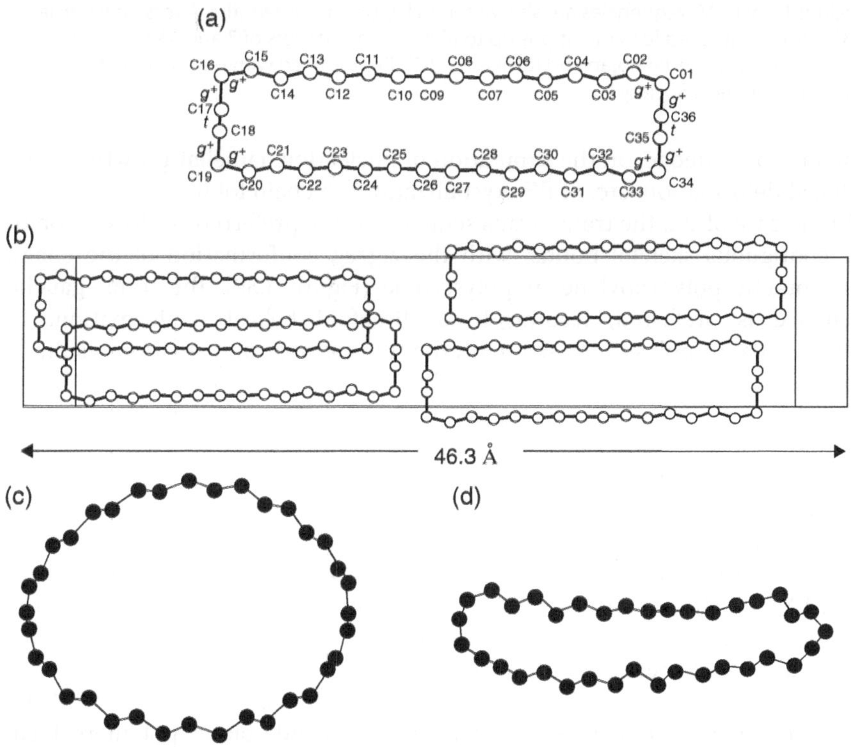

Figure 3.28 (a) The conformation of $(CH_2)_{36}$ in the crystal structure, (b) packing of four molecules in the unit cell, (c) model used for simulations, and (d) folded conformation obtained from MD simulations. (Source: (a) and (b) Trzebiatowski et al. [51]. Reproduced with permission of Wiley.)

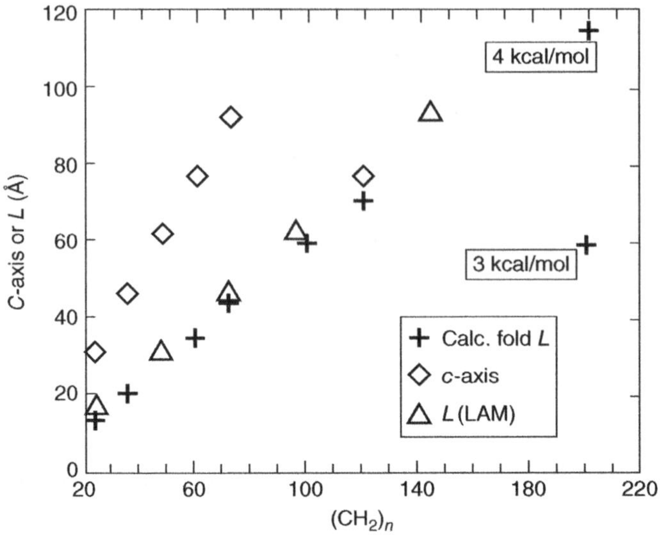

Figure 3.29 Calculated fold lengths, crystallographic c-axis dimensions, and the long spacings from LAM frequencies are shown for $(CH_2)_n$ of various lengths. A torsional barrier of 3 kcal/mol was used for simulations up to $(CH_2)_{100}$ and barriers of 3 and 4 kcal/mol for $(CH_2)_{200}$. (Source: Sundararajan and Kavassalis [52]. Reproduced with permission of American Chemical Society.)

Yamamoto, Rutledge, on the simulation of nucleation, lamellar growth, shear-induced deformation, etc. of PE crystallization via chain folding.

In the case of PE, the trans–trans sequence is the preferred conformation of the contiguous skeletal bonds, with the zigzag conformation of the chain. For isotactic polypropylene or polystyrene (Figure 3.30), the trans–gauche sequence is preferred, leading to a threefold helical conformation. In Figure 3.30a for polystyrene segment, the trans–trans conformation results in close contacts between the adjacent phenyl groups, as shown by the dotted line. This is a short-range interaction, which depends on the rotations φ and ψ about two contiguous skeletal bonds. Following Flory [68], we define the trans–trans conformation to correspond to φ and $\psi = 0°$. Changing one of the rotations, for example, ψ to gauche (Figure 3.30b) alleviates the phenyl–phenyl interaction, and the rotations $\varphi = 0°$ and $\psi = 120°$ would lead to the threefold helical conformation, with a pitch of 6.6 Å. Isotactic poly(methyl methacrylate) (PMMA) which is a disubstituted vinyl chain forms a double helix in the crystalline state, with each strand consisting of 10 monomer units [69–71]. This is due to the preference for the trans–trans sequence of the skeletal bonds, as suggested by Sundararajan and Flory [72]. It should be noted that PMMA is a unique case in which the isotactic and syndiotactic polymers form stereocomplex. It is well known that the α-helix of the poly(amino acids)

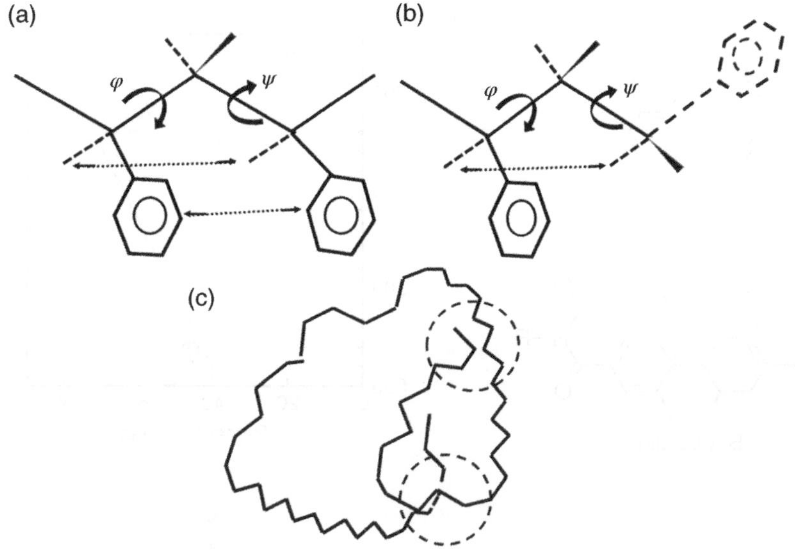

Figure 3.30 (a) Illustration of short-range interaction in terms of the rotations φ and ψ in the trans–trans conformation resulting in phenyl–phenyl proximity (indicated by the dotted line) in a diad segment of isotactic polystyrene; (b) the trans–gauche conformation; and (c) long-range interaction arising between remote segments, denoted by dashed circles.

consists of a nonintegral helix, stabilized by intramolecular hydrogen bonds between adjacent turns of the helix. It is an axiom that if the rotational states of the bonds between contiguous repeat groups are identical, the polymer would adopt a helical shape. Given the geometry of the repeat unit and the interunit torsion angles, the number of repeat units (n) per turn of the helix and the pitch can be exactly calculated, using the method developed by Ramachandran et al. [73]. Although the repeat units would fold to generate the helix, this is not considered chain folding nor as foldamers in the modern context, as it is due to short-range interactions [74]. Figure 3.30c schematically shows the long-range interaction. Here, long-range does not mean a large distance between the atoms, but close proximity between different segments along the chain, that is, long range along the chain sequence, for example, between atoms attached to a skeletal bond n and those of a skeletal bond $n+x$, where x is large. This is the basis of the excluded volume and the theta temperature concepts of Flory [75, 76].

The long-range van der Waals interactions between the stems stabilize the chain-folded structure of PE discussed earlier. The folding of PE and cyclic PE involves only van der Waals interactions. If a polymer is made up of, for example, an aromatic moiety linked through a flexible segment (not necessarily showing liquid crystalline behavior) folding of the latter could occur, facilitated

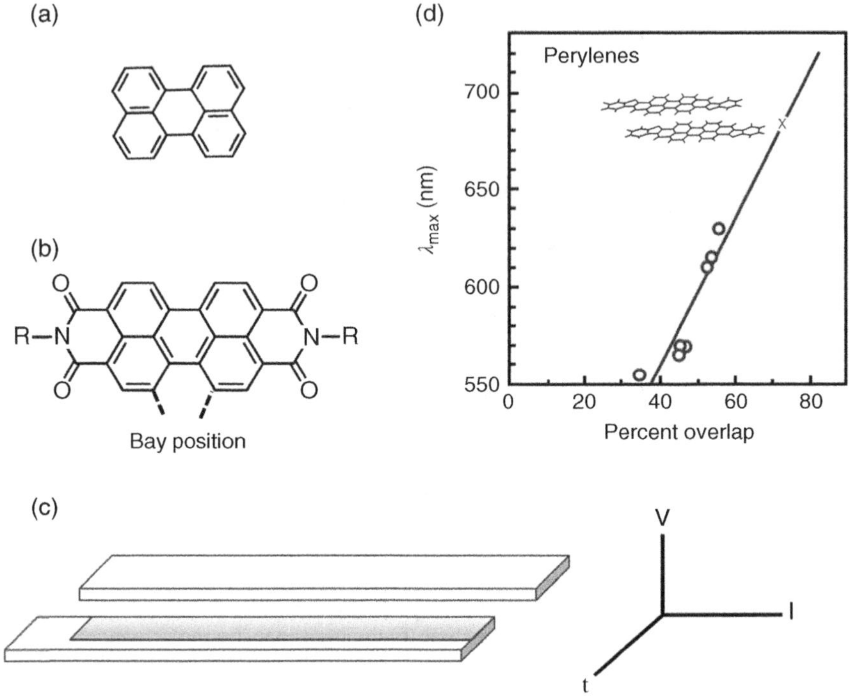

Figure 3.31 (a) Perylene. (b) Illustration of substitutions R on the imide nitrogens and at the bay positions. (c) A sketch of the stacking of perylene diimides, with the gray area showing the projected aromatic overlap. v, vertical; t, transverse; and l, longitudinal displacements. (d) Correlation between percent overlap and the maximum absorption wavelength λ_{max}.

by the tendency of the aromatic groups to associate via π-interactions. The photophysics and photochemistry of systems based on perylenes, phthalocyanines, and porphyrins have been studied extensively in the literature. Figure 3.31 shows the schematics of perylene and perylene diimide. The influence of several types of substituents such as alkyl chains, oligo(ethylene oxide), and amides at the imide nitrogen as well as at the bay positions (Figure 3.31b), on the self-assembly of the aromatic core, has been investigated. In their pioneering studies, Hädecke and Graser [77–79] reported the crystal structures of several perylene derivatives and discussed the effect of the substituents on the π-stacking and the color of the dye. Figure 3.31c shows the schematic of the stacking of two aromatic cores, for example, those of perylene diimide. Apart from the vertical packing distance, which is typically about 0.35 nm, the aromatic planes would also be shifted with respect to each other in the longitudinal and transverse directions, depending on the nature of substituents. The projected overlap of one unit over the other is illustrated by the gray

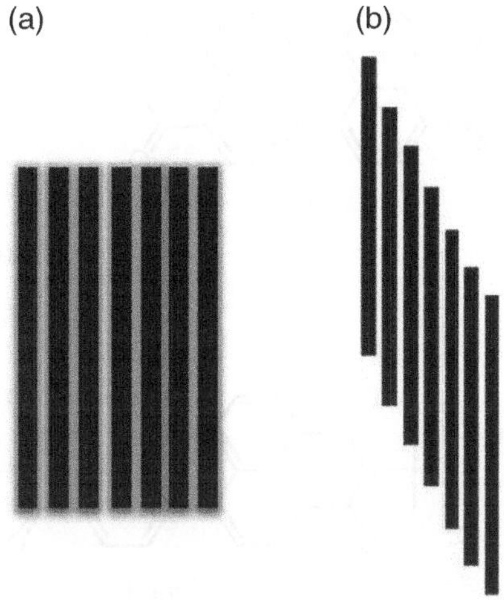

Scheme 3.4 (a) H-type and (b) J-type stacking of aromatics.

area in Figure 3.31c. Bluhm and Sundararajan (unpublished), using some of the crystal structure results of Hädecke and Graser, calculated the projected overlaps (% area) in some of the structures and plotted them against the experimentally observed λ_{max}, as shown in Figure 3.31d. A linear relationship is seen between these two, that is, the larger the overlap, the higher the absorption wavelength. The data marked by X in Figure 3.31d correspond to that of benzamidazole perylene [80]. The self-assembly of these molecules in solution and their stacking mode are often characterized by the effect of concentration of the solute on the absorption and fluorescence spectra. The generic modes of stacking are the H- and J-type, as illustrated in Scheme 3.4. A blue shift in the absorption spectra and fluorescence quenching with concentration are indicative of H-stacking. The J-stacking would cause a red shift in the absorption spectra and fluorescence enhancement with concentration.

Using perylenetetracarboxylic dianhydride and aliphatic diamines (with $(CH_2)_5$ to $(CH_2)_{12}$), Wang et al. [81] synthesized polymeric PERYs shown in Scheme 3.5. The original intent was not studying self-assembly or chain folding but to evaluate these as potential candidates for electro-photographic applications. These were soluble only in concentrated H_2SO_4 and partially soluble in *m*-cresol. The crystallinity of the solids improved upon annealing at temperatures as high as 320°C. Although the chain folding could not be studied, X-ray diffraction showed that prolonged annealing at such high temperatures led to a crystal–smectic transition [82]. MD simulations on **3f** (Scheme 3.5) showed [83] that theoretically chain folding could occur with intramolecular self-assembly of the perylene diimide segment as shown in Figure 3.32.

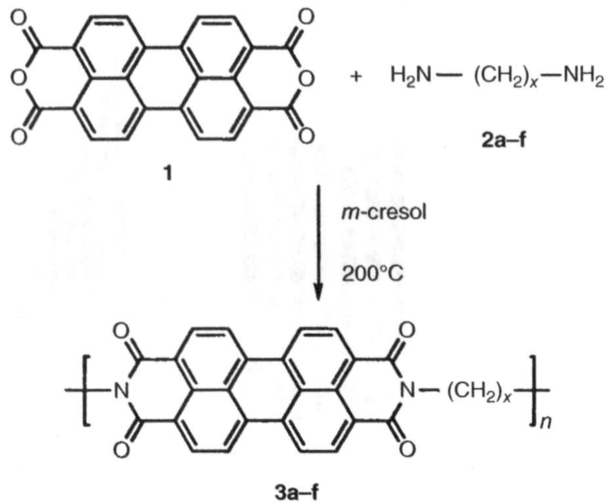

	a	b	c	d	e	f
$x =$	5	7	8	9	10	12

Scheme 3.5 Illustration of polymeric perylene diimides with alkyl spacers [81].

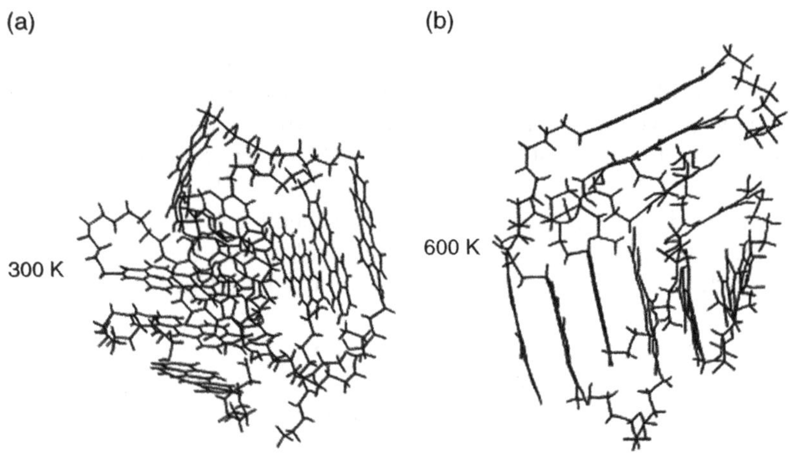

Figure 3.32 Folded chain conformation of **3f** (Scheme 3.5) after 200 ps of molecular dynamics simulations at 300 K and 600 K. (Source: Sundararajan et al. [83]. Reproduced with permission of Elsevier.)

The π-interaction between the perylene diimide segments and the van der Waals forces between the short alkyl chains essentially rendered the aforementioned molecules **3a–f** (Scheme 3.5) insoluble in common solvents. Later studies with hydrophilic tetra(ethylene glycol) or poly(tetrahydrofuran)

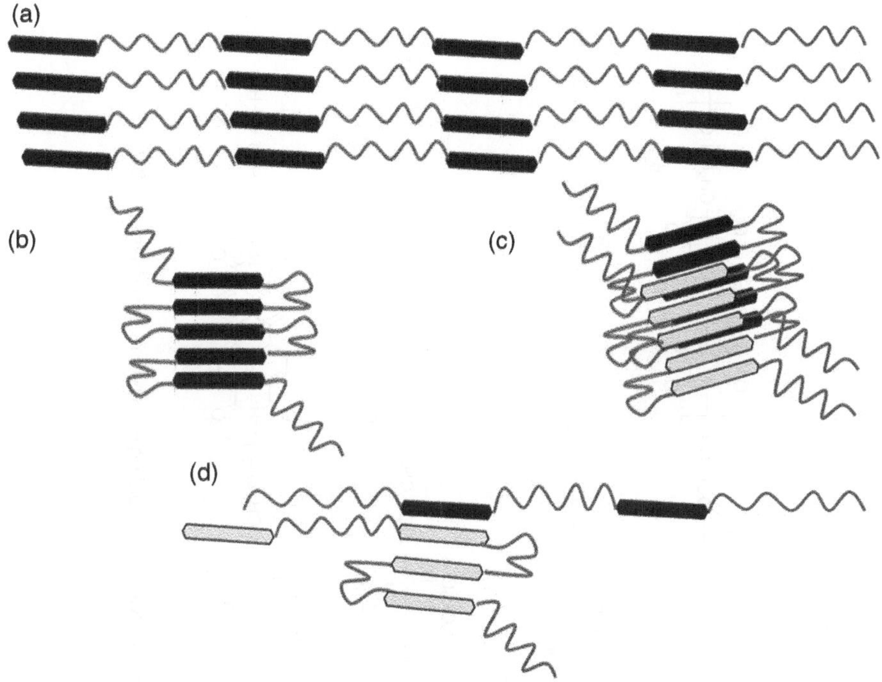

Figure 3.33 (a)–(d) Various possible modes of association of the aromatic segment.

spacers enabled observations of both intramolecular self-assembly via chain folding and intermolecular self-assembly. A schematic of the various scenarios is shown in Figure 3.33.

Intermolecular self-assembly of the aromatic segments in the extended conformation of the chain is shown in Figure 3.33a, and the intramolecular self-assembly leading to chain folding is illustrated in Figure 3.33b. The folded chains could associate to form stacks as shown in Figure 3.33c or they could mix with the extended chains (Figure 3.33d). Li et al. [84] synthesized polymeric PERYs with tetra(ethylene glycol) (Scheme 3.6) and studied the chain folding and self-assembly. Their model of association of the perylene segments, shown in Figure 3.34, did not explicitly consider the intermolecular assembly of the extended chains shown in Figure 3.33a. Since perylene is a chromophore, optical absorption spectra can be used for studying the self-assembly. The typical absorption spectra are shown in Figure 3.35 in which the peaks at about 531, 498, and 472 nm correspond to S_{0-0}, S_{0-1}, and S_{0-2} transitions. The positions of these peaks would vary slightly depending on the substituent as well as the solvent medium. In the work of Islam et al. [85] on PERYs substituted with an ethylene oxide–propylene oxide copolymer, these peaks were observed

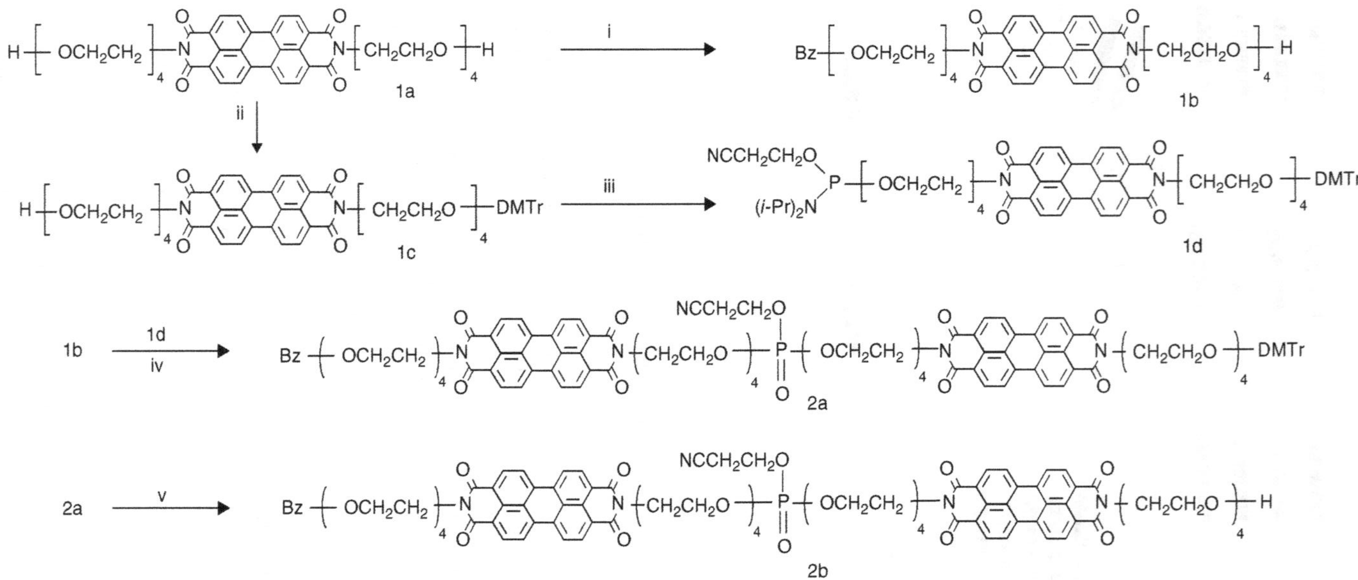

Scheme 3.6 Schematics of some of the molecules prepared by Li et al. for studying chain folding of perylene-based systems [84].

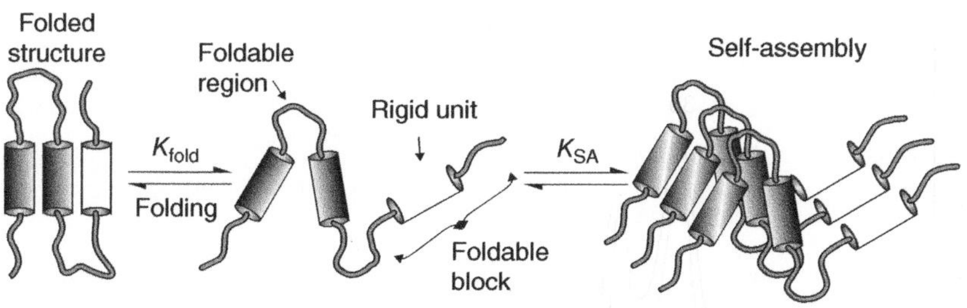

Figure 3.34 Illustration of chain folding leading to self-assembly [84].

at 453, 482, and 519 nm in various organic solvents with some bathochromic shift and at 466, 497, and 531 nm in water.

In the absence of association of the perylene segments the relative intensities would be such that $S_{0-0} > S_{0-1} > S_{0-2}$. This is seen in Figure 3.35a for the free monomer **1b**. The spectra for the dimer and trimer (**2b** and **3b**, respectively, in Figure 3.35a show intensity reversal between 0–0 and 0–1 peaks, indicating aggregation due to folding and self-assembly. Figure 3.35b shows that a certain minimum concentration is required for such self-assembly to occur.

Neuteboom et al. [86] studied the chain folding and self-assembly of perylene diimide–poly(tetrahydofuran) copolymers, with different chain lengths of poly(THF) shown in Scheme 3.7. Alkyl-substituted perylene diimide (PERY) was used as the reference compound. In *o*-dichlorobenzene, the π-stacking was impeded in the case of PERY due to the alkyl substituents. With the polymers, it was found that the absorption peak at 493 nm (0–1 transition) was most intense with the shorter poly(THF) segment (P1 in Scheme 3.7), indicating that the aggregation was stronger than in the case of P2 and P3. Based on absorption and fluorescent spectra, models similar to those shown in Figure 3.33b and d were proposed.

3.8 Foldamers

In the aforementioned examples of perylene-based polymers, the long-range π-interaction between the aromatic groups facilitated by the flexible linkers enabled chain folding. Biological molecules, for example, proteins fold in specific ways to imprint functionality such as binding in the grooves, etc. The amino acid sequences (primary structure) determine the secondary structure, which would consist of (depending on the preferred (φ, ψ) conformation) sequences of α-helices, β-sheets, and nonhelical segments. Such segmented secondary structure then leads to a folded tertiary structure, stabilized by hydrogen bonds, disulfide bridge, etc., between nonadjacent monomers. This

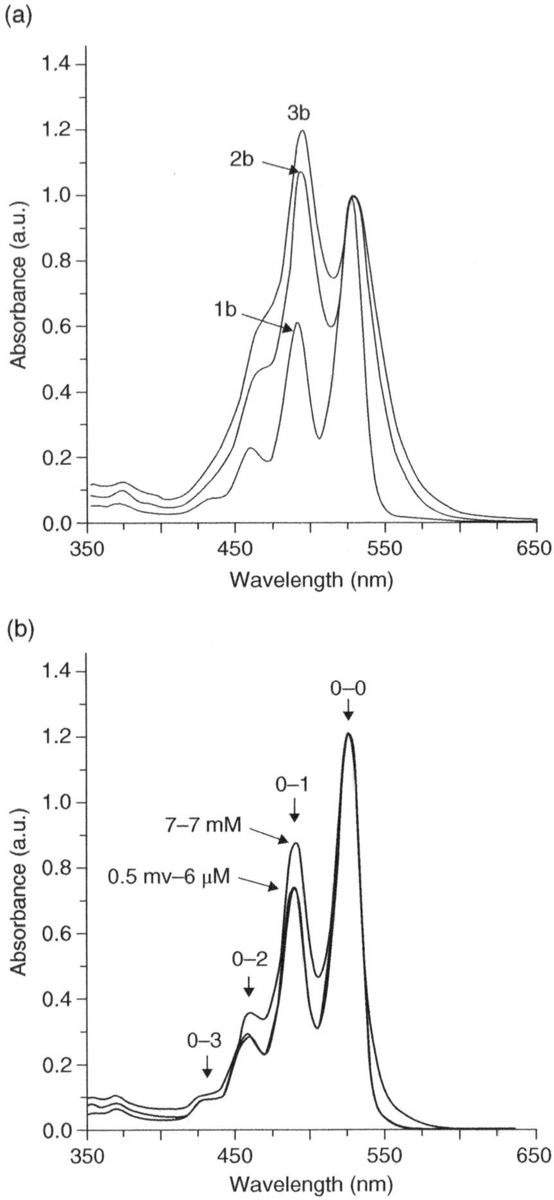

Figure 3.35 (a) Normalized optical-absorption spectra of folded chromophoric dimer **2b** and trimer **3b** that have an intensity reversal between the $0 \rightarrow 0$ and $0 \rightarrow 1$ transitions when compared with the free monomer **1b**. The chromophoric unit concentrations are 6.6 μM for all oligomers. (b) Normalized optical-absorption spectra of chromophoric monomer **1b** and its self-assembled products. At concentrations less than ~1 mM (0.5 mM and 6 μM shown), few self-assembled products are formed, as indicated by the absence of perturbation of the optical spectra. At concentrations above ~1 mM (7.7 mM shown), the $0 \rightarrow 1$ transition intensity begins to increase, indicating self-organization in solution. (Source: Li et al. [84]. Reproduced with permission of Wiley.)

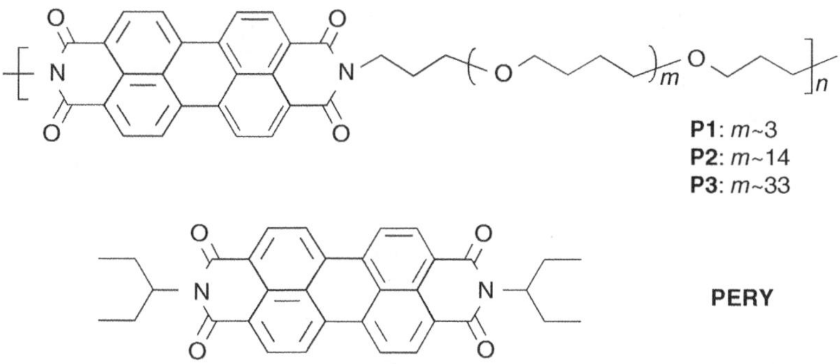

Scheme 3.7 Structures of alternating perylene bisimide–polyTHF copolymers and the reference compound [86].

occurs consistently for a given protein, unless the amino acid sequence is altered by mutation, etc.

Such precise folding to yield consistently compact structures is hard to achieve in the case of synthetic polymers, although in the examples discussed earlier, PE chains fold with integral number of folds (quantization) and perylene-based polymers show H- or J-stacking of the aromatic groups. In the case of PE, the chain segments which are not part of the stems are relegated to the amorphous interface, and consist of distribution of trans and gauche conformations. Similar is the case of perylene polymers, in which the flexible linkers fold with nonuniform conformations (see, e.g., Figure 3.32).

In an effort to design molecular segments that would enable precise folding to create compact tertiary structures, Gellman [87, 88] introduced the concept of "foldamers," which are oligomers that would adopt desirable secondary structure conformations through hydrogen bonding between nonadjacent groups. These were based on β-peptides, as shown in Figure 3.36. Six possible compact helical structures were identified, as shown by the helix designations. For example, the 18-helix means 18 atoms between the hydrogen bond donor and acceptor. Various secondary structures as foldamers were discussed by Moore et al. [74], as shown in Figure 3.37. A foldamer is a molecule which would take a folded shape, as a result of strategically placed hydrogen bonding groups, or supramolecular guest molecule association, or metal binding. Most of the early works on foldamers were aimed at mimicking biopolymers. Moore and coauthors [74] define the requirements for foldamers as follows:

1) A foldamer is an oligomer, which, through noncovalent interactions between *nonadjacent monomer units*, folds into an ordered structure in solution.
2) Solvent hence plays a role.

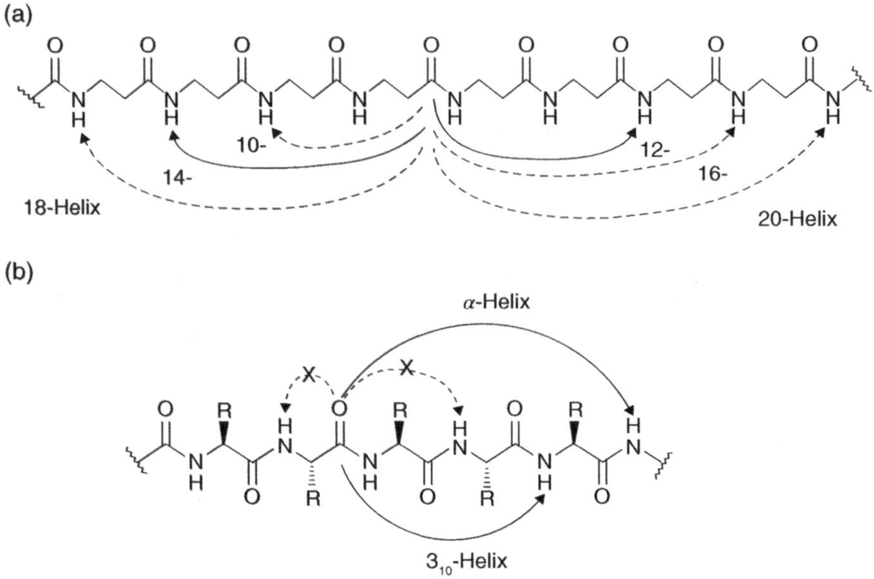

Figure 3.36 (a) Poly(β-amino acid) backbone. The dashed arrows indicate H-bonds associated with the six tightest helices available to β-peptides. The solid arrows indicate the two helices that have been documented in short β-peptides. (b) Poly(α-amino acid) backbone is also shown for comparison [88].

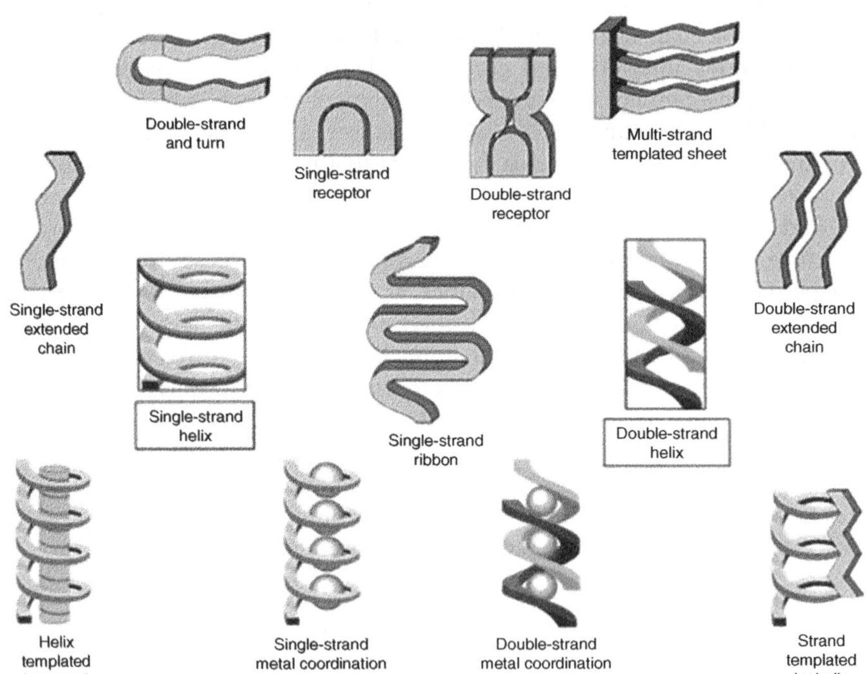

Figure 3.37 Illustrations depicting the different types of foldamer secondary structures. (Source: Hill et al. [74]. Reproduced with permission of American Chemical Society.)

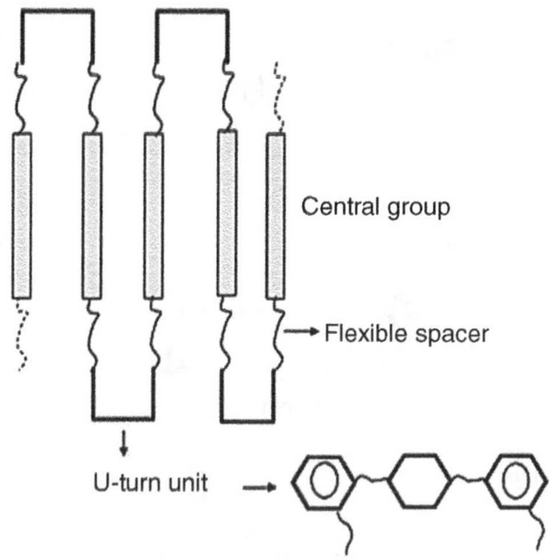

Figure 3.38 Schematic of a rigid-rod U-turn polyamide [89].

3) It could be single-stranded or multistranded. In the latter case, the individual strands associate between themselves and the fold.
4) In the presence of a solvent, folding is a dynamic phenomenon.

Just as a protein is a sequence of different secondary structures (α-helix, zigzag, and nonhelical), a polymer containing foldamers with different secondary structures is called "tyligomer," meaning an object consisting of folded parts.

The strategy to design a foldamer is to position the hydrogen bonding (or any such noncovalent bonding) groups at appropriate positions in the oligomer or to incorporate a group that would twist the direction of the chain. One of the earliest such attempt [89] was to insert U-turns in poly(*p*-phenylene terephthalamide) (Kevlar) which is a rigid rod polymer (Figure 3.38). A similar approach was followed by Heckel and Dervan [90]. They considered two methods to designing a pyrrole–imidazole polyamide motif for recognition of the DNA minor groove. As shown schematically in Figure 3.39a, they distinguished between a hairpin, a U-pin, and an H-pin. Along with the hairpin molecule **2** (Figure 3.39c) as a reference, they varied the *n* and R in the U-pin molecule **1** shown in Figure 3.39b and studied their binding to DNA.

Using such a hairpin model, Ajayaghosh et al. [91] created toroidal nano-objects. It is known that melamine (1,3,5-triazine-2,4,6-triamine) and cyanuric acid (1,3,5-triazine-2,4,6-triol) can form a complex with a network of hydrogen bonds as shown in Figure 3.40a. Melamine was attached to oligo(*p*-phenyleneethynylene) (OPE) to serve as the hairpin (Molecule **1** in Figure 3.40b). Each segment of the OPE in **1** bears three $OC_{12}H_{25}$ tails and the cyanurate (Molecule **2**) bears one $C_{12}H_{25}$ tail. With an equimolar mixture of **1** and **2** in $CDCl_3$, co-assembly as discrete rosettes resulted (Figure 3.40b), as confirmed

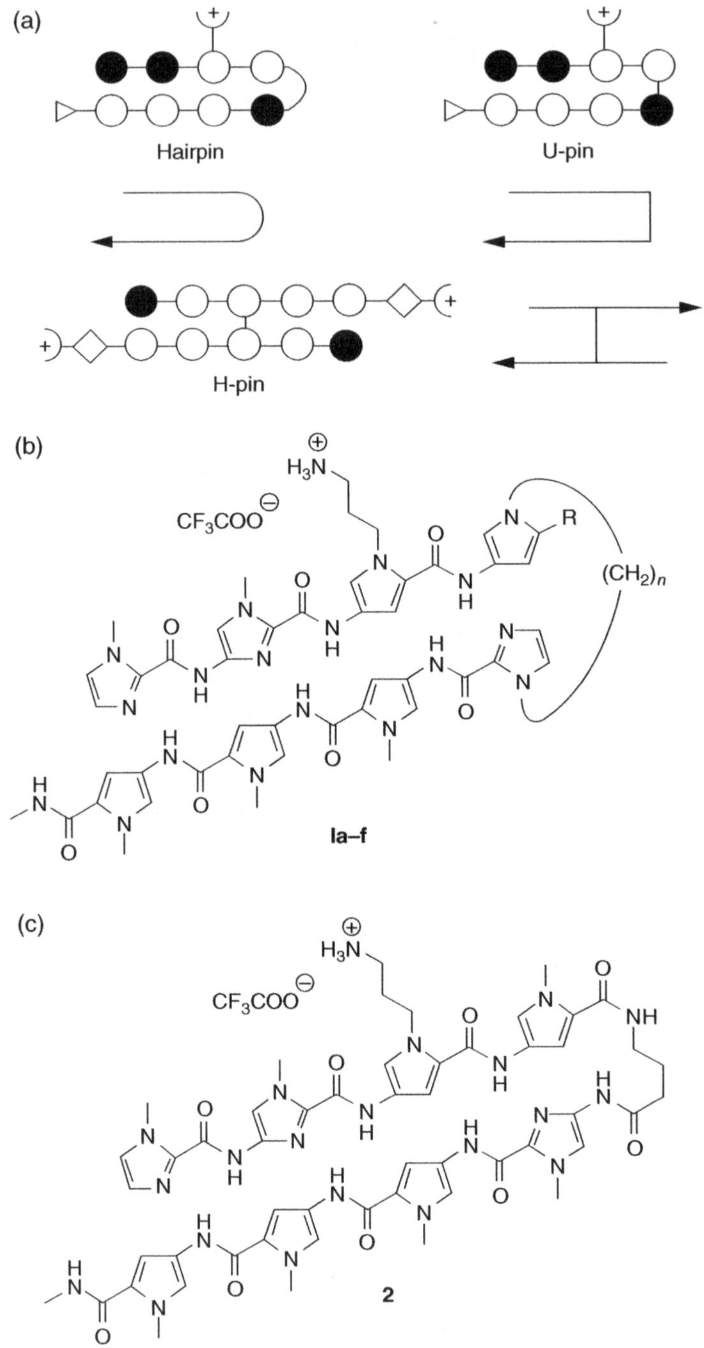

Figure 3.39 (a) Schematic of hairpin, U-pin, and H-pin turns; (b) the U-pin (**1a–f**); and (c) hairpin (**2**) molecules studied [90]. In **1a–d**, n=3, 4, 5, and 6, respectively, and R=CONHMe; **1e**: n=6 and R=$CONH_2$; **1f**: n=6 and R=H.

by NMR spectra. The absorption spectrum of the mixture in $CHCl_3$ showed that the OPE segments were H-stacked in the hydrogen bonded rosette assembly. Dynamic light scattering experiments indicated an average hydrodynamic diameter of 8 nm, corresponding to the rosette assemblies with extended alkyl chains. On the other hand, much larger hydrodynamic diameter of 50 nm was obtained in decane, resulting from the formation of nano-objects, due to the organization of the rosettes. The absorption spectra showed extended H-stacked aggregation.

An atomic force microscopy image (Figure 3.40d) of a spin-coated film of the decane solution showed toroidal nano-objects, formed by the stacking of individual rosettes. The circumference of the toroid was calculated to be about 67 nm. It was concluded that 191–134 rosettes are in a toroid with an average inclination of 1.9–2.7 between them, as shown in Figure 3.40d. Although the objective of this work was not stated to be targeted toward a foldamer, it is the folding of the melamine-linked OPE that gave rise to the morphology that was seen.

We distinguished the short- and long-range interactions earlier. The foldamers that were illustrated earlier were based on the latter. However, the role of short-range interactions cannot be neglected. A classic case is the difference in the conformations of cellulose and amylose (component of starch). The β-1,4 linkage between the D-glucose residues in cellulose results in an extended zigzag conformation of the chain, stabilized by a hydrogen bond between the OH-3 hydroxyl group of one D-glucose unit to the O-5 ring oxygen in its contiguous unit, which was also modeled by molecular mechanics simulations [92–94]. The α-1,4 linkage between two contiguous D-glucose units in amylose leads to folding of the chain into a flat helical conformation, with hydrogen bonds between the OH-2 group of one glucose unit and OH-3 of the contiguous residue and with a large cavity, again confirmed by molecular modeling [92–94]. Although both polymers are made up of D-glucose units with 1,4 linkage, the difference in the type of linkage (α versus β) causes the difference. A folded flat helical conformation of a cellulose chain is stereochemically unfavorable with large steric hindrance. Likewise, a zigzag extended conformation of the amylose chain does not exist. The α-1,4 linkage between two contiguous D-glucose units allows for the formation of cyclic dextrins (e.g., α, β, and γ cyclodextrins), but the β-1,4 linkage as in cellulose does not allow the formation of cyclic celluloses.

It was discussed earlier that the Moore definition [74] of a foldamer requires that "it is an oligomer, which, through noncovalent interactions between *nonadjacent monomer units*, folds into an ordered structure in solution." The examples of cellulose and amylose discussed earlier show that near-neighbor short-range interaction also could determine the shape of the polymer, as is well known even with respect to the α-helix, 3_{10} helix, and β-structures of polypeptides. Designing oligomers with a conformational twist resulting from

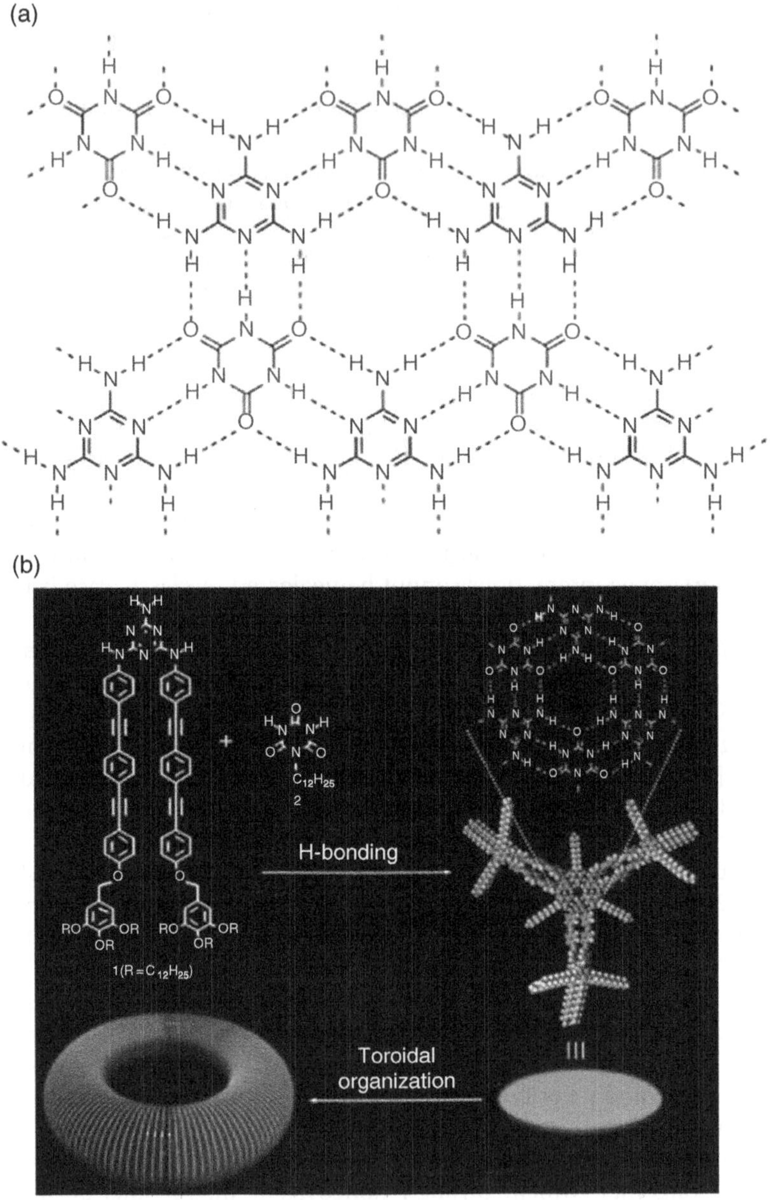

Figure 3.40 (a) Schematic of the hydrogen bond network between melamine and cyanuric acid, (b) schematic of the molecules 1 and 2 studied and their co-assembly, (c) atomic force microscopy images of the torroids formed by the co-assembly of 1 and 2, and (d) illustration of the geometry of the structure of the torroids. (Source: Yagai et al. [91]. Reproduced with permission of Wiley.)

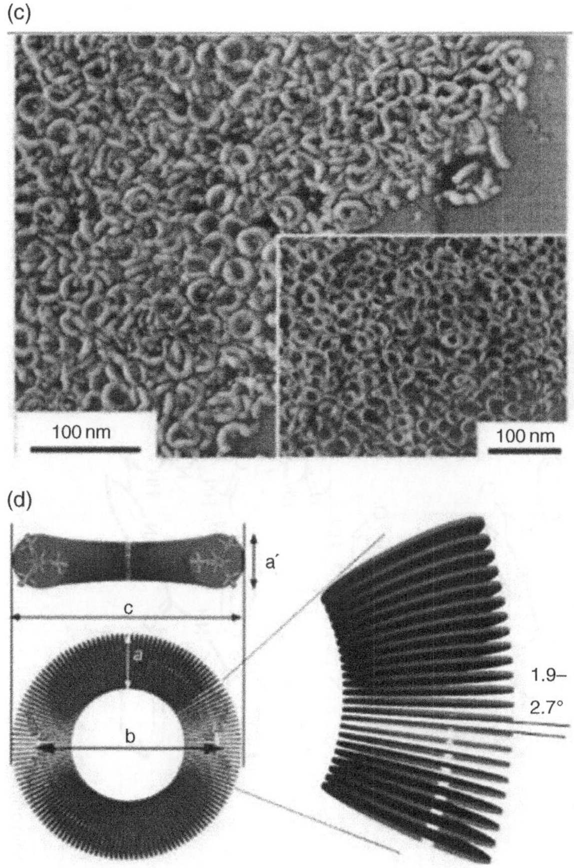

Figure 3.40 (Continued)

hydrogen bonding between adjacent units (to cause folding) has been demonstrated by Lehn et al. They [95] constructed an oligomeric molecule shown in Figure 3.41, consisting of alternating 2,6-diaminopyridine and 2,6-pyridinedicarbonyl units. Molecular modeling showed formation of hydrogen bond between the nitrogen atom and the next amide hydrogen that would cause a twist, leading to a bent conformation with an angle of 39° between the atoms 6 and 6′ as illustrated in Figure 3.41a. The molecule would then fold into a single-strand helix with one and a half turns, as in Figure 3.41b and c. Since it was difficult to obtain single crystals of **1** in Figure 3.41, molecules without the long alkyl side chains were synthesized and the molecular structures of pentamers (**12**, **13**, and **14**), an undecamer (**15**), and a heptamer (**20**) are shown in Scheme 3.8. Crystal structures of **12**, **14**, **15**, and **20** showed that these four oligomers formed single helical structures as predicted from molecular

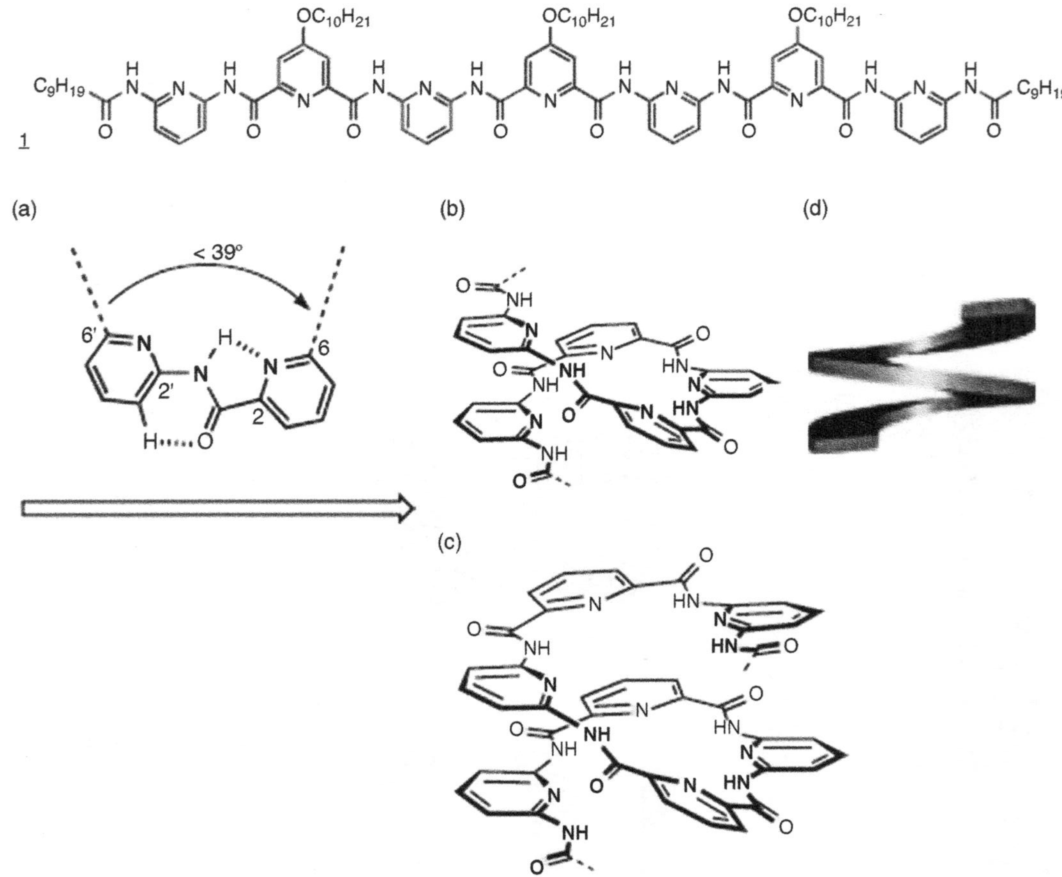

Figure 3.41 A heptamer strand of alternating 2,6-diaminopyridine and 2,6-pyridinedicarbonyl units (1) which bends as shown in (a) due to hydrogen bond formation, leading to helical shapes (b), (c), and (d). (Source: Berl et al. [95]. Reproduced with permission of Wiley.)

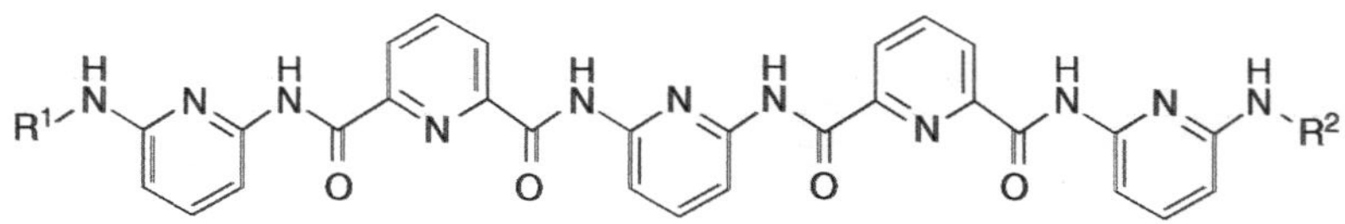

12: $R^1 = R^2 = COCH_3$; **13**: $R^1 = H$; R^2 = Fmoc
+**14**: $R^1 = R^2$ = Fmoc

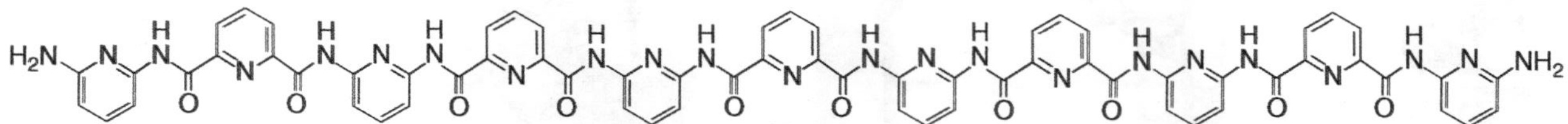

15

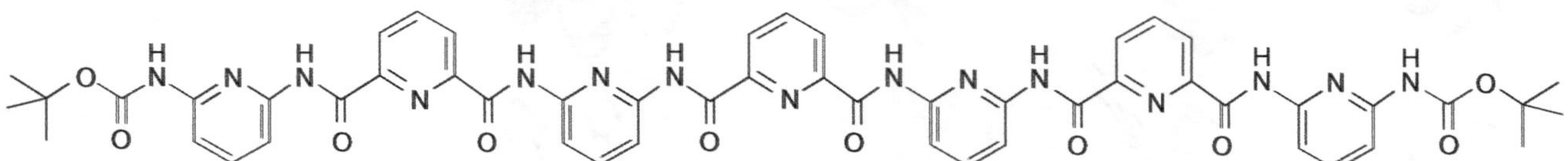

20

Scheme 3.8 Molecular strands of pentamers (**12**, **13** and **14**), undecamer (**15**), and a heptamer (**20**) synthesized by Lehn et al. for folding studies [95]. Fmoc, fluorenylmethoxycarbonyl amino-protecting group; Boc, tert-butoxycarbonyl.

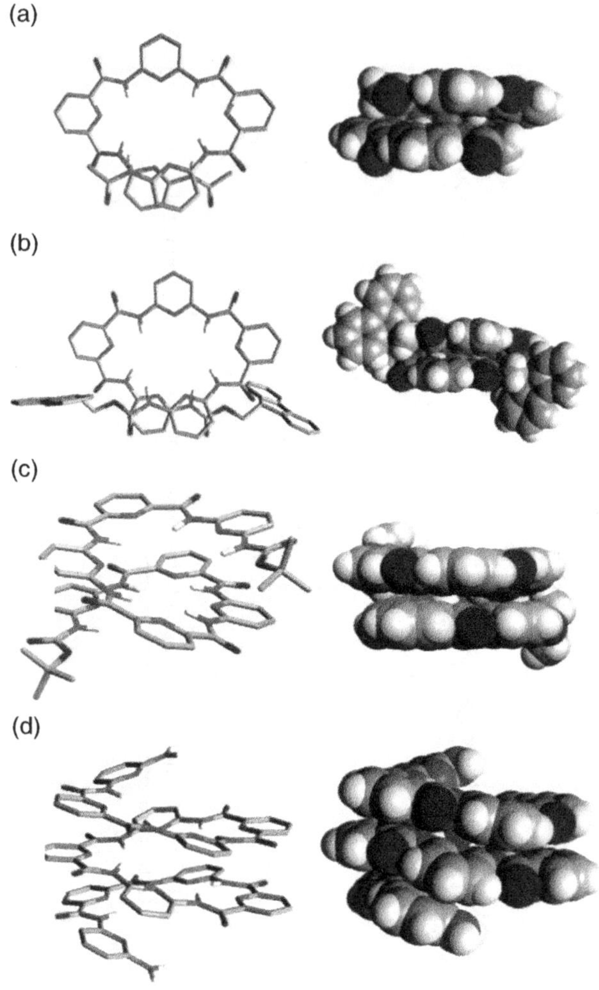

Figure 3.42 Projections (left, wire model) and side views (space-filling models) of (a) pentamer **12**, (b) pentamer **14**, (c) heptamer **20**, and (d) undecamer **15**. (Source: Berl et al. [95]. Reproduced with permission of Wiley.)

modeling. As mentioned earlier, intramolecular hydrogen bonds between amide N—H bonds and the pyridine nitrogen lead to bending of the molecules, resulting in helical motifs. With 4.5 rings per turn of the helix, the heptamer **20** would consist of one and a half turns, and the undecamer **15** two and a half turns. The models of these structures are shown in Figure 3.42. The aromatic ring of unit n overlapped with that of unit $n+4$. The authors compared the crystal structures of the four oligomers in detail. Folding of the oligomers into helices in solutions was also studied using NMR.

In a further study [96], it was found that the pentamers and heptamers formed supramolecular double helices in solution and solid state. A detailed discussion of crystal structures of these foldamers was published by Schmitt et al. [97] Thus, the short-range interactions between neighboring units that cause a twist can also be used to generate foldamers. In fact, using the methods developed by Ramachandran [73], it is possible to calculate the type of helix that the molecules in Figure 3.41 and Scheme 3.8 would adopt with various stereochemically possible twists between the repeat units.

3.9 Single Chain Polymer Crystals and Nanoparticles

As the studies on folding-induced self-assembly progressed, the next step was to control the folding of a single chain, so that the polymer collapses into nanoparticles. Although these investigations are recent, the first ever attempt to prepare single-chain polymer nanoparticles (SCPNs) is attributed to Boyer and Heidenreich [98] in 1945. Dissolving a sample of polychlorostyrene ($M_w = 40\,000$) in benzene and adding a nonsolvent, they obtained particles with circular cross section, measuring a diameter of 1.5–50 nm, as seen in an electron microscope. Interestingly, they were not realizing the potential of their work, but they were simply using the technique to fractionate and measure the molecular weight of the particles. Much later, Wunderlich et al. [99, 100] published pioneering work on growing single crystals of single polymer chains, to understand the structure of a single chain in a crystal. They prepared single-chain single crystals of PE, with samples of various molecular weights, using a stock solution of materials from preliminary crystallization and then diluting these and recrystallizing with low and high super cooling. An electron micrograph image of these single crystals is shown in Figure 3.43a. The melting temperatures of these single crystals containing a single or just a few polymer chains were about 135°C, similar to that of the bulk high density PE. They also were successful in creating SCPNs of poly(oxyethylene) (PEO, $M_n = 6.7 \times 10^5$). In this case, they isolated the polymer into single molecules using a Langmuir film balance, which consisted of a PTFE trough, a surface measuring unit, and a moving barrier. Since PEO is soluble in cold water but not hot water, Wunderlich et al. [99] spread a dilute solution (2×10^{-4} wt%) in benzene, drop by drop, on to a hot water surface kept at 80°C. Ultrafine particles formed and when sufficient number had formed, they were compressed for collecting the particles on TEM grids. The single-molecule single crystals of PEO are shown in Figure 3.43b. The number average particle diameter was 7.6 nm. Confirmation of these being single crystals of single chains was obtained from electron

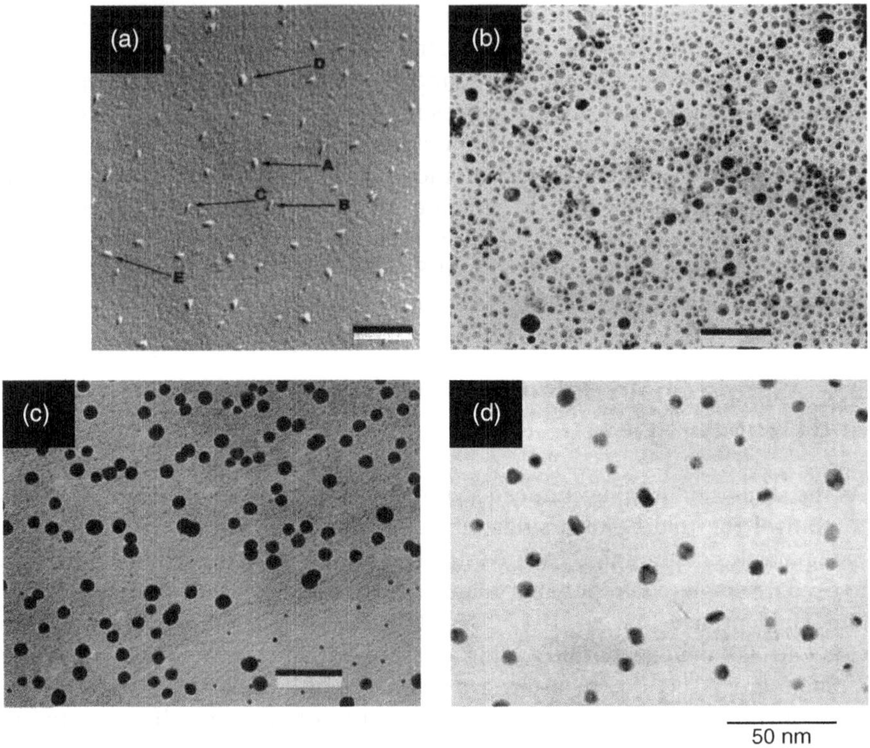

Figure 3.43 Electron micrograph images of (a) single molecule or a few molecules of single crystals of PE, with A–E containing one molecule of mass 4.2×10^7, 2.1×10^7, 2.0×10^7, 2.9×10^7, and 2.2×10^7, respectively. Scale bar: 1 μm. (b) Single-molecule single crystals of poly(oxyethylene) grown as described in the text. Scale bar: 0.1 μm. (c) Single-molecule particles of amorphous polystyrene. Scale bar: 0.1 μm. (d) Single-molecule single crystals of isotactic polystyrene crystallized at 442.8 K for 8 h. (Source: (a)–(c) Bu et al. [99]. Reproduced with permission of Wiley; (d) Bu et al. [100]. Reproduced with permission of Wiley.)

diffraction, with reflections corresponding to the monoclinic unit cell of PEO. The same method of Langmuir film balance and casting drops of solutions on hot water surface was used to prepare single-molecule particles of amorphous polystyrene and single-molecule single crystals of isotactic polystyrene [100]. The electron micrographs of the particles are shown in Figure 3.43c and d, respectively. Whereas the amorphous polystyrene SCPNs are spherical, those corresponding to isotactic polystyrene have a definite shape and the angles between the edges are close to 120°.

In a recent article, Altintas and Barner-Kowollik [101] summarized the approaches to collapsing single polymer chains into nanoparticles using noncovalent and covalent interactions. The noncovalent route involves

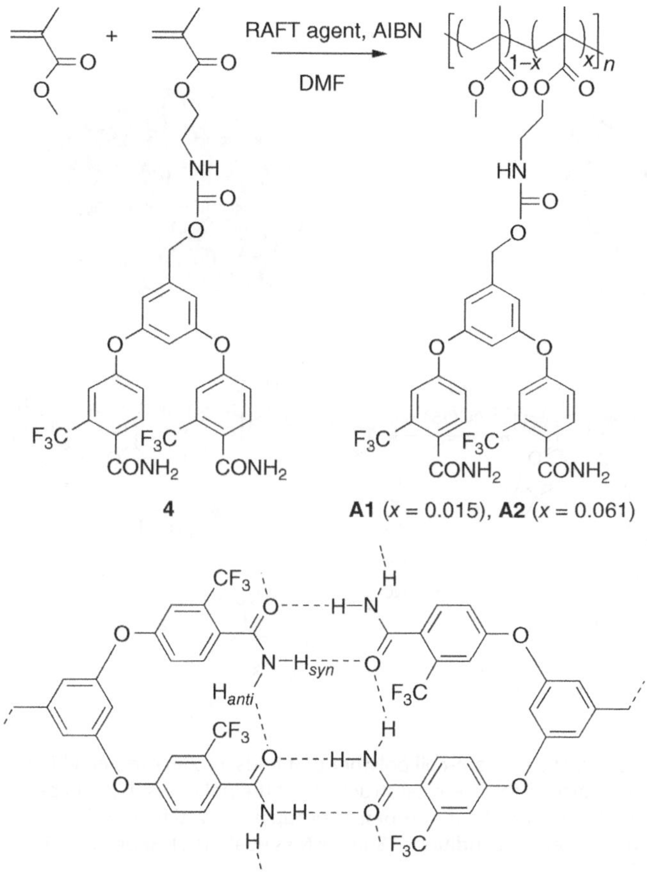

Scheme 3.9 Random copolymers of PMMA and PMMA-dendrons A1 and A2. The proposed hydrogen bonding scheme is also shown [102].

introducing hydrogen bonding groups, usually randomly, along the chain. It is thus the synthesis of random copolymers, with one of the components bearing the hydrogen bonding motifs. The covalent method includes, for example, disulfide bridge formation and photodimerization. The folding of the chain and collapse into an SCPN were followed using techniques, such as size exclusion chromatography, CD in the case of chiral substituents, atomic force microscopy, and transmission and scanning electron microscopies. Seo et al. [102] used self-complementary hydrogen bonding (SCHB) units in a PMMA chain to prepare nanoparticles via chain collapse. They synthesized copolymers of PMMA with a benzamide dendron as shown in Scheme 3.9, with 1.5 and 6.1% of SCHB (**A1** and **A2** in Scheme 3.9). The proposed hydrogen bonding

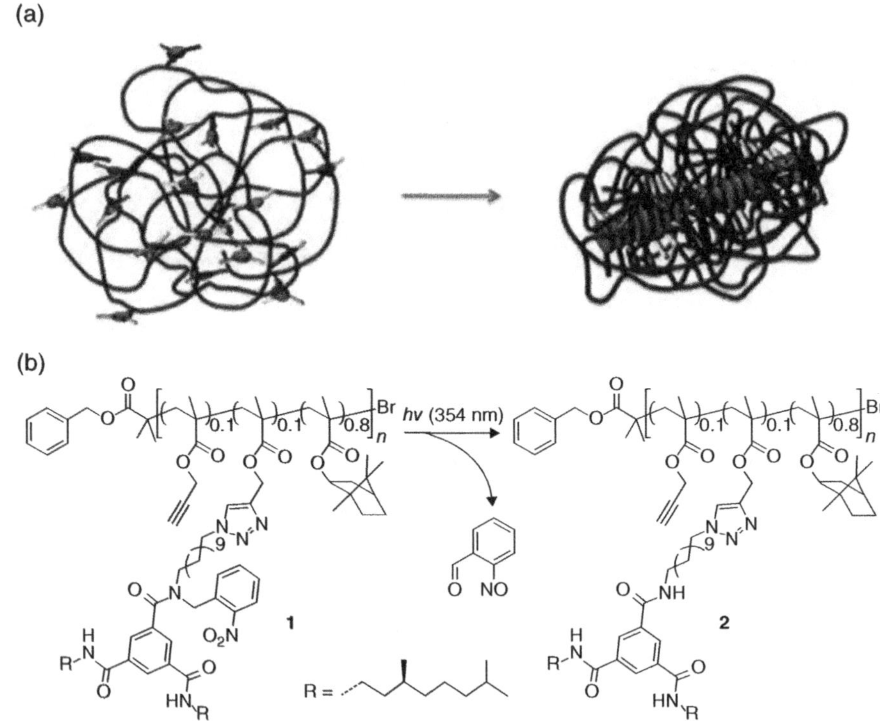

Figure 3.44 (a) Representation of a random-coil polymer that folds into an ordered chiral single-chain polymeric nanoparticle. (b) Chemical structures of the polymers that can be triggered to fold. Upon irradiation of **1**, the *o*-nitrobenzyl group is cleaved to afford **2**, allowing self-assembly under selected conditions. (Source: Mes et al. [103]. Reproduced with permission of Wiley.)

scheme is also shown in Scheme 3.9. Upon dissolving **A2** in a mixture of toluene and THF followed by the evaporation of the latter, uniformly spherical nanoparticles formed, which was confirmed by scanning force microscopy and dynamic light scattering experiments, with a mean diameter of 24 nm. With the copolymer **A1**, with a smaller fraction of dendron substitution, large irregular aggregates with diameters of over 100 nm was observed. This shows that while **A2** collapsed with predominantly intramolecular hydrogen bonding, smaller fraction of dendrons in **A1** leads to interchain association.

Using hydrogen bonding motifs, Meijer et al. [103] showed that SCPN can be prepared in a controlled manner. In one of their investigations, a copolymer of silyl-protected propargyl methacrylate and isobornyl methacrylate was synthesized and azide-functionalized "caged" chiral benzene-1,3,5-tricarboxamide (BTA) was attached to the alkyne functionalized polymer to yield polymer 1 in Figure 3.44. The BTAs act as supramolecular cross-linking units. Irradiation with 354 nm UV radiation for 20 min, in a mixture of 1,2-dichloromethane (DCE) and methyl cyclohexane (MCH), cleaved the O-nitrobenzyl group, enabling polymer 2 to form hydrogen bonding between the BTA groups. While BTA by itself forms helical aggregates in MCH, the methacrylate polymers were insoluble in this solvent. Hence, the aforementioned solvent mixture was used. The self-assembly was followed by the appearance of Cotton effect in CD spectroscopy. The Cotton effect was proportional to the amount of deprotected BTA and was typical of helical hydrogen bonded aggregates of BTA. The composition of the solvent mixture also affects the magnitude of the Cotton effect. It was found that for a BTA concentration of 7.5×10^{-5} M, the maximum Cotton effect was found with a DCE/MCH: 70/30 mixture.

It was concluded that the aggregation is due predominantly to a single-chain folding since the extent of aggregation depended on the local concentration of BTA and not the overall concentration of the polymer. In another study [104], the Meijer group prepared a random copolymer of oligo(ethylene glycol) methacrylate (*o*-EGMA) (to provide water solubility) and BTA-methacrylate (BTAMA) (for supramolecular cross-linking). A molar ratio of BTAMA/*o*-EGMA: 10/90 was used. Intramolecular folding occurred in dilute aqueous solutions resulting in SCPNs.

Meijer et al. also used quadruple hydrogen bonding interactions between uriedo-pyrimidinone (UPy) groups to prepare SCPNs. In one of their publications [105], a norbornene imide-based polymer was used, with 10% of the monomers functionalized with Upy units which were protected with an O-nitrobenzyl group which could be photo-cleaved (Scheme 3.10). Polymers **1a–f** had M_n ranging from 52 to 1200 kg/mol. The long alkyl chains were introduced in the structure to provide solubility. Irradiating a dilute solution of the protected polymer **1a–f** in chloroform with UV light (350 nm) cleaved the nitro group and resulted in **2a–f** forming the SCPN. The M_n of **2a–f** decreased by 23–29% upon irradiation of **1a–f**. SEC analysis of fractionated samples showed that the size of the nanoparticles depended on the molecular weight, as shown in Figure 3.45. It was also shown that the inclusion of urea or urethane in the molecule is necessary to provide orthogonal hydrogen bonding to the quadruple hydrogen bonding of the Upy and stabilize the nanoparticles. Polymers related to **1a–f** in Scheme 3.10 but without the urea or urethane segment did not form SCPN. Meijer et al. [106] also used PMMA-based copolymers, with Upy and urethane units to prepare SCPN's supramolecular cross-linking of these groups.

Tuten et al. [107] designed poly(norbornene-*exo*-anhydride), a polymer with a reactive functional group, which folded by the introduction of *p*-aminophenyl disulfide, a difunctional cross-linker. Chain expansion and collapse as nanoparticles in THF were reversibly controlled by oxidation and reduction.

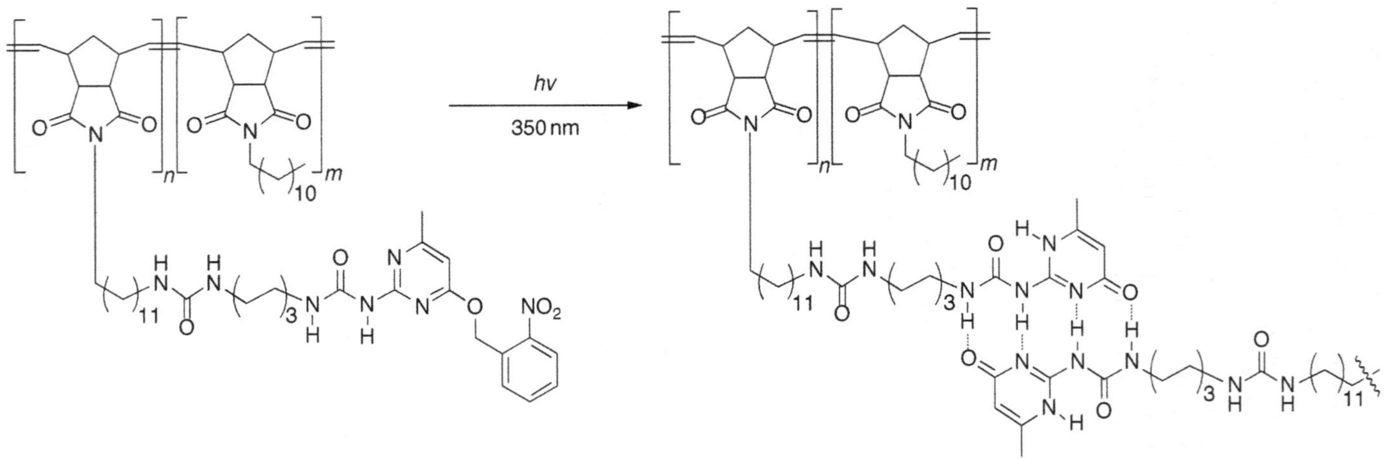

Scheme 3.10 Photo-induced folding of polymers **1a–f** leading to nanoparticles **2a–f**. **1a–f** correspond to polymers with M_n varying from 52 to 1200 kg/mol (2-nitrosobenzaldehyde side product is not shown) [105].

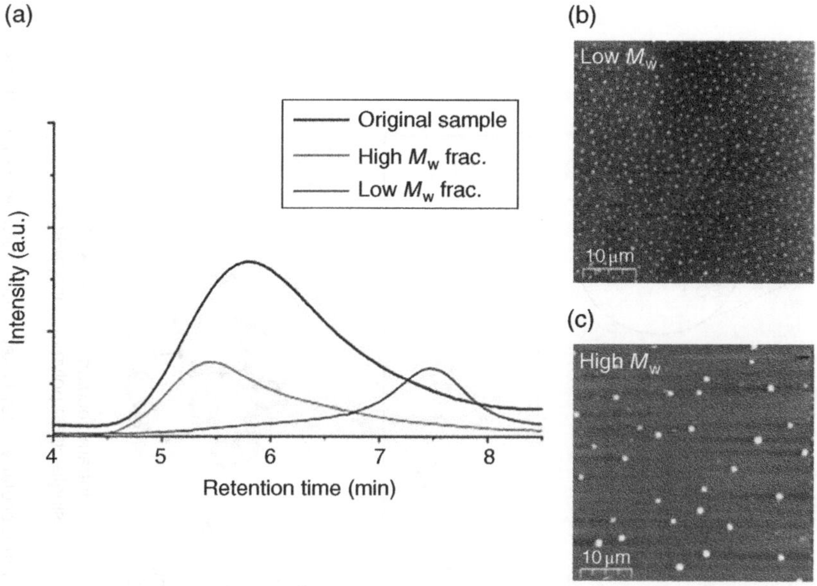

Figure 3.45 (a) SEC traces from the fractionation of a nanoparticle sample; (b) and (c) AFM images of low- and high-molecular-weight samples, respectively. (Source: Foster et al. [105]. Reproduced with permission of Wiley.)

Such reversible cross-linking has been named "dynamic covalent bonds," Whitaker et al. [108] used copolymerization of oligo(ethyleneglycol) methyl ether methacrylate ($OEGMA_{300}$, $M_n \sim 300$ g/mol) with *p*-(2-methacryloxyethoxy) benzaldehyde (MAEBA) to form in reactive thermoresponsive copolymer scaffolds (Figure 3.46). Addition of adipic acid dihydrazide (in 0.1 M NaCl, pH 4.5), caused intramolecular cross-linking through the formation of dynamic covalent acylhydrazone bonds resulting in chain folding and SCPNs.

Frank et al. [109] used copolymers of PPMA and poly(9-anthracenyl methyl methacrylate) (AMMA), with 10, 20, and 46% of the latter, as shown in Figure 3.47. When dilute solutions of THF were irradiated with 350 nm UV light, photo dimerization of anthracene occurred, forming SCNP of sub-20 nm size. While predominantly single chain particles were detected at low concentrations of 10% AMMA in the copolymer, some interchain dimerization was observed with higher AMMA content. The SCNP appeared ellipsoidal in TEM images with 15–20 nm in length and 7–10 nm in width.

Thus, while the traditional methods [98–100] depended on routes such as nonsolvent induced precipitation for creating SCPN, incorporation of noncovalent interactions such as hydrogen bonds along the chain has been successful in preparing SCPN of desired dimensions and with reproducibility. Application of stimuli such as UV radiation, thermal treatments and pH to

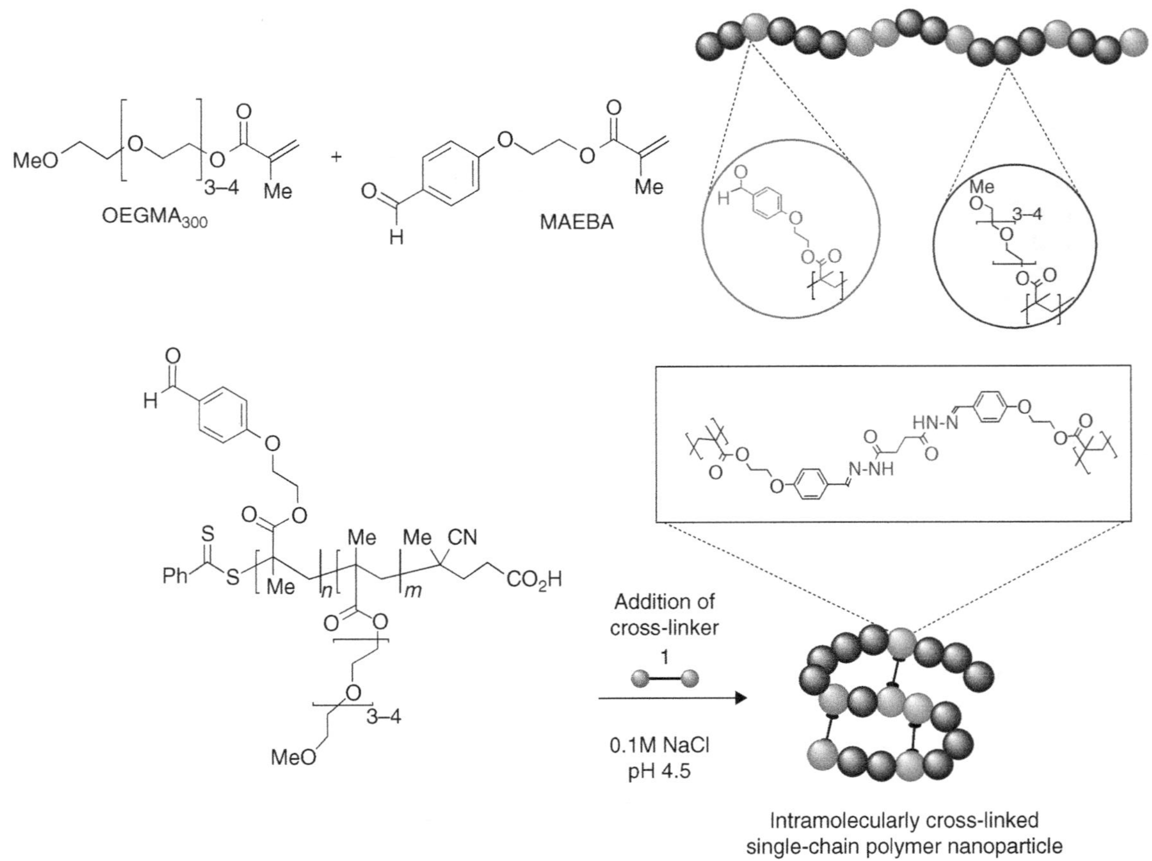

Figure 3.46 Conjugation of OEGMA/MAEBA copolymer with dihydrazide to form intramolecularly cross-linked single polymer nanoparticles. (Source: Whitaker et al. [108]. Reproduced with permission of Wiley.)

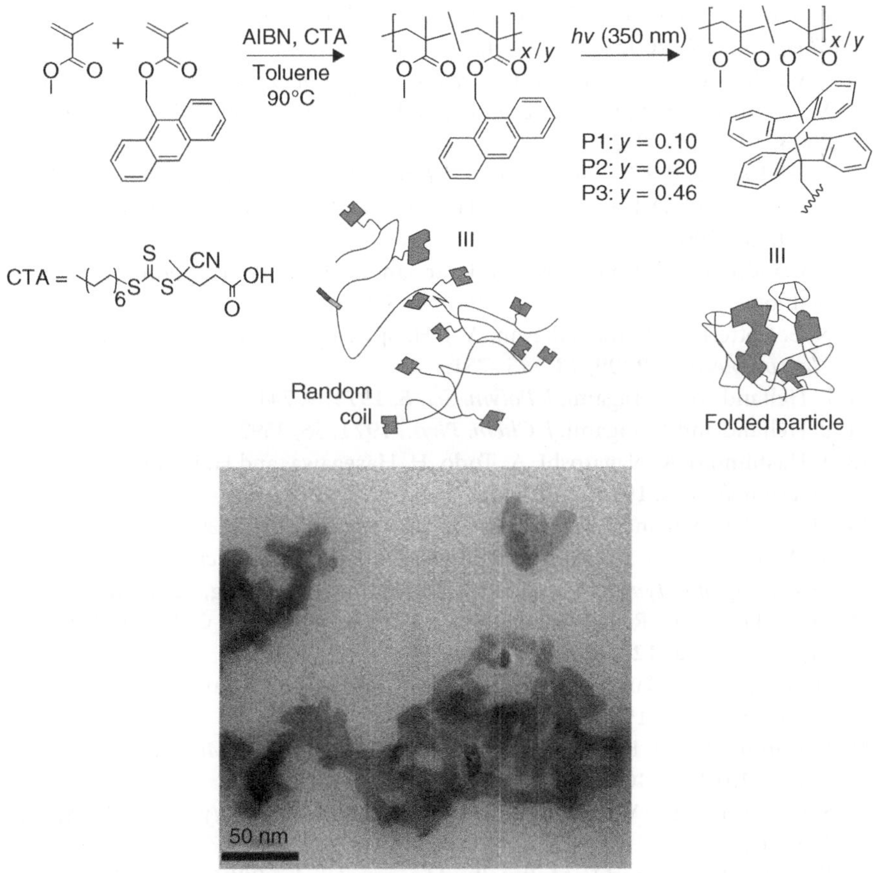

Figure 3.47 Schematic of the synthesis and folding of anthracene-functionalized polymers (top) and the TEM images of the SCNPs from P1. (Source: Frank et al. [109]. Reproduced with permission of Wiley.)

reversibly collapse the single chains to nanoparticles and to expand them again would have several applications such as delivery of cargo for targeted drug delivery [110, 111].

References

1 R. Krämer, J.-M. Lehn, and A. Marquis-Rigault, *Proc. Natl. Acad. Sci. U. S. A.*, **1993**, *90*, 5394–5398.

2 J.-M. Lehn, *Science*, **2002**, *295*, 2400–2403.

3 A. Wu and L. Isaacs, *J. Am. Chem. Soc.*, **2003**, *125*, 4831–4835.

4 A. Wu, A. Chakraborty, J. C. Fettinger, R. A. Flowers Ii, and L. Isaacs, *Angew. Chem. Int. Ed.*, **2002**, *41*, 4028–4031.
5 P. Mukhopadhyay, A. Wu, and L. Isaacs, *J. Org. Chem.*, **2004**, *69*, 6157–6164.
6 M. Moniruzzaman, B. Goodbrand, and P. R. Sundararajan, *J. Phys. Chem. B*, **2003**, *107*, 8416–8423.
7 M. K. Khan and P. R. Sundararajan, *J. Phys. Chem. B*, **2008**, *112*, 4223–4232.
8 B. Goodbrand, D. Boils, P. R. Sundararajan, and R. Wong, US Patent 6,414,051, **2002**.
9 B. Goodbrand, D. Boils, P. R. Sundararajan, R. Wong, M. P. Breton, S. Malhotra, and C. Tripp, US Patent 6,334,890, **2002**.
10 N. E. Botterhuis, S. Karthikeyan, A. J. H. Spiering, and R. P. Sijbesma, *Macromolecules*, **2009**, *43*, 745–751.
11 E. Helfand and Y. Tagami, *J. Polym. Sci. B*, **1971**, *9*, 741.
12 E. Helfand and Y. Tagami, *J. Chem. Phys.*, **1971**, *56*, 3592.
13 T. Hashimoto, K. Nagatoshi, A. Todo, H. Hasegawa, and H. Kawai, *Macromolecules*, **1974**, *7*, 364–373.
14 M. M. Coleman and P. C. Painter, *Prog. Polym. Sci.*, **1995**, *20*, 1.
15 M. M. Coleman, J. F. Graf, and P. C. Painter, *Specific Interactions and the Miscibility of Polymer Blends*, **1991**, Technomic Publishing, Lancaster, PA.
16 M. J. Mayoral, C. Rest, J. Schellheimer, V. Stepanenko, and G. Fernández, *Chem. Eur. J.*, **2012**, *18*, 15607–15611.
17 H. Asanuma, T. Ban, S. Gotoh, T. Hishiya, and M. Komiyama, *Macromolecules*, **1998**, *31*, 371–377.
18 P. Szatkowska, M. Koba, P. Koslinski, and M. Szablewski, *Mini-Rev. Org. Chem.*, **2013**, *10*, 400–408.
19 S. Li, S. Cao, M. J. Whitcombe, and S. A. Piletsky, *Prog. Polym. Sci.*, **2014**, *39*, 145–163.
20 F. Ilhan, M. Gray, and V. M. Rotello, *Macromolecules*, **2001**, *34*, 2597–2601.
21 F. Ilhan, L. Diamondis, L. Gautreau, and V. M. Rotello, *Chem. Commun.*, **2000**, 447–448.
22 C. Burd and M. Weck, *Macromolecules*, **2005**, *38*, 7225–7230.
23 M. M. Green, J.-W. Park, T. Sato, A. Teramoto, S. Lifson, R. L. B. Selinger, and J. V. Selinger, *Angew. Chem. Int. Ed.*, **1999**, *38*, 3138–3154.
24 M. M. Green, in *Circular Dichroism: Principles and Applications*, 2nd Ed., **2000**, N. Berova, K. Nakanishi, and R. W. Woody, Eds., Chapter 17, John Wiley & Sons, Inc., New York.
25 M. M. Green, M. P. Reidy, R. D. Johnson, G. Darling, D. J. O'Leary, and G. Wilson, *J. Am. Chem. Soc.*, **1989**, *111*, 6452.
26 S. Mayer and R. Zentel, *Prog. Polym. Sci.*, **2001**, *26*, 1973–2013.
27 S. Lifson, C. E. Felder, and M. M. Green, *Macromolecules*, **1992**, *25*, 4142–4148.
28 Y. Nagata, T. Yamada, T. Adachi, Y. Akai, T. Yamamoto, and M. Suginome, *J. Am. Chem. Soc.*, **2013**, *135*, 10104–10113.

29 R. Fiesel and U. Scherf, *Acta Polym.*, **1998**, *49*, 445–449.
30 A. Ajayaghosh, R. Varghese, S. Mahesh, and V. K. Praveen, *Angew. Chem. Int. Ed.*, **2006**, *45*, 7729–7732.
31 S. Khanna, M. K. Khan, and P. Sundararajan, *Langmuir*, **2009**, *25*, 13183–13193.
32 M. M. Green, B. A. Garetz, B. Munoz, H. Chang, S. Hoke, and R. G. Cooks, *J. Am. Chem. Soc.*, **1995**, *117*, 4181–4182.
33 B. M. W. Langeveld-Voss, R. J. M. Waterval, R. A. J. Janssen, and E. W. Meijer, *Macromolecules*, **1999**, *32*, 227–230.
34 (a)A. Keller, *Philos. Mag.*, **1957**, *2*, 1171. (b)L. H. Sperling, *An Introduction to Physical Polymer Science*, 3rd Ed., **2001**, Wiley Interscience, New York.
35 D. M. Sadler and A. Keller, *Polymer*, **1976**, *17*, 37.
36 S. J. Spells and D. M. Sadler, *Polymer*, **1984**, *25*, 739.
37 D. M. Sadler and R. J. Harris, *J. Polym. Sci. Polym. Phys. Ed.*, **1982**, *20*, 561.
38 M. L. Mansfield, *Macromolecules*, **1983**, *16*, 914.
39 J. D. Hoffman, *Polymer*, **1983**, *24*, 3.
40 P. J. Flory, *J. Am. Chem. Soc.*, **1962**, *84*, 2857.
41 P. J. Flory and D. Y. Yoon, *Nature*, **1978**, *272*, 226.
42 D. Y. Yoon and P. J. Flory, *Polymer*, **1977**, *18*, 509.
43 P. J. Flory and D. Y. Yoon, *Discuss. Faraday Soc.*, **1979**, *68*, 288.
44 T. A. Kavassalis and P. R. Sundararajan, *Macromolecules*, **1993**, *26*, 4144–4150.
45 P. R. Sundararajan and T. A. Kavassalis, *J. Chem. Soc. Faraday Trans.*, **1995**, *91*, 2541–2549.
46 G. Ungar, J. Stejny, A. Keller, I. Bidd, and M. C. Whiting, *Science*, **1985**, *229*, 386–389.
47 G. Ungar and A. Keller, *Polymer*, **1986**, *27*, 1835–1844.
48 K. S. Lee and G. Wegner, *Makromol. Chem. Rapid Commun.*, **1985**, *6*, 203.
49 K. S. Lee, G. Wegner, and S. L. Hsu, *Polymer*, **1987**, *28*, 889.
50 W. M. Leung, R. St. John Manley, and A. R. Panaras, *Macromolecules*, **1985**, *18*, 746, 753, 760.
51 T. Trzebiatowski, M. Drager, and G. R. Strobl, *Makromol. Chem.*, **1982**, *183*, 731.
52 P. R. Sundararajan and T. A. Kavassalis, *Macromolecules*, **1997**, *30*, 5172–5174.
53 H. Takeshita, M. Poovarodom, T. Kiya, F. Arai, K. Takenaka, M. Miya, and T. Shiomi, *Polymer*, **2012**, *53*, 5375–5384.
54 R. Napolitano and B. Pirozzi, *Macromol. Theory Simul.*, **2001**, *10*, 532–541.
55 H. Meyer and F. Muller-Plathe, *Macromolecules*, **2002**, *35*, 1241–1252.
56 L. Toma, S. Toma, and J. A. Subirana, *Macromolecules*, **1998**, *31*, 2328–2334.
57 C. Liu and M. Muthukumar, *J. Chem. Phys.*, **1998**, *109*, 2536–2542.
58 M. Muthukumar and P. Welch, *Polymer*, **2000**, *41*, 8833–8837.
59 M. Muthukumar, *Eur. Phys. J. E*, **2000**, *3*, 199.
60 P. Welch and M. Muthukumar, *Phys. Rev. Lett.*, **2001**, *87*, 218302-1–218302-4.

61 I. Dukovski and M. Muthukumar, *J. Chem. Phys.*, **2003**, *118*, 6648–6655.
62 S. Liu, B. Ashok, and M. Muthukumar, *Polymer*, **2004**, *45*, 1383–1389.
63 T. Yamamoto, *J. Chem. Phys.*, **2001**, *115*, 8675–8680.
64 T. Yamamoto, *Polymer*, **2004**, *45*, 1357–1364.
65 T. Yamamoto, *J. Chem. Phys.*, **1997**, *107*, 2653–2661.
66 H. Takeuchi, *J. Chem. Phys.*, **1998**, *109*, 5614–5621.
67 M. S. Lavine, N. Waheed, and G. C. Rutledge, *Polymer*, **2003**, *44*, 1771–1779.
68 P. J. Flory, *Statistical Mechanics of Chain Molecules*, **1969**, John Wiley & Sons, Inc., New York (Reprinted by Hanser Publishers 1989).
69 H. Tadokoro, Y. Chatani, H. Kusanagi, and M. Yokoyama, *Macromolecules*, **1970**, *3*, 441–447.
70 H. Kusanagi, H. Tadokoro, and Y. Chatani, *Macromolecules*, **1976**, *9*, 531–532.
71 H. Kusanagi, Y. Chatani, and H. Tadokoro, *Polymer*, **1994**, *35*, 2028–2039.
72 P. R. Sundararajan and P. J. Flory, *J. Am. Chem. Soc.*, **1974**, *96*, 5025–5031.
73 G. N. Ramachandran, C. M. Venkatachalam, and S. Krimm, *Biophys. J.*, **1966**, *6*, 849.
74 D. J. Hill, M. J. Mio, R. B. Prince, T. S. Hughes, and J. S. Moore, *Chem. Rev.*, **2001**, *101*, 3893–4012.
75 P. J. Flory, *Principles of Polymer Chemistry*, **1969**, Cornell University Press, Ithaca, NY.
76 P. R. Sundararajan, "Theta Temperatures", in *Physical Properties of Polymers Handbook*, **2007**, J. E. Mark, Ed., Springer, Chapter 15.
77 E. Hädicke and F. Graser, *Acta Crystollagr.*, **1986**, *C42*, 189–195.
78 F. Graser and E. Hädicke, *Liebigs Ann. Chem.*, **1980**, *12*, 1994–2011.
79 G. Klebe, F. Graser, E. Hädicke, and J. Berndt, *Acta Crystallogr.*, **1989**, *B45*, 69–77.
80 R. O. Loutfy, P. M. Kazmaier, A.-M. Hor, and R. A. Burt, US Patent 4,808,506, **1989**.
81 Z. Y. Wang, Y. Qi, J. P. Gao, G. G. Sacripante, P. R. Sundararajan, and J. D. Duff, *Macromolecules*, **1998**, *31*, 2075–2079.
82 D. Yao, Z. Y. Wang, and P. R. Sundararajan, *Polymer*, **2005**, *46*, 4390–4396.
83 P. R. Sundararajan, G. Sacripante, and Z. Y. Wang, *Comput. Theor. Polym. Sci.*, **2000**, *10*, 219–220.
84 A. D. Q. Li, W. Wang, and L-Q Wang, *Chem. Eur. J.*, **2003**, *9*, 4594–4601.
85 M. R. Islam, E. Dahan, S. Saimani, and P. Sundararajan, *Eur. Polym. J.*, **2012**, *48*, 1538–1554.
86 E. E. Neuteboom, S. C. J. Meskers, E. W. Meijer, and R. A. J. Janssen, *Macromol. Chem. Phys.*, **2004**, *205*, 217–222.
87 D. H. Appella, L. A. Christianson, I. L. Karle, D. R. Powell, and S. H. Gellman, *J. Am. Chem. Soc.*, **1996**, *118*, 13071–13072.

88 S. H. Gellman, *Acc. Chem. Res.*, **1998**, *31*, 173–180.
89 P. Yarborough and D. Sogah, *Polym. Mater. Sci. Eng.*, **1997**, *76*, 395–396.
90 A. Heckel and P. B. Dervan, *Chem. Eur. J.*, **2003**, *9*, 3353–3366.
91 S. Yagai, S. Mahesh, Y. Kikkawa, K. Unoike, T. Karatsu, A. Kitamura, and A. Ajayaghosh, *Angew. Chem. Int. Ed.*, **2008**, *47*, 4691–4694.
92 P. R. Sundararajan and V. S. R. Rao, *Biopolymers*, **1969**, *8*, 313–323.
93 V. S. R. Rao, N. Yathindra, and P. R. Sundararajan, *Biopolymers*, **1969**, *8*, 325–333.
94 P. R. Sundararajan and V. S. R. Rao, *Biopolymers*, **1969**, *8*, 305–312.
95 V. Berl, I. Huc, R. G. Khoury, and J.-M. Lehn, *Chem. Eur. J.*, **2001**, *7*, 2798–2809.
96 V. Berl, I. Huc, R. G. Khoury, and J.-M. Lehn, *Chem. Eur. J.*, **2001**, *7*, 2810–2820.
97 J.-L. Schmitt, A.-M. Stadler, N. Kyritsakas, and J.-M. Lehn, *Helv. Chim. Acta*, **2003**, *86*, 1598–1624.
98 R. F. Boyer and R. D. Heidenreich, *J. Appl. Phys.*, **1945**, *16*, 621.
99 H. Bu, Y. Pang, D. Song, T. Yu, T. M. VolL, G. Czornyj, and B. Wunderlich, *J. Polym. Sci. B Polym. Phys.*, **1991**, *29*, 139–152.
100 H. Bu, E. Chen, S.-Y. Xu, K. Guo, and B. Wunderlich, *J. Polym. Sci. B*, **1994**, *32*, 1351–1357.
101 O. Altintas and C. Barner-Kowollik, *Macromol. Rapid Commun.*, **2012**, *33*, 958–971.
102 M. Seo, B. J. Beck, J. M. J. Paulusse, C. J. Hawker, and S. Y. Kim, *Macromolecules*, **2008**, *41*, 6413–6418.
103 T. Mes, R. van der Weegen, A. R. A. Palmans, and E. W. Meijer, *Angew. Chem. Int. Ed.*, **2011**, *50*, 5085–5089.
104 P. J. M. Stals, M. A. J. Gillissen, T. F. E. Paffen, T. F. A. de Greef, P. Lindner, E. W. Meijer, A. R. A. Palmans, and I. K. Voets, *Macromolecules*, **2014**, *47*, 2947–2954.
105 E. J. Foster, E. B. Berda, and E. W. Meijer, *J. Polym. Sci. A Polym. Chem.*, **2011**, *49*, 118–126.
106 E. B. Berda, E. J. Foster, and E. W. Meijer, *Macromolecules*, **2010**, *43*, 1430–1437.
107 B. T. Tuten, D. Chao, C. K. Lyon, and E. B. Berda, *Polym. Chem.*, **2012**, *3*, 3068–3071.
108 D. E. Whitaker, C. S. Mahon, and D. A. Fulton, *Angew. Chem. Int. Ed.*, **2013**, *52*, 956–959.
109 P. G. Frank, B. T. Tuten, A. Prasher, D. Chao, and E. B. Berda, *Macromol. Rapid Commun.*, **2014**, *35*, 249–253.
110 I. Odriozola, M. K. Aiertza, G. Cabanero, H.-J. Grande, and I. Loinaz, in *Handbook of Harnessing Biomaterials in Nanomedicine: Preparation,*

Toxicity, and Applications, **2012**, D. Peer, Ed., Pan Stanford Publishing, Chapter 1.
111 M. K. Aiertza, I. Odriozola, G. Cabanero, H.-J. Grande, and I. Loinaz, *Cell. Mol. Life Sci.*, **2012**, *69*, 337–346.

Further Reading

M. M. Safont-Sempere, G. Fernández, and F. Würthner, *Chem. Rev.*, **2011**, *111*, 5784–5814.
E. Yashima, K. Maeda, H. Iida, Y. Furusho, and K. Nagai, *Chem. Rev.*, **2009**, *109*, 6102–6211.

4

Supramolecular Macromolecules and Polymers

Mutual molecular recognition between two or more molecules via noncovalent interactions leads to their self-assembly. Reversibility of the association is an attractive feature as is the ease of changing the molecular architecture at will. If a molecule A noncovalently associates with its own kind to form a dimer, by adding a stronger, competing ligand B, the A···A association could be replaced by A···B pair to form a hetero-dimer. Such a modification can be achieved without any "bond-breaking" chemistry as in the case of covalently bonded systems. It was mentioned in Chapter 1 that the International Union of Pure and Applied Chemistry (IUPAC) does not consider the definition of a polymer to be the same as that of a "macromolecule." Association of molecules by noncovalent interactions via self-assembly would lead to "supermolecules" or "macromolecules" through "supramolecular" assembly. The synthetic chemistry protocol then should be such that the molecular recognition leading to supramolecular assembly could be achieved by strategically placing the recognition elements in the molecule. In his book *Supramolecular Chemistry*, Lehn [1] explained the whole concept as that of extending the chemistry of molecules to the association of many molecules to make supermolecules. In order to build a molecule with specific properties from two or more smaller molecules or fragments, conventional molecular chemistry would employ covalent bonds. Functionalizing a molecule with, for example, specific side groups to impart or modify certain properties (i.e., solubility) is usually done by covalent attachment. Achieving the same by using recognition elements based on noncovalent intermolecular forces is the tenet of supramolecular chemistry.

4.1 Supramolecular Macromolecules

Different shapes of the same molecule, brought about, for example, by rotations around single bonds (Figure 1.8c) are known as "conformational isomers." Flory called them "rotational isomers" and developed the rotational isomeric

Physical Aspects of Polymer Self-Assembly, First Edition. P. R. Sundararajan.

state (RIS) theory for studies on statistical conformational behavior of chain molecules. Conventional chemistry would use methods such as attaching a bulky or an articulated side group to restrict the conformational flexibility and to force the molecule to a desired conformation. Lehn called the different spatial arrangements of a molecule driven by the recognition element/receptor interactions via supramolecular association as "supramolecular isomers." We will illustrate this with one of the publications of Lehn et al. [2] as an example.

Figure 4.1 shows a molecular strand of an oligo-isophthalamide **1** with alkyl substituents R^1 and R^2. The molecule has a sequence of four donor–acceptor–donor (DAD: NH—N—NH) hydrogen bonding groups. With the rotations around the bond linking C=O/aromatic ring, it was determined that 36 different conformations are possible for **1**. With the addition of a Janus-type cyanurate (**4** in Figure 4.1) that has the ADA/ADA (C=O, NH, C=O, NH, C=O) recognition sequence, the binding leads to a C-shaped, or an S-shaped, or a helical strand. Up to two equivalents of **4** were added to **1** in $CDCl_3$. ^{1}H NOE experiments confirmed that the **1**:$\mathbf{4}_2$ helical strand was more predominant than the C- or S-shaped conformers. The self-assembly of the helical strands led to helical columns as the second-level aggregation entity. This was confirmed by the observation that addition of a hydrocarbon solvent such as cyclohexane to a 20–40 mM solution of **1**:$\mathbf{4}_2$ in chloroform led to strong viscoelasticity. This confirmed the formation of long and entangled fibers. The transmission electron microscopy (TEM) images showed the fibers to be 3–6 μm long, with about 50 Å in diameter (cryo imaging), and they aggregated further to form large bundles and two-dimensional sheets. The molecules **1** or **4** by themselves did not form such fibers. The binding of **4** to **1**, causing template-induced hierarchical self-assembly, is required to form such long fibers.

Combining the hydrogen bonding, π-interaction as well as chiral centers, the groups of Schenning and Meijer [3, 4] synthesized oligo(*p*-phenylenevinylene)s with three, four, and five phenylenevinylene units (**OPV3**, **OPV4**, and **OPV5**, respectively), as shown in Figure 4.2. All three molecules were functionalized with chiral side chains, three aliphatic dodecyloxy side chains as the end groups, and an ureido-*s*-triazine hydrogen bonding group. The self-assembly in chloroform, heptane, and dodecane were studied using ^{1}H NMR and ultraviolet–visible (UV–vis) and fluorescent spectra. The supramolecular association as dimers occurred first via hydrogen bonding, as shown in Figure 4.2. Note that the hydrogen bonding moiety in OPV3, OPV4, and OPV5 has the donor (NH)–acceptor (N)–donor (NH)–acceptor (C=O) sequence, as illustrated in Figure 4.2b. The dimerization would involve association that requires an inversion along the long axis of the molecule. The OPV molecules studied here do not have complimentary hydrogen bonding groups in the same molecule, as described in Section 4.2. Thus, hydrogen bonding in this case would lead only to supramolecular dimerization, not polymerization. The dimers were found to molecularly dissolve in chloroform. The optical spectra showed

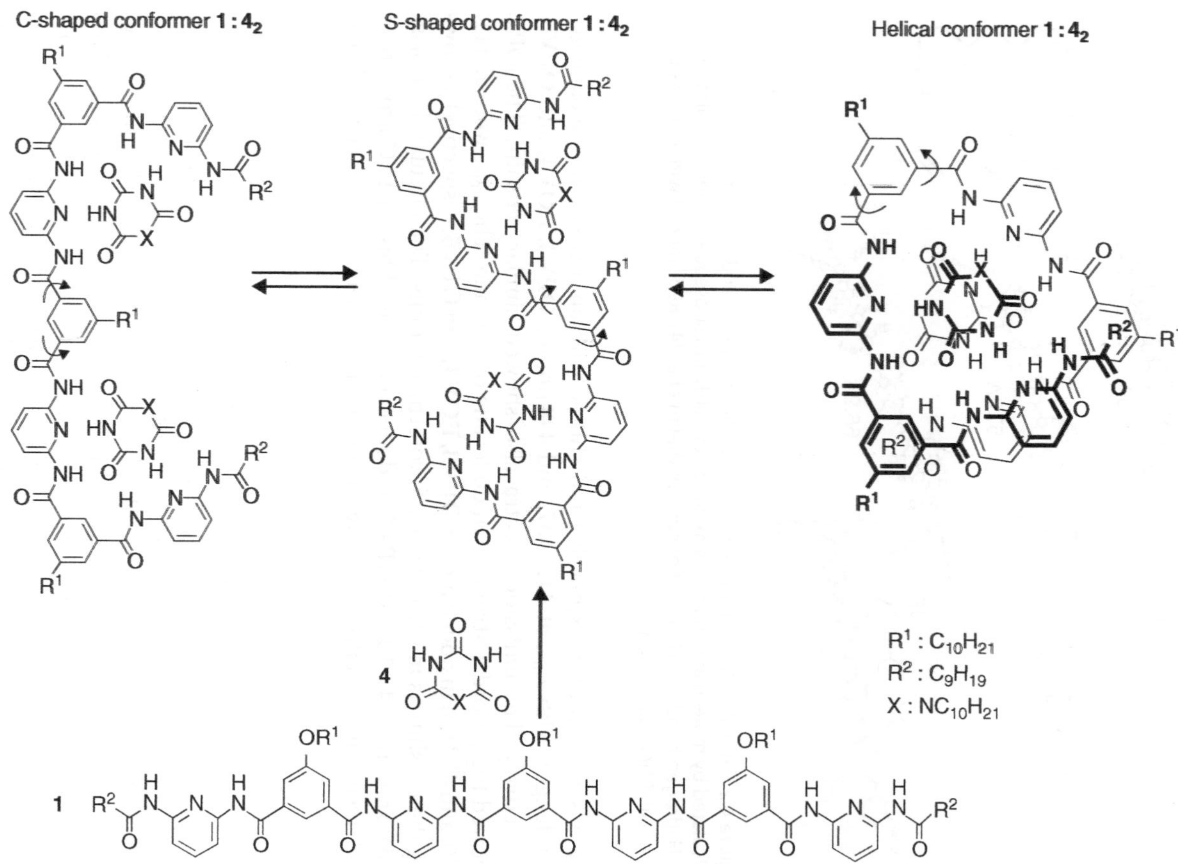

Figure 4.1 Binding of the two-equivalent cyanurate template (4) to the molecular strand (1) leads to C- or S-shaped conformers and predominant helical assembly. (Source: Berl et al. [2]. Reproduced with permission of Wiley.)

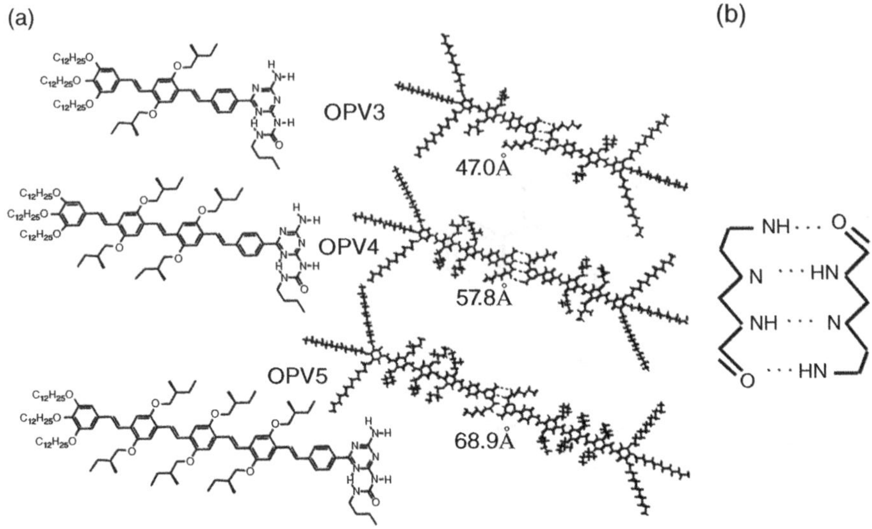

Figure 4.2 (a) Molecular structures of OPV3, OPV4, and OPV5, and the dimerization as calculated by molecular dynamics simulation at 300 K and subsequent energy minimization. (b) Hydrogen bonding scheme. (Source: Jonkheijm et al. [4]. Reproduced with permission of American Chemical Society.)

UV–vis absorption maxima that red-shifted with an increase in the conjugation length, with λ_{max} at 412, 437, and 444 nm for **OPV3, POV4,** and **OPV5,** respectively. The emission spectra also showed similar redshift. The OPV showed higher order structure in dodecane or heptane. The λ_{max} were at 411, 440, and 453 nm in dodecane for **OPV3, POV4,** and **OPV5,** respectively, showing the redshift as in the case of chloroform solutions. The circular dichroism (CD) spectra of 4×10^{-5} M **OPV5** in dodecane showed bisignate Cotton effect and the negative exciton coupling suggested the presence of a left-handed helical columnar arrangement. This is the result of the hierarchical arrangement which originates with the hydrogen bonded dimers, which further self-assemble into stacks via π-interaction between the phenylenevinylene units, with the helicity of the stacks promoted by the chiral side groups. The calculated lengths of the dimers (not including the dodecyloxy side chains) using molecular dynamics simulations as marked in Figure 4.2a, were 47, 58, and 69 Å for **OPV3, POV4,** and **OPV5,** respectively. Further simulations led to the conclusion that two hydrogen bonded **OPV4** dimers stack with a distance of 0.35 nm between them and an angle of rotation of 12°, leading to a helical pitch of 6 nm (see figure 5 of Jonkheijm et al. [4] for the simulated helical structure). These results from simulations were in agreement with scattering data. Small-angle neutron scattering from **OPV4**/dodecane solutions led to a length of 150 nm for the helical columns and a diameter of 5.8 nm. While the helical structure

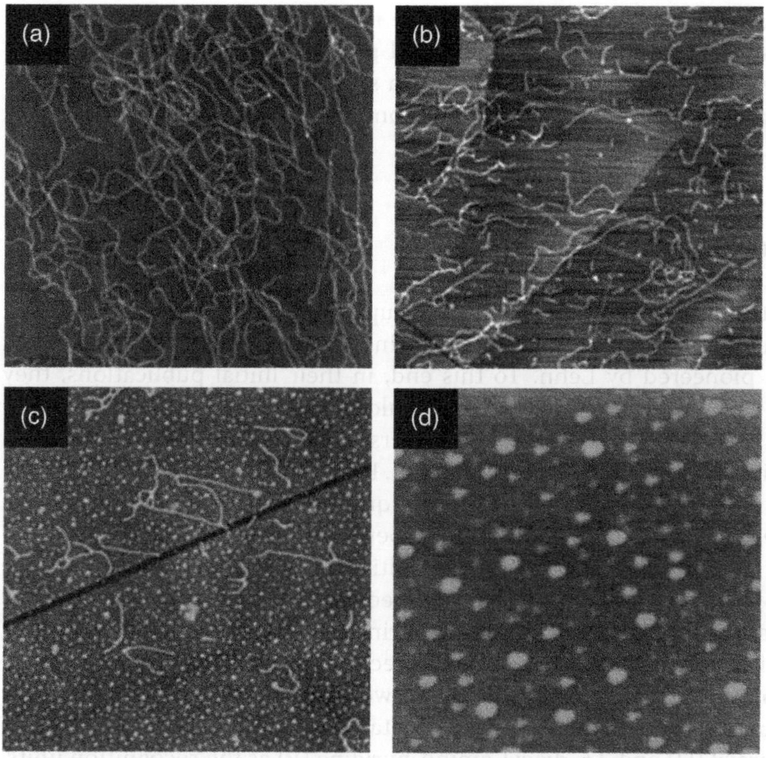

Figure 4.3 Concentration-dependent morphology (AFM images) of drop-cast solutions of OPV4 on graphite surface. (a) 1.3×10^{-4} M, (b) 1.3×10^{-5} M, (c) 1.3×10^{-6} M, and (d) 1.3×10^{-7} M. The surface shown in (a) is $10 \times 10\,\mu m^2$, and it is $7 \times 7\,\mu m^2$ in (b)–(d). (Source: Jonkheijm et al. [4]. Reproduced with permission of American Chemical Society.)

was stable at room temperature, "melting" of the stacks occurred in a narrow temperature range around 343 K, which would indicate that the transition was cooperative.

Jonkheijm et al. [4] pointed out the difficulty in transferring the self-assembled structures from solution on to solid substrates. The atomic force microscopy (AFM) images of the drop-cast films of **OPV4** on to a graphite surface are shown in Figure 4.3 for various concentrations of the solution. With a high concentration of 1.3×10^{-4} M, several intertwined strands of the aggregate structure were seen as in Figure 4.3a. As the concentration was reduced, short strands along with globules were seen (Figure 4.3b and c), and only globules were seen at a concentration of 1.3×10^{-7} M (Figure 4.3d). The globular structures were attributed to uncontrolled self-assembly of molecularly dissolved material. The morphology of the drop-cast stacks also depended on the type of substrate. Single fibers were seen on hydrophobic surfaces such as silicon oxide wafers and graphite, whereas lamellar morphology was seen with the glass and mica surfaces. It was emphasized that transferring the supramolecular stacks from solution to solid substrates is a delicate process.

4.2 Supramolecular Polymers

Whereas the aforementioned examples of supramolecular approach were used for building "macromolecules," synthesizing polymers with regular repeat units was pioneered by Lehn. To this end, in their initial publications, they designed receptors and wedges that could mutually associate with three hydrogen bonds. Supramolecular polymers built through hydrogen bonds often exhibited liquid crystalline properties. If one of the components is a large aromatic molecular fragment, discotic liquid crystalline behavior can be anticipated. Naturally, the more the number of hydrogen binds between the repeating units, the more robust the resulting polymer. The use of multiple hydrogen bonds to build such supramolecular polymers was reported by Fouquey et al. [5]. Figure 4.4a shows the principle of the approach, in which a template is functionalized with a ditopic receptor and another with a ditopic wedge, and the mutual recognition of the two would lead to a supramolecular polymer chain. To this end, they chose a tartaric acid derivative as the template (T), and uracil (U) and 2,6-diacyl-amino-pyridine (P) as the recognition units. Chirality could be introduced by choosing the L, D, or *meso* isomer of tartaric acid. The self-assembly of these units is illustrated in Figure 4.4b. Three hydrogen bonds between the TP_2 and TU_2 units stabilize the chain. Thermal analysis was performed of the 1/1 mixtures ($LP_2 + LU_2$), ($DP_2 + LU_2$), and ($MP_2 + MU_2$) in tetrahydrofuran (THF) after solvent evaporation. The letters L and D denote the isomers of T. Thermotropic mesophase was observed from below room temperature to above 200 °C.

X-ray diffraction studies indicated the structure to be composed of hexagonal columnar mesophases. A model for the helicoidal arrangement was derived from the X-ray data. The TEM images [6] (Figure 4.5a and b) showed helical fibers of opposite chirality for ($LP_2 + LU_2$) and ($DP_2 + DU_2$), whereas the meso ($MP_2 + MU_2$) did not show any helicity (Figure 4.5c).

With a view to increase the number of hydrogen bonds as well as to design molecular fragments that could be used for selective purposes, Berl et al. [7] designed a receptor and a wedge with complementary hydrogen bonding sites, such as, for example, the receptor possessing a sequence of donor–acceptor–donor (D–A–D) moieties and the wedge with A–D–A groups. The basic components of such a strategy are shown in Figure 4.6a.

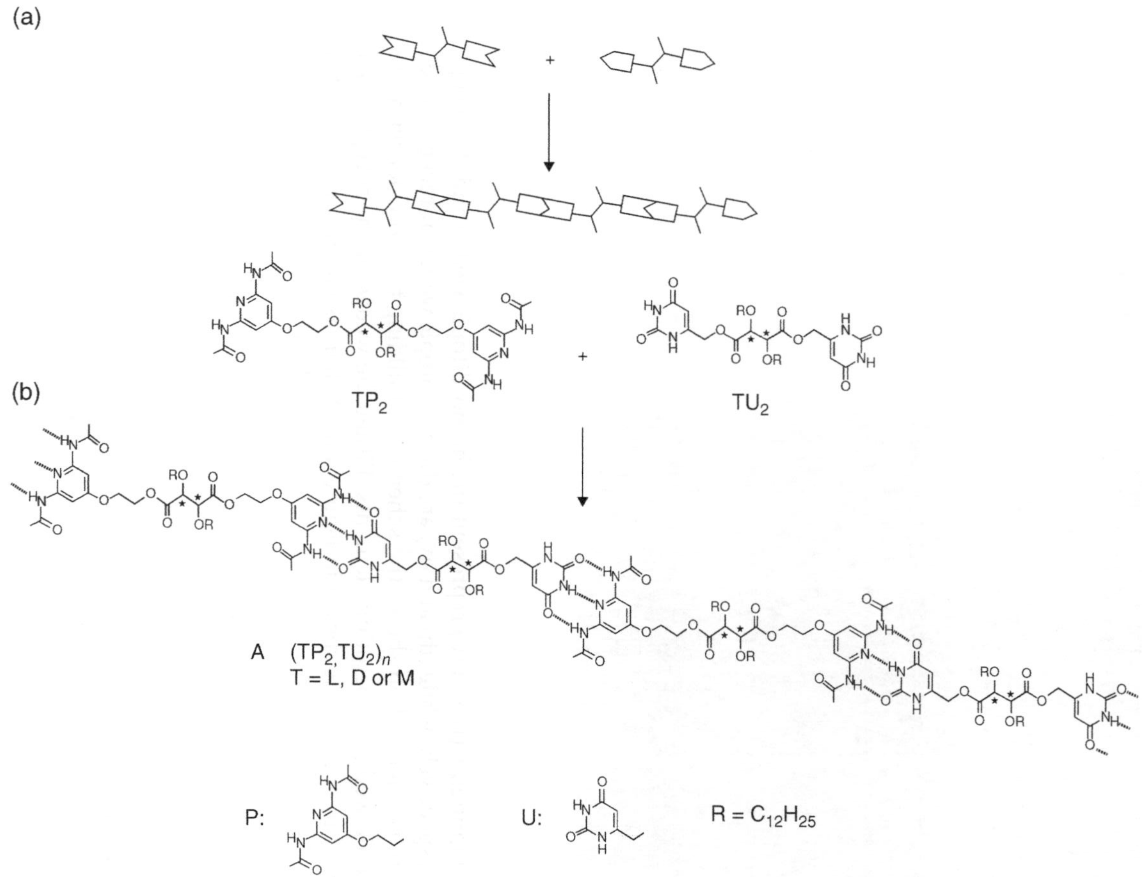

Figure 4.4 (a) Schematic illustration of the supramolecular assembly of templates functionalized with hydrogen bond recognition fragments. (b) Supramolecular assembly of tartaric acid (T) functionalized with uracil (U) and amino-pyridine (P), with three hydrogen bonds between the recognition units. (Source: Fouquey et al. [5]. Reproduced with permission of Wiley.)

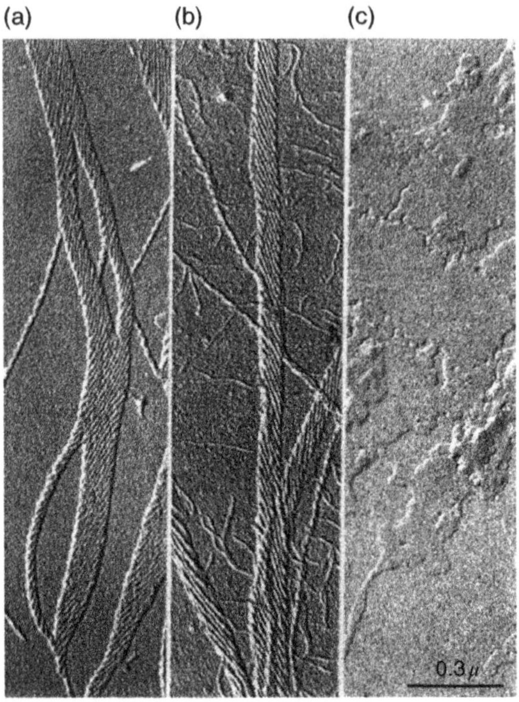

Figure 4.5 TEM images of the helical structures formed by the mixtures: (a) LP_2+LU_2, (b) DP_2+DU_2, and (c) MP_2+MU_2. (Source: Lehn [6]. Reproduced with permission of Wiley.)

Components (i) and (ii) represent the homo-ditopic wedge and the receptor, respectively, while (iii) and (iv) are the monotopic wedge or receptor end-cap units. Figure 4.6b shows the scheme for building the polymer with the receptor/wedge recognition elements. Either the wedge or the receptor can be used to end-cap, as shown in Figure 4.6c. Whereas Figure 4.6b illustrates the linear polymerization, cross-linked network is schematically shown in Figure 4.6d.

Berl et al. [7] used a cyanuric wedge with ADA–ADA (Janus-type) hydrogen bond array and a diaminopyridine-substituted isophthalamide bis-receptor with the DAD–DAD sequence, shown as **3BB** and **1AA** in Figure 4.7. Six hydrogen bonds are possible between these units, affording a strong supramolecular association. Their self-assembly is also illustrated in the figure. The choice of Janus-type **AA** and **BB** units as opposed to the AB type was rationalized on the basis of the ease of synthesis as well as stoichiometric control in the mixtures. A tris-receptor was used for cross-linking (shown illustrated in Figure 4.6a(v) and d). The association constants for the association of the components in chloroform were determined using titration studies. Detailed nuclear magnetic resonance (NMR) studies were conducted for the determination of

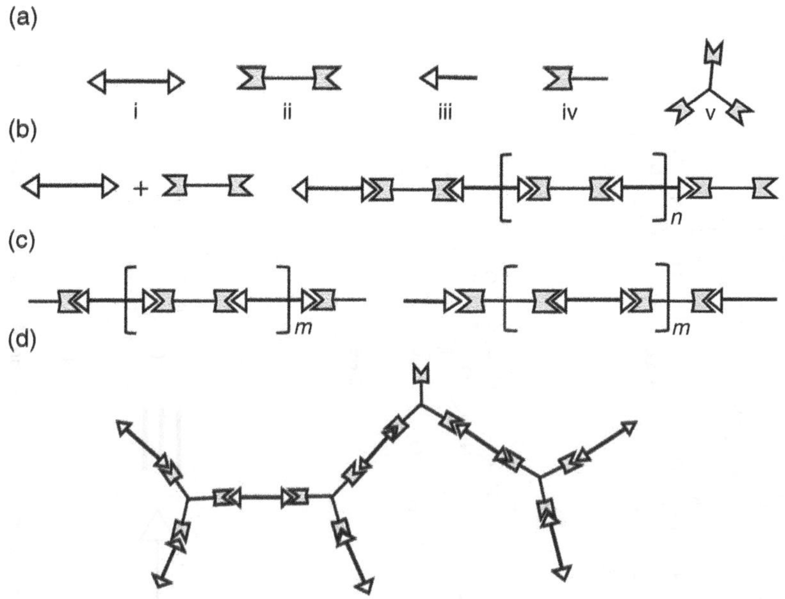

Figure 4.6 Basic components to building supramolecular polymers using receptor–wedge design [7] (see text).

supramolecular linear polymerization, branched structures, and end-capping. The linear supramolecular polymer thus synthesized is shown in Figure 4.8.

Rheological measurements of highly viscous toluene solutions showed a large drop in viscosity with applied shear. This was attributed to the mechanical stress-induced disruption of the hydrogen bonded assembly. However, regeneration of the structure was achieved upon equilibrium conditions, illustrating the dynamic nature of these supramolecular polymers. The high viscosity of the solutions arises due to the formation of fibers. Such highly viscous solutions resulted when a nonpolar solvent such as heptane was added to the chloroform mixtures. The addition of such nonpolar solvent reduces the polarity and strengthens the hydrogen bond association. Figure 4.9a and b shows the directly observed and cryo-fractured TEM images of drop-cast solutions (2.5 mM) of stoichiometric mixtures of **1AA** and **3BB** $\{[1:3]\}_n$ in chloroform/heptane (1 : 4) mixtures. Dense fiber network is seen. However, no fibers were observed when chloroform alone was used. Tetrachloroethane (TCE) was found to be a good solvent for observing fiber formation. Solutions of **1**, **3** as well as other components by themselves did not show any fibers. When stoichiometric mixtures of **1** and **3** were used, well-defined helical fibers were observed. Another interesting observation was the difference in the self-assembly behavior depending on the procedure used for the preparation of the

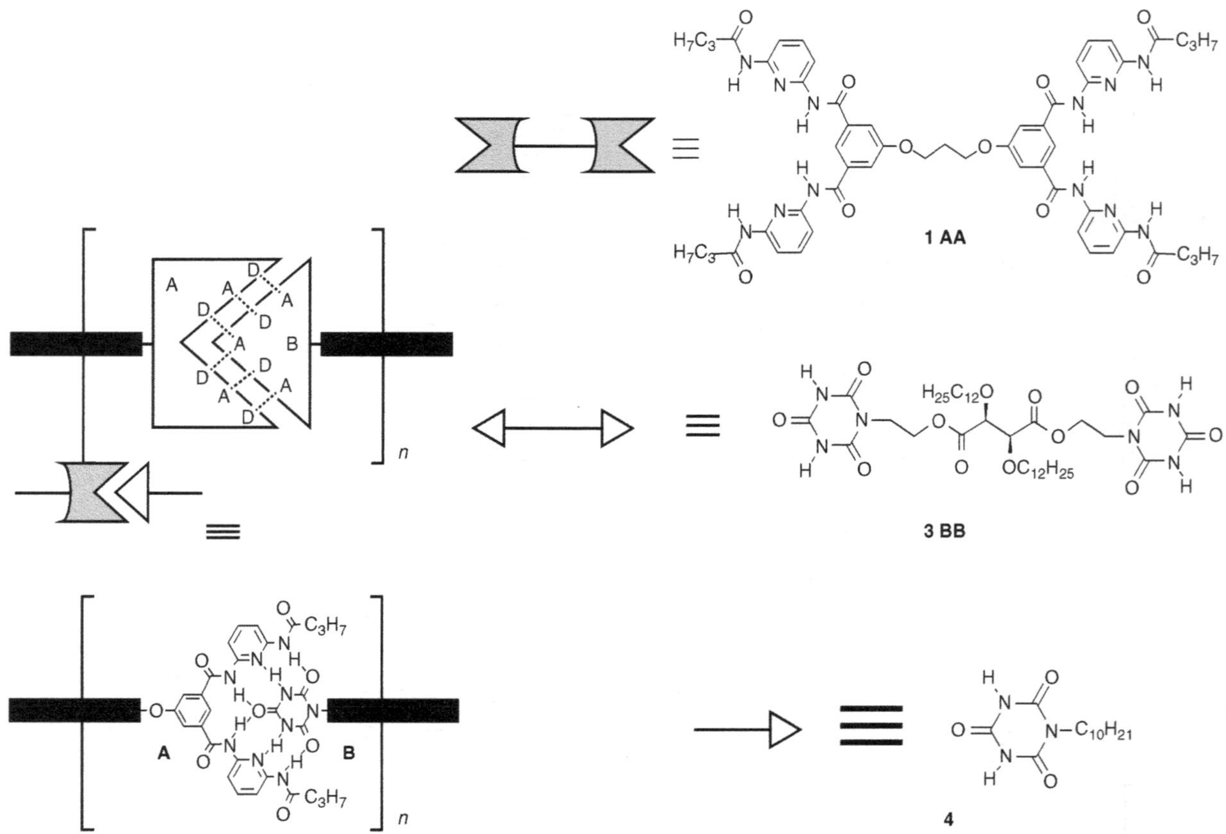

Figure 4.7 The wedge and receptor units as well as their hydrogen bonding pattern are shown. The structure **4** is the stopper wedge. (Source: Berl et al. [7]. Reproduced with permission of Wiley.)

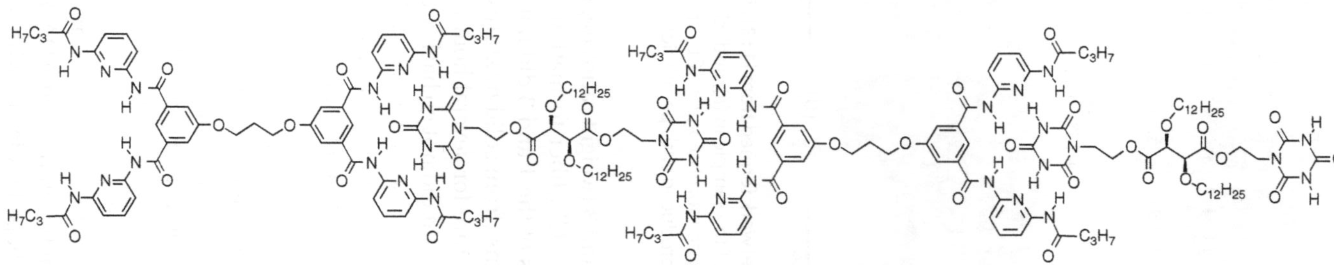

Figure 4.8 Linear supramolecular polymer formed by molecular recognition involving six hydrogen bonds per supra-link between **1AA** and **3BB**. (Source: Berl et al. [7]. Reproduced with permission of Wiley.)

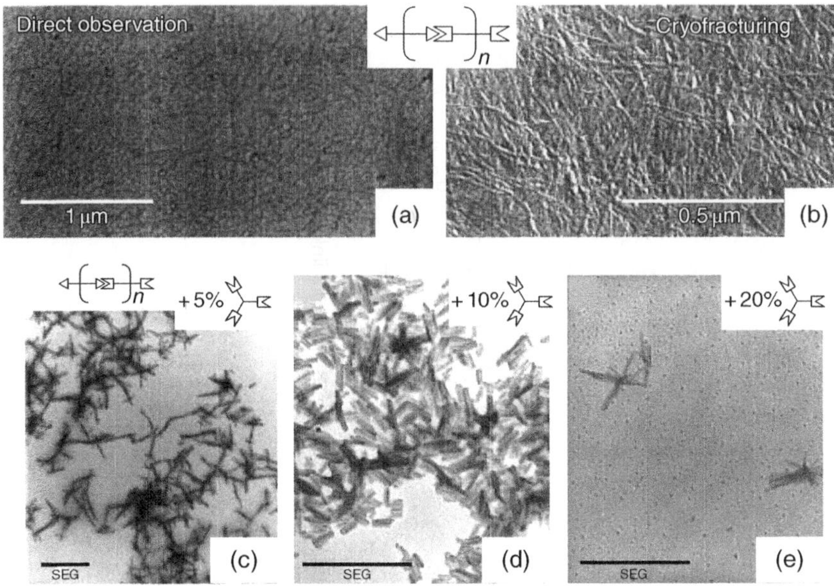

Figure 4.9 (a) A TEM image of fiber network formed from $[\mathbf{1:3}]_n$ in chloroform/heptane solutions; (b) a cryo-fractured image of the sample in (a); and (c)–(e) effect of adding the cross-linking agent on the aggregate morphology in TCE for various ratios of the cross-linker. (Source: Berl et al. [7]. Reproduced with permission of Wiley.)

mixtures. If solid mixtures of **1** and **3** (weight ratios depending on the stoichiometric ratio) were dissolved in TCE, fiber formation was observed, whereas mixing premade stock solutions of the **1** and **3** did not show any fibers. It was also found that the TCE solutions (from solid mixtures) developed cloudiness upon standing at room temperature for several hours and a white precipitate formed after several days. This solid did not go back into solution in TCE even after rigorous heat treatments. It did so only upon adding a polar solvent and subsequent heat treatment. Aggregate fiber formation beyond a critical size upon standing was thought to be a reason for the "irreversibility" of the supramolecular polymerization. It is likely that standing at room temperature for several hours or days promotes entanglement formation (see Chapter 1). Such irreversibility of the supramolecular association was likened by Berl et al. [7] to the irreversible association of proteins to form aggregates in the brain which leads to Alzheimer's and similar diseases. Figure 4.9c–e shows the influence of adding the cross-linking agent. The fibers are shortened significantly with an increase in the amount of the cross-linker.

The system described earlier with ADA–ADA Janus receptor and DAD–DAD Janus wedge has spacers between the two ADA and the two DAD units. Ikeda et al. [8] designed a bow-shaped molecule with four self-complementary

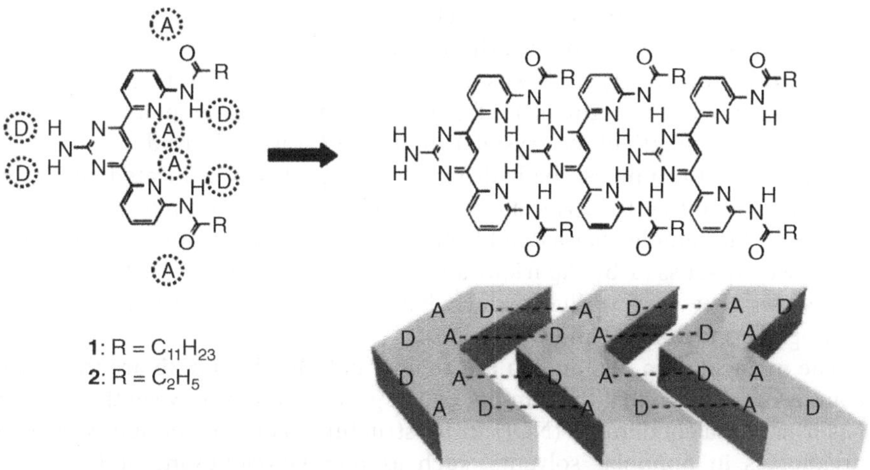

1: R = $C_{11}H_{23}$
2: R = C_2H_5

Figure 4.10 Bow-shaped monomer bearing self-complementary hydrogen bonding sites on either side, the self-assembly of which leads to ribbon-like supramolecular polymer. The donor and acceptor sites are marked by D and A, respectively [8].

hydrogen bonding sites on either side of the molecule as shown in Figure 4.10, without a spacer. The donor (D) and acceptor (A) sites are marked in the figure, as well as the hydrogen bonding scheme. The self-assembly in chloroform was investigated using ^{1}H NMR spectroscopy.

With a concentration of 1.0 mM of **1**, the spectrum corresponded to that of a molecular solution. Upon increasing the concentration up to 250 mM, both amide NH and pyrimidine-NH_2 peaks showed significant downfield shifts, showing that self-assembly had taken place, corresponding to the quadruple hydrogen bonding pattern shown in Figure 4.10. Since the hydrogen bond strength would be expected to be enhanced in nonpolar solvents, these authors prepared gels of **1** in solvents, such as decalin, toluene, *p*-xylene, and paraffin. Similar to the procedure for making organo- or polymer gels discussed in Chapter 7, solid powder of 1 was dispersed (10.0 mM) in the chosen solvent, heated to dissolution, and cooled back to room temperature. While gels were obtained by this procedure with **1**, compound **2** did not form a gel, suggesting that the long alkyl side chains of **1** aided in the gelation, suppressing the tendency to crystallize. The gelation was reversible. The hydrogen bond-mediated aggregation was also confirmed by UV–vis spectra of the gels in decalin which showed absorption peaks that were red-shifted with respect to those recorded for chloroform solutions. The fluorescence spectra also showed similar redshift, and it was proposed that the J-type stacking was present. Variable temperature NMR spectra showed that the gels were stable up to 80°C, although some aggregates persisted after the gel–sol transition.

The TEM image of the **1**/decalin gel is shown in Figure 4.11. Lamellar morphology was seen [8], with a width of 1.0 μm and a thickness of about 30 nm. The highly ordered structure was confirmed by the sharp diffraction pattern shown in the inset to Figure 4.11, from which unit cell parameters of $a = 9.6$ Å (parallel to the length of the crystals) and $b = 16.1$ Å were derived.

In the earlier examples, the hydrogen bonding sites were located along the molecules such that the association led to a linear extension into a polymeric structure. Supramolecular assembly along the lateral association of the molecules was discussed by Aparicio and Sánchez [9]. Figure 4.12 shows the molecules that they studied, which have aromatic segments and laterally hydrogen bondable NH and CO groups.

The molecules **1a, 2a,** and **3a** are achiral and **1b, 2b,** and **3b** are the chiral counterparts. Note that the number of NH groups differs between them and **3** has a naphthalenediimide (NDI) in its structure. These compounds formed organogels in nonpolar solvents, such as methylcyclohexane and toluene, which was attributed to supramolecular polymerization. Although fibrillar morphologies have been routinely seen in the gelation of self-assembling organic small molecules, such behavior is not generally termed "supramolecular polymerization." Although it can be called supra "macromolecules," we go along with the term "polymerization" here since these authors have used it. In an earlier publication, Aparicio et al. [10] showed that the self-assembly of **2** led to helical, columnar supramolecular structures that associated further to form bundles of fibers. The AFM images of the gel fiber morphologies are shown in Figure 4.13. Achiral **1a** forms long rods of several micrometers (Figure 4.13a), whereas Figure 4.13b shows intertwined fibrils with several twists for chiral **1b**. The morphologies seen in Figure 4.13c and d for achiral **3a** with the NDI unit consist of long nanowires (4 nm height) which form intertwined thicker wires of 12–16 nm in height. It also appears from Figure 4.13d that these nanowires are composed of a mixture of right- and left-handed helices (this is indicated by black lines in Figure 4.13d). The chiral **3b** forms filaments of 1.5 nm height and aggregates into thicker fibers. It was found that the critical concentration required for gelation (CGL) differed markedly, with 7, 2.5, 2.8, 7, and 34 mg/ml for **1a, 2a, 2b, 3a,** and **3b**, respectively, indicating that the type and strength of the noncovalent interactions play an important role in the self-assembly. Such factors were also seen in the CD experiments that were performed with mixtures of **2a** and **2b** as well as **3a and 3b**, to examine the sergeants–soldiers effect. Addition of only 10% of **3b** to **3a** was sufficient to achieve maximum chiral response, whereas 30% of **2b** was needed for the same result with **2a**. This again shows that intercalation of **2b** with the achiral **2a** was impeded to some extent, as compared with the case of **3b** and **3a**. Nuclear Overhauser effect (NOE) experiments were combined with ^{1}H NMR spectral data to derive a model for the packing of the molecules **2** and **3**. Figure 4.14 shows the difference in the packing of these molecules. In the case of **2**,

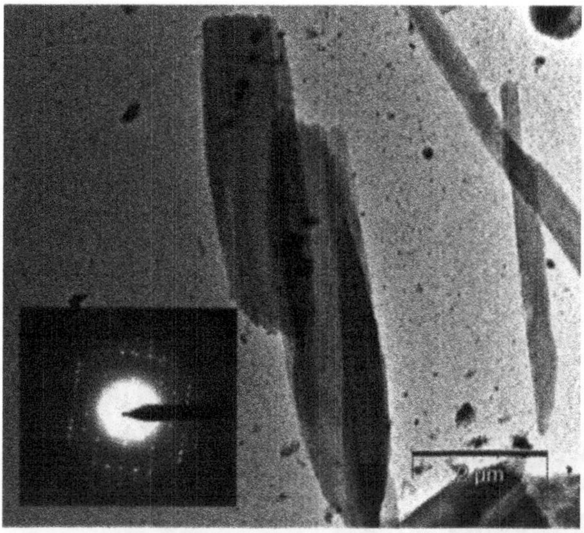

Figure 4.11 Transmission electron micrograph of **1** in decalin (1.0 mM) with the selected area diffraction pattern in the inset. (Source: Ikeda et al. [8]. Reproduced with permission of Wiley.)

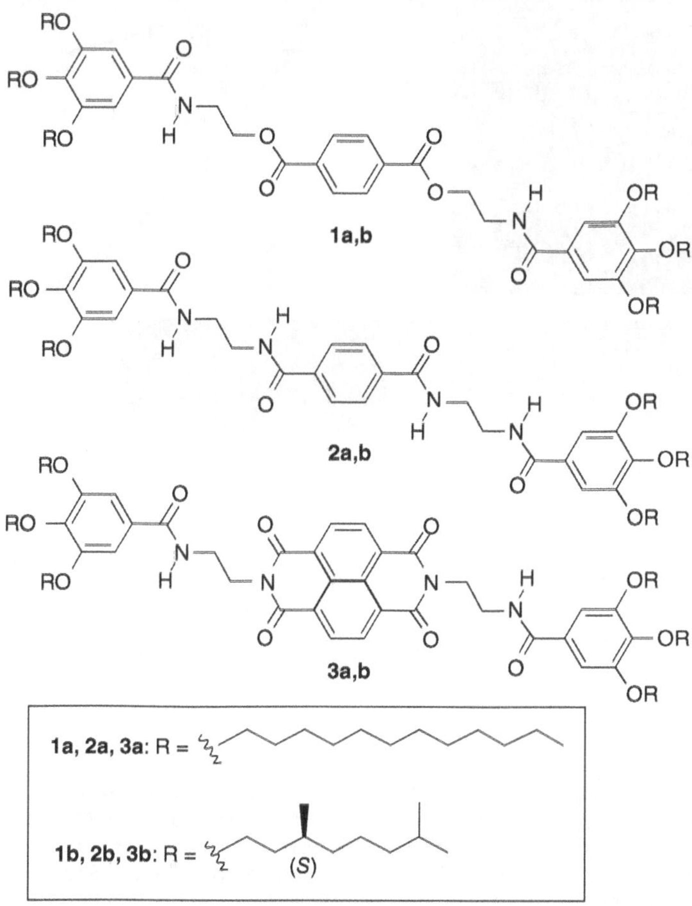

Figure 4.12 Chemical structures of self-assembling molecules investigated by Aparicio and Sánchez [9].

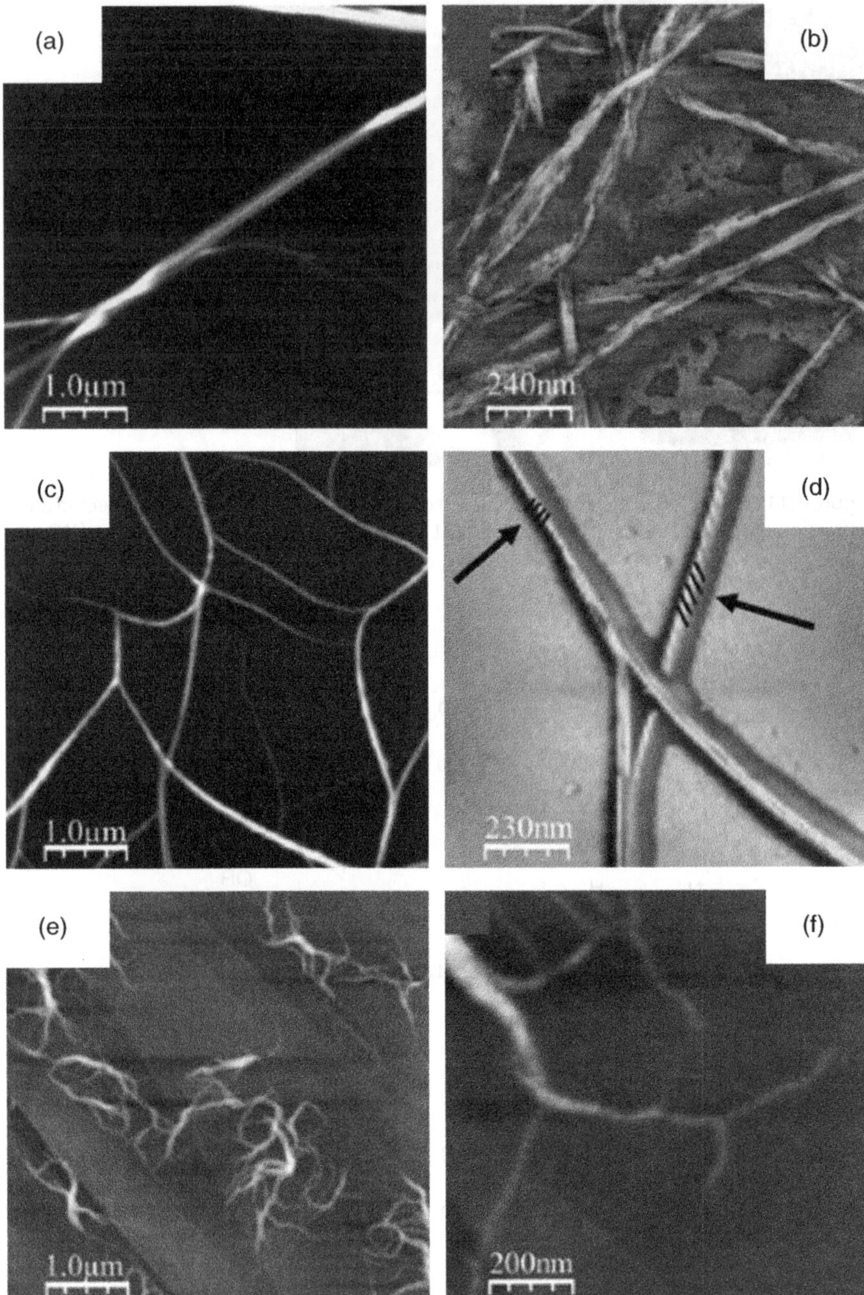

Figure 4.13 AFM height images of (a) achiral **1a**, *z*: 25 nm; (b) chiral **1b**, *z*: 50 nm; (c, d) achiral **3a**, *z*: 35 and 20 nm; and (e) and (f) chiral **3b**, z: 20 and 8 nm. The surface was highly oriented pyrolytic graphite (HOPG), solution was 1×10^{-5} M in toluene, and *z* is the AFM height scale. The arrows in (d) point to the black lines that were drawn as a guide to the eye to indicate the chirality. (a) and (b) were given in color in the original publication. (Source: Aparicio and Sánchez [9]. Reproduced with permission of Wiley.)

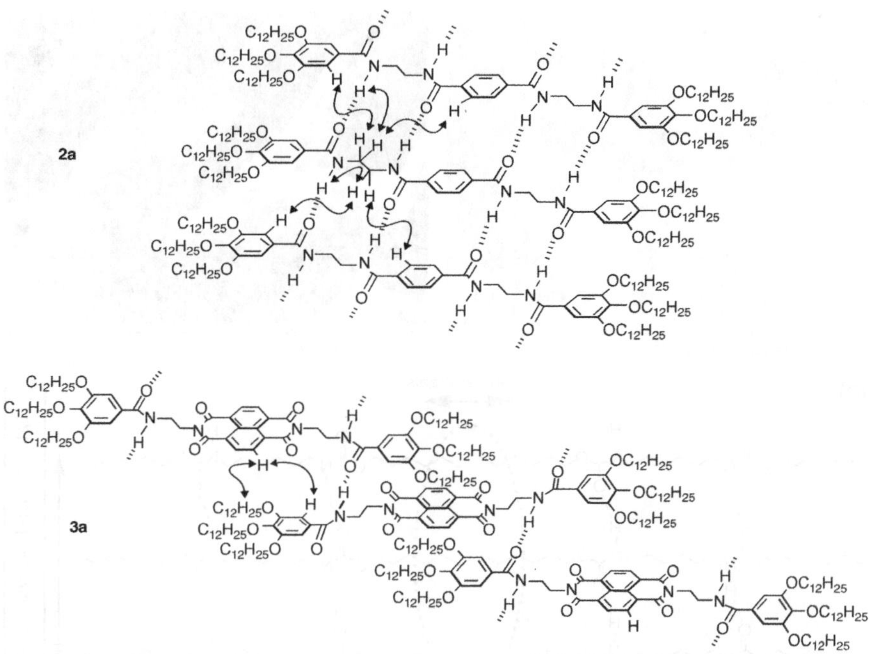

Figure 4.14 Packing modes of **2** and **3** as deduced from NMR experiments conducted at 298 K with 20 mM concentration in $CDCl_3$. (Source: Aparicio and Sánchez [9]. Reproduced with permission of Wiley.)

supramolecular assembly is predominantly facilitated by the N—H···O═C hydrogen bonds and with very little influence of the π-stacking. The packing of the molecule **3** is staggered such that a near head to tail arrangement is seen. The π-stacking occurred between the terminal phenyl group of one molecule and the NDI of the next one but not between two of the neighboring NDI segments. Such differences influenced the CGL, the morphology, and the sergeant–soldier behavior.

The morphologies discussed earlier invariably involve twisted or helical fibers. Self-assembly of chiral molecules usually results in helical twist leading to hollow tube formation. Such helical structures leading to nanotubular morphology are also known in the case of surfactant-like peptides [11, 12], diacetylenic lipids, [13, 14] and with a two-component system with amphiphilic groups [15]. The possible mechanism of such tube formation and related theories were discussed by Schnur [16]. Khanna et al. [17] presented the first case of such tube formation during the gelation of biscarbamates (model compounds for polyurethanes), which are neither chiral nor amphiphilic. Figure 4.15 shows the morphology of the gel fibers of a biscarbamate with a C_{12} side chain. Figure 4.15a for the gel prepared with benzonitrile shows twisted

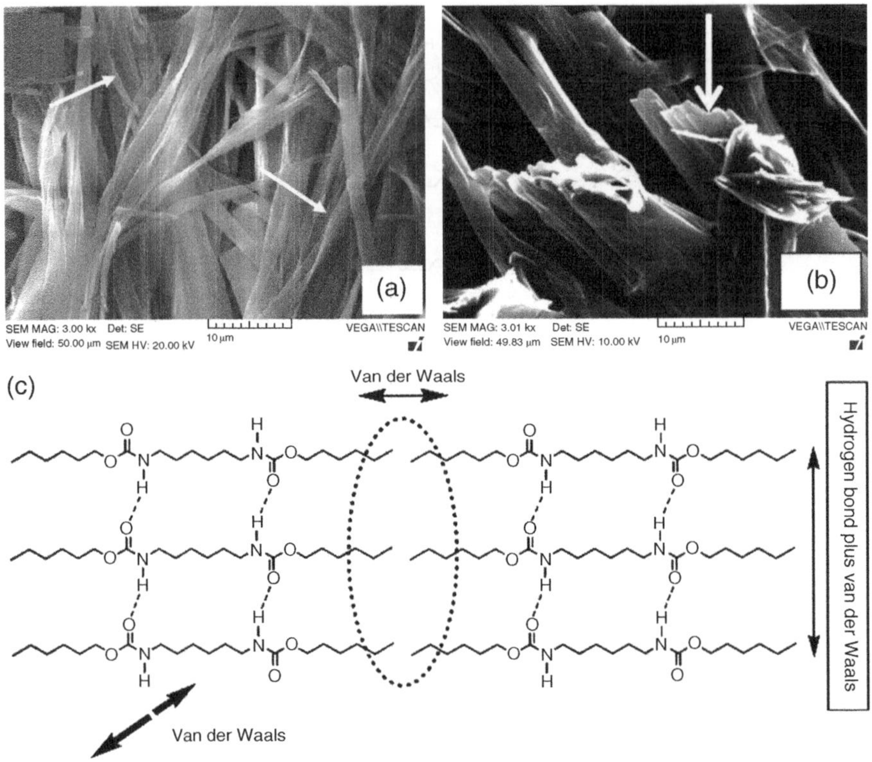

Figure 4.15 (a) and (b) SEM images of xerogels of a biscarbamate with C_{12} side chain, using benzonitrile and 1-octanol, respectively. The arrows point to the eaves troughs and hollow tubes. (c) The packing scheme of the biscarbamates (C_6 side chains) and the directions of the noncovalent interactions that lead to an asymmetry in the molecular interactions. (Source: Khanna et al. [17]. Reproduced with permission of American Chemical Society.)

fibers which also exhibit folds along their length resembling an eaves trough (indicated by the arrows). The latter is simply an incomplete tube. Distinct hollow tubes are seen in Figure 4.15b. As the molecule is neither chiral nor amphiphilic, the eaves trough and tube formation was rationalized on the basis of asymmetric interactions along the three directions, as illustrated in Figure 4.15c. The packing mode of the biscarbamate was determined using X-ray diffraction data and molecular modeling. The primary plane of interaction is the one along the length of the page of this book, with two N—H···O=C hydrogen bonds. In addition, this plane also involves van der Waals interactions between the C_6 side chains as well as the C_6 spacer (between the two NH groups along the molecule). In the horizontal direction as well as the depth of the page, only van der Waals interactions are present. Thus, the interactions between any two molecules of the biscarbamate are not equal in the three directions. Helfrich and Prost [18] concluded that amphiphilic mono- and bilayers are not at their minimal energy when they are flat and prefer a bent state. Similar was the observation of Chappell and Yager [19] that the hollow tube formation of chiral lipids was due to anisotropy of intermolecular interactions. We argued [17] that similar anisotropy of interactions allows the planes of the biscarbamate molecules to bend (rather than be flat) that would lead to the tube formation when the gel was still wet. This is similar to the example of a sheet of ordinary bond paper, which is also anisotropic due to the inherent machine direction and the additives mostly on one side (felt side). It is a common observation that if a sheet of paper is left on a table top on a humid day, the paper would tend to curl. This is due to the anisotropy. Thus, the twist and tube formation in the case of gels of nonchiral, nonamphiphilic systems can be attributed to the asymmetry of the noncovalent interactions along the three directions.

4.3 Modular Supramolecules

Stubbs and Weck [20] developed a method to design a "universal polymer scaffold," which is a polymer or copolymer that is functionalized with recognition units on their side chains, to which a receptor could supramolecularly bind itself. Their concept is illustrated in Figure 4.16a. Here, the black regular rectangle and the snipped rectangle denote the recognition sites of the functionalized monomers. Upon polymerization, the (green, blue, red, or orange) receptors can supramolecularly (and reversibly) self-assemble to bind with the corresponding recognition units on the polymer side chain. They used norbornenes functionalized with diaminotriazene or diaminopyridine and alkyl spacers as monomers and polymerized by ring opening. The M_n of the polymers ranged from 8665 to 22 700. The alkyl chains showed a strong effect on the reactivity of these monomers. Self-association constants for the monomers and the polymers due to dimerization were determined using NMR spectroscopy. The acylated diaminotriazenes showed strong tendency toward dimerization. On the other hand, diaminopyridines showed lower dimerization and higher association with receptors, such as uracil and thymine. It was found that the diaminotriazene-based polymers precipitated from solution due to self-association as shown in Figure 4.16b (marked insoluble). This is a case of reversible cross-linking by hydrogen bonds. By adding *N*-butyl thymine in the presence of heat and ultrasonification, solubilization was achieved, as the butyl thymine breaks up the self-association and formed triple hydrogen bonds with diaminotriazene as shown in Figure 4.16b (marked soluble). Hence, butylthymine was added (1–3 equivalents) during polymerization of diaminotriazane-based monomers. This suppressed the self-dimerization since butylthymine self-assembled with the backbone recognition units. It was also found that

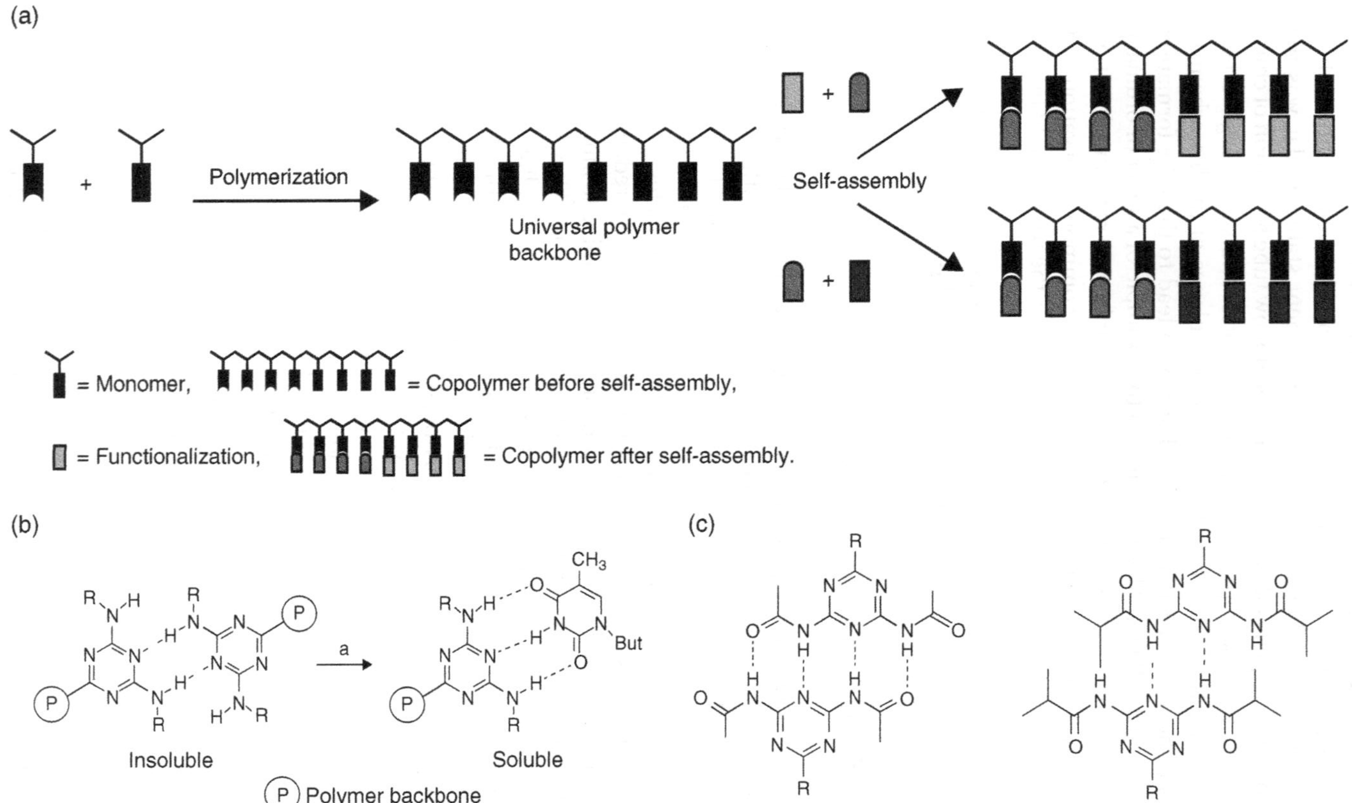

Figure 4.16 (a) Illustration of the design of the "universal polymer backbone"; (b) insolubility of the polymer due to self-association and the addition of butyl thymine to inhibit self-association and render the polymer soluble; and (c) quadruple hydrogen bonds between dimers of *N*-acetyl-substituted diaminotriazene monomers (requiring three equivalent butyl thymine to solubilize the polymer) and two hydrogen bonds with isobutyryl substitution [20].

polymers based on *N*-acetyl-substituted diaminotriazene monomers required three equivalents of butylthymine to dissolve whereas polymers of *N*-isobutyryl-substituted monomers required only one equivalent. This was rationalized on the basis of four hydrogen bonds in the case of dimer association of the *N*-acetyl monomers (Figure 4.16c) compared to only two hydrogen bonds with the *N*-isobutyryl substitution, which causes steric hindrance. Similar noncovalently cross-linked polymers via metal coordination in addition to hydrogen bond functionalization were described by Pollino et al. [21]. A tutorial review on the various designs of such side chain functionalized polymers leading to supramolecular assembly was given by Pollino and Weck [22].

Supramolecular chain-extended polymers or block copolymers could be prepared by end-functionalizing two homopolymers with complementary charged groups or hydrogen bonding moieties. Folmer et al. [23] used quadruple hydrogen-bonded ureido-pyrimidinone (UPy) units to functionalize OH-telechelic polymers and built supramolecular chain-extended polymers. The UPy was linked to a reactive isocyanate group, and this was reacted with the telechelic polymer. In one of their studies, they used telechelic poly(ethylene/butylene) functionalized with the UPy. Their approach is shown in Figure 4.17. The dynamical mechanical thermal analysis showed that the "virtual" molecular weight of the resulting supramolecular chain-extended polymer was very high. Figure 4.18 compares the flow behavior of the telechelic poly(ethylene/butylene) with OH end groups and the one functionalized with the hydrogen bonding UPy units. The telechelic polymer is a viscous liquid, whereas the supramolecular polymer behaves as an elastic solid. A significant advantage is that the large drop in melt viscosity with an increase in temperature is due to the reversibility of the hydrogen bonded association.

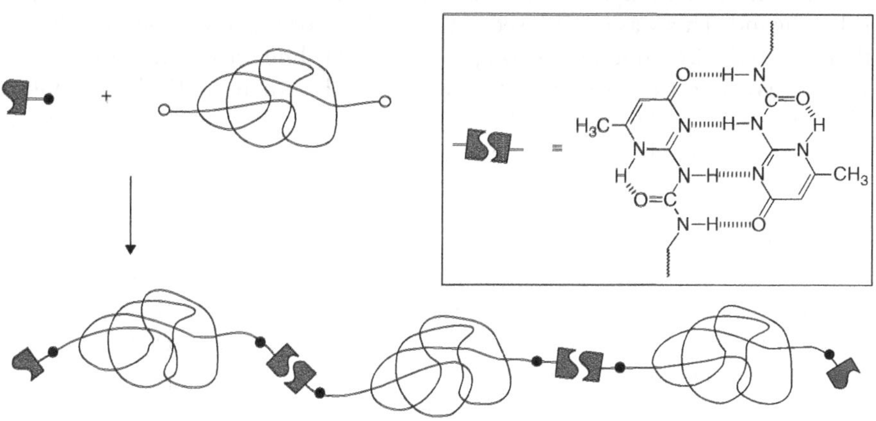

Figure 4.17 Building supramolecular chain-extended polymers by functionalization of telechelic polymers with quadruple hydrogen bonded ureido-pyrimidinone units. (Source: Folmer et al. [23]. Reproduced with permission of Wiley.)

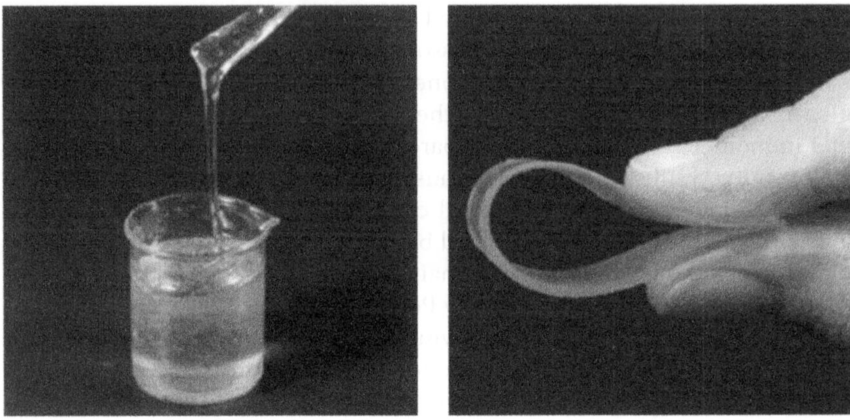

Figure 4.18 The telechelic poly(ethylene/butylene) is a viscous liquid (left), whereas the supramolecular chain-extended polymer is an elastic solid. (Source: Folmer et al. [23]. Reproduced with permission of Wiley.)

Using the supramolecular approach, Dionisio et al. [24] made compatible mixtures of two otherwise immiscible polymers. Polystyrene (PS) and poly(butyl methacrylate) (PBMA) are normally immiscible. These authors functionalized PS and PBMA with a tetraphosphonate cavitand as the host and methylpyridinium as the guest, respectively. The complexation led to a miscible blend characterized by a single glass transition temperature. The reversibility of the miscibility was demonstrated by adding a competing guest.

Another approach presented by Schubert et al. [25, 26] involves metal–ligand interactions. A hydroxyl-functionalized polymer was end-capped with a terpyridine ligand. This was complexed with $RuCl_3$ to form a monocomplex. It was then reacted with another (uncomplexed) terpyridine terminated polymer to yield an asymmetrical AB diblock copolymer with ruthenium (II) bis-complex. These were referred to by the authors as "metallo-supramolecular block copolymer." Some of the block copolymers that were constructed by this procedure are shown in Figure 4.19. Similar to the micelle formation of amphiphilic block copolymers, the block copolymers formed via Ru complex with poly(ethylene oxide) (PEO) and PS blocks, for example, formed "metallo-supramolecular micelles," as shown in Figure 4.20A. The PEO forms the corona and the hydrophobic block forms the core. The terpyridine (shown as brackets)–Ru complex is at the interface. The terpyridine and Ru are shown as brackets and large black dots, respectively. Apart from the stability of these supramolecular block copolymers in various solvents and a range of temperatures, the reversibility is also an advantage. It was found that by adding a large excess of a competing ligand (B in Figure 4.20), the complex was released and the core with the terpyridine ligand (C) and the corona separated. Another hydrophilic polymer,

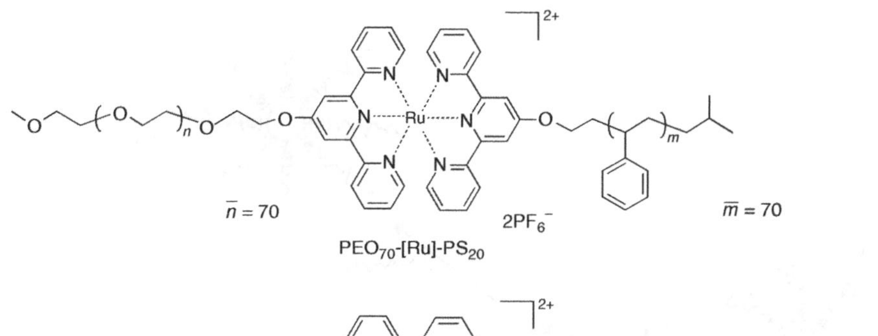

Figure 4.19 Molecular structures of metallo-supramolecular block copolymers. (Source: Gohy et al. [25]. Reproduced with permission of Wiley.)

with another metal to complex with the terpyridine can be added to make a new metallo-supramolecular block copolymer, as shown in the path (a) in Figure 4.20. Another opportunity would be to add, via an AFM tip for example, to modify the terpyridine-functionalized hydrophobic core (path (b) in Figure 4.20).

4.4 Solvent Influence

It is well known that the solvent medium plays an important role in the self-assembly. It is a common practice to add a nonsolvent to a solution to precipitate the solute. In the case of amphiphilic (block) polymers, which are analogous

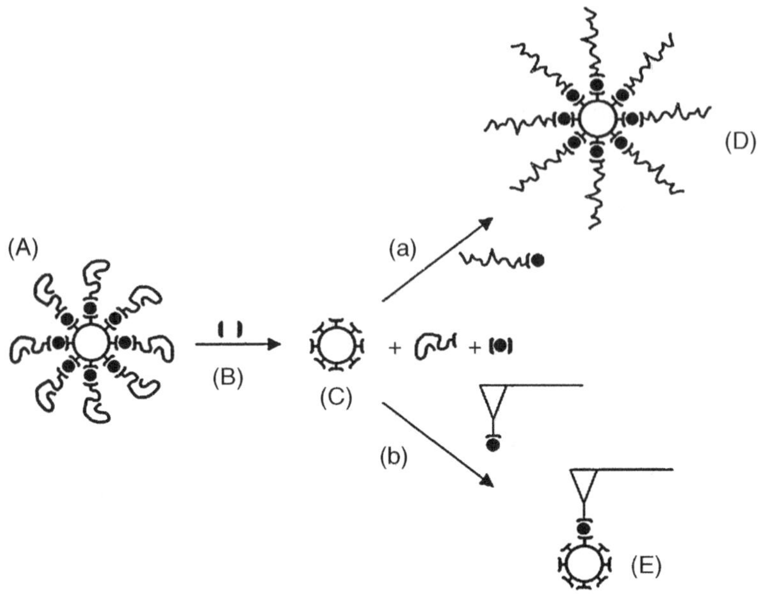

Figure 4.20 (A) Micellar morphology of the block copolymer with PEO corona and the hydrophobic block as the core. The terpyridine–ruthenium complex forms the interface. Addition of a competing ligand (B) displaces the complex. The core (C) and corona are separated. (a) Postmodification of the terpyridine-functionalized core with another polymer to form a corona to make a new block copolymer (D). (b) Nano-manipulation of the metallo-supramolecular nano-objects (E). (Source: Gohy et al. [25]. Reproduced with permission of Wiley.)

to soaps, one part of the molecule would be soluble in a solvent (e.g., water) and the other part, not. The insoluble part would aggregate (self-assemble) to avoid the solvent, and this is the process by which micelles and vesicles are formed. The morphology would depend on the relative lengths of the soluble and insoluble segments as well as the solvent [27–36].

We describe a few cases of supramolecular assembly of π-interaction-based systems with oligomer substitution that depend on the type of solvent [37–41]. These are molecules with perylene tetracarboxylic diimide (PTCDI) as the large aromatic segment either mono- or disubstituted with oligostyrene, poly(dimethyl siloxane) (PDMS), or poly(ethylene/propylene glycol), as shown in Figure 4.21. Also included is a terthiophene substituted with oligostyrene.

With PTCDI substituted with the flexible oligomers, these are rod–coil macromolecular surfactants. The molecules shown in Figure 4.21a, c, and f with disubstituted chains are Gemini surfactants, applicable to nonaqueous systems (except (f) which can be both aqueous and nonaqueous). Those shown in Figure 4.21d and e are inverse surfactants in that the head group (PTCDI) is

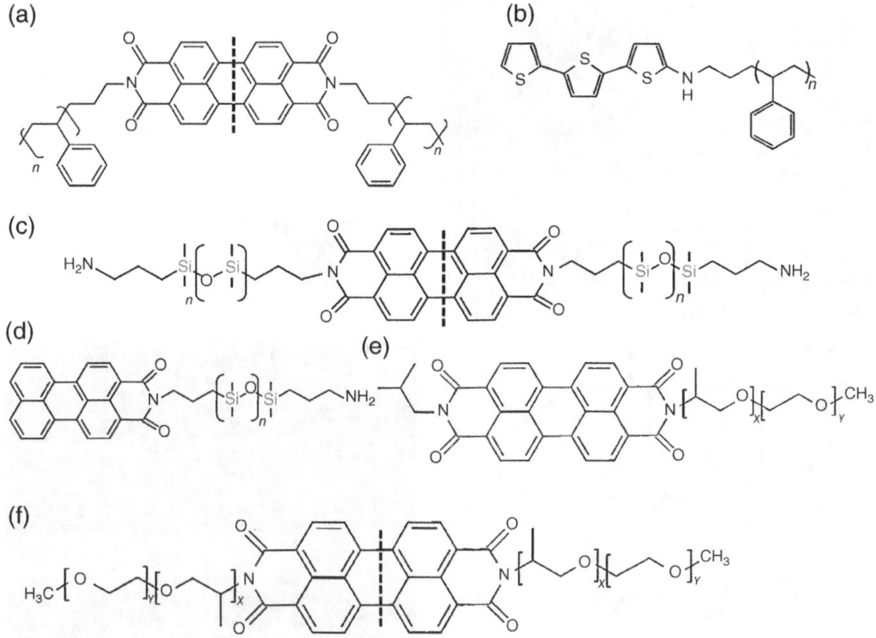

Figure 4.21 Chemical structures of (a) PS–PTCDI–PS (Gemini surfactant), (b) terthiophene–PS, (c) PDMS–PTCDI–PDMS (Gemini surfactant), (d) inverse macromolecular surfactant PDI–PDMS, (e) inverse macromolecular surfactant PTCDI–poly(ethylene/propylene glycol) (MJ–PTCDI), and (f) poly(ethylene/propylene glycol)–PTCDI–poly(ethylene/propylene glycol) (DJ–PTCDI) (Gemini surfactant). The dashed vertical lines in (a), (c), and (f) indicate the symmetry on either side that render these molecules Gemini surfactants.

not soluble in common solvents but the tail is. Yao et al. [37] found that in the case of perylene–PDMS (Figure 4.21d), the self-assembly of the perylene segment occurred in chloroform, dichloromethane, and carbon tetrachloride as shown by UV–vis spectra, and the peak positions blue-shifted with an addition of hexane. With dichloromethane, the peaks were seen at 481 and 503 nm, with $I_{481} > I_{503}$, confirming the aggregation. In solvent mixtures, with an increase in the concentration of hexane, these peaks gradually blue-shifted to 467 and 495 nm, respectively. The spectral behavior was rationalized on the basis of solubility parameters of the solvents and that of PDMS as well as the χ parameters between these solvents and PDMS. TEM images of drop-cast films showed a trilayer vesicular morphology, with crystalline perylene domains, and the size of the vesicles depended on the solvent composition. With the Gemini molecule PDMS–PTCDI–PDMS (Figure 4.21c), Yao et al. [38] found that the morphologies of samples from chloroform and hexane differed significantly. Figure 4.22 shows the cryo-SEM and TEM images of drop-cast films from these two solvents. With chloroform, Figure 4.22a shows a nanoweb

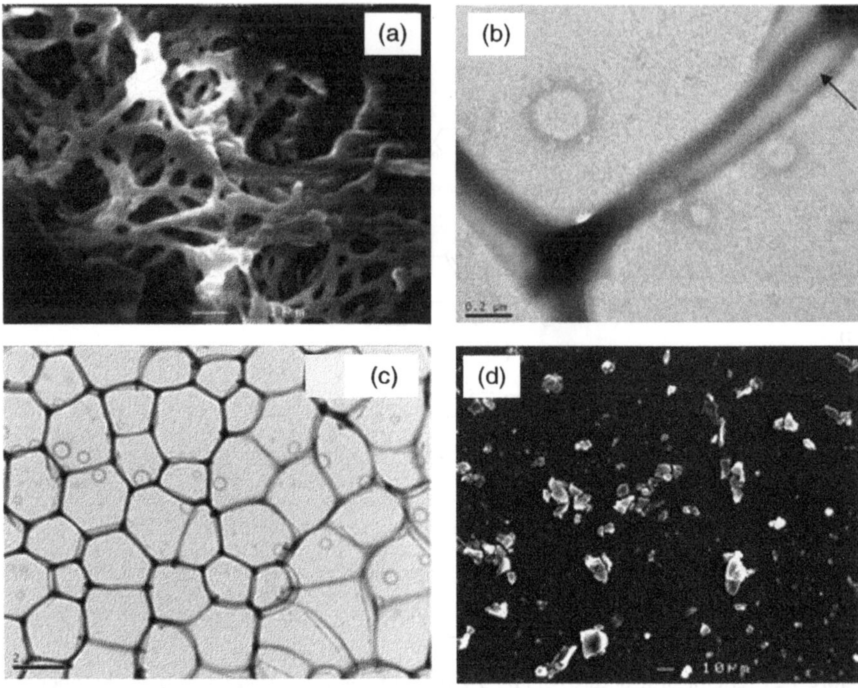

Figure 4.22 (a) Cryo-SEM image of PDMS–PTCDI–PDMS from $CHCl_3$; (b) and (c) TEM images of PDMS–PTCDI–PDMS from $CHCl_3$; and (d) SEM image of PDMS–PTCDI–PDMS from hexane. (Source: Yao et al. [37]. Reproduced with permission of American Chemical Society.)

morphology in the cryo-SEM image. An enlargement of one of the branches is shown in Figure 4.22b. The TEM image in Figure 4.22c confirms the nanoweb morphology. The sample from hexane shows single crystals rather than vesicles, micelles, or fibrillar morphology.

In the molecules MJ–PTCDI (Figure 4.21e) and DJ–PTCDI (Figure 4.21f), the oligomer was Jeffamine®, which is a copolymer of poly(ethylene/propylene glycol) [42] (EO/PO: 31/10 mol ratio; $M_w = 2000$). Islam et al. [39] found that the side chain (EO/PO copolymer) is soluble in nonaqueous solvents such as acetone, chloroform, dichloromethane, methanol, THF, and dimethyl formamide, and the PTCDI is not. One would then expect aggregation of PTCDI in these solvents. However, as shown in Figure 4.23a and b, UV–vis spectra exhibited absorption patterns corresponding only to monomeric form, irrespective of the concentration. Thus, there was no self-assembly of PTCDI in these block-selective solvents, as would be expected from analogy to amphiphilic block copolymers or soaps. Islam et al. [39] attributed this result to the solvation of the PTCDI π-system by these solvents. Several authors [43–49] have studied the phenomenon of solvation of the π-system theoretically and

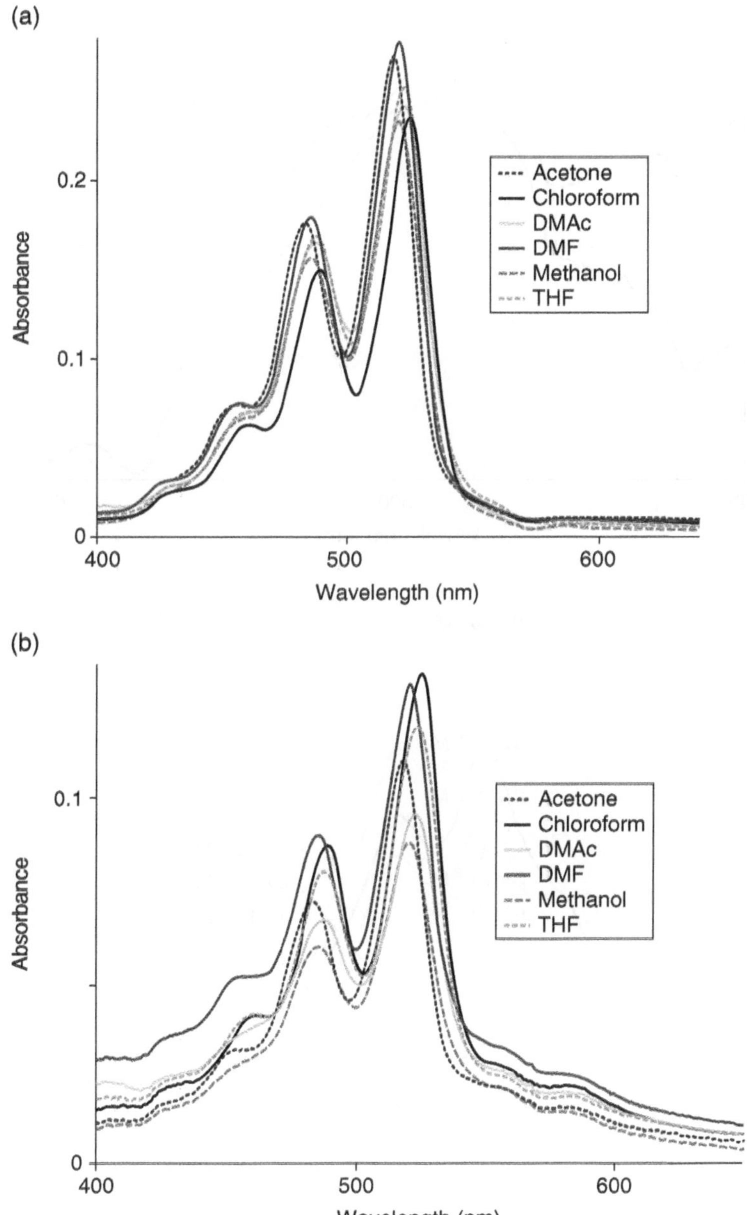

Figure 4.23 UV–vis spectra of (a) DJ–PTCDI and (b) MJ–PTCDI in various nonaqueous solvents. (c) Drop-cast films of DJ–PTCDI and MJ–PTCDI from chloroform and (d) DJ–PTCDI in water at various concentrations. (Source: Islam et al. [39]. Reproduced with permission of Elsevier.)

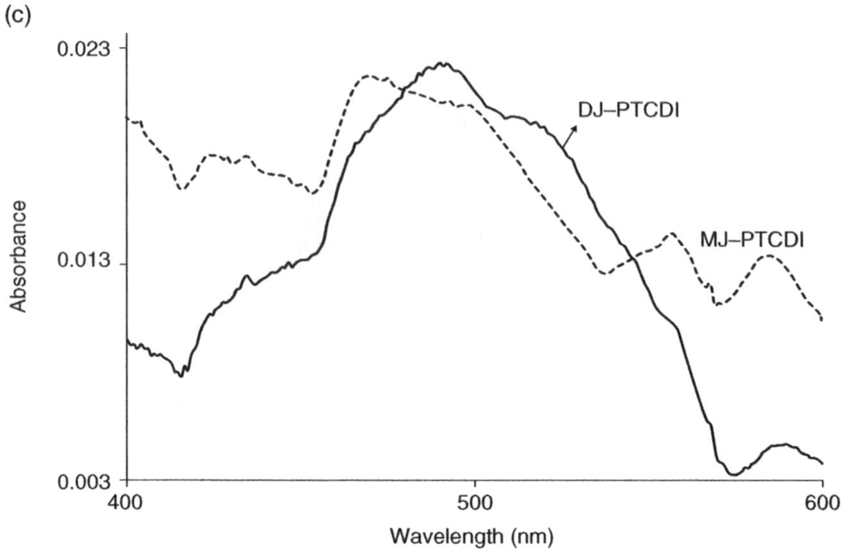

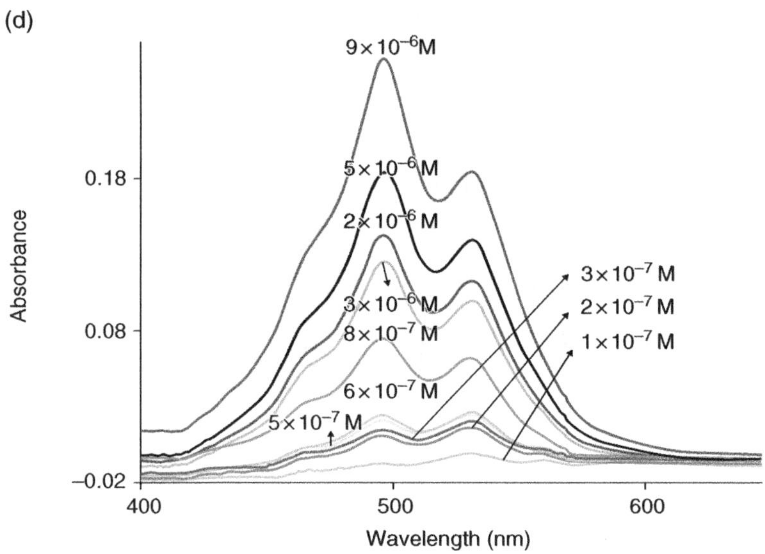

Figure 4.23 (Continued)

experimentally. Sugiura and Ogawa [45] reported such an effect with cyclohexane, heptane, benzene, 1,4-dioxane, and chloroform. The anisotropic solvation of aromatic rings in aqueous and mixed solvent systems was studied by Sindkhedkar et al. [46]. With a system of peripherally substituted perylene bisimides investigated by Ghosh et al. [49], no aggregation occurred in chloroform due to such solvation of the π-system. Similar to the aforementioned cases, the solvation of the PTCDI segment by the nonaqueous solvents inhibit the self-assembly of PTCDI, although these are good solvents for the EO/PO copolymer. However, the UV–vis spectra of drop-cast films from chloroform (Figure 4.23c) showed the self-assembly of PTCDI. Upon evaporation of the solvent, the solvation effect vanished and the aggregation ensued. The crystalline X-ray diffraction of these films confirmed the aggregate formation of PTCDI.

Both MJ–PTCDI and DJ–PTCDI showed self-assembly of PTCDI in water as well as aqueous solvent mixtures. Figure 4.23d shows the UV–vis spectra of DJ–PTCDI in water for various concentrations. While monomeric absorption pattern is seen with very low concentrations, self-assembly of PTCDI begins with a concentration of 10^{-7} M. Addition of a nonaqueous solvent gradually disrupts the self-assembly. Figure 4.24 shows the mole fraction of aggregates in water/acetone mixtures. The total aggregation with water was disrupted by the addition of acetone, and the extent of self-assembly decreases gradually with an increase in acetone content. A detailed study of the effect of varying the ratio of good and poor solvents on the self-assembly of oligo(*p*-phenylene vinylene) (OPV) (see Figure 4.2) as well as a few other aromatic systems was reported by Korevaar et al. [50].

As observed from optical, SEMs as well as dynamic light scattering data, MJ–PTCDI and DJ–PTCDI form spherical aggregates of about 10–15 μm diameter in water. Gemini surfactants (e.g., DJ–PTCDI) can self-assemble with various types of morphologies, such as micelles and vesicles [51, 52]. In the nonaqueous solvents, in which self-assembly occurred only during the evaporation of the solvent, discrete crystalline aggregates of PTCDI was seen, similar to the case of PDMS–PTCDI–PDMS from hexane (Figure 4.22d). The optical microscopy (OM) and SEM images of drop-cast films of DJ–PTCDI and MJ-PTCDI from THF are shown in Figure 4.25. The OM images show a tree-like morphology in both cases. A number of discrete crystals are also seen in the background in Figure 4.25c for MJ–PTCDI. The SEM images in Figure 4.25 show that these tree-like formations are aggregates of crystals of these molecules. It would appear that the crystals, while still in the solvent medium, aggregate together through surface interactions.

While the self-assembly of DJ–PTCDI and MJ–PTCDI was precluded in nonaqueous solvents, their supramolecular assemblies in water were found to be time-evolutionary and the prolonged self-assembly led to various types of supramolecular structures. Manners and Winnik [53–56] reported the slow

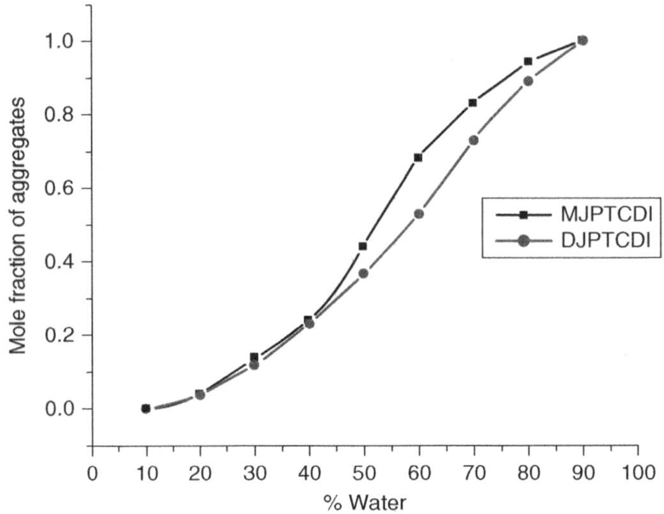

Figure 4.24 Mole fraction of aggregates as a function of water/acetone composition for MJ–PTCDI and DJ–PTCDI (Source: Islam et al. [39]. Reproduced with permission of Elsevier.)

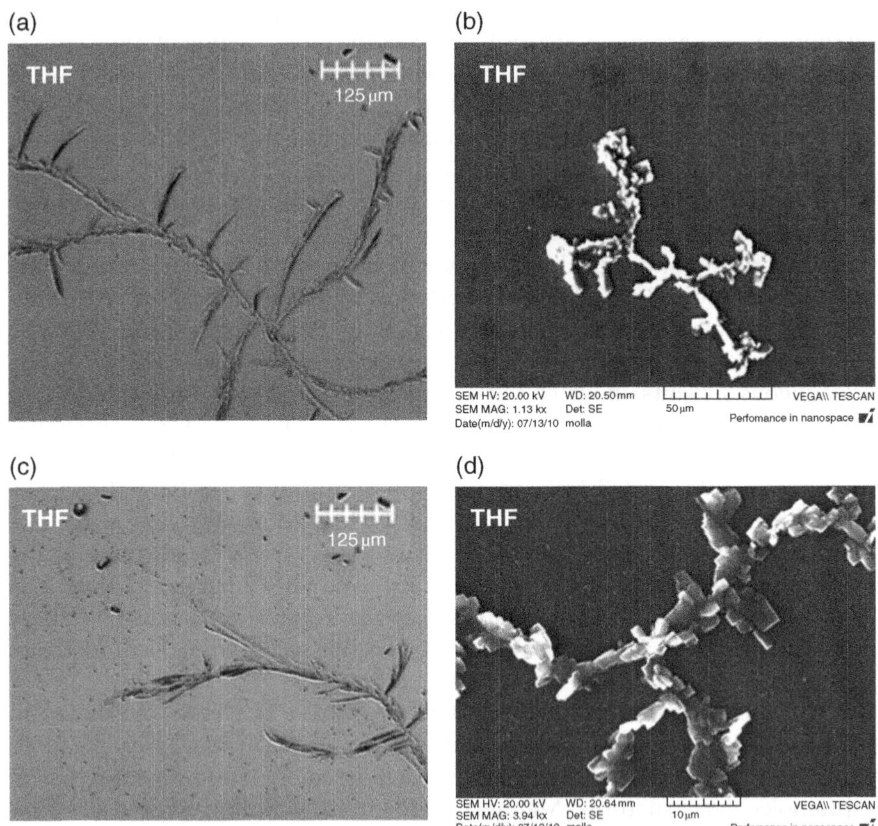

Figure 4.25 (a) and (c) OM images of DJ–PTCDI and MJ–PTCDI, respectively, of drop-cast films from THF. The scale bar is 125 μm. (b) and (d) Corresponding SEM images. The scale bar is 10 μm. (Source: Islam et al. [39]. Reproduced with permission of Elsevier.)

aggregation of micelles, which was called "living self-assembly" into hierarchical flower-like structures in the di- and triblock copolymers with crystalline cores. Varying the rate of addition of nonsolvents to copolymer solutions to trap "kinetically controlled" morphologies was discussed by Pochan et al. [57, 58]. It was mentioned earlier that DJ–PTCDI and MJ–PTCDI films drop-cast from aqueous solutions showed spherical morphology. However, when the drop-cast solutions were covered with an aluminum foil and dried slowly over an extended period of time, bundles of fibers were seen, as if the spherical entities coalesced to form higher order morphologies. Islam et al. [39] examined the aqueous solutions of DJ–PTCDI and MJ–PTCDI after standing for 2 days to 1 year. Figures 4.26 and 4.27 show the SEM images of the supramolecular morphologies of these two molecules after 2, 6, and 7 days and after 1 year. These were from drop-cast solutions aged over the aforementioned time periods. It is seen in Figure 4.26 that the DJ–PTCDI initially forms spheres and then fuses to form short rod-like aggregates after 6 days. These then assemble after 7 days into "floral" entities which resemble spherulites. After 1 year, cobweb-like morphology is seen, with long intertwined fibers. That these are composed of smaller spherical assemblies is seen from Figure 4.27f. The MJ–PTCDI also forms flower-like assemblies that grow to be long fibers. A magnified image of a section of a fiber (Figure 4.27f) confirms that these consist of spherical assemblies. Although the details were not examined, it is likely that slow diffusion and surface interaction of the spherical entities grow to such cobweb structures via living epitaxial assembly of primary spheres.

It was mentioned in Chapter 2 that the C—H···π interaction, although usually weak, plays a role on molecular recognition and self-assembly. In the case of the Gemini surfactant polystyrene–PTCDI–polystyrene (PS–PTCDI–PS) (for simplicity, we use the abbreviation PS although oligostyrene was employed here), Islam and Sundararajan [40] interpreted the experimental observations to imply that the C—H···π interaction inhibited the self-assembly of the PTCDI. The oligostyrene segment is soluble in THF, chloroform, and toluene but precipitated from methanol and hexane. The UV–vis spectra showed no self-assembly of PTCDI in the solvents in which the PS segment is soluble (acetone, benzene, chloroform, cyclohexane, THF, and toluene). The λ_{max}, however, shifted with the polarity of the solvent. The S_{0-0}, S_{0-1}, and S_{0-2} transition peaks were in the range of 519–525, 484–490, and 456–459 nm, respectively. However, as observed before in the case of MJ–PTCDI and DJ–PTCDI, drop-cast films showed the supramolecular assembly, as the solvent evaporated, with a strong J-stacking. In addition, the absorption maxima remained at 542, 496, and 467 nm, irrespective of the solvent used for drop-casting, whereas these varied in the case of the spectra of the solutions.

Islam and Sundararajan [40] attributed the aforementioned lack of self-assembly to the C—H···π interaction between the phenyl side group of the PS and the perylene core. In a good solvent, the PS chain would be extended

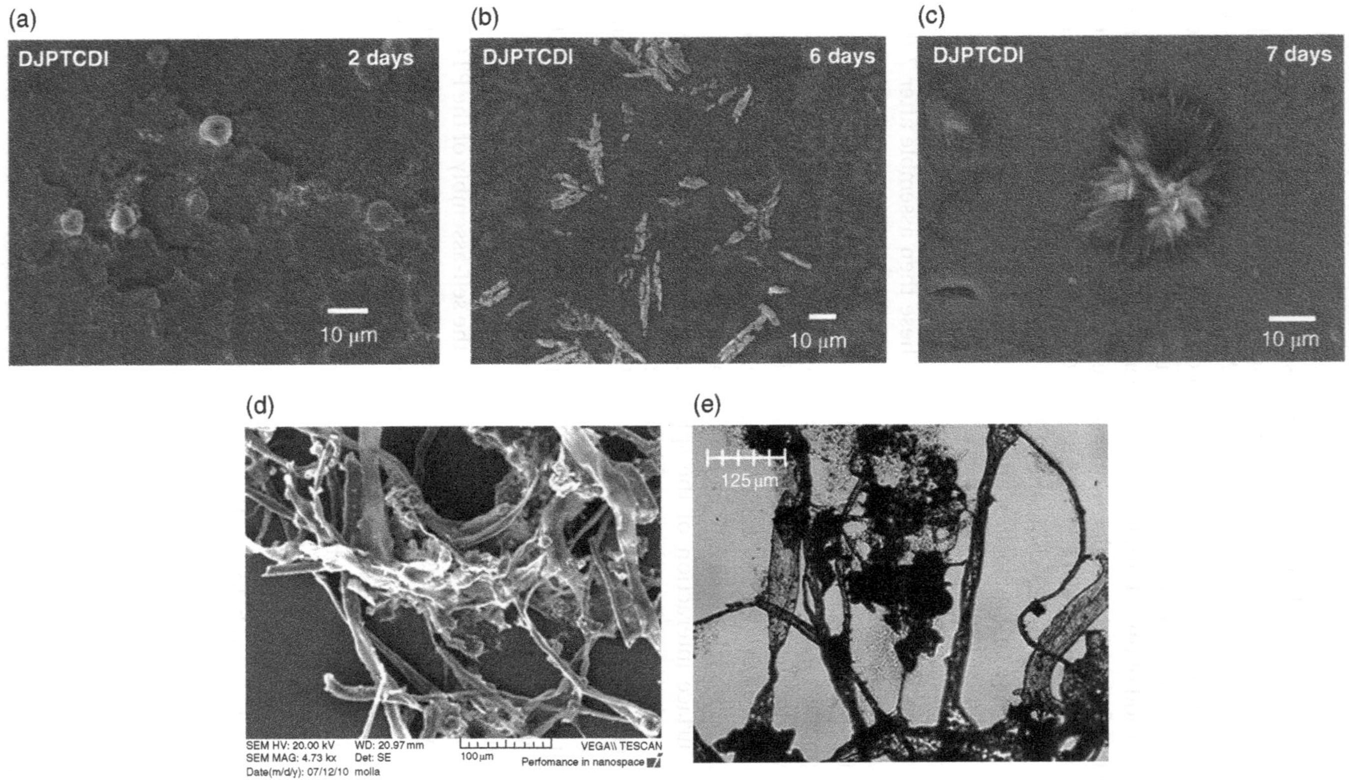

Figure 4.26 (a)–(c) SEM images of drop-cast films of DJPTCDI showing evolution of self-assembly in aged aqueous solutions after 2, 6, and 7 days; (d) SEM of drop-cast film after aging the solution for 1 year; and (e) OM after aging the solution for 1 year. (Source: Islam et al. [39]. Reproduced with permission of Elsevier.)

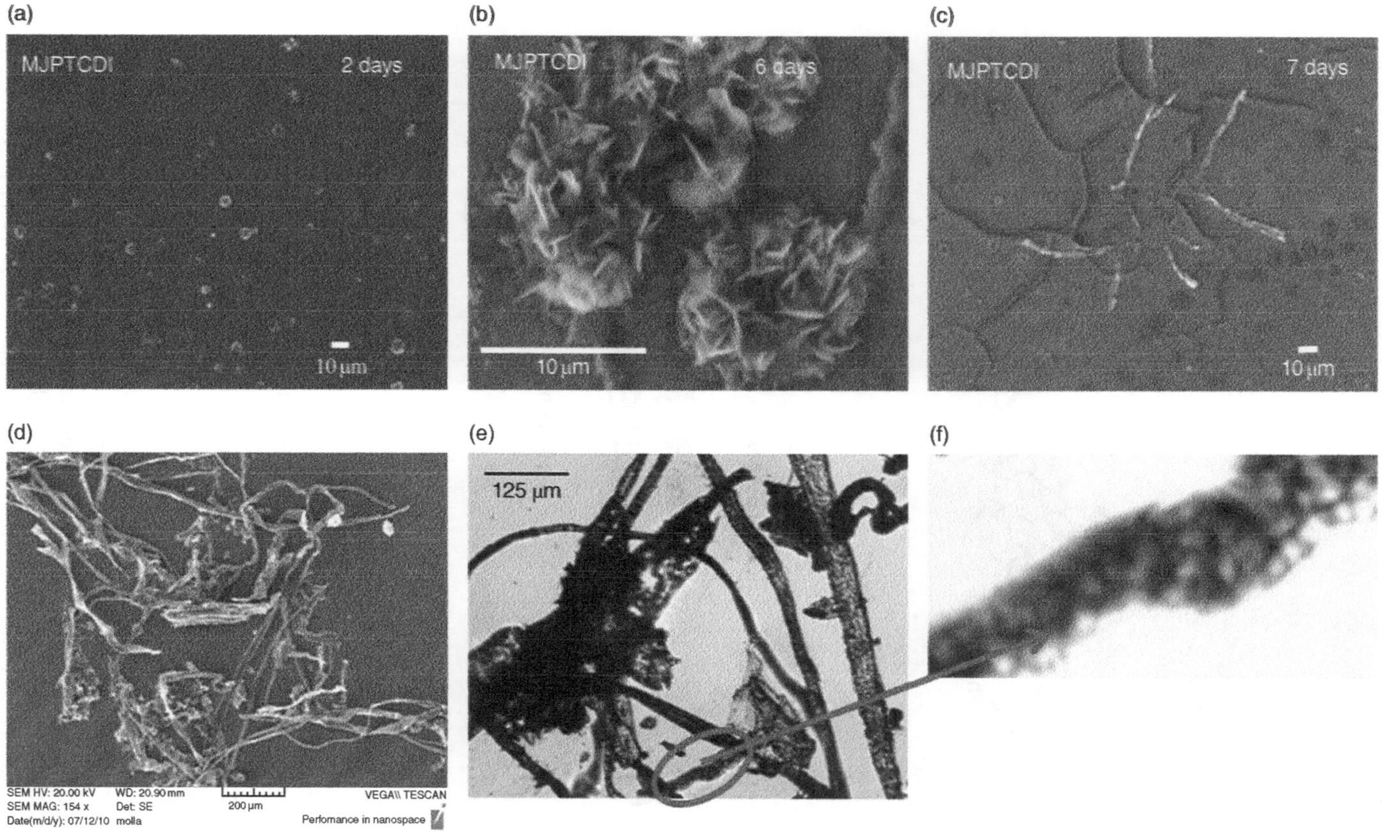

Figure 4.27 (a)–(c) SEM images of drop-cast films of MJPTCDI showing secondary assembly in aged aqueous solution after 2, 6, and 7 days; (d) SEM of drop-cast film after aging the solution for 1 year; (e) OM after aging the solution for 1 year; and (f) a magnified image of a section in (e) showing the assembly of spheres along one direction. (Source: Islam et al. [39]. Reproduced with permission of Elsevier.)

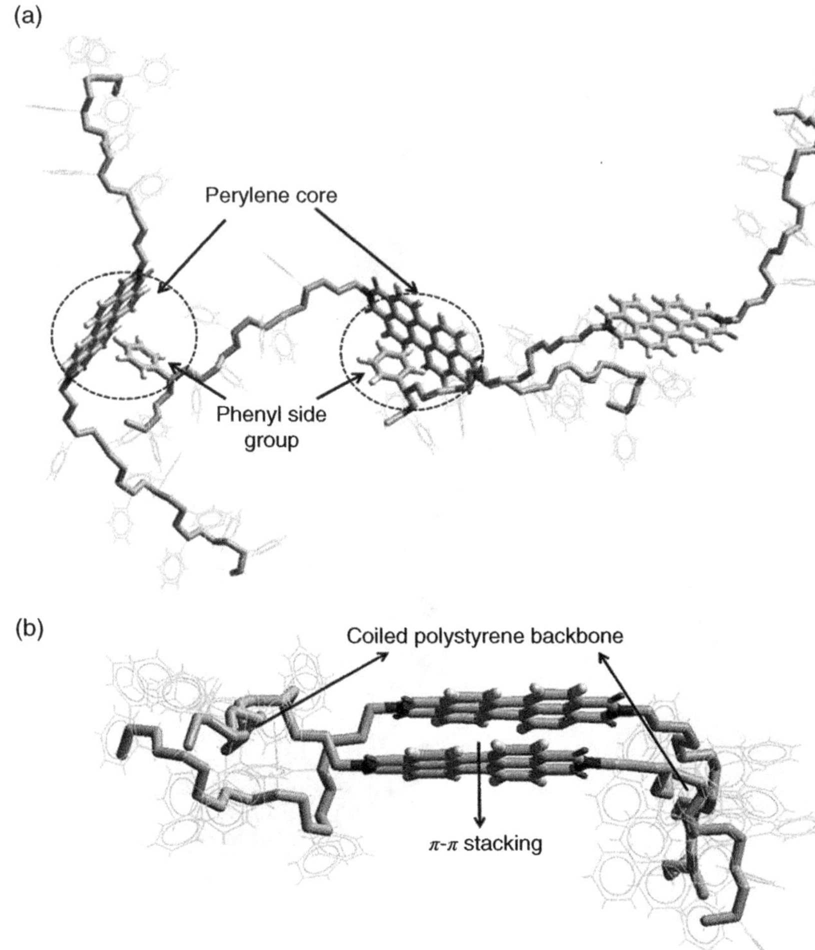

Figure 4.28 Illustration of (a) the CH···π interaction between the phenyl group of oligostyrene and perylene core when the PS chain is in an extended conformation in a good solvent. This prevents the self-assembly of PTCDI. (b) Collapsed, coiled conformation of oligostyrene which allows self-assembly of PTCDI in solvent/nonsolvent mixtures or in drop-cast films. (Source: Islam and Sundararajan [40]. Reproduced with permission of Elsevier.)

(stretched out), so that the phenyl side groups of neighboring molecules would be in a favorable position to form C—H···π interaction with the perylene core as shown in Figure 4.28a. This would prevent the self-assembly of the PTCDI. In the case of drop-cast films, as the solvent evaporates, the PS would collapse and the PTCDI would self-assemble (Figure 4.28b). This interpretation was confirmed by ^{1}H NMR spectra using different concentrations of PS–PTCDI–PS in $CDCl_3$ and acetone-d_6. Although the self-assembly did not occur in the

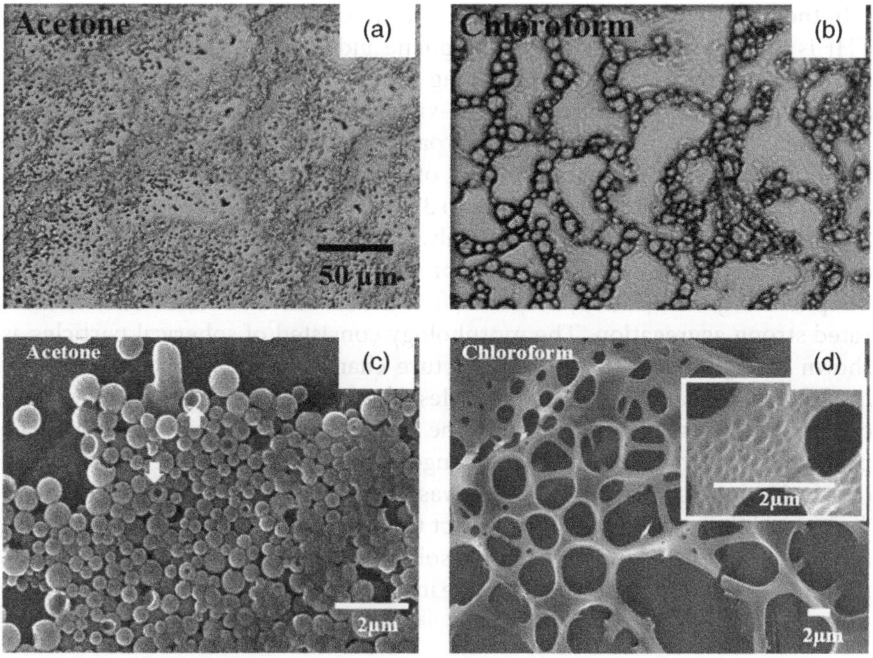

Figure 4.29 (a) and (c) Optical and SEM micrographs of drop-cast films of PS–PTCDI–PS from acetone, and (b) and (d) from chloroform. (Source: Islam and Sundararajan [40]. Reproduced with permission of Elsevier.)

solvents in which the PS was soluble, addition of a nonsolvent led to the supramolecular association. For example, chloroform is a good solvent for PS, but methanol is a nonsolvent for both PS and PTCDI. UV–vis spectra recorded with different relative concentrations of chloroform and methanol showed that with a high methanol concentration (1/9), intensity reversal of the peaks occurred, confirming the self-assembly of PTCDI. On the other hand, with the acetone/water mixture, the monomeric form was seen only with acetone/water: 9/1 composition, and aggregation of PTCDI was seen with all other ratios. Thus, addition of a nonsolvent in this case leads to self-assembly. The morphology of the drop-cast films from acetone and chloroform are shown in Figure 4.29. The OM of the acetone films in Figure 4.29a shows small spherical objects, which are hollow spheres of this Gemini surfactant as seen in the corresponding SEM image (Figure 4.29c). Interconnected lacy formation is seen in the OM image of the film cast from chloroform, which is seen more clearly in the corresponding SEM image (Figure 4.29d) to be a complex organization, with nanoweb similar to that in Figure 4.22a and c and spherical aggregates are seen within these nanowebs.

In the case of oligostyrene attached to terthiophene (Ter-OS, Figure 4.21b), THF is a good solvent for both terthiophene and oligostyrene. Self-assembly of the terthiophene was achieved by using solvent/nonsolvent mixtures. Deepak and Sundararajan [41] recorded UV–vis spectra for Ter-OS in THF–water mixtures. Ter-OS in THF showed a broad peak with λ_{max} at 379 nm. With the addition of water, there was a blueshift of the absorption peak as well as sharpening. The λ_{max} shifted from 379 nm to 377, 375, 371, 367, and 365 nm with 10, 20, 30, 40, and 50% of water, respectively. Such a blueshift from the molecularly dissolved state was assigned to H-type self-assembled aggregates of the terthiophene segment [59, 60]. A blueshift of 14 nm with 50% water content indicated strong aggregation. The morphology consisted of spherical particles as shown in Figure 4.30, but their structure changed with water content. The models for the packing of the molecules are also shown. With just 10% water content, the OS segment, which has the identical solubility parameter as THF, points outward to form a micellar arrangement. With an increase in water content to 20%, a vesicular morphology was seen, with the terthiophene segment forming a bilayer. This is due to the fact that terthiophene is hydrophilic. With a further increase in water content, solid spheres are seen again. Thus, the morphology changes just with a change in the relative ratios of solvent mixtures.

4.5 Comb Polymers

Cao and Smith [61] found that polyaniline, when complexed with camphorsulfonic acid, forms lyotropic phase in solutions in *m*-cresol, and it was attributed to increased rigidity of the chain conformation resulting from such complexation. Following this observation, the theoretical work of Fredrickson [62] showed that the association of the head group of oligomeric surfactants with flexible polymers, with an associative interaction, would lead to steric stiffening of the latter. The persistence length would increase. The number of surfactant molecules bound to the polymer would be σN, where σ is defined as the number of bound surfactant molecules and N is the degree of polymerization of the flexible polymer. If M is the degree of polymerization (e.g., the number of alkyl groups in the tail) of the surfactant, for low coverage, the polymer would be a flexible coil and $\sigma \ll M^{-3/5}$. For higher coverage, $\sigma \gg M^{-3/5}$ and with the tails of the surfactant molecules sticking out, the system would resemble a bottlebrush. Essentially, the flexibility of a chain depends on the rotations around the skeletal bonds shown in Figure 4.31a (see also Figure 1.8c). An all-trans conformation of the chain, in isotactic form, is shown for convenience. The conformation of the chain and the disposition of R would depend on the tacticity. If the barrier to rotation is small (as, e.g., in the case of poly(dimethyl siloxane)) the chain would be highly flexible, with low energy barrier for conformational transitions between the rotational isomers (e.g., tt and tg). If the

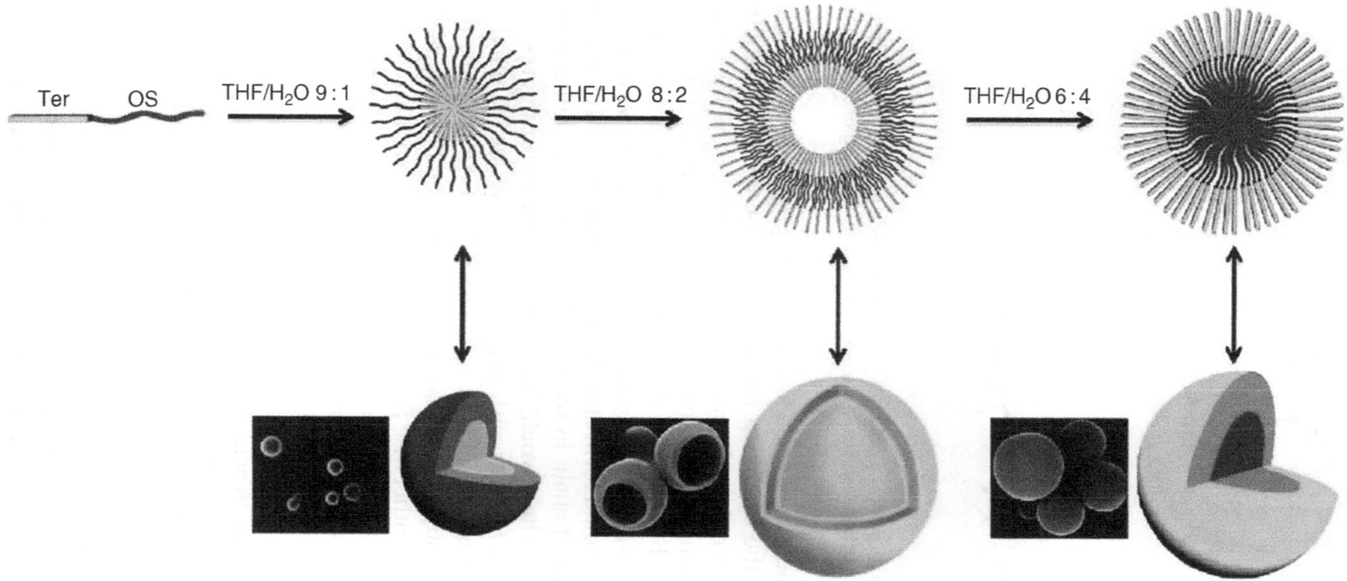

Figure 4.30 SEM images and the models for the morphologies of Ter-OS with solvent mixtures. The terthiophene segment is light gray and the OS is dark gray in these images. (Source: Deepak and Sundararajan [41]. Reproduced with permission of American Chemical Society.)

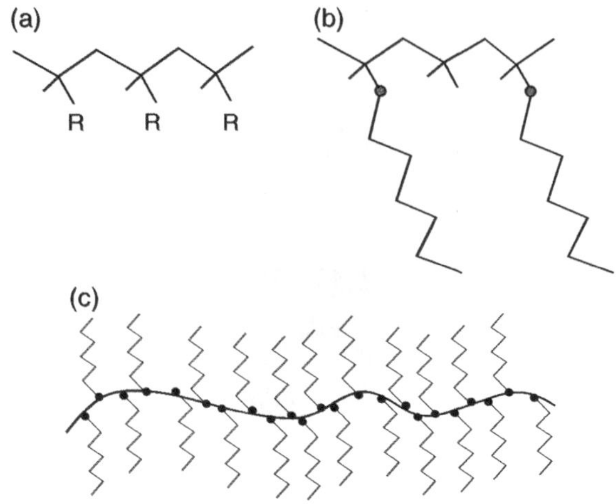

Figure 4.31 Schematics of (a) a polymer chain, with a side group R; (b) with a surfactant bound to R; and (c) bottlebrush structure upon significant binding of surfactant (see text).

head of the surfactant associates with the side group of the polymer as shown in Figure 4.31b, the barrier to rotation would be increased, leading to chain stiffening. A preferred conformation would dominate and the conformational transitions would be reduced or inhibited, depending on the number of surfactant molecules bound along the chain. In a good solvent for the alkyl chain of the surfactant, with the stretched conformation of the alkyl chain, a "bottle-brush" aggregate would form at high coverage (Figure 4.31c). Fredrickson [62] modeled such a stiff chain in the wormlike framework rather than as a rigid rod. The wormlike chain would lead to mesomorphic structures.

In the case of surfactants with longer alkyl chains, as the number of bound molecules increases, the alkyl chains would begin to pack themselves. This would lead to the supramolecular analogue of a comb copolymer. Binding of the surfactant to an otherwise flexible coil would stiffen the polymer chain and lead to a mesomorphic state. The formation of such mesomorphic structures with polymer–surfactant supramolecular complexes has been discussed in a series of papers by ten Brinke et al. [63–70] Using poly(4-vinyl pyridine) (P4VP) and poly(2-vinyl pyridine) (P2VP) with *p*-dodecylbenzenesulfonic acid (DBSA), it was found [63] that an acid–base complex was formed between the nitrogen of the pyridine ring and SO_3H of DBSA resulting in protonation of the nitrogen, as shown schematically in Figure 4.32. Note that in addition to being a flexible coil, the P4VP and P2VP were also atactic, that is, there is no stereoregularity. The bulk samples of polymer/DBSA were prepared from dimethyl formamide, and for solution studies xylene, which is a good solvent for the alkyl segment, was used. Infrared (IR) spectroscopy showed that protonation of the nitrogen of the pyridine ring occurred, and it was complete at a 1 : 1 ratio

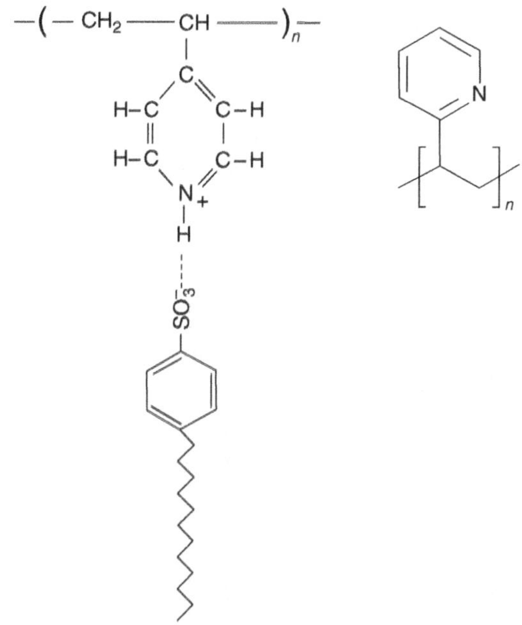

Figure 4.32 Protonation of poly(4-vinylpyridine) with *p*-dodecylbenzenesulfonic acid and (right) schematic of poly(2-vinylpyridine). (Source: Ikkala et al. [63]. Reproduced with permission of American Chemical Society.)

of the DBSA and the vinyl pyridine monomer. Mesomorphic layer structure was formed in the bulk, with an X-ray long period of 2.93 nm for a P4VP–DBSA complex upon complete protonation, which would indicate that the alkyl chains are stretched, with possible interdigitation with those of neighboring chains. The domain size was determined to be of 26 nm from line broadening of the X-ray diffraction peak. This corresponds to nine layers in a domain. With P2VP/DBSA complex (1 : 1), the long period and the domain size were 2.82 and 18 nm, respectively, with seven layers in a domain. Liquid crystallinity was observed in xylene solutions with a weight fraction of 50% or higher of P4VP–DBSA (1 : 1) and with 70% or higher in the case of P2VP/DBSA (1 : 1). Since xylene is a good solvent for the alkyl tail, a bottlebrush conformation could be expected. Small angle X-ray scattering of the solutions showed that the long spacing, which was 2.94 nm for the bulk, increased to 3.44 and 4.75 nm for weight fractions of 70 and 50% of P4VP–DBSA (1 : 1), respectively, in the solvent. A similar mesomorphic structure was also studied with metal coordination, using P4VP and zinc dodecyl benzene sulfonate.

Extending these studies beyond the aforementioned complexes, ten Brinke et al. [65, 68, 70] used surfactants with a hydrogen bonding group (Figure 4.33) to associate with flexible polymers to create mesomorphic structures.

Hydrogen bonding was confirmed by IR spectra. In the case of P4VP–pentadecylphenol (PDP), stretching and packing of the alkyl chains leads to a microphase separated lamellar morphology as shown in Figure 4.34, similar to

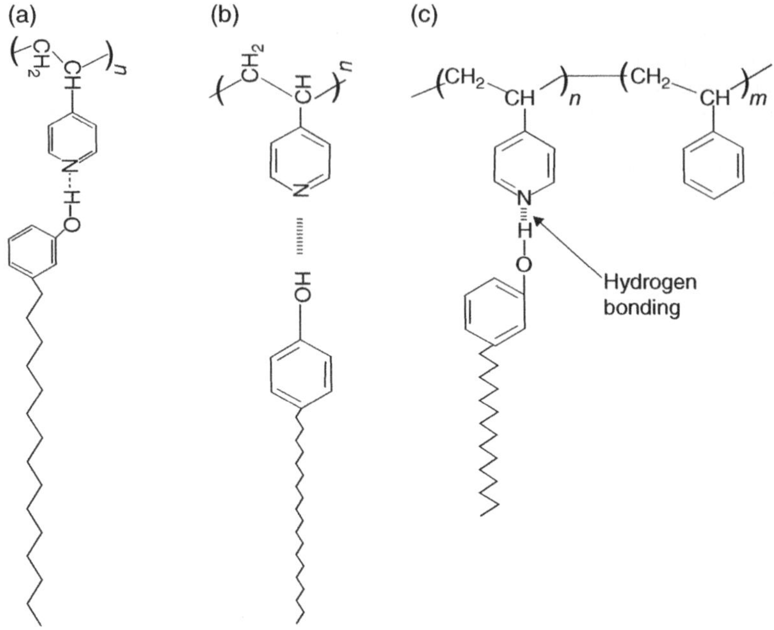

Figure 4.33 Association of surfactants with P4VP via hydrogen bonding: (a) pentadecylphenol (PDP), (b) 4-nonadecylphenol (NDP), and (c) PDP hydrogen bonded to the P4VP block of P4VP-*b*-PS. (Source: Ruokolainen et al. [65]. Reproduced with permission of American Chemical Society.)

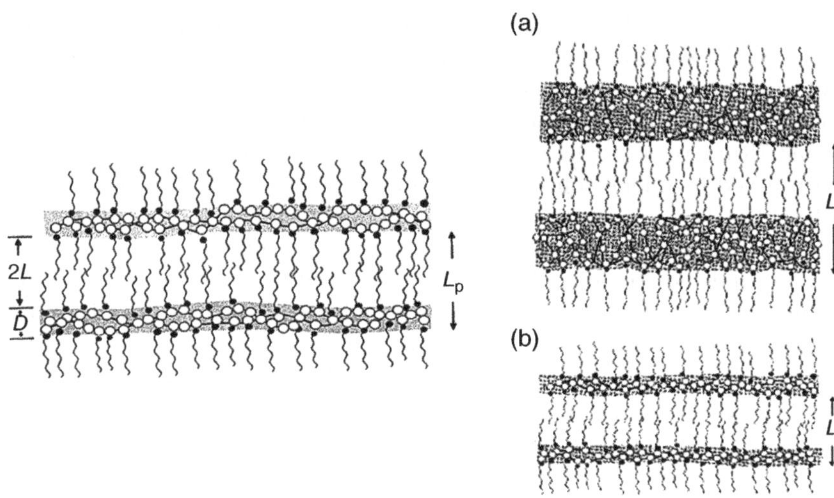

Figure 4.34 Lamellar model of the polymer/PDP system. (a) and (b) illustrate the reduction in the long spacing L_p with an increase in the mole fraction of the surfactant. The parameters D (polymer layer thickness) and $2L$ (surfactant layer thickness) are defined in the schematic on the left. (Source: Ruokolainen et al. [65, 66]. Reproduced with permission of American Chemical Society.)

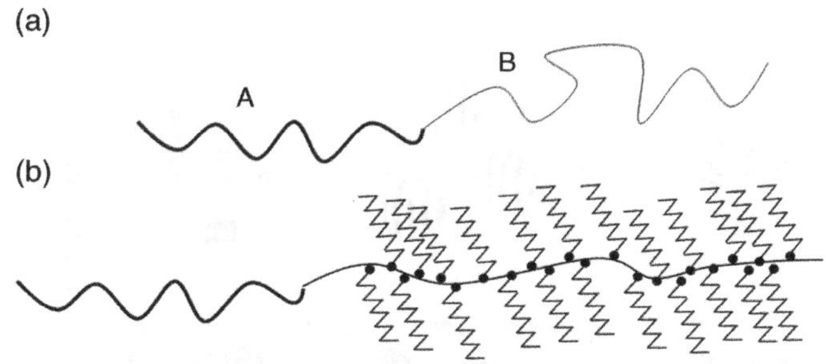

Figure 4.35 Illustration of (a) an AB block copolymer in which both blocks are flexible coils and (b) the block B becoming a supramolecular comb polymer, when associated with surfactant molecules.

that of block copolymers. X-ray scattering showed that the long period L_p decreased with an increase in the mole fraction x of the surfactant, from 5 to 3.8 nm as x increased from 0.15 to 1.0. Theoretical considerations showed that the surfactant layer thickness $2L$ remained unchanged with an increase in x, whereas the thickness D of the polymer layer was proportional to $1/x$. Crystallization of the bound and free PDP also occurred depending on the sample treatment. Similar lamellar morphology with a long period of about 4 nm was observed for P4VP–4-nonadecylphenol (NDP) system. While the domain sizes in the case of block copolymers are about 1000–2000 Å, the mesomorphic periods discussed earlier are about a few tens of angstroms.

The aforementioned process of creating mesomorphic structures was extended to designing comb–coil diblock copolymers, which would have microphase separated morphologies at two length scales, that is, the domains formed by the phase separation of the blocks and the organization of the surfactant molecules within one of the blocks [70]. Consider an AB block copolymer in which both blocks are flexible coils as shown in Figure 4.35a. If surfactants preferentially hydrogen bond with the block B, it would then become a supramolecular comb block. In such a system, the AB block copolymer would show a phase separated morphology that is typical of block copolymers, and the comb block would show submorphology within its domain, thus leading to two length scales.

ten Brinke et al. [69, 70] used polystyrene-*b*-poly(4-vinyl pyridine) (PS-P4VP) block copolymers with the surfactant PDP or NDP, which preferentially forms hydrogen bond with the pyridine group of P4VP. This results in a block copolymer, with a flexible block (PS) and a supramolecular comb polymer block. They used stoichiometric amount of PDP or NDP (one surfactant molecule for each monomer unit of P4VP: $(\text{P4VP/PDP})_{1.0}$), and varied the relative ratios of the PS and the P4VP–NDP (or P4VP–PDP) blocks to study the morphology. Figure 4.36 shows that with a P4VP–NDP comb fraction of $f_{comb} = 0.48$,

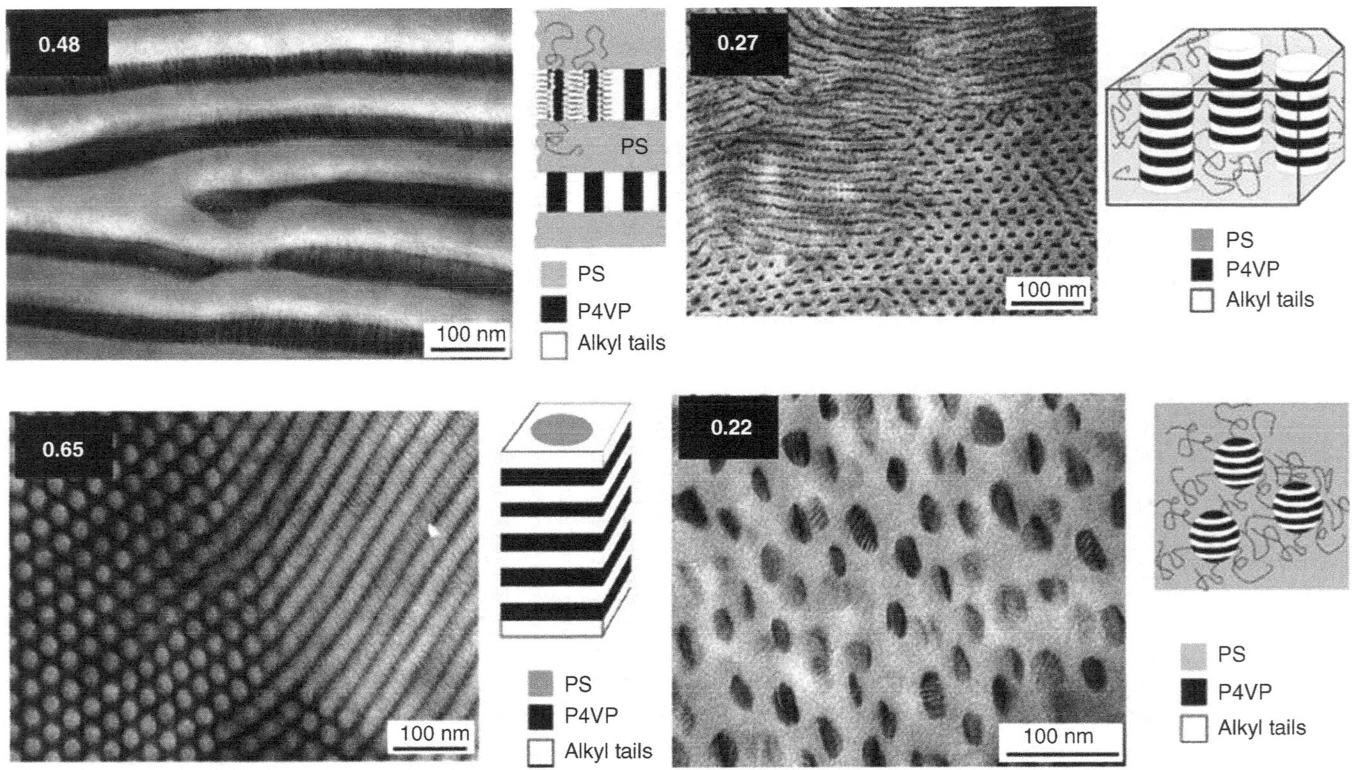

Figure 4.36 Morphologies of PS-$(P4VP/PDP)_{1.0}$ coil/comb block copolymers with various fractions (*f*) of the comb block. The values of *f* are noted on the figures. (Source: Ruokolainen et al. [69]. Reproduced with permission of Wiley.)

a lamellar–within-lamellar structure was obtained. The phase separated block copolymer morphology was lamellar, due to the approximately equal amount of PS and P4VP–NDP comb. The long period of the block copolymer was 90 nm. Within the lamellae of the P4VP–NDP block, another set of lamellae were seen, corresponding to the organization of the NDP, as illustrated in the schematic in this figure. The long period of these lamellae was 4.5 nm, which is the same as that of the size observed for the case of homopolymer of P4VP/NDP supramolecular complex. Thus, hierarchical organization with two length scales was achieved using supramolecular approach. Figure 4.36 also shows that with $f_{comb} = 0.27$ the morphology consisted of lamellae within the cylindrical structure and with $f_{comb} = 0.22$, lamellae within spherical domains are seen. A morphology with cylinders within lamellar structure was seen with $f_{comb} = 0.65$. Similar variations in the morphology with the compositions of the blocks were reported for the case of PS–P4VP–PDP system also.

Ikkala and ten Brinke [71] presented a "View Point" article on the utilization of such supramolecular assemblies as discussed earlier to fabricate functional materials. Their pictorial routes to achieve such structures are shown in Figure 4.37. Since a variety of morphologies can be tuned, as discussed earlier, by the supramolecular association of surfactants with homopolymers or block copolymers, postprocessing of such morphological structures could lead to useful nanoscale objects. Figure 4.37a shows the simple case of a homopolymer turned into a comb polymer similar to the case of hydrogen bonded association between P4VP and PDP. The organization of the tails leads to a lamellar structure. It was pointed out by Antonietti et al. [72] that the use of fluoroalkyl side chains would lead to low surface energy materials with low friction. If two or more types of binding molecules are used with polymer, for example, a surfactant such as PDP and a dye molecule (Figure 4.37b), the latter would be intercalated along the structure, and hence the dye would be dispersed rather than aggregated. Such a system could be useful for photonic applications, as was demonstrated by Wang et al. [73] with the supramolecular association of DNA/cationic surfactant complexes which included functional dye molecules. In yet another design (Figure 4.37c), a charged molecule such as camphorsulphonic acid (CSA) or methanesulfonic acid (MSA) can be used as an intermediate, with which a hydrogen bonding molecule could associate. For example, electrical conductivity could be achieved if CSA or MSA is used with a polymer such as polyaniline. Zheng et al. [74] used DBSA with polyaniline and the resulting supramolecular complexation led to a highly soluble and processable electroactive material. Kosonen et al. [75] treated polyaniline with CSA and 4-hexylresorcinol (which formed two hydrogen bonds with CSA) to fabricate nanoscale conducting cylinders. Selectively removing the alkyl side chains from the lamellar structures shown in Figure 4.37a–c would lead to functional films (Figure 4.37d).

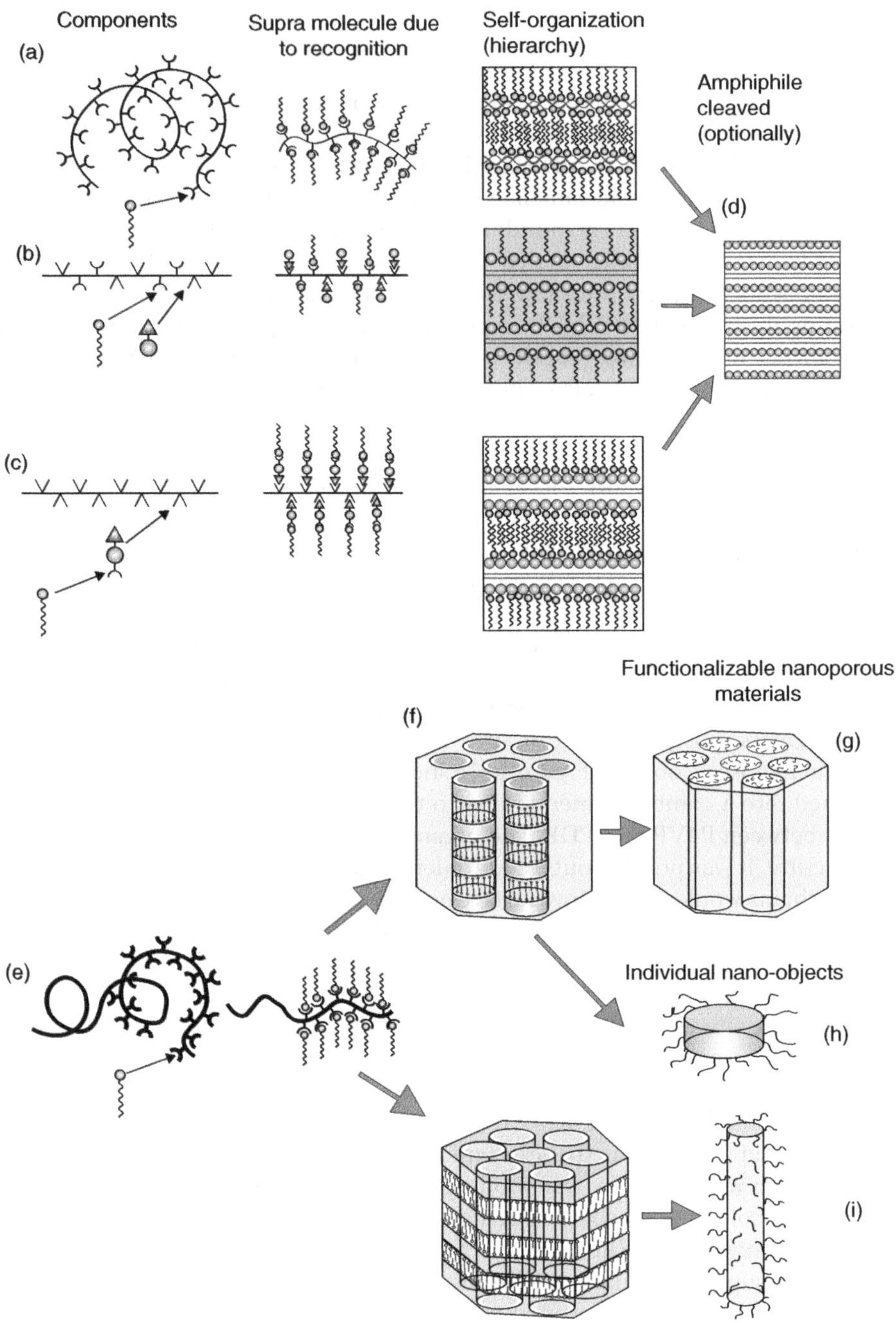

Figure 4.37 (a)–(d) Illustration of supramolecular association of surfactants by hydrogen bonding (or metal complexation) with (a) flexible or (b) and (c) rigid-rod polymers. The resulting hierarchical self-organization and films of such organized structures (d) are shown schematically. (e) and (f) Illustration of the supramolecular association of surfactants with one of the blocks of a diblock copolymer and the resulting domain morphologies. Selective removal of the components could lead to mesoporous films or hairy nano-objects as in (g), (h), and (i). (Source: Ikkala and ten Brinke [71]. Reproduced with permission of American Association for Advancement of Science.)

With the comb–coil block copolymer (Figure 4.37e), phase separation of the blocks would lead to a morphology in the length scale of about 1000– 2000 Å, whereas the surfactant self-organization would be in the range of 10–60 Å. Thus two length scales are combined in one system with the supramolecular comb–coil block copolymer. Morphologies such as lamellae-within-cylinders (Figure 4.37f) also allow selective cleaving of the constituents (as in the case of Figure 4.37d) that form the supramolecules. For example, starting from PS-block-P4VP/PDP, removal of PDP would result in a structure in which empty cylindrical pores with poly(4-vinylpyridine) brushes at the walls are left in the glassy PS matrix (Figure 4.37g). Using such a PS-block-P4VP/PDP system, Saito [76] cross-linked the P4VP, removed the PDP, and obtained "slices," leading to a method of preparing nanoscale colloidal disks (Figure 4.37h). Other shapes of nano-objects, such as polymeric rods and hairy rods, as shown in Figure 4.37i can also be prepared by selective removal of constituents from predesigned morphologies.

References

1 J.-M. Lehn, *Supramolecular Chemistry: Concepts and Perspectives*, **1995**, Wiley-VCH Verlag GmbH, Weinheim.
2 V. Berl, M. J. Krische, I. Huc, J.-M. Lehn, and M. Schmutz, *Chem. Eur. J.*, **2000**, *6*, 1938–1946.
3 A. P. H. J. Schenning, P. Jonkheijm, E. Peeters, and E. W. Meijer, *J. Am. Chem. Soc.*, **2001**, *123*, 409–416.
4 P. Jonkheijm, F. J. M. Hoeben, R. Kleppinger, J. van Herrikhuyzen, A. P. H. J. Schenning, and E. W. Meijer, *J. Am. Chem. Soc.*, **2003**, *125*, 15941–15949.
5 C. Fouquey, J.-M. Lehn, and A.-M. Levelut, *Adv. Mater.*, **1990**, *2*, 254–257.
6 J.-M. Lehn, *Polym. Int.*, **2002**, *51*, 825–839.
7 V. Berl, M. Schmutz, M. J. Krische, R. G. Khoury, and J.-M. Lehn, *Chem. Eur. J.*, **2002**, *8*, 1227–1244.
8 M. Ikeda, T. Nobori, M. Schmutz, and J.-M. Lehn, *Chem. Eur. J.*, **2005**, *11*, 662–668.
9 F. Aparicio and L. Sánchez, *Chem. Eur. J.*, **2013**, *19*, 10482–10486.
10 F. Aparicio, E. Matesanz, and L. Sánchez, *Chem. Commun.*, **2012**, *48*, 5757–5759.
11 S. Vauthey, S. Santoso, H. Gong, N. Watson, and S. Zhang, *Proc. Natl. Acad. Sci.*, **2002**, *99*, 5355.
12 C. Valéry, M. Paternostre, B. Robert, T. Gulik-Krzywicki, T. Narayanan, J.-C. Dedieu, G. Keller, M.-L. Torres, R. Cherif-Cheikh, P. Calvo, and F. Artzner, *Proc. Natl. Acad. Sci.*, **2003**, *100*, 10258.
13 D. A. Frankel and D. F. O'Brien, *J. Am. Chem. Soc.*, **1994**, *116*, 10057.

14 J. M. Schnur, B. R. Ratna, J. V. Selinger, A. Singh, G. Jyothi, and K. R. K. Easwaran, *Science*, **1994**, *264*, 945.

15 H. Y. Lee, S. R. Nam, and J.-I. Hong, *J. Am. Chem. Soc.*, **2007**, *129*, 1040.

16 J. M. Schnur, *Science*, **1993**, *262*, 1669.

17 S. Khanna, M. K. Khan, and P. Sundararajan, *Langmuir*, **2009**, *25*, 13183–13193.

18 W. Helfrich and J. Prost, *Phys. Rev. A*, **1988**, *38*, 3065.

19 J. S. Chappell and P. Yager, *Chem. Phys. Lipids*, **1991**, *58*, 253.

20 L. P. Stubbs and M. Weck, *Chem. Eur. J.*, **2003**, *9*, 992–999.

21 J. M. Pollino, K. P. Nair, L. P. Stubbs, J. Adams, and M. Weck, *Tetrahedron*, **2004**, *60*, 7205–7215.

22 J. M. Pollino and M. Weck, *Chem. Soc. Rev.*, **2005**, *34*, 193–207.

23 B. J. B. Folmer, R. P. Sijbesma, R. M. Versteegen, J. A. J. van der Rijt, and E. W. Meijer, *Adv. Mater.*, **2000**, *12*, 874–878.

24 M. Dionisio, L. Ricci, G. Pecchini, D. Masseroni, G. Ruggeri, L. Cristofolini, E. Rampazzo, and E. Dalcanale, *Macromolecules*, **2014**, *47*, 632–638.

25 J.-F. Gohy, B. G. G. Lohmeijer, and U. S. Schubert, *Chem. Eur. J.*, **2003**, *9*, 3472–3479.

26 B. G. G. Lohmeijer and U. S. Schubert, *Angew. Chem. Int. Ed.*, **2002**, *41*, 3825–3829.

27 I. Astafieva, X. Zhong, and A. Eisenberg, *Macromolecules*, **1993**, *26*, 7339.

28 L. Zhang and A. Eisenberg, *Science*, **1995**, *268*, 1728.

29 L. Zhang and A. Eisenberg, *J. Am. Chem. Soc.*, **1996**, *118*, 3168.

30 L. Zhang, K. Yu, and A. Eisenberg, *Science*, **1996**, *272*, 1777.

31 X. S. Wang, M. A. Winnik, and I. Manners, *Macromol. Rapid Commun.*, **2002**, *23*, 210.

32 J. Raez, I. Manners, and M. A. Winnik, *J. Am. Chem. Soc.*, **2002**, *124*, 10381.

33 B.-S. Kim, D.-J. Hong, J. Bae, and M. Lee, *J. Am. Chem. Soc.*, **2005**, *127*, 16333.

34 L. Jiang, R. C. Hughes, and D. Y. Sasaki, *Chem. Commun.*, **2004**, *8*, 1028.

35 H. Wang, H. H. Wang, V. S. Urban, K. C. Littrell, P. Thiyagarajan, and L. Yu, *J. Am. Chem. Soc.*, **2000**, *122*, 6855.

36 M. Lee, C.-J. Jang, and J.-H. Ryu, *J. Am. Chem. Soc.*, **2004**, *126*, 8082.

37 D. Yao, T. P. Bender, P. J. Gerroir, and P. R. Sundararajan, *Macromolecules*, **2005**, *38*, 6972–6978.

38 D. Yao, B. Tuteja, and P. R. Sundararajan, *Macromolecules*, **2006**, *39*, 7786–7788.

39 M. R. Islam, E. Dahan, S. Saimani, and P. R. Sundararajan, *Eur. Polym. J.*, **2012**, *48*, 1538–1554.

40 M. R. Islam and P. R. Sundararajan, *Eur. Polym. J.*, **2013**, *49*, 2042–2051.

41 V. D. Deepak and P. R. Sundararajan, *J. Phys. Chem. B*, **2011**, *115*, 8458–8464.

42 Jeffamine® M-2070 Polyetheramine, Technical Bulletin, **2008**, Huntsman Corporation. www.huntsman.com (accessed August 18, 2016).

43 M. Allesch, F. C. Lightstone, E. Schwegler, and G. Galli, *J. Chem. Phys.*, **2008**, *128*, 014501/1–014501/9.

44 M. Allesch, E. Schwegler, and G. Galli, *J. Phys. Chem. B*, **2007**, *111*, 1081.
45 T. Sugiura and H. Ogawa, *J. Chem. Thermodyn.*, **2009**, *41*, 1297.
46 M. D. Sindkhedkar, H. R. Mulla, and A. Cammers-Goodwin, *J. Am. Chem. Soc.*, **2000**, *122*, 9271.
47 N. Ben-Horin, U. Even, and J. Jortner, *Chem. Phys. Lett.*, **1991**, *177*, 153.
48 T. W. Bentley, I. S. Koo, and S. J. Norman, *J. Org. Chem.*, **1991**, *56*, 1604.
49 S. Ghosh, X.-Q. Li, V. Stepanenko, and F. Wurthner, *Chem. Eur. J.*, **2008**, *14*, 11343.
50 P. A. Korevaar, C. Schaefer, T. F. A. de Greef, and E. W. Meijer, *J. Am. Chem. Soc.*, **2012**, *134*, 13482–13491.
51 F. M. Menger and C. A. Littau, *J. Am. Chem. Soc.*, **1993**, *115*, 10083–10090.
52 F. M. Menger and J. S. Keiper, *Angew. Chem. Int. Ed.*, **2000**, *39*, 1906–1920.
53 J. B. Gilroy, S. K. Patra, J. M. Mitchels, M. A. Winnik, and I. Manners, *Angew. Chem. Int. Ed.*, **2011**, *50*, 5851.
54 F. He, T. Gadt, I. Manners, and M. A. Winnik, *J. Am. Chem. Soc.*, **2011**, *133*, 9095.
55 T. Gadt, F. H. Schacher, N. McGrath, M. A. Winnik, and I. Manners, *Macromolecules*, **2011**, *44*, 3777.
56 J. B. Gilroy, T. Gadt, G. R. Whittell, L. Chabanne, J. M. Mitchels, R. M. Richardson, M. A. Winnik, and I. Manners, *Nat. Chem.*, **2010**, *2*, 566.
57 J. Zhu, K. Zhang, S. Zhang, Y. Lin, K. Wooley, and D. J. Pochan, *Polym. Prepr.*, **2010**, *51*, 436.
58 H. Cui, Z. Chen, S. Zhong, K. Wooley, and D. J. Pochan, *Science*, **2007**, *317*, 647.
59 D. G. Whitten, *Acc. Chem. Res.*, **1993**, *26*, 502.
60 F. J. M. Hoeben, P. Jonkheijm, E. W. Meijer, and A. P. H. J. Schenning, *Chem. Rev.*, **2005**, *105*, 1491.
61 Y. Cao and P. Smith, *Polymer*, **1993**, *34*, 3139–3143.
62 G. H. Fredrickson, *Macromolecules*, **1993**, *26*, 2825–2831.
63 O. Ikkala, J. Ruokolainen, G. ten Brinke, M. Torkkeli, and R. Serimaa, *Macromolecules*, **1995**, *28*, 7088–7094.
64 J. Ruokolainen, J. Tanner, G. ten Brinke, O. Ikkala, M. Torkkeli, and R. Serimaa, *Macromolecules*, **1995**, *28*, 7779–7784.
65 J. Ruokolainen, G. ten Brinke, O. Ikkala, M. Torkkeli, and R. Serimaa, *Macromolecules*, **1996**, *29*, 3409–3415.
66 J. Ruokolainen, M. Torkkeli, R. Serimaa, S. Vahvaselkä, M. Saariaho, G. ten Brinke, and O. Ikkala, *Macromolecules*, **1996**, *29*, 6621–6628.
67 J. Ruokolainen, M. Torkkeli, R. Serimaa, E. Komanschek, G. ten Brinke, and O. Ikkala, *Macromolecules*, **1997**, *30*, 2002–2007.
68 J. Ruokolainen, J. Tanner, O. Ikkala, G. ten Brinke, and E. L. Thomas, *Macromolecules*, **1998**, *31*, 3532–3536.
69 J. Ruokolainen, G. ten Brinke, and O. Ikkala, *Adv. Mater.*, **1999**, *11*, 777–780.
70 J. Ruokolainen, M. Saariaho, O. Ikkala, G. ten Brinke, E. L. Thomas, M. Torkkeli, and R. Serimaa, *Macromolecules*, **1999**, *32*, 1152–1158.

71 O. Ikkala and G. ten Brinke, *Science*, **2002**, *295*, 2407–2409.
72 M. Antonietti, S. Henke, and A. Thünemann, *Adv. Mater.*, **1996**, *8*, 41–45.
73 L. Wang, J. Yoshida, N. Ogata, S. Sasaki, T. Kajiyama, *Chem. Mater.*, **2001**, *13*, 1273–1281.
74 W.-Y. Zheng, R.-H. Wang, K. Levon, Z. Y. Rong, T. Taka, and W. Pan, *Makromol. Chem. Phys.*, **1995**, *196*, 2443–2462.
75 H. Kosonen, J. Ruokolainen, M. Knaapila, M. Torkkeli, K. Jokela, R. Serimaa, G. ten Brinke, W. Bras, A. P. Monkman, and O. Ikkala, *Macromolecules*, **2000**, *33*, 8671–8675.
76 R. Saito, *Macromolecules*, **2001**, *34*, 4299–4301.

Further Reading

J. L. Atwood, J. E. D. Davies, D. D. MacNicol, and F. Vögtle, Eds. *Comprehensive Supramolecular Chemistry*, Vols. *1–11*, **1996**, Pergamon, Oxford.
A. Ciferri, Ed., *Supramolecular Polymers*, **2005**, CRC Press Taylor & Francis.
T. F. A. De Greef, M. M. J. Smulders, M. Wolffs, A. P. H. J. Schenning, R. P. Sijbesma, and E. W. Meijer, *Chem. Rev.*, **2009**, *109*, 5687–5754.
V. Rotello and S. Thayumanavan, Eds., *Molecular Recognition and Polymers*, **2008**, John Wiley & Sons, Inc., Hoboken.
J. W. Steed and J. L. Atwood, *Supramolecular Chemistry*, 2nd Ed., **2009**, John Wiley & Sons, Inc., New York.
L. Yang, X. Tan, Z. Wang, and X. Zhang, *Chem. Rev.*, **2015**, *115*, 7196–7239.

5

Block Copolymers

A block copolymer (BCP), as the name implies, consists of two or more homopolymers (poly(A) with monomer of type A and poly(B) with monomer of type B) and which are usually immiscible with each other, end-linked covalently. The idea of linking two polymers covalently and synthesizing "block copolymers" dates back to 1952. In a communication to *Nature*, Dunn and Melville [1] introduced the concept of a block polymer, which would be expected to have properties different from a random copolymer of two monomers or a physical mixture of two types of polymers. They reported [1, 2] the synthesis of polystyrene-*b*-poly(methyl methacrylate) (PS-*b*-PMMA). At about the same time, Herman Mark discussed the synthesis of PS-*b*-PMMA, by using hydroxyl-terminated PS and PMMA and treating them with diisocyanate to make a BCP. In a footnote to his paper [3], Mark credits Dr. G. F. Tuckett with suggesting the term "block copolymer" in a lecture at the *Gordon Research Conference on Textiles* in 1952 (also see Patent Appeal No. 7115 United States Court of Customs and Patent Appeals, April 9, 1964: http://openjurist.org/329/f2d/977/application-of-robert-p-zelinski).

5.1 Theoretical Aspects

Figure 5.1 illustrates a diblock copolymer with $(A)_n$ and $(B)_m$ blocks, consisting of two types of monomers: a symmetric $(A)_n$–$(B)_m$–$(A)_n$ triblock copolymer and an asymmetric $(A)_n$–$(B)_m$–$(C)_k$ triblock copolymer. Apart from such linear sequences, other architectures such as star copolymers with BCP arms are also possible [4]. The self-assembly of the individual blocks gives rise to various interesting phase separated morphologies, which can be tailored by the chemistry and processing conditions. In the case of physical blends of two homopolymers A and B, the entropy of mixing is small and the interaction parameter χ_{AB} is usually small and positive, leading to macrophase separation, although completely miscible pairs are known. The miscibility in the case of polymer

Physical Aspects of Polymer Self-Assembly, First Edition. P. R. Sundararajan.

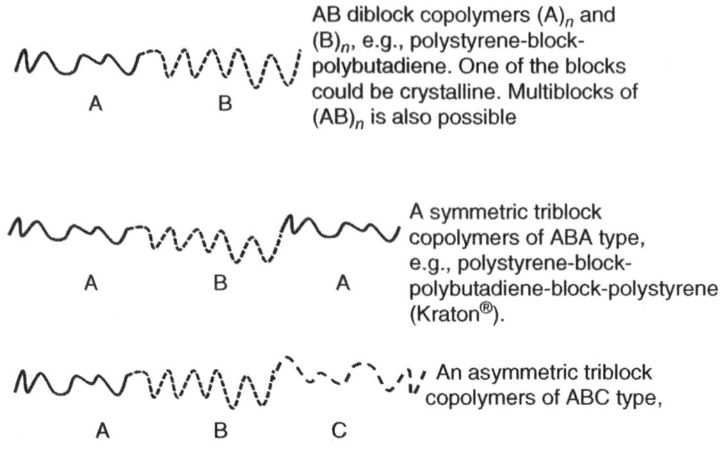

Figure 5.1 Illustration of AB diblock, ABA symmetric triblock, and ABC asymmetric triblock copolymers.

blends is usually characterized by a single glass transition temperature (T_g), which would be intermediate between those of the two homopolymers depending on the relative composition. Immiscibility leads to two distinct T_g's corresponding to those of the individual components. Even with immiscible blends, if the blending leads to enhancement of, for example, mechanical properties, the term "compatible" is used. In contrast to the macrophase separation in blends, the entropic force that constrains the constituent blocks to the covalent link leads to mesoscopic phase separation with dimensions of the order of 10 nm. When we consider the self-assembly of BCPs, it should be noted that it is not driven by hydrogen bond or similar specific interactions. For example, in the case of the widely studied BCPs such as PS-*b*-PMMA, polystyrene-*b*-polybutadiene (PS-*b*-PBt), and polystyrene-*b*-polyisoprene (PS-*b*-PI), there are no groups that lead to hydrogen bonding or polar interactions. It is the immiscibility of the blocks and their self-sorting that leads to the phase separated morphologies.

The physics of phase morphologies of the BCP based on mean field theories has been discussed by Bates and Fredrickson [4, 5] and others. The morphology in the melt depends on N_A and N_B; the number of monomers in the A and B blocks, respectively; the overall degree of polymerization N (= $N_A + N_B$); and the interaction parameter χ_{AB}. Typical morphologies are spheres and cylinders of one component in the matrix of the other and lamella. As an example [6], the transmission electron microscopy (TEM images and the corresponding schematics of the morphologies of PS-*b*-PBt diblock copolymers are shown in Figure 5.2, for various concentration of PS. It is usual to stain PBt or PI with osmium tetroxide or similar agents, to provide contrast in the TEM images.

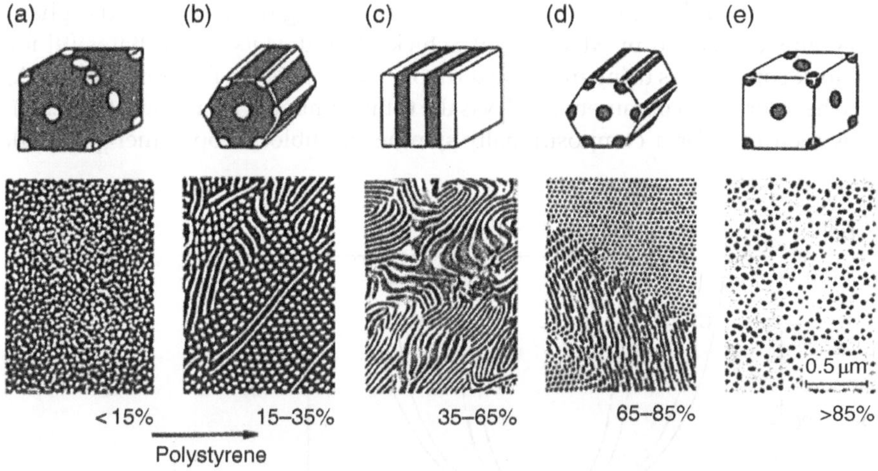

Figure 5.2 Transmission electron micrographs, along with the corresponding schematics, showing the changes in the morphology of polystyrene-*b*-polybutadiene (PS-*b*-PBt) diblock copolymer with increasing PS content. (Source: Schmitt [6]. Reproduced with permission of Wiley.)

When the fraction f_{PS} of PS is small (<15%), PS adopts spherical domains (shown in white) in the PBt (black) matrix (Figure 5.2a). This is shown as body-centered cubic (bcc) phase in the schematic. With an increase in concentration, the domains change to cylinders (Figure 5.2b). Here again, the black matrix corresponds to PBt. Note that this image consists of "dots" which look like spheres as well as lines. This demonstrates that in order to verify the cylindrical morphology, the image should be recorded with different orientations of the sample with respect to the electron beam. If the beam is exactly normal to the cylinders in the sample, it will see the end-on cross section of the cylinders and it would appear as the projection of spheres. With different angles of recording, the cylindrical appearance can be confirmed. When f_{PS} and f_{PBt} of PBt are about equal (a symmetric diblock) lamellar morphology occurs, as seen in Figure 5.2c. In the TEM image, the dark lines correspond to the stained PBt. With a further increase in f_{PS} phase inversion occurs, with PBt forming cylinders in the PS matrix. Finally, with $f_{PS} > 85\%$, PBt forms spheres in the PS matrix.

For a copolymer with a small N and χ_{AB}, the repulsion between the blocks will be small. In this case, discrete self-sorting would not be present and there will be disorder due to the partial mixing of the blocks. With an increase in N and χ_{AB}, self-sorting would become stronger and microphase separation would lead to well-ordered domains. The theory of Leibler [7] predicted that $N\chi_{AB}$ should be greater than 10.5 for this to occur. If the blocks are short and the interaction parameter χ_{AB} is also favorable such that $N\chi_{AB} < 10.4$, mixing of the blocks would occur in the melt leading to disorder. Thus, $N\chi_{AB} = 10.5$ marks the order–disorder transition (ODT) condition. Figure 5.3 shows the phase diagrams developed by Matsen and Schick [8] and Matsen and Bates [9] for diblock copolymers using mean field theory. Here, f_A is the fraction of block A, χ is the interaction parameter, and N is the total number of monomers ($N_A + N_B$). It is seen that for a compositionally symmetric diblock copolymer ($f_A = 0.5$),

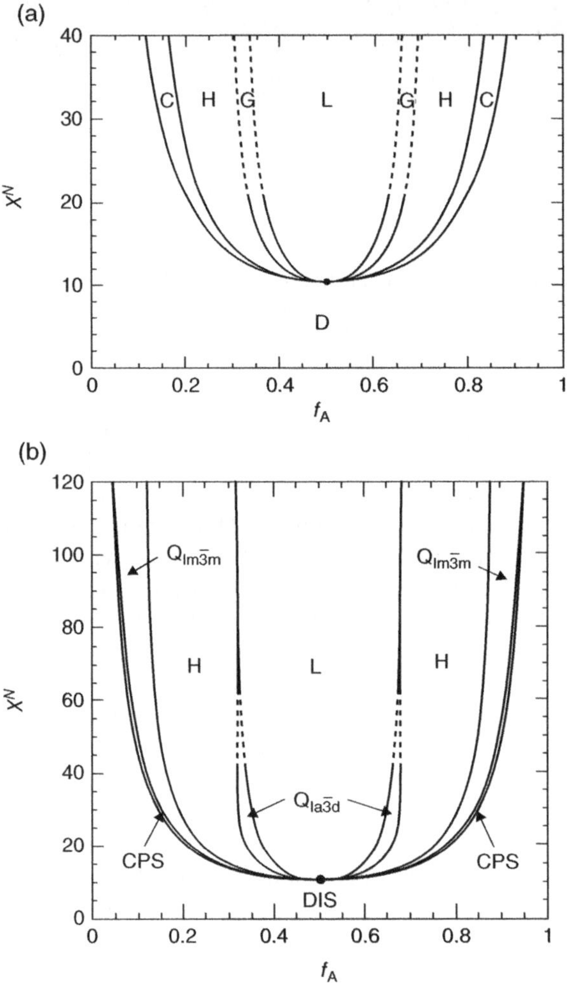

Figure 5.3 Theoretically predicted microphase separated morphologies of A–B diblock copolymers in terms of volume fraction f_A, interaction parameter χ, and N the total degree of polymerization. (a) $\chi N = 0$–40; (b) $\chi N = 0$–120. (Source: (a) Matsen and Schick [8]. Reproduced with permission of American Physical Society and (b) Matsen and Bates [9]. Reproduced with permission of American Chemical Society.)

with $N\chi_{AB} < 10.4$ the morphology is disordered. The "dot" at the bottom of the curves in Figure 5.3 is the "mean field critical point," at which there is weak segregation. When f_A deviates from 0.5, the microphase transition point moves to higher $N\chi_{AB}$. With large $N\chi_{AB}$ well-defined order–order boundaries are found. For the symmetric diblock copolymer, with f_A between 0.4 and 0.6, lamellar morphology is predicted. Hexagonally packed cylindrical morphology occurs on either side of the lamellar domains, and the bcc phase (spheres, $Im\bar{3}m$ space group) occurs on either side of the H phase.

Note that the ordinate in Figure 5.3a shows a small range of $N\chi_{AB}$ up to 40. In this weak segregation regime, a bicontinuous gyroid morphology between the lamellar and cylindrical phases was predicted [8, 9] and has been observed experimentally. Figure 5.4 shows the schematic of such a morphology in which each of the blocks forms continuous gyroid, a labyrinth with threefold connectors [10]. Schoen [11], a NASA scientist, is credited with naming these structures "gyroid" as he developed the mathematics of "three-dimensional Euclidean space into two interpenetrating labyrinths by intersection-free infinite periodic minimal surfaces." These have $Ia\bar{3}d$ space group symmetry. Gyroids are based on *Haniwa*, the terracotta clay figures (hollow sculptures) which were made for ritual use and as funerary objects during the Kofun period (third to sixth century AD) of the history of Japan. The gyroid morphology is seen in a narrow region between the lamellae and hexagonally packed cylinders in Figure 5.3b for $N\chi_{AB}$ less than 60 and this phase ends at $N\chi_{AB} = 11.14$, $f_A = 0.452$ and 0.548. Later work by Cochran et al. [12] indicated that the gyroid morphology could exist even in the strong segregation limit, that is, when $N\chi_{AB} \rightarrow \infty$. A narrow region corresponding to close-packed spheres (CPS) is also shown

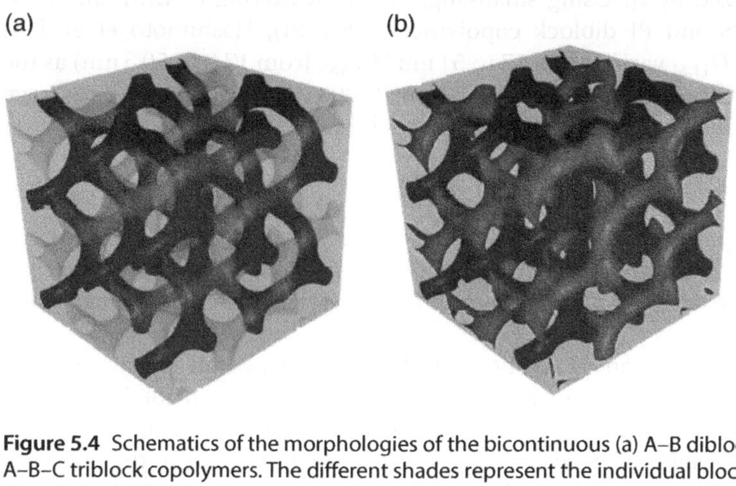

Figure 5.4 Schematics of the morphologies of the bicontinuous (a) A–B diblock and (b) A–B–C triblock copolymers. The different shades represent the individual blocks. See Vignolini et al. [10] for the corresponding color image. (Source: Vignolini et al. [10]. Reproduced with permission of Wiley.)

between the bcc sphere phase and the disordered phase. This phase terminates at $N\chi_{AB} = 17.67, f_A = 0.235$ and 0.765.

The phase diagrams shown in Figure 5.3 are based on the assumption that the A and B segments are conformationally symmetric, that is, their statistical lengths are the same. Matsen and Bates [13] discussed the roles of the interfacial tension between the blocks and the entropic penalty (arising from the stretching of the segments to occupy the domain space) in determining the shape of the interface and the morphology. When the stretching entropies of the blocks are balanced, flat interfaces and lamellar morphology result. An imbalance of the stretching of the blocks leads to curved interfaces and spherical morphology as the volume fraction f_A of the A block approaches 0 or 1. Matsen and Bates [13] showed that when the segment lengths a_A and a_B of the blocks are unequal, the ratio a_A/a_B had a significant effect on the theoretical order–order phase boundaries and the mean field critical points.

5.2 Diblock Copolymers

As mentioned before, when $N\chi_{AB}$ is greater than 10.5, strong segregation would lead to distinct lamellae of blocks A and B. In this case, the theory of phase separation predicted that the lamellar dimension L would be proportional to $N^{2/3}$. The dimension of the interphase will be much smaller than the size of the lamellae, as shown schematically in Figure 5.5. In the figure, DA and DB denote the lamellar spacing of blocks A and B, and the overall dimension is given by D. The interphase thickness is indicated by t and the diffuseness of the boundary is characterized by t_1. Using small-angle X-ray scattering (SAXS) and TEM images of PS and PI diblock copolymers (PS-*b*-PI), Hashimoto et al. [14] showed that D_{TEM} varied from 17 to 51 nm (D_{SAXS} from 17.2 to 50.3 nm) as the total M_n of the polymer was increased from 21 000 to 102 000. Thus the average domain size increased with the total molecular weight according to

$$\bar{D} = 0.024\bar{M}_n^{\ 2/3} \tag{5.1}$$

Thus, the experimental result that the lamellar dimension scales with $N^{2/3}$ is in accord with the theoretical prediction. Their work also showed that the interphase thickness t was about 2 nm and was invariant with the total molecular weight.

In a later publication, Hashimoto et al. [15] tabulated an extensive list of PS-*b*-PI diblock copolymers with various compositions and total molecular weights and the morphology of toluene-cast films from such samples. Toluene is a neutral solvent for PS and PI blocks. They concluded that (i) the microdomain morphologies were dependent on the volume fractions of the blocks and not on the total molecular weight, (ii) the transition from one morphology

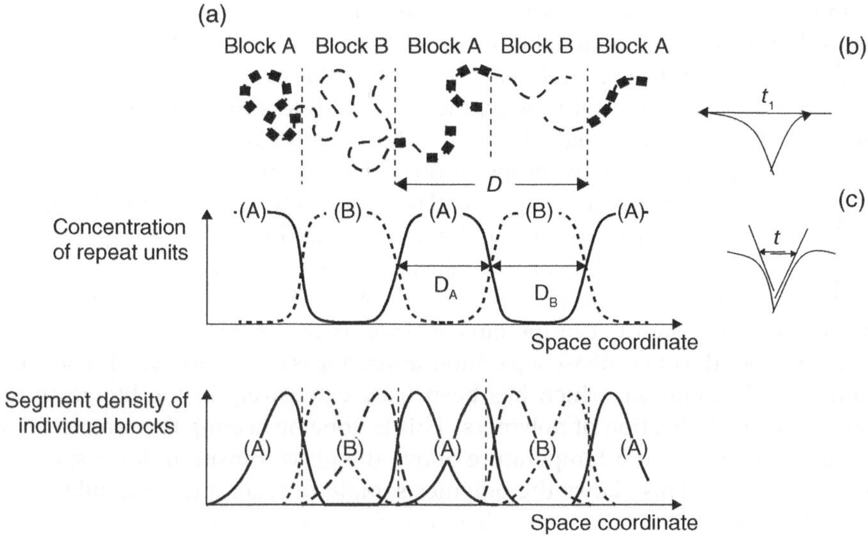

Figure 5.5 Schematic of the interphase between phase separated blocks. (a) Phase separated domains of the two blocks A and B; (b) diffuseness of the boundary t_1; (c) interphase thickness t. (Source: Schmitt [6]. Reproduced with permission of Wiley.)

Table 5.1 Microdomain morphologies for the various volume fractions of polystyrene in the PS-*b*-PI diblock copolymers as observed by Hashimoto et al. [15].

PS volume fraction (Φ_{PS})	Morphology
>0.76	Spherical PI domains in PS matrix
0.76–0.67	PI cylinders in PS matrix
0.66–0.62	PI tetrapod networks in PS matrix
0.61–0.32	Alternating PI–PS lamellae
0.30–0.18	PS cylinders in PI matrix
0.17	PS spheres in PI matrix

to the other was sharp, and (iii) the dimensions of the domains obeyed the 2/3 power law. The morphologies determined by these authors for the various volume fractions of PS are listed in Table 5.1. Note that the tetrapod network, which is a bicontinuous structure (gyroid), was observed in a narrow Φ_{PS} range between the cylindrical and lamellar morphologies.

Sample preparation is an important step in the characterization of the BCP morphologies. Since TEM is often used to visualize the features, most often thin films are cast from solution, with a careful selection of the solvent. Ideally, the solubility of the individual blocks should be similar, which means that the solvent should not have preference for one block over the other, that is, it should be nonselective. Otherwise, the block with less solubility would phase separate first and this could influence the morphology during the solvent evaporation. Controlling the evaporation of the solvent is also necessary (to be slow) since a fast evaporation could lead to "kinetically trapped" morphologies. The surface onto which to cast the film depends on the aim of the experiment. Most authors cast the films on glass surface. A glass treated with dichlorodimethylsilane as a release agent allows peeling of the film without much difficulty. Solution casting on water or mercury surfaces is also preferred.

To enhance the microphase separation, annealing is often performed. Thermal annealing is common, which has been used extensively in the literature to study the crystallization of polymers. This is done by heating the sample (in a vacuum chamber) at a temperature above the glass transition, for a specific period of time. This allows the polymer sample to relax into an equilibrium morphology. Note that the phase diagrams shown in Figure 5.3 are based on f_A and $N\chi_{AB}$, but the temperature is not included as an explicit parameter. The interaction parameter χ would change with temperature. Although N would a constant for a given BCP system, the coil dimension would also be temperature-dependent. Hence, order–order transitions (OOTs) or ODTs could be induced by thermal annealing.

Solvent annealing is similar except that the sample is exposed to a solvent vapor instead of a thermal medium. This method has its origin in the studies on the phenomenon of solvent-induced crystallization (SINC) which was studied by a few groups [16–19]. When an amorphous but crystallizable polymer is exposed to a solvent vapor (at room temperature), the diffusion of the solvent results in the plasticization of the polymer and the glass transition temperature is depressed significantly, often to below room temperature. The gap between the T_g and T_m is widened, and this results in the crystallization at room temperature. The SINC of several polymers, such as poly(ethylene terephthalate) and isotactic PS, has been studied [16–18]. Due to the depression of T_g, solvent vapor exposure would also cause phase separation and crystallization of plasticizers from the polymer films. Solvent vapor-induced crystallization of cyclic oligomers from films of poly(ethylene terephthalate) was reported by Sundararajan and Perovic [19]. Thus, solvent vapor exposure or solvent "annealing" has been used in several studies on BCP morphologies. The intent is to allow for the mobility of the chains to enhance the phase separation.

The spherical, cylindrical, and lamellar morphologies shown in Figure 5.2 are known as "classic morphologies." Guided by theoretical predictions, other types of morphologies (i.e., gyroid) have been discovered by synthesis of BCPs with defined volume fractions of the blocks, thermal and solvent annealing, blending two BCPs, etc. These additional morphologies, shown in Figure 5.6,

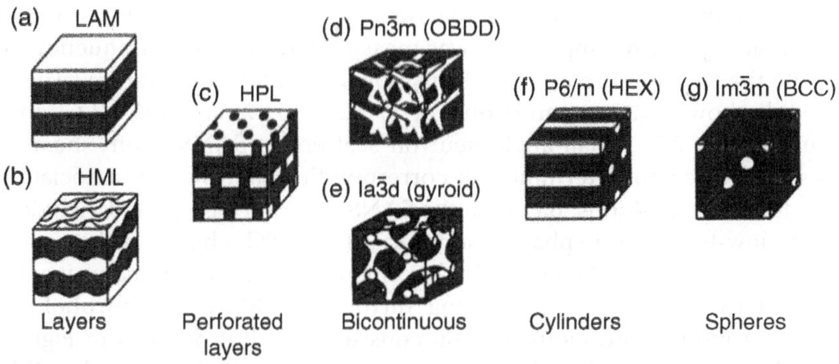

Figure 5.6 Schematics of various morphologies of diblock copolymers which depend on microstructure, annealing, and other conditions. (a) lamellar (LAM), (b) hexagonally modulated lamellae (HML), (c) hexagonally perforated layers (HPL), (d) ordered bicontinuous double diamond (OBDD), (e) cubic bicontinuous network (gyroid), (f) hexagonally packed cylinders (HEX), and (g) spheres (body-centered cubic, bcc). The space groups corresponding to the structures are shown. (Source: Schulz and Bates [20]. Reproduced with permission of Springer.)

occur between the classic ones near the ODTs [20]. While the classic morphologies depend on the relative volume fractions of the blocks, annealing a BCP with a given volume fraction can give rise to OOTs, leading to intermediate morphologies prior to the ODTs. There have been several studies in this regards, especially related to identifying the gyroid morphology. Some examples will be discussed later.

Hajduk et al. [21] reported the OOT that was thermoreversible in the case of a PS-*b*-poly(ethene-*co*-butene) (PS–EB) diblock copolymer, which was fractionated from the commercial Kraton G-1726 to remove the triblock polymer which was present as a minor component. The volume fraction of PS was 0.29. Films for the measurements were made using tetrahydrofuran (THF), which is a nonselective solvent for both blocks. The as-cast films showed lamellar morphology. Upon annealing the film at 130°C, a modified lamellar morphology appeared, which consisted of layers of cylinders, but the cylinders in adjacent layers were out of registry between one another. Annealing at a higher temperature of 175°C resulted in cylindrical morphology of the PS domains. These transformations were found to be reversible. When the films were cast from preferential solvents for the EB block, such as hexane or cyclohexane, only cylindrical morphology of the PS block in the matrix of EB was seen since the latter preferred contact with the solvent. Annealing at temperatures up to 200°C did not change the morphology.

Hamley and Bates [22] discovered two intermediate phases between the lamellar and hexagonal phases upon heating a poly(ethylene–propylene)-*b*-poly(ethylethylene) (PEP–PEE) diblock copolymer with a PEP volume fraction of 65%. Using dynamic mechanical spectroscopy (DMS), the variation of the moduli G' and G'' with temperature was measured at a constant frequency to determine the temperature of phase transitions. With the temperature sweep, the moduli showed three transitions, at 87 ± 2, 136 ± 2, and 147 ± 2°C, and disorder at 178 ± 2°C. Small-angle neutron scattering (SANS) from sheared samples were used to determine the corresponding morphology associated with the thermal transitions. Based on the SANS experiments, it was concluded that (i) the low-temperature phase is lamellar; (ii) at 120°C, a hexagonally modulated phase as in Figure 5.6b forms. In this schematic, the dark domains would correspond to PEP; (iii) upon heating further to 140°C, the morphology changes to a mono-continuous domain consisting of PEP holes as in Figure 5.6c; further heating to the high-temperature phase and shearing, the PEE component forms cylinders in the PEP matrix (Figure 5.6f). Thus, in the case of the PEP–PEE diblock copolymer with a PEP volume fraction of 65%, with increasing temperature an intermediate hexagonally modulated lamellar phase at low temperature and a layered hexagonally packed channels at high temperature are seen between the classical lamellar and cylindrical morphologies.

By annealing diblock copolymers at higher temperatures, Hajduk et al. [23] identified that the gyroid morphology could exist in equilibrium in the intermediate to weak segregation limit, at temperatures of about 50°C below the order–disorder temperature (ODT). They used a diblock copolymer of PS-*b*-polyisoprene (PS–PI) with a 37 wt% of PS and a total molecular weight of 27 400. From the work of Hashimoto et al. [24], it was determined that $N\chi$ for this copolymer ranges from 22.4 to 16.4 between the temperatures of 90 and 160°C, respectively. According to theory [25], the upper bound for the weak segregation limit corresponds to $f = 0.34$ and $N\chi = 16$ so that the copolymer is in the weak-to-intermediate segregation limit. Films were cast from toluene solution on to water surface. The as-cast films showed lamellar morphology with a spacing of 20 ± 0.5 nm, which increased to 22.6 ± 0.2 nm upon annealing at 90°C for 2 h. There was no change in the morphology upon several hours of annealing at temperatures up to 120°C. However, annealing at 160°C resulted in the gyroid phase. This transformation was thermo-reversible, since annealing the sample at 100°C restored the lamellar morphology. By adding short-chain homopolymers, it was determined that the gyroid phase occurs with styrene volume fractions of 0.31–0.42 in a limited temperature range.

Bates group studied [26] the phase behavior of polyethylene-*b*-PEE (PE–PEE) diblock copolymers (Scheme 5.1) with the PE volume fraction (f_{PE}) varying from 0.25 to 0.46. In addition to using small-angle X-ray and SANS and TEM for morphology analysis, Bates et al. used DMS to determine the temperatures of ODTs for various f_{PE}. The samples were shear oriented to enhance the initial morphology and then the PEE block was stained with ruthenium tetraoxide. With an increase in the PE fraction f_{PE}, the morphology changed from PE spheres to cylinders, gyroid, and lamellae as in Figure 5.6g, f, e, and a

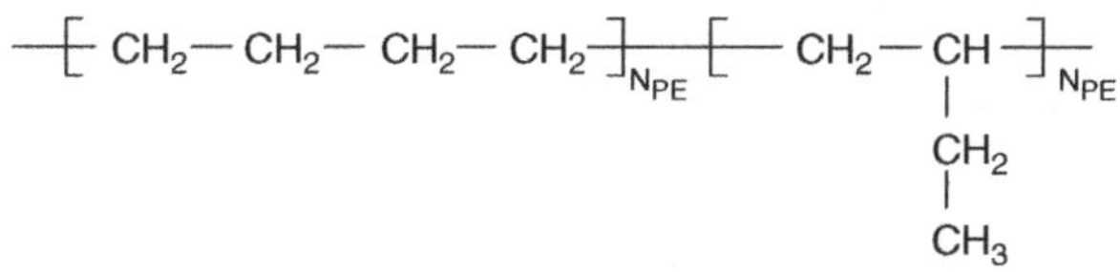

Poly(ethylene)–poly(ethylethylene)

Scheme 5.1 Schematic of the PE–PEE diblock copolymer [26].

Table 5.2 Temperature-dependent morphologies for various volume fractions of polyethylene block in the poly(ethylene)-*b*-poly(ethylethylene) diblock copolymers.

Sample[a]	Volume fraction (f_{PE})	Phases
1H	0.50_0	Lamellar → 136°C → disorder
2D	0.49_0	Lamellar → 120°C → disorder
3H	0.45_6	Lamellar → 276°C → disorder
4D	0.43_0	Hexagonally perforated layer (HPL) → 154°C → gyroid → 207°C → disorder
5D	0.40_0	Cylinders → 176°C → gyroid → 250°C → disorder
6D	0.37_4	Cylinders → 294°C → disorder
7D	0.25_2	Cylinders → 238°C → bcc spheres → 273°C → disorder

Source: Zhao et al. [26]. Reproduced with permission of American Chemical Society.
[a] H and D denote the hydrogenated and deuterated precursor polymers, respectively. See Zhao et al. [26] for the original sample identifications.

respectively. For studying the temperature-dependent morphologies, the samples were annealed at the temperature of phase transitions detected by the variations in the modulus G' and quenched in liquid nitrogen to freeze the ordered phase and crystallize the PE block. The PEE block was then stained. Measurement of G' with temperature showed ODT at 294 and 276°C, respectively, with $f_{PE} = 0.37$ and 0.46. However, with $f_{PE} = 0.25$, 0.40, and 0.43, a sharp increase in G' was observed before ODT was reached, which would suggest an OOT. The phase transitions and temperature-dependent morphologies are listed in Table 5.2, and the various annealing experiments and resulting morphologies are summarized in Figure 5.7. Thus, in this system, the gyroid phase is seen for f_{PE} between 0.37 and 0.43. In a similar system of PS–PI studied by these authors [27], the bicontinuous gyroid phase occurred for the volume fraction of PI between 0.36 and 0.39. They constructed a phase diagram of f_{PI} versus χN. The gyroid phase was found between $\chi N = 20$ and 27, bound by cylinders, lamellae, and HPL phases. It was mentioned earlier that sample preparation protocol would have an influence on the morphology. Lipic et al. [28]

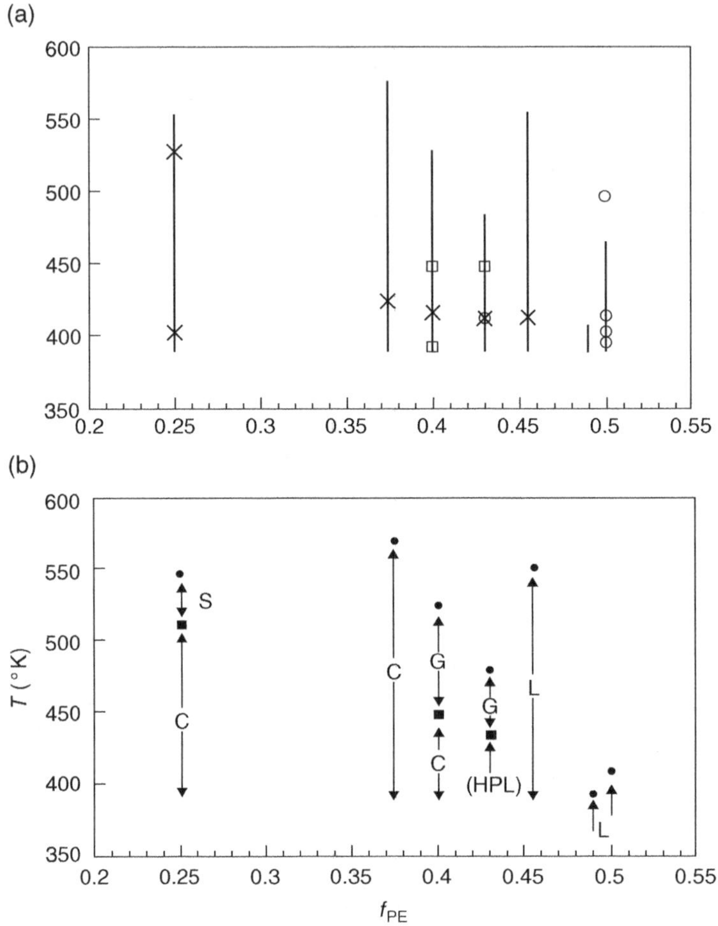

Figure 5.7 (a) Summary of experiments performed with the pure PE–PEE diblock copolymers as a function of temperature and the minority PE block volume fraction. Vertical lines identify the temperature range for dynamic mechanical spectroscopy (DMS) measurements; □, SAXS on unsheared specimens; ○, SANS on sheared and unsheared specimens; ●, annealing temperature prior to liquid nitrogen quenching for TEM analysis. (b) A "phase diagram" for PE–PEE diblock copolymers, constructed based on the DMS, SAXS, SANS, and TEM results. ■ and × symbols mark OOT and ODT temperatures, respectively, determined from isochronal G′(T) measurements. The S, C, G, and L morphologies were clearly identified while the HPL state was a tentative assignment. (Source: Zhao et al. [26]. Reproduced with permission of American Chemical Society.)

found differences in the morphologies of the samples of the PE–PEE diblock copolymers and their blends that were prepared using two different methods. In one, they poured the hot solution of the samples in cyclohexane or toluene into a methanol bath (nonsolvent) and the polymer precipitated. It was collected, dried, and films were made by compressing between Teflon sheets at 120°C. In the other method, solutions were cast on glass slides at high temperature and were subsequently quenched in liquid nitrogen. The films were then allowed to warm up to room temperature.

In a subsequent publication, Hajduk et al. [29] summarized the phase behavior of the following diblock copolymers: PE–PEE, PI–PS, PS-poly(2-vinylpyridine) (PS–P2VP), PEP–poly(dimethylsiloxane) (PEP–PDMS), and poly(ethylene oxide)–PEE (PEO–PEE) with respect to the hexagonally modulated layer and hexagonally perforated layer morphologies. They showed that the HML and HPL are long-lived nonequilibrium morphologies, which transition to bicontinuous gyroid with isothermal annealing.

5.3 Organic/Inorganic Diblocks

While most of the earlier studies were concerned with BCPs with a hard segment and a rubbery segment (e.g., PS/PBt, PS/PI, and those with PMMA and the rubbery blocks), those with organic and inorganic blocks have been studied in later years. For example, BCPs with poly(ferrocenosilane) as one of the blocks since the organometallic segment would lead to magnetic, electrical, and similar properties and also serve as precursors for ceramic materials [30].

Vancso group studied [31] the morphologies of PS-*b*-poly(ferrocenyldimethylsilane) diblock copolymer (PS-*b*-PFS, Scheme 5.2) which exhibited periodic organic and organometallic microdomains.

It was mentioned earlier that the theoretical phase diagrams such as those in Figure 5.3 for diblock copolymers are based on the assumption of conformational symmetry of the blocks and that Matsen and Bates [13], Vavasour and Whitmore [32], and Matsen and Schick [33] discussed the effect of conformational asymmetry on the shape of the order–order and order–disorder boundary contours. Such asymmetry could arise from differences in Kuhn segment lengths

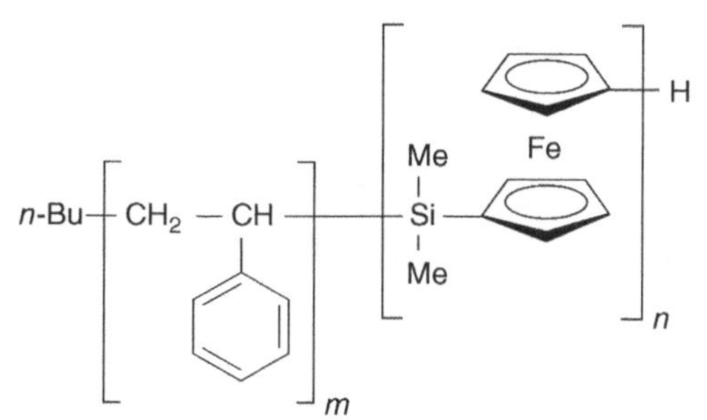

Scheme 5.2 Schematic of the polystyrene-*b*-poly(ferrocenyldimethylsilane) diblock copolymer [31].

and/or densities of the pure components. With the organic/organometallic blocks, the latter could contribute to asymmetry ($\rho_{PS} = 1.05$; $\rho_{PFS} = 1.26\,g/cm^3$). The PS block is amorphous with a T_g of about 100°C, while the PFS block is crystallizable, with a low T_g of 33°C and a melting temperature of 142°C.

Vancso et al. [31] used PS-*b*-PFS copolymers with M_w ranging from 27 300 to 43 500 g/mol and PFS volume fractions of 12–84%. Films were cast from THF solutions, dried overnight, annealed for a week at 165°C, and quenched in ice water bath (to prevent crystallization of the PFS block). SAXS was used to confirm the morphologies observed with TEM, and rheological measurements were used to follow OOT and ODT with temperature. The copolymer with M_n = PS/PF: 27 000/12 000 (denoted as SF 27/12) and $\Phi_{PFS} = 27\%$ showed hexagonally packed PFS cylinders in PS matrix as seen in Figure 5.8a. The dark regions correspond to the PFS block. Small-angle X-ray scattering (SAXS) confirmed the cylindrical morphology and showed the interdomain distance and the radius of the PFS cylinders to be 26.1 and 7.1 nm, respectively. Figure 5.8b shows that increasing Φ_{PFS} to 47% (with SF 18/19, total $M_n = 36.8$) leads to alternating lamellar morphology with spacings of 12.9 and 11.5 nm for the PS and PFS blocks, respectively, and an interdomain distance of 24.4 nm. The copolymer SF 12/25 also had the same M_n as SF 18/19 with a total M_n of 36.8,

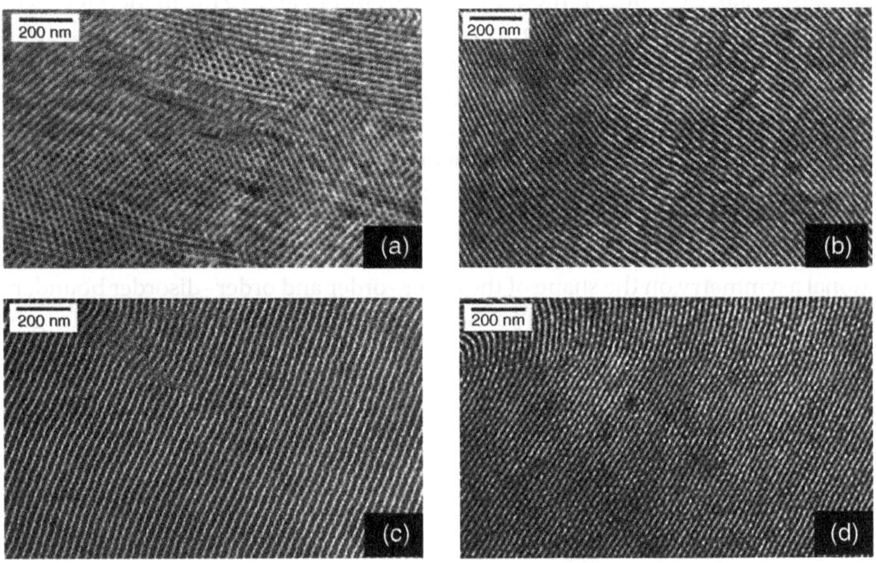

Figure 5.8 Bright-field TEM micrographs of (a) SF 27/12 (27 vol% PF) displaying hexagonally packed PF cylinders in a PS matrix; (b) SF 18/19 (47 vol% PF) showing an alternating lamellar morphology; (c) SF 9/19 (64 vol% PF) shows the regular lamellar morphology; and (d) perforated lamellar (PL) structure present in SF 9/19 at 200°C. In the identification of the samples, for example, SF 27/12 denotes the copolymer with a M_n of 27 000 for PS and 12 000 for PFS. (Source: Lammertink et al. [31]. Reproduced with permission of Wiley.)

but with a much larger Φ_{PFS} of 64%. Lamellar morphology was seen in this case also, but the PFS domain size was significantly larger (16.8 nm) than that of PS (9.4 nm). Another copolymer SF 9/19, which also contained Φ_{PFS} of 64%, but of lower molecular weight ($M_n = 28.5$), showed lamellar morphology (Figure 5.8c), with a larger PFS lamellar spacing of 13.7 nm and a PS spacing of 7.7 nm. This polymer showed a change in the G′ modulus at 200°C, which was attributed to an OOT. Upon annealing the sample at 200°C, a lamellar morphology was still seen (Figure 5.8d), but the PS layers were perforated with PF. There was, however, no long-range order in these perforations.

Kloninger et al. [34] synthesized PFS-*b*-PMMA diblock copolymers with narrow polydispersity and molecular weights ranging from 23 000 to 159 000 g mol^{-1}. The PFS volume fraction was varied from 0.25 to 0.48. Films were prepared using THF, methylene chloride and chloroform by evaporating the solvent slowly (as mentioned above) by keeping the sample at room temperature in a desiccator for a week. The trace solvent was then removed by slowly heating to above the T_g of PMMA to prevent bubble formation due to the evaporating solvent. They were then annealed at various temperatures ranging from 165 to 250°C and quenched in ice to prevent PFS crystallization. For a sample with $M_n = 68\,300$ and Φ_{PFS} of 0.25 (annealed at 190°C), sphere-like domains were found in the TEM images. While typically cubic bcc structure is seen in the case of the spherical morphology upon phase separation of the blocks, lack of well-ordered domains in this case was attributed to the high M_n and high melt viscosity of the PFS block which impede well-defined ordering of the domains. The authors noted that the glass transition and melting temperatures of PFS are 33 and 143°C, respectively, well below the annealing temperatures used. Increasing the volume fraction of PFS resulted in well-defined hexagonally packed cylinders for $\Phi_{PFS} = 0.25–0.37$ (annealed at 165°C). These cylindrical domains exhibited long range order of 3–4 μm in the case of samples of low M_n (38 200) and less so in the case of higher M_n. Lamellar morphology was seen with $\Phi_{PFS} = 0.40$ (film annealed at 165 or 190°C), although the coexistence of the cylindrical morphology was also found in the TEM images. The authors attributed this to the presence of about 1 w/w% of the PFS homopolymer in the samples during synthesis, which caused concentration fluctuations of the PFS domains in the melt. Upon increasing the annealing temperature to 220°C, interdiffusion of the homo- and copolymers led to uniformity of the PFS concentration in the sample and only lamella were observed. The fact that direct transitions from spherical to cylindrical to lamella with PFS volume fraction, without any intermediate morphology such as the gyroid, confirmed that the phases existed in the strong segregation limit due to the incompatibility of PFS and PMMA. The gyroid phase is seen in a very narrow composition range in the weak segregation region. As discussed earlier, the addition of homopolymer of one of the blocks leads to gyroid morphology. The presence of 1 w/w of the homopolymer PFS did not lead to such transition in this case.

In the aforementioned case of the dimethyl substitution at the silicon atom, the PFS would tend to crystallize and the authors quenched the annealed samples in ice to prevent crystallization. In the case of a block copolymer in which both blocks are amorphous, the self-assembly would be driven by the incompatibility of the blocks without any contribution from the crystallization of one of the blocks. To this end, Manners et al. [35] synthesized PFS-based diblock copolymers, with an asymmetric (ethyl/methyl) substitution at the silicon atom. They synthesized PS-*b*-poly(ferrocenylethylmethylsilane) (PFEMS) with M_n ranging from 38 700 to 149 000 and PFEMS ratio of 0.07–0.68, enabling tunability of the iron content of the BCP. Living anionic polymerization was used which led to very low polydispersity.

Bulk samples for physical studies were prepared by solvent casting in toluene and thermally annealed. Solvent annealing prior to thermal treatment was also used in some cases. Both TEM and SAXS were used to characterize the morphology. The amorphous nature of the samples was confirmed by DSC and X-ray diffraction. A glass transition at 15°C was recorded. With a volume fraction Φ of 10% of PFEMS, disordered spherical domains of PFEMS were observed in the PS matrix. The spheres were about 10 nm in size. Increasing Φ_{PFEMS} to 36% led to highly ordered, hexagonally packed cylindrical domains with a diameter of about 20 nm and a periodicity of 44 nm. With a similar Φ of both PS/PFEMS blocks (0.47/0.52), a highly ordered lamellar morphology with an average domain spacing of 88 nm was observed, consisting of alternating lamellae of PFEMS (46 nm) and PS (42 nm). In a later publication [36], with the intent of significantly increasing the iron content high-molecular-weight PS-*b*-PFEMS polymers (M_n = 50 000–250 000) were studied, with the PFEMS volume fraction ranging from 0.69 to 0.87. Lamellar, cylindrical, and spherical morphologies were observed. The lamella were asymmetric in the case of Φ_{PFEMS}, with the PFEMS domain thickness of 18 nm, and the PS domain size of 8 nm. A glass transition at 31°C was recorded for a BCP sample with M_n = 74 000 and Φ_{PFEMS} = 0.69.

Although the aforementioned ethyl/methyl substituted PFS were amorphous, the glass transition temperatures were close to or below room temperature, which makes it undesirable for most applications. To overcome this limitation, Gwyther and Manners [37] increased the size of one of the substituents to an isopropyl group to increase the steric bulk and thereby increase the T_g. A T_g of 60°C was recorded for a homopolymer of PFiPMS with M_n = 25 000. Diblock polymers polystyrene-*b*-poly(ferrocenylisopropylmethylsilane) were prepared with M_n = 42 000–51 000 and Φ_{PFiPMS} ranging from 0.26 to 0.77. For the diblock copolymer with M_n = 51 000 and Φ_{PFiPMS} = 0.26, TEM images showed hexagonally packed cylinders of PFiPMS in the PS matrix with an intercylinder distance of approximately 25 nm. This is in accord with the theoretical phase diagrams relating the volume fraction and morphology. Upon increasing the PFiPMS volume fraction to 0.38 (M_n = 45 000), a gyroid morphology (which usually occurs in a small section of the phase diagram between the lamellar and cylindrical morphologies) in the sample, along with regions of lamellar morphology, was seen. With a further increase of Φ_{PFiPMS} to 0.77 (M_n = 42 000), phase inversion occurred, with PS cylinders in the PFiPMS matrix. Thus, BCPs with a high volume fraction of the inorganic component, with a glass transition well above room temperature, were successfully prepared.

While most of the efforts to create gyroid morphology focussed on organic BCPs or organic/inorganic blends, Manners et al. [35–38] created such morphology using the organic/inorganic BCPs. As mentioned earlier, the PFEMS diblock copolymers showed gyroid phases along with lamellar domains. In a later publication [38], they synthesized PS-block-PFEMS (PS-*b*-PFEMS) diblock copolymers with a molecular weight of 64 000 and Φ_{PFEMS} = 0.39. The M_n of PS and PFEMS were 341 and 110, respectively. Melts were quenched in liquid nitrogen and thin microtomed sections of 50 nm thickness were analyzed with TEM. The images shown in Figure 5.9a and b show the double-wave

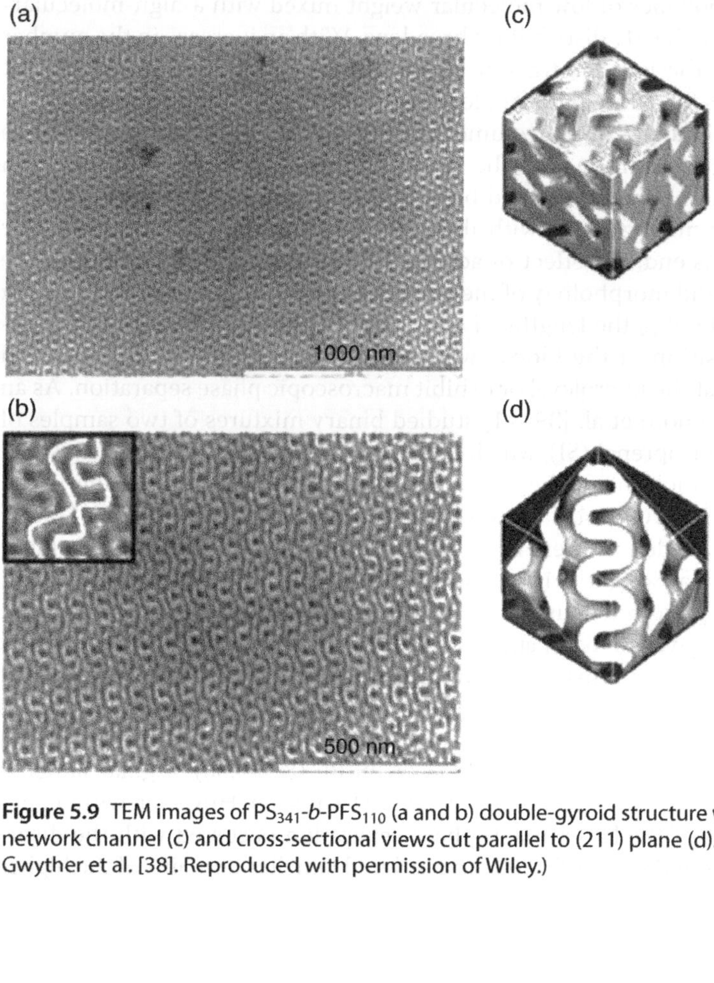

Figure 5.9 TEM images of PS_{341}-*b*-PFS_{110} (a and b) double-gyroid structure with void network channel (c) and cross-sectional views cut parallel to (211) plane (d). (Source: Gwyther et al. [38]. Reproduced with permission of Wiley.)

pattern that is characteristic of the gyroid morphology. This morphology was present exclusive of others, and over a long range. Figure 5.9c and d shows the $Ia\bar{3}d$ space group symmetry and a cross section along the [211] crystallographic plane, respectively.

5.4 Blends of Diblock Copolymers

The field of polymer blends has been well explored and is mature. The purpose of adding one polymer to another type is to modify the properties mutually to the desired end. Tuning the glass transition temperature and mechanical properties are the common pursuit with miscible or compatible polymer blends. In the case of the latter, while mixing at the molecular level might not be present, the components of the blends could effectively modify the properties mutually. When the components of the blends are of the same type of polymer, for example, if a polymer of low molecular weight mixed with a high-molecular-weight polymer, the M_w distribution broadens. With an increase in the number of chain ends, the low M_w fraction could act as a plasticizer. In the case of BCPs, the phase separated morphology depends on the volume fraction of the blocks. Subtle changes in the volume fraction, along with annealing, give rise to morphologies additional to the classical ones. Synthesis of blocks with fine-tuned volume fraction would be a tedious task, and this effort could be augmented by mixing BCPs with different volume fractions and molecular weights. To this end, the effect of adding one type of BCP to another on the self-assembly and morphology of the pair was studied. In a mixture of a BCP α with another BCP β, the lengths of the chains N_α and N_β as well as the fractional composition of the blocks would determine if the two BCPs would self-assemble at the microlevel or exhibit macroscopic phase separation. As an example, Hashimoto et al. [39–41] studied binary mixtures of two samples of poly(styrene-*b*-isoprene) (SI), which differed in total molecular weights as well as the weight fractions of the PS blocks. Seven copolymers, ranging in M_n from 8.5×10^{-3} to 1030×10^{-3} and weight fraction of PS from 35 to 69% were prepared. We will denote these BCPs as PS/PI: (M_n, Φ) where M_n is the number average molecular weight ($\times 10^{-3}$) and Φ is the weight fraction of the PS block. These authors preferred to use weight fraction rather than volume fraction, but these values are given in their paper. Five mixtures from the set of seven polymers were examined using films cast from toluene which is a nonselective solvent for the blocks. The polymer PS/PI: (8.5, 0.5) showed disordered, featureless morphology in the TEM due to its very low M_n. When mixed with PS/PI: (81, 0.63), at low concentrations (20%) of PS/PI: (8.5, 0.50), regular lamellar microdomains with a single identity spacing D was seen. However, with higher concentrations of the latter, macrophase separation occurred with the disordered domains of PS/PI: (8.5, 0.50) and ordered lamellar domains of the higher

M_n component. Note that the ratio of M_n of the two BCPs in this mixture is about 9.6. In the case of mixtures of PS/PI: (32, 0.35) and PS/PI: (81, 0.63), lamellar morphology with a single D was seen, which varied with the weight fraction of the higher M_n polymer. In this case, the ratio of the M_n of the two components in the mixture is 2.5. Interestingly, with 80 and 50 wt% of the lower M_n copolymer in the mixture, a well-ordered bicontinuous microdomain was seen. This was not identified with a gyroid morphology since this work was reported in 1993, before such a morphology was predicted and verified. Mixtures of PS/PI: (31, 0.48) and PS/PI: (164, 0.69) (with a ratio of 5.2 of the M_n of the two BCPs) also showed a single, composition-dependent lamellar identity period. In the case of mixtures of PS/PI: (1030, 0.52) and PS/PI: (81, 0.63) a single composition-dependent lamellar identity period was seen with 90 and 80% of the larger M_n polymer in the mixture. However, for other compositions lamellar but macrophase separated morphology was seen.

These observations led to the conclusions that the PS–PI BCPs of different molecular weights are completely miscible at the molecular level, showing single microdomain, composition-dependent lamellar period if the ratio of the molecular weights $N\alpha/N\beta$ of the two BCPs α and β was less than about 5. With this ratio greater than about 10, macroscopic phase separation occurred, with α and β forming lamellar domains, each with different periodicity, although there was partial miscibility. In the case of partial miscibility, the BCP with the higher M_n solubilized the lower M_n copolymer up to about 30 wt%. With the miscible blends, for example, PS/PI: (31, 0.48) and PS/PI: (164, 0.69), the lamellar spacing D varied almost linearly with the weight fraction of the higher M_n copolymer, from about 30 to 80 nm. If an effective M_n is defined as

$$M_n = x_\alpha M_{n,\alpha} + x_\beta M_{n,\beta} \tag{5.2}$$

where x_α and x_β are the mole fractions of BCPs α and β, respectively; it was found that with the miscible blends, the lamellar spacing scaled as

$$D = {M_n}^{2/3} \tag{5.3}$$

In another publication, Hashimoto et al. [41] studied the macrophase separated morphology of immiscible mixtures. As an example, Figure 5.10 shows the TEM images of the thin sections of the 50/50 wt% mixture of PS/PI: (31, 0.48) and PS/PI: (534, 0.49) in which two macrophase separated lamella with distinctly different D spacings. There is no mixing of the lamellae and a sharp interface is seen. In later studies, Court and Hashimoto [42, 43] blended a long asymmetric PS-*b*-PI with three short symmetric PS-*b*-PI and concluded that addition of a small amount of the short PS-*b*-PI to the asymmetric BCP enlarged the composition range of lamellar microdomain formation. Apart from the relative volume fractions of the blocks (PS and PI) the ratio of the degree of polymerizations of the asymmetric and symmetric diblock copolymers

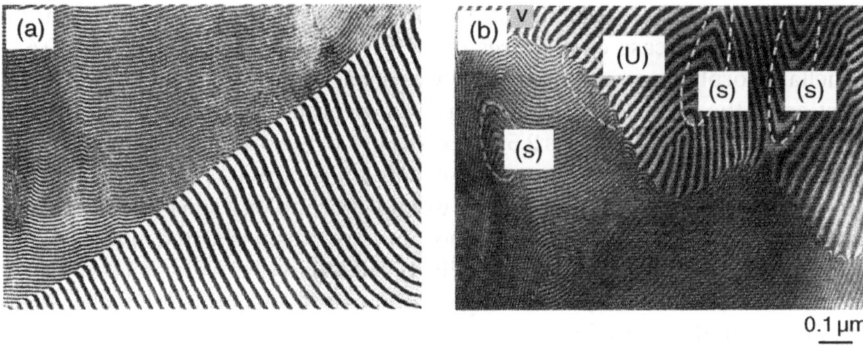

Figure 5.10 TEM images of thin sections of the 50/50 wt% mixtures of PS/PI: (31, 0.48) and PS/PI: (534, 0.49) stained with osmium tetraoxide. (a) Two different, macroscopically phase separated thick and thin lamellar domains with sharp commensuration along an inclination. (b) U- and V-shaped commensurations as well as grain boundary (S) are marked. (Source: Hashimoto et al. [41]. Reproduced with permission of American Chemical Society.)

also influences the phase diagram, and in shifting the phase boundaries between adjacent morphological domains. Blending two diblock copolymers of different lengths and volume fractions of the blocks provides the means to alter the thermodynamically stable morphology. The role of the short symmetric diblock copolymer in controlling the morphology in the blends is attributed [44–46] to the "co-surfactant effect."

Previously, we discussed the work of Bates group on the PE–PEE diblock copolymer, which showed bicontinuous gyroid morphology with a PE volume fraction of 0.40 and 0.43 upon annealing (see Table 5.2). To fine-tune the range of PE volume fractions over which such morphology occurs, they studied the blends of the BCPs with f_{PE} 0.37 and 0.46 using DMS, SANS, and TEM. The overall volume fraction of PE in the blends was denoted by $\langle f_{PE} \rangle$. In the temperature sweep of the DMS experiments, a single thermally induced OOT was observed at about 210, 180, and 150°C for $\langle f_{PE} \rangle = 0.39$, 0.395, and 0.40, respectively. Two OOT transitions were seen for $\langle f_{PE} \rangle = 0.405$, 0.410, and 0.415. While the gyroid phase was observed for the PE–PEE BCP for f_{PE} in the range of 0.37–0.43, this range was narrowed to $\langle f_{PE} \rangle$ from 0.390 to 0.420 in the case of blends. Thus, the blends were used to identify the narrow range of composition in which the bicontinuous morphology occurs.

5.5 Diblock/Homopolymer Blends

Blends of diblock copolymers and homopolymers have been studied to examine the effect of the relative molecular weight and volume fractions on the microphase morphology [47, 48]. This was also driven by the curiosity to identify the conditions for the formation of the gyroid structures. While the volume fraction could be tuned to attain a desired morphology, the additives also serve as sacrificial components. As will be discussed later in this chapter, after designing the required morphology (lamellar, cylindrical, or gyroid), one of the blocks could be selectively degraded to yield porous materials or nanochannels, which could be used as nanoreactors, functional membranes or similar devices. Hashimoto et al. [49] used a diblock copolymer of PS-*b*-PI with a total M_n of 88 000 with a PS weight fraction of 0.51. Homo-PS with a M_n of 7500 was added to create a blend with total PS volume fraction of 0.66. When cast from toluene, the films showed a bicontinuous (gyroid) phase separated morphology, as in Figure 5.11. Ozonolysis was used to selectively degrade the PI block to leave

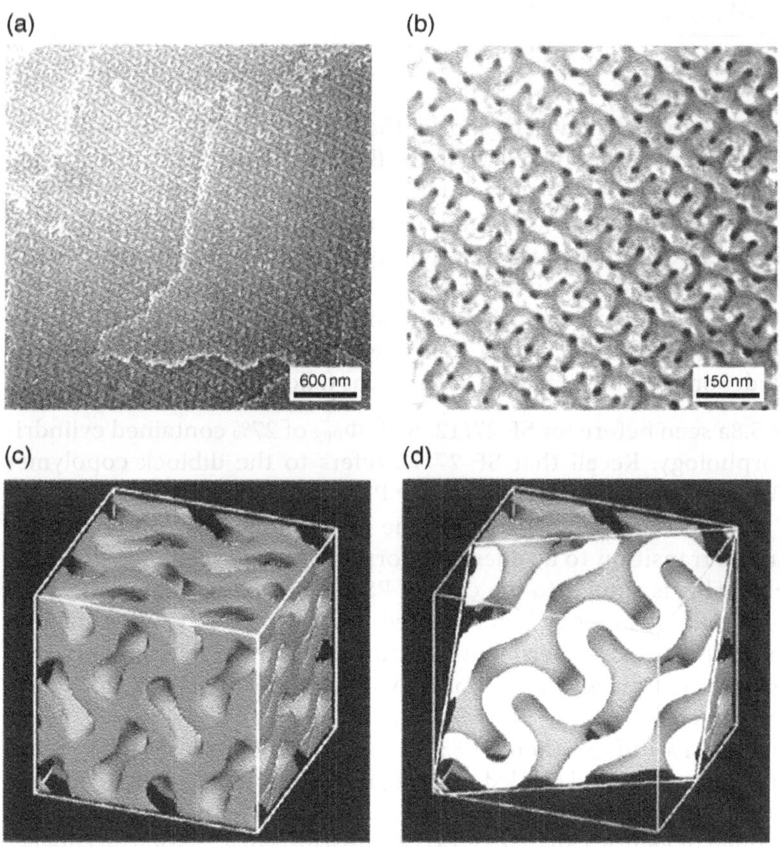

Figure 5.11 (a) and (b) A bicontinuous nanochannel in the matrix of PS with two different magnifications is seen in SEM images. Computer graphics of a double-gyroid network are shown in (c) and (d). A 3D view (c) and a 2D intersection cut along the (211) direction (d) are shown. (c) A solid model in which only the matrix phase, corresponding to the PS matrix is shown. In (d), the bright domain corresponds to the PS matrix and the gray and dark phases correspond to the degraded PI phase. (Source: Hashimoto et al. [49]. Reproduced with permission of American Chemical Society.)

nanochannels in the PS matrix. The channels were plated with nickel, using a nonelectrolytic method. It was suggested that such metal coated nanochannels in a glassy matrix could serve as membrane reactors.

In the section on diblock copolymer blends, it was noted that homogeneous microphase separation occurred only if the ratio N_α/N_β of the two BCPs α and β was less than about 5. In the case of diblock/homopolymer blends, the relative molecular weight of the homopolymer with respect to that of the corresponding block influences the morphology. Following Vancso et al. [31], let us define the ratio of molecular weights of the A block of an A–B BCP and the corresponding homopolymer of A as

$$\alpha = \frac{M_{\mathrm{A,\ homo}}}{M_{\mathrm{A,\ diblock}}} \tag{5.4}$$

If $\alpha < 0.5$, the homopolymer will be shorter than the corresponding block in the diblock copolymer, will be solubilized in the BCP, and the chains will distribute themselves uniformly in the domains [50–53]. Otherwise, the addition of the homopolymer could cause disorder or macrophase separation. Following the studies described earlier on the morphology of PS-*b*-PFS diblock copolymers, Vancso et al. [31] blended homopolymers of PS (h-PS) or PFS (h-PF) with the BCPs. Two h-PS samples with $M_n = 4000$ and 7600 and three h-PF samples with $M_n = 6500$, 11 400, and 24 200 were used, and a total of 30 blends were investigated. Some of the highlights are given as follows:

1) Figure 5.8a seen before for SF 27/12, with Φ_{PFS} of 27% contained cylindrical morphology. Recall that SF 27/12 refers to the diblock copolymer with $M_n = 27\,000$ for PS and 12 000 for PFS. With the addition of 25% of hPS-7.6 (homo-PS with $M_n = 7600$), the PFS volume fraction reduces to 20%, and a transition to a spherical morphology was seen (see Ref. [31], figure 8). This is similar to the case of PS-*b*-PI listed in Table 5.1, where the spherical morphology was recorded with a PS or PI volume fraction of about 20%. SAXS experiments confirmed that the PFS spheres seen belonged to BCC packed morphology with a radius of 9.5 nm, in a cell of 32.9 nm dimension.
2) Addition of 11% hPF-6.5 (homo PFS with $M_n = 6500$) increases the PFS volume fraction to 34% and this led to hexagonally packed PFS cylinders in the PS matrix (see Ref. [31], figure 10).
3) It was noted in Figure 5.8b that SF 18/19 with a Φ_{PFS} of 47% showed lamellar morphology. No change in the morphology was found when the homopolymer hPF was added up to 26%, leading to a total Φ_{PFS} of 60%. However, adding just a 9% hPS-7.6 (changing Φ_{PFS} to 43%) led to a biphasic morphology with PFS cylinders and lamellae (see Ref. [31], figure 11). No gyroid structure was observed in this blend.

4) As seen in Figure 5.8c, the neat diblock copolymer SF 9/19 ($\Phi_{PFS} = 64\%$) showed lamellar morphology. Upon adding 6% of hPF-11.4 to this copolymer, a PFS-rich double-gyroid morphology was observed (see Ref. [31], figure 13). Similar gyroid morphology was observed with SF 9/19 + 12% hPF-11.4 as well as SF12/25 + 6% hPF-24.2 blends.

Based on these blend studies in which the volume fractions of PS and PFS as well as the number of monomer units were varied, Vancso et al. [31] concluded that (i) the phase diagram was asymmetric with respect to the volume fraction $\Phi_{PFS} = 50\%$, (ii) the order–order phase transitions in the phase diagram were almost vertical lines for $\Phi_{PFS} > 50\%$, and (iii) OOTs and their locations with respect to PFS volume fractions are located at approximately BCC: 0.22; hexagonal: 0.43, lamellae: 0.65, double gyroid, hexagonal: 0.68, and BCC: 0.79.

The work of Kloninger et al. [34] on the diblock copolymer of ferrocenyl dimethylsilane-*b*-methyl methacrylate was discussed earlier. With some samples of these PFS-*b*-PMMA diblock copolymers, with a PFS volume fraction (Φ_{PFS}) of 0.4, hexagonally packed cylinders and lamellae were found to coexist in films cast from methylene chloride solutions, even with annealing at 165°C. However, lamellar phase was exclusively seen when the annealing temperature was increased to 220°C. This behavior was attributed to the presence of a small amount of PFS homopolymer. It was concluded that with high-molecular-weight diblock copolymers, the cylindrical morphology changes to lamellar with an increase in Φ_{PFS}. To explore [54] the morphological features further with Φ_{PFS} of about 0.40, homopolymers of PFS or PMMA were blended with the BCPs of various molecular weights, to achieve $\Phi_{PFS} \approx 0.4$. It was concluded that with annealing at the temperatures listed before, the transition from cylindrical to lamellar morphology occurred at $\Phi_{PFS} = 0.4$. Using a diblock copolymer of lower molecular weight and upon adding PMMA homopolymer to meet the condition $\Phi_{PFS} = 0.4$, the gyroid morphology was achieved with a high annealing of 190°C. The appearance of the gyroid phase in this system was interpreted as due to such a high annealing temperature effectively reducing the incompatibility between the PFS and PMMA blocks, lowering the χ and hence χN. This moves the system to the weak segregation domain.

5.6 BCP/Small-Molecular Supramolecular Association

Supramolecular assembly of BCPs with small molecules via hydrogen bonding has been reported in a number of publications [55–66]. A BCP with poly(4-vinyl pyridine) (P4VP) as one of the blocks is often used, so that the nitrogen atom would form hydrogen bond with the small-molecule additive. As will be discussed in Chapter 8, complexation of the small molecule with one of the blocks would lead to an expansion of the coil. The complexed block would appear as a bottlebrush. Nandan et al. [60] used PS-*b*-P4VP with 2-(4′-hydroxybenzeneazo)benzoic acid (HABA) with different molar ratios (X) of HABA with respect to the 4VP monomer unit. In the bulk samples, with X up to 0.25, the VP block formed bcc spheres in the PS matrix; with a further increase in X, hexagonally close packed cylinders were formed. In thin films which were prepared on silicon wafers by dip-coating, the orientation of the cylindrical domains was dependent on the solvent annealing procedure. Annealing in 1,4-dioxane, which is a selective solvent for the PS block, led to cylindrical domains oriented normal to the substrate (the atomic force microscope (AFM) images were recorded after washing off HABA). The films which were pre-annealed in 1,4-dioxane were then annealed with chloroform for different periods of time. After 14 min of annealing, the orientation changed and the cylinders became parallel to the surface. Extended annealing led to perpendicular orientation again, but with poor long-range order. The AFM images of the perpendicular and parallel orientation and a schematic illustration of the switching of the orientation with solvent annealing are shown in Figures 5.12 and 5.13. The transformations were discussed in terms of the solvent selectivity for the domains and HABA. When annealing with chloroform, the solvent enters both PS and P4VP/HABA domains resulting in increased mobility of the chains. With this process creating sufficient mobility in the perpendicular metastable cylindrical structures, undulations developed as in Figure 5.13b. This causes the development of ellipsoids, which then coalesce with the ellipsoids a long the neighboring cylinders as in Figure 5.13c–e. The coalescence step was verified by AFM. Further coalescence of the ellipsoid lead to fully developed cylinders which were oriented parallel to the surface (Figure 5.13f).

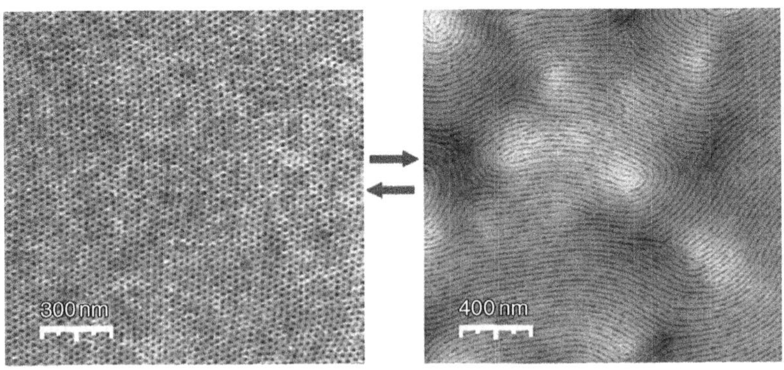

Figure 5.12 AFM images of perpendicular and parallel orientation of cylindrical morphology of PS-*b*-P4VP/HABA supramolecular assembly. (Source: Nandan et al. [60]. Reproduced with permission of American Chemical Society.)

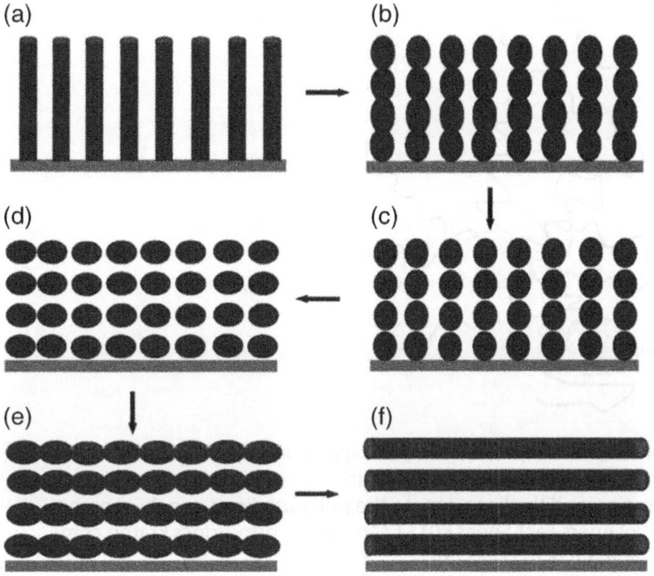

Figure 5.13 A schematic illustration of switching of the orientation of the cylindrical domains of PS-*b*-P4VP/HABA with chloroform annealing (see text). (Source: Nandan et al. [60]. Reproduced with permission of American Chemical Society.)

5.7 Triblock Copolymers

The simplest form of triblock copolymer is of the A–B–A type. The most common and well-studied polymers are styrene–butadiene–styrene (SBS) and styrene–isoprene–styrene (SIS) copolymers, which were pioneered by Shell Chemical Company. These were named "Kraton" after the Greek god Kratos, the one for strength and the brother of Nike. The very first application of Kraton was in footwear industry as a replacement for PVC soles. Other applications include adhesives, as modifiers for asphalt paving, etc. These polymers are now marketed by Kraton Performance Polymers, Inc. (Kraton.com), in various types, depending on the composition of the copolymer and end-use. Some of these types are Kraton D (SBS), Kraton D SIS/SIBS, and Kraton G (hydrogenated midblock of styrene–ethylene/butylene–styrene (SEBS) or styrene–ethylene/propylene–styrene (SEPS)) and Kraton FG which are SEBS polymers with maleic anhydride (MA, 1.0–1.7 wt%) grafted onto the rubber midblock.

The Kraton-type SBS or SIS triblock copolymers consist of about 30 wt% styrene. They behave as a strong rubber at room temperature, can be processed at high temperatures and show recoverable deformations (as with vulcanized rubber), but without any *chemical* cross-linking. Such a behavior

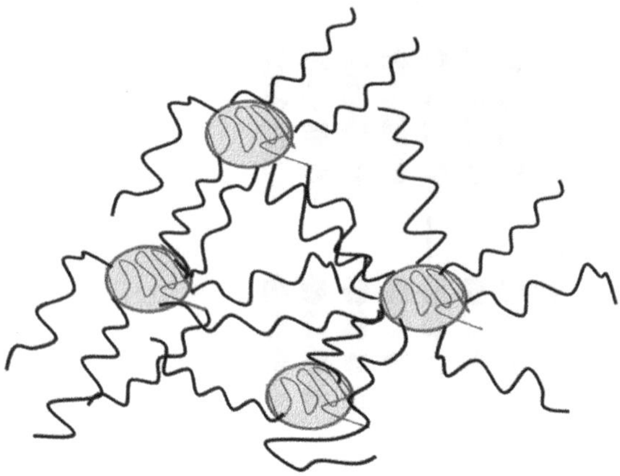

Figure 5.14 A model for the phase separated morphology for polystyrene-*b*-polybutadiene-*b*-polystyrene or polystyrene-*b*-isoprene-*b*-polystyrene triblock copolymers. The gray ellipses denote polystyrene domains, and the lines denote the polybutadiene or polyisoprene chains. See also Kraton Polymers Fact Sheet K0151 (Americas). http://docs.kraton.com/kraton/attachments/downloads/82021AM.pdf

is accounted for by the model [67] shown in Figure 5.14. With the phase separation of the blocks, the PS chains self-assemble into glassy domains, which act both as a reinforcing filler and as *physical* cross-links between the rubber domains. Although the commercial Kraton polymers typically have 30 wt%, PS, SBS, and SIS polymers with differing styrene compositions and molecular weights have been synthesized and their properties characterized. For example, Uchida et al. [68] prepared SIS polymers with styrene content ranging from 9.5 to 72 wt%. The variation of the morphology of these A–B–A triblock copolymers with the composition of the blocks follows the behavior of the A–B diblock copolymers shown in Figure 5.2 with five characteristic domain structures. With an increase in the PS weight fraction, the morphology changed from PS spheres in PI matrix, PS cylinders, PS–PI lamella, PI cylinders in PS matrix, and PI spheres. However, the solvents used for casting the films result in significant changes in morphology depending on whether they are nonselective or selective for one of the blocks. Figure 5.15 shows the TEM images of S–I–S copolymer films (with 72 wt% PS) cast from four different solvents. Methyl ethyl ketone (MEK) is a good solvent for PS blocks but is a poor solvent for PI. Figure 5.15a shows spherical domains of PI in the PS matrix. Toluene is a good solvent for both PS and PI, and in this case, the PI domains become elongated to rod-like, interconnected shape (Figure 5.15b). Similar is the case with the films cast from carbon tetrachloride. With cyclohexane (which is a poor solvent for PS), PI forms the matrix, with dispersed PS domains as in Figure 5.15d.

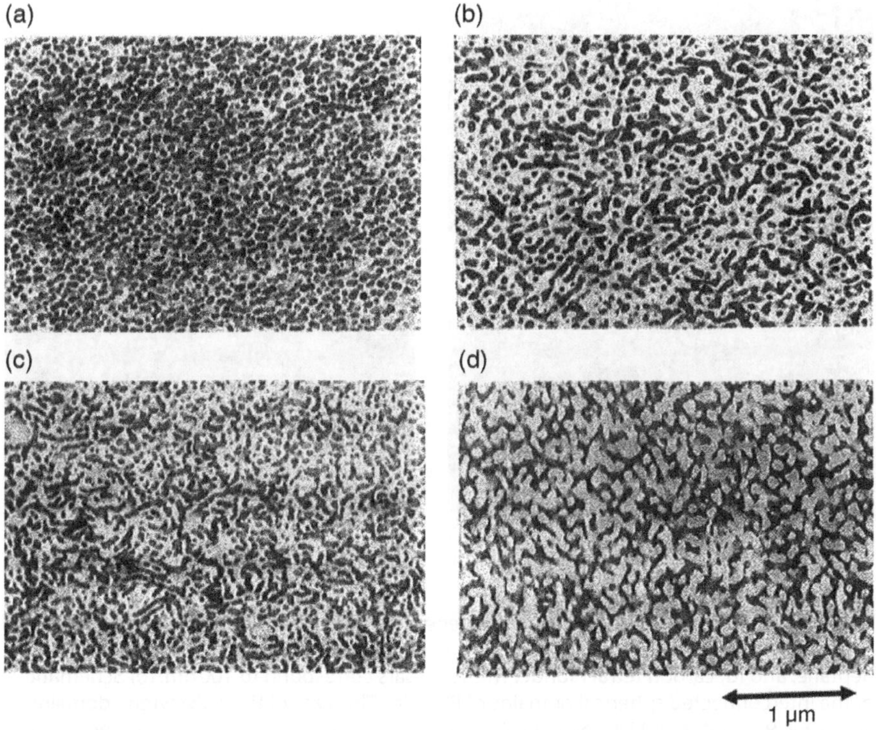

Figure 5.15 TEM images of ultrathin sections of films cast from 5% solutions of SIS triblock copolymer with 72 wt% polystyrene (total $M_n = 70.2 \times 10^{-4}$). (a) MEK, (b) toluene, (c) carbon tetrachloride, and (d) cyclohexane. (Source: Uchida et al. [68]. Reproduced with permission of Wiley.)

The SIS triblock copolymer discussed before consisted of 72 wt% PS. A PS/rubber triblock copolymer with smaller PS content was studied by Aggarwal et al. [69]. They studied the solvent effect on the morphology and mechanical properties of commercial Kraton 101, a PS-*b*-PBt–PS triblock copolymer with about 28 wt% PS. Figure 5.16 shows the TEM images of films cast from three different solvent/mixtures.

With THF/MEK: 90/10 mixture as the casting solvent, both are good solvents for the PS blocks; THF is a good solvent for the PB block, but MEK is a poor solvent. In this case, the PS forms spherical domains in the PB matrix (Figure 5.16a), but these domains tend to be interconnected, as shown schematically in Figure 5.16b and form a long-range network. Due to the network of the PS spheres, films cast from this solvent mixture exhibited high modulus. The films cast from benzene/heptane: 90/10 solution (heptane is a good solvent for PB) show smaller spherical domains of PS (9.4 ± 2.4 nm), without interconnections. Since PB is the continuous phase in these films, the modulus

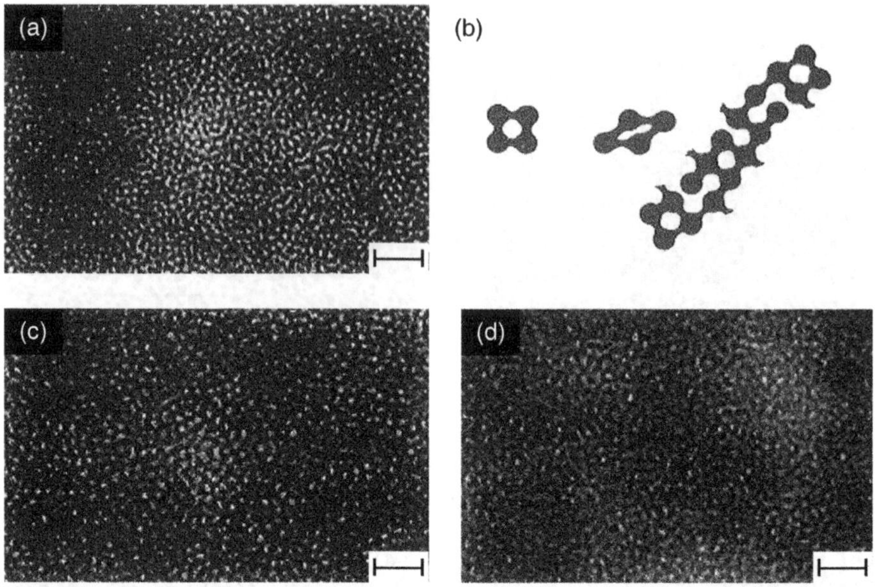

Figure 5.16 TEM images of films of polystyrene-*b*-polybutadiene-*b*-polystyrene triblock copolymer (with 28 wt% PS content) cast from (a) 90/10 THF/MEK, (c) 90/10 benzene/heptane, and (d) carbon tetrachloride. The scale bars correspond to 100 nm. (b) Schematic of the interconnected spherical domains of PS in (a). The sizes of the polystyrene domains are 12 ± 2, 9.4 ± 2.4, and 11 ± 2.5 nm in (a), (c), and (d), respectively. (Source: Beecher et al. [69]. Reproduced with permission of Wiley.)

was lower than the films cast from THF/MEK mixture. In the case of films cast from carbon tetrachloride, no discrete phase boundary was seen and the PS was dispersed in the PB matrix giving rise to a "mottled" appearance of the PB matrix.

As discussed earlier, the morphologies of diblock copolymers depend on N, N_A, N_B, and χ_{AB}, and the main architectures are lamellar, cylindrical, spherical, and gyroid. In the case of triblock polymers of the type ABC, the morphology would depend on many more variables which include N, N_A, N_B, N_C, χ_{AB}, χ_{BC}, and χ_{AC} that would lead to a rich spectrum of self-assembly. In addition, the sequence of the polymer blocks could be ABC, ACB, and BAC. Zheng and Wang [70] and Stadler et al. [71] presented theoretical treatments of the phase diagrams of linear ABC triblock copolymers in the strong segregation limit. Figure 5.17 shows the schematics of the morphologies from the theoretical work of Zheng and Wang (the gyroid morphology (*l*) of a triblock copolymer is also shown in the figure; see Figure 5.28). According to Helfand and Sapse [72], the interfacial tension σ in the strong segregation limit is approximately equal to $(\chi)^{1/2}$. Strong incompatibility and

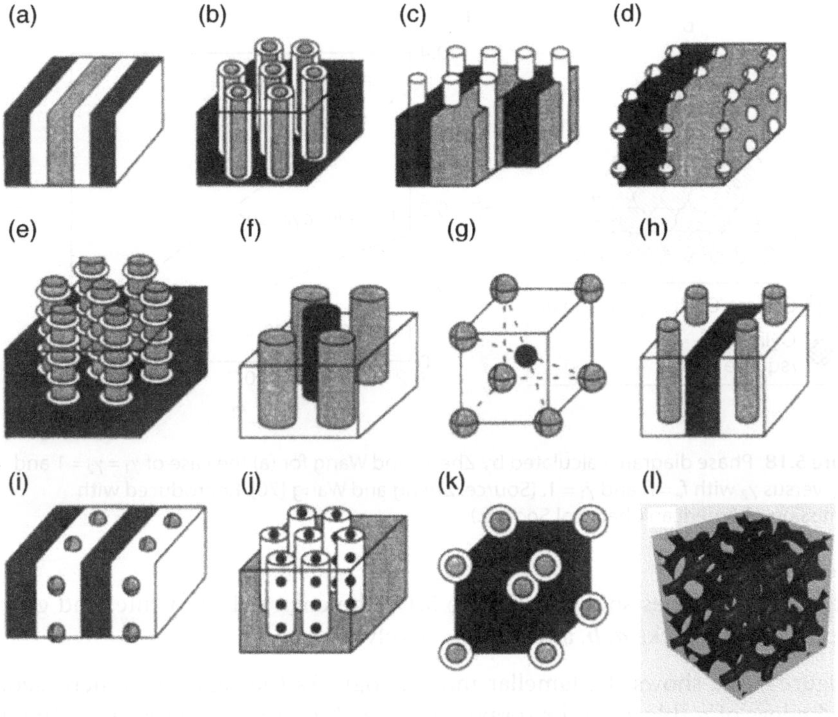

Figure 5.17 Morphologies of linear ABC triblock copolymers (denoted by black (A block), white (B block), and gray (C block), respectively). (Source: Zheng and Wang [70]. Reproduced with permission of American Chemical Society.) The gyroid morphology (l) is also shown (see Figure 5.28). (Source: Vignolini et al. [10]. Reproduced with permission of Wiley.) See text for the descriptions.

hence large positive χ leads to large interfacial tension. With the interfacial tensions σ_{ac}, σ_{ab}, and σ_{bc} between the blocks, the relative strengths of the interfacial tensions were defined as $\gamma_1 = \sigma_{bc}/\sigma_{ab}$ and $\gamma_2 = \sigma_{ac}/\sigma_{ab}$. Zheng and Wang generically defined the blocks as ***a***, ***b***, and ***c*** such that ***b*** is the middle block. In the actual polymer, the middle block may be different depending on the sequence ABC, BCA, or BAC. Six different cases were considered, in which (i) $\gamma_1 = 1$, $\gamma_2 < 1$; (ii) $\gamma_1 = 1$, $\gamma_2 = 1$; (iii) $\gamma_1 = 1$, $\gamma_2 > 1$; (iv) $\gamma_1 > 1$, $\gamma_2 < 1$; (v) $\gamma_1 > \gamma_2 > 1$; and (vi) $\gamma_2 > \gamma_1 > 1$. The volume fractions of ***a***, ***b***, and ***c*** were defined as f_a, f_b, and f_c, respectively.

Zheng and Wang [70] calculated triangular phase diagrams in terms of compositions f of ***a***, ***b***, and ***c*** for several combinations of γ_1 and γ_2. Figure 5.18a is the case with $\gamma_1 = \gamma_2 = 1$, in which the interaction parameter and hence the interfacial tension is taken to be equal for the blocks. Each step in the triangle corresponds to 0.1 in the fractions of blocks ***a***, ***b***, and ***c***. These relate as follows

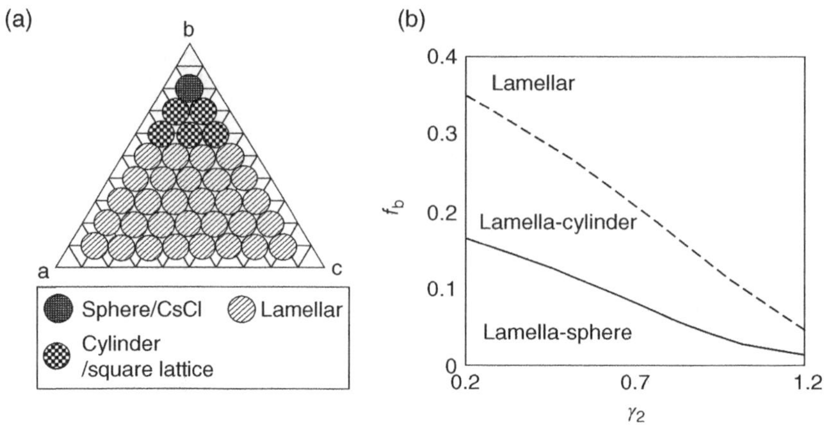

Figure 5.18 Phase diagram calculated by Zheng and Wang for (a) the case of $\gamma_1 = \gamma_2 = 1$ and (b) f_b versus γ_2 with $f_a = f_c$ and $\gamma_1 = 1$. (Source: Zheng and Wang [70]. Reproduced with permission of American Chemical Society.)

to the morphologies shown in Figure 5.17 (the colors black, white, and gray correspond to blocks ***a***, ***b***, and ***c***, respectively):

- Figure 5.17a shows the lamellar morphology. As the value of f_b increased, cylinders of ***a*** (black) and ***c*** (gray) are formed in the square matrix of the ***b*** block (white) (Figure 5.17f). With a further increase in f_b, the ***a*** and ***c*** blocks form spheres in the ***b*** matrix, with a structure similar to that of cesium chloride (Figure 5.17g).
- At the edge of the triangular phase diagram Figure 5.18a, where f_b is small (e.g., $f_b = 0.1$) and $f_a \approx f_c$ lamellar-cylinder and lamellar-sphere morphologies have free energies similar to that of the lamellar. In these cases, the block ***b*** (white) forms cylinders (Figure 5.17c) or spheres (Figure 5.17d), and these are located at the interface of the lamellae of the ***a*** (black) and ***c*** (gray) blocks.
- At the ***ab*** edge of the phase diagram Figure 5.18a, if f_c is small and $f_a \approx f_b$, the blocks ***a*** and ***b*** form lamellae and the minority block ***c*** forms cylinders (gray) inside the lamellae of the ***b*** block (Figure 5.17h). This was called lamellar-cylinder II. Or, the block ***c*** forms hexagonally packed spherical domains inside the ***b*** block (Figure 5.17i). This was called lamellar-sphere II. Likewise, at the ***bc*** edge of the phase diagram, if f_a is small and $f_b \approx f_c$, the block ***a*** forms cylinders or spheres inside the ***b*** block. In the case of the lamellar-cylinder (Figure 5.17c) morphology (or lamellar-sphere, Figure 5.17d) cylinders (or spheres) of the center block are located at the interface of the lamellae of the two outer blocks. On the other hand, in the lamellar-cylinder II (or lamellar-sphere II) morphology, the cylinders (or spheres) of the minority outer block is located inside of the lamellae of the center block (Figure 5.17h and i).

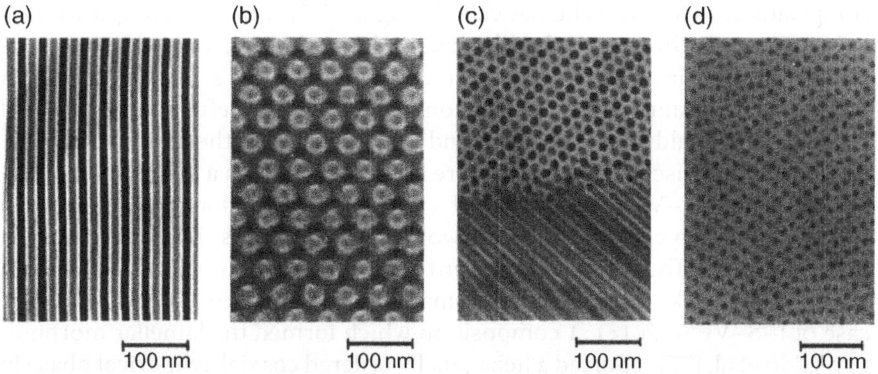

Figure 5.19 TEM images of films of poly(isoprene-*b*-styrene-*b*-2-vinylpyridine) (poly I-S-VP) with the following volume fractions: (a) 0.36/**0.31**/0.33; (b) 0.26/**0.48**/0.26; (c) 0.14/**0.68**/0.18; and (d) 0.11/**0.80**/0.09. (Source: Mogi et al. [73]. Reproduced with permission of American Chemical Society.)

- At the ***a*** corner of the phase diagram where f_b and f_c are small (or the ***c*** corner where f_a and f_b are small), coaxial morphologies of the minor phases were found to be stable. Corresponding to the ***a*** corner, as seen in Figure 5.17b, the minority ***c*** phase (gray) forms cylinders and the ***b*** phase (white) forms coaxial cylinders enveloping the ***c*** phase in the ***a*** (black) matrix. Similarly, in Figure 5.17j, spheres of ***c*** block are embedded in the cylinders of the ***b*** block, with ***a*** forming the matrix.

Figure 5.18b shows the phase diagram for f_b versus γ_2 ($=\sigma_{ac}/\sigma_{ab}$) with $f_a = f_c$ and $\gamma_1 = 1$.

The aforementioned morphologies of triblock copolymers have been observed experimentally, depending on the type of polymers constituting the blocks and their composition. Mogi et al. [73] studied the morphologies of poly(isoprene-*b*-styrene-*b*-2-vinylpyridine) (Poly I-S-VP) by varying the volume fraction of PS (middle block) from 0.3 to 0.8, while the volume fractions of the PI and poly(VP) were kept equal. Films were prepared from solutions with THF, which is a good solvent for the three blocks and were stained with OsO_4. The TEM images of the films are shown in Figure 5.19, for I–**S**–VP volume fractions as follows (the PS fraction is in bold): (a) 0.36/**0.31**/0.33, (b) 0.26/**0.48**/0.26, (c) 0.14/**0.68**/0.18, and (d) 0.11/**0.80**/0.09. With OsO_4 staining, the PI block appears black in these images, PS white, and P2VP gray in these images.

With about equal volume fraction of the blocks, Figure 5.19a shows lamellar morphology. It is seen that the repeat sequence consists of four layers in the sequence: PI (black), PS (white), P2VP (gray), and PS (white). This is the four-layered lamellar morphology of the three phase polymer with 1:1:1

composition, as shown schematically in Figure 5.17a. With a composition of 1 : 2 : 1, Figure 5.19b shows a double-diamond morphology, formed by two end-block polymers in a matrix of the middle block. Mogi et al. called this an "ordered tricontinuous double-diamond structure." Careful examination of Figure 5.19b would show the black and gray domains in the white PS matrix. Mogi et al. [74] discussed this structure further in detail in a subsequent publication. With I–**S**–VP composition of 1 : 4 : 1, cylindrical domains are seen. In this case, the two end-blocks form two types of cylinders (black and gray) in the PS matrix. With a further increase in the PS volume fraction (0.11/**0.80**/0.09), the two end-blocks form spherical domains in the PS matrix. In contrast to the case of I–**S**–VP with 1 : 1 : 1 composition which formed the lamellar morphology, Gido et al. [75] reported a hexagonally ordered coaxial cylindrical phase in the PS matrix in a triblock polymer with the sequence VP–I–S.

Auschra and Stadler [76] studied the triblock copolymer system consisting of PS and PMMA as end-blocks and poly(ethylene-*co*-butylene) as the middle block (P(S-*b*-EB-*b*-MMA)). The fraction of the EB block was varied from 38 to 6 wt%, while keeping the PS and PMMA blocks to be of equal fraction. Films were cast from chloroform solutions and thin sections were stained with RuO_4 vapor. With weight fractions of 0.24 : 0.38 : 0.38 for PS : EB : MMA, lamellar morphology was seen as shown schematically in Figure 5.20a. This is similar to the four-layered morphology observed by Mogi et al. [73, 74] for the poly I–S–VP triblock polymer. Figure 5.20a shows the lamellae with the sequence PS, EB, PMMA, EB. The total long period was determined to be 90–95 nm, with the individual dimensions of 23 nm (PS), 35 nm (PMMA), and 2 × 18 nm (EB).

When the weight fraction of the middle (EB) block was reduced to 0.17, lamellar morphology was seen for the PS and PMMA blocks as in Figure 5.20b, but the EB block formed cylinders at the interface of the PS/PMMA lamellae. Note the sequence of dots seen in Figure 5.20b. The schematic of the cylinder in lamellae morphology is shown in Figure 5.20c which is similar to that shown in Zheng and Wang model Figure 5.17c. With a further reduction of the EB ratio to 6%, cylindrical morphology of PS in PMMA matrix was seen although the equal weight fraction of PS and PMMA would be expected to lead to a lamellar morphology. In addition, the EB fraction was found to form rings around the PS cylinders, as shown schematically in Figure 5.20d. Auschra and Stadler [76] rationalized these morphologies on the basis of the strengths of the interfacial tensions between the blocks. They took these values to be $\gamma = 5.0$ and 9.5 dyn/cm for PS/EB and EB/PMMA, respectively, but only $\gamma = 1.5$ dyn/cm for the PS/PMMA interface. Due to the large interfacial tension with EB, the system minimized the overall interface free energy by reducing the surface of the EB domains. Their calculations on the system with 0.17 EB fraction showed that the morphology with cylinders along the lamellar boundaries had lower overall interfacial free energy than the four-layered lamellar morphology. In a

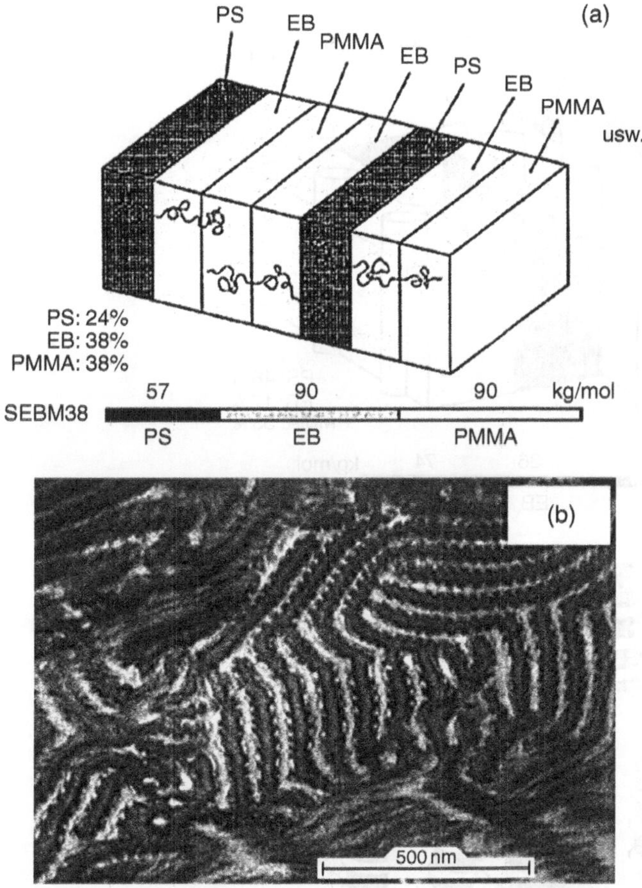

Figure 5.20 Morphologies of P(S-*b*-EB-*b*-MMA) triblock copolymers with varying weight fractions of the EB block (a), (c), and (d) show the schematics based on the TEM images in Ref. [76]. (b) TEM image with EB fraction of 0.17 (see text). (Source: Auschra and Stadler [76]. Reproduced with permission of American Chemical Society.)

later publication [71], these authors used PBt as the middle block, and found that with a weight fraction of 6%, spheres of PB formed along the interface between the PS and PMMA lamellae (which they called "ball at the wall" morphology). This is similar to the schematic shown in Figure 5.17d.

Stadler's group published a detailed study [77–79] of the cylindrical morphologies in the case of asymmetric ABC triblock copolymers. When the central block B is the major component, the end-blocks would form cylinders in the matrix of B. If one of the end blocks, let us say, C is the major component forming the matrix, and Φ_A and Φ_B are the volume fractions of the A and B blocks, the morphology would depend on the sum of the volume fractions of

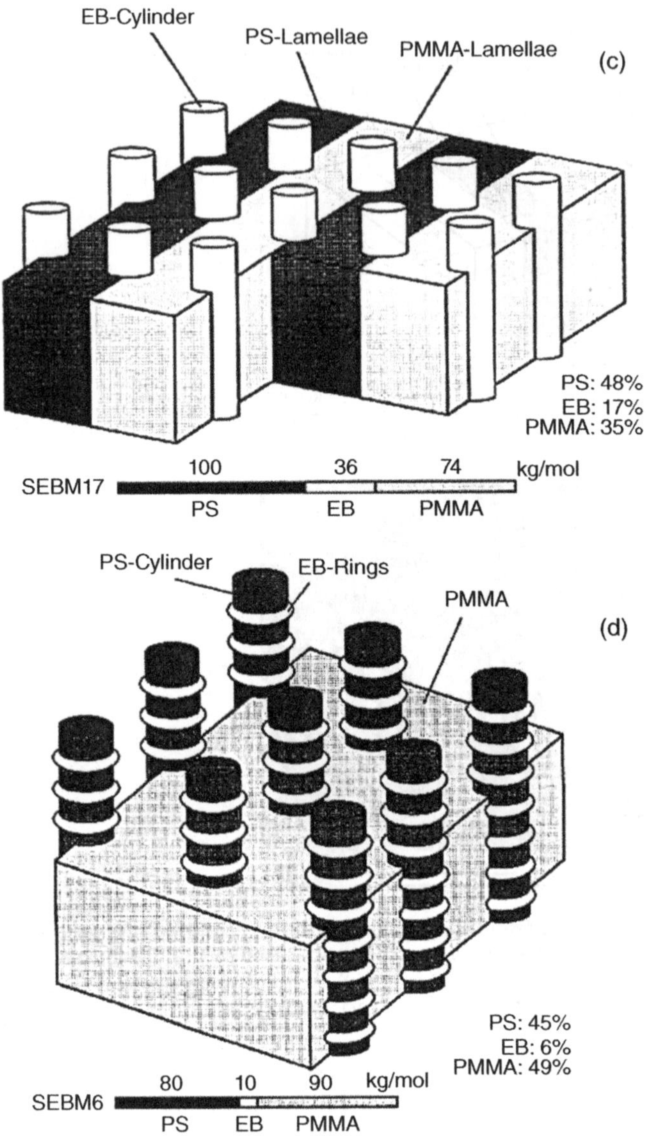

Figure 5.20 (Continued)

A and B $[\Phi_A + \Phi_B]$ as well as the ratio Φ_A/Φ_B. These authors used triblock copolymers of PS–PBt-PMMA]as well as PS–PEB–PMMA with various volume fractions of the blocks as well the total molecular weight. They considered the cases in which (a) $1.4 < [\Phi\ (PB)/\Phi\ (PMMA)] < 2.5$, (b) $0.1 < [\Phi\ (PB)/\Phi$

(PMMA)] < 1.0, and (c) 0.3 < [Φ (PB)/Φ (PS)] < 0.6. We will briefly describe the morphologies that Stadler et al. [79] found for the (chloroform) solution cast and annealed (at 185°C) for these copolymer compositions.

Case 1: Φ (PB)/Φ (PMMA) = 2.5 The polymer $S_{72}B_{18}M_{10}^{(120)}$ was used. (Note: (120) here is not a reference number. It is a sample code. See the following text). This sample code was used where weight percent of the PS, PB, and PMMA blocks were 72, 18, and 10, respectively. The total molecular weight was 120 000. In terms of volume fractions, Φ (PS) = 0.72; Φ (PB) = 0.2; Φ (PMMA) = 0.08. Hence, [Φ (PB)/Φ (PMMA)] = 2.5 and [Φ (PB) + Φ (PMMA)] = 0.28. In the TEM images, PB which was stained with OsO_4, appeared as black rings surrounding the clear cylindrical PMMA cores suggesting a core/shell cylinder-in-cylinder (CiC) morphology. The core was PMMA and the shell was PB. A schematic of this morphology is shown in Figure 5.21a and b. This is similar to the morphology shown in Figure 5.17b.

Case 2: Φ (PB)/Φ (PMMA) = 1.4 The polymer $S_{64}B_{19}M_{17}^{(94)}$ was used in this case. The subscripts denote the weight percent of the blocks and the total M_w is shown in the superscript. The volume fractions are Φ (PS) = 0.64, Φ

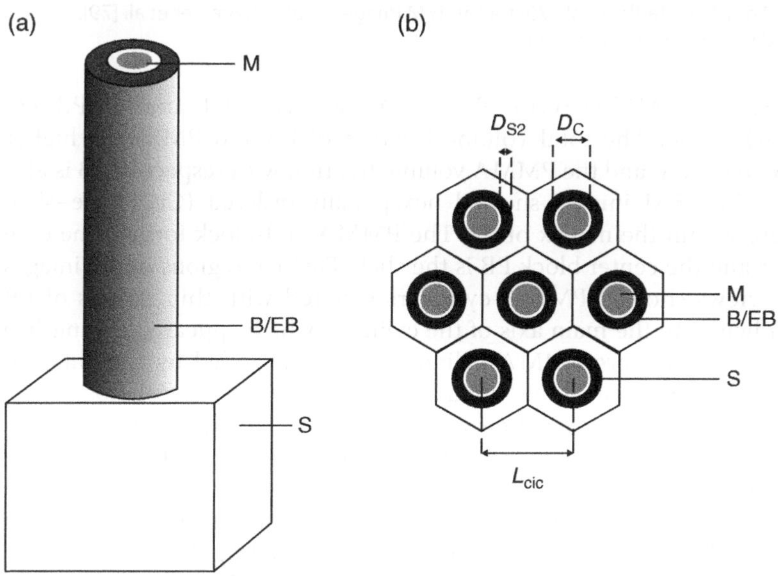

Figure 5.21 Schematics of the cylinder-in-cylinder morphology of $S_{72}B_{18}M_{10}^{(120)}$ triblock copolymer based on TEM observations. The projection is shown in (b). S, B/EB, and M in (a) and (b) denote PS, PB/PEB, and PMMA blocks, respectively. In the projection (b), the dimensions were L_{cic} = about 40 nm with no long-range order, D_c = 10.8 ± 1.9 nm, and D_s = 5.4 ± 1.9 nm. See Stadler et al. [79] for the TEM images. (Source: Breiner et al. [79]. Reproduced with permission of Wiley.)

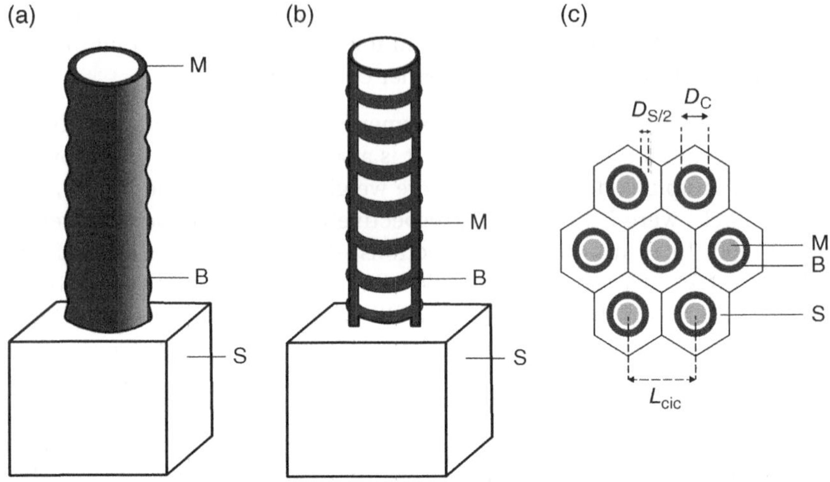

Figure 5.22 The schematics of the undulated/perforated cylinder-in-cylinder morphology morphologies observed for the $S_{64}B_{19}M_{17}^{(94)}$ triblock copolymer are shown in (a), (b), and the projection (c). M, B, and S denote the PMMA, PB, and PS blocks, respectively. The dimensions of the features shown in (c) are $L_{cic} = 41.0 \pm 2.8$ nm, $D_c = 17.5 \pm 1.8$ nm, and $D_s = 5.2 \pm 1.3$ nm. See Stadler et al. [79] for the TEM images. (Source: Breiner et al. [79]. Reproduced with permission of Wiley.)

(PB) = 0.21, Φ (PMMA) = 0.15, [Φ (PB)/Φ (PMMA)] = 1.4, and [Φ (PB) + Φ (PMMA)] = 0.36. The total volume fraction of PB and PMMA is higher relative to Case 1, and the PMMA volume fraction with respect to PB is also higher. The TEM images showed hexagonally ordered (CiC) core–shell structure within the matrix of PS. The PMMA end-block formed the core cylinder and the center block PB is the shell. Further, regions of the images of side views showed PMMA cylinders covered with thin stripes of PB perpendicular to the main axis of the cylinder, which appeared as ring-like objects, as if the core PMMA cylinders are surrounded by cage-like PB cylinders. This was called the "undulated/perforated CiC morphology." This is similar to the morphology shown in Figure 5.17e. In this sample, a homogeneous shell is not formed due to the small volume fraction. It was reasoned that a rupture of the PB shell occurs as a consequence **of** the unfavorable interactions between the center block and the end-blocks. A more favorable interface is formed with cylindrical objects surrounding the main cylinder. Schematics of the different morphologies observed in this case are shown in Figure 5.22a–c.

Case 3: Φ (PB)/Φ (PMMA) = 0.84 A high-molecular-weight $S_{65}B_{14}M_{21}^{(129)}$ was used here. The volume fractions are Φ (PS) = 0.65, Φ (PB) = 0.16, Φ (PMMA) = 0.19 leading to [Φ (PB)/Φ (PMMA)] = 0.84, and [Φ (PB) + Φ (PMMA)] = 0.35. The TEM images showed that in the gray PS is the matrix,

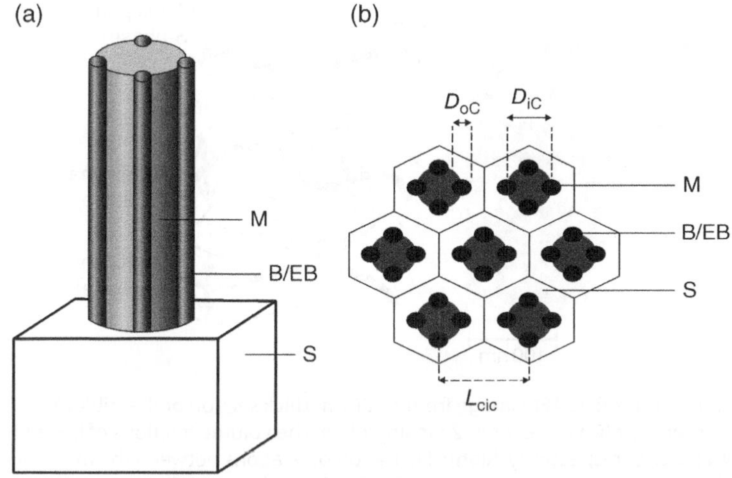

Figure 5.23 (a) Schematics of the cylinder-at-cylinder morphology of $S_{65}B_{14}M_{21}^{(129)}$ triblock copolymer. PS, PB/PEB, and PMMA are denoted by S, B/EB, and M, respectively. (b) The distance between the cylinders (L_{CaC}) was 60.5 ± 5.0 nm and the diameters of the inner (D_{iC}) and outer (D_{oC}) cylinders were 32.6 ± 6.4 and 9.7 ± 2.6 nm, respectively. See Stadler et al. [79] for the TEM images. (Source: Breiner et al. [79]. Reproduced with permission of Wiley.)

unstained (PMMA) cylinders were surrounded by four strongly stained objects. Further features seen in the TEM images lead to the model in which the central PMMA core was surrounded by four parallel PB cylinders. These cylinders were packed hexagonally in the PS matrix, although the PB cylinders surrounding the PMMA core are not hexagonally arranged, but tetragonally. This morphology was called "cylinder-at-cylinder" (CaC). The schematics of the morphology are shown in Figure 5.23. In this case, the distance between the cylinders (L_{CaC}) was 60.5 ± 5.0 nm and the diameters of the inner (DiC) and outer (DoC) cylinders were 32.6 ± 6.4 and 9.7 ± 2.6 nm, respectively. Comparing the three cases, the change from the CiC to CaC morphology occurred when the volume fraction Φ (PB)/Φ (PMMA) was decreased (from 2.5 to 0.84).

Case 4: 0.3 < Φ (PB)/Φ (PS) < 0.6 Analysis of the morphologies of several triblock copolymers with various ratios of the volume fractions of the center and end-blocks showed that in some cases, the CaC structure actually exhibited a helical twist, with four PB chains wrapping around a cylindrical core of PS to form a fourfold helix. One such case is that of a high-molecular-weight polymer $S_{25}B_{12}M_{63}^{(218)}$ with a large volume fraction of the PMMA end-block. Here, Φ (PS) = 0.27, Φ (PB) = 0.14, Φ (PMMA) = 0.59, Φ (PB)/Φ (PS) = 0.52. [Φ (PB) + Φ (PS) = 0.41]. The TEM image of such a helically twisted morphology can be seen in Ref. [79].

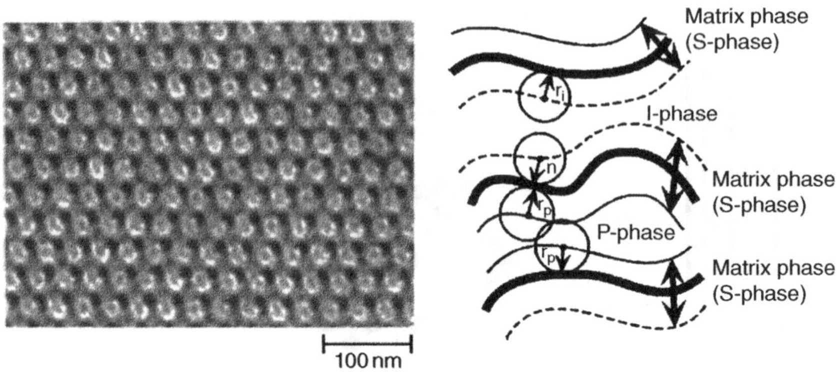

Figure 5.24 Left: The bright-field TEM image from an 80 nm thick section of the triblock copolymer polyisoprene–polystyrene–poly-2-vinylpyridine. The volume fractions of I, S, and P were 0.22, 0.59, and 0.19, respectively. Right: Illustration of relations between the minimal surface and parallel surfaces in a gyroid structure. The bold solid lines are the native minimal surface. The regular solid lines and the broken lines are parallel surfaces which correspond to I/S and P/S interfaces, respectively. r_i is not equal to r_p for the present triblock polymer, because the volume fractions of I and P phases are not equal. (Source: Suzuki et al. [80]. Reproduced with permission of American Institute of Physics.)

Suzuki et al. [80] used a symmetric triblock copolymer of polyisoprene-*b*-PS-*b*-P2VP with volume fractions of 0.22—0.59—0.19 and observed a tricontinuous double-gyroid morphology, with two parallel surfaces to the gyroid surface. The space group of this structure was determined to be $I4_132$. The TEM image is shown in Figure 5.24. The minimal surface and the parallel surfaces of the double-gyroid morphology are illustrated in the figure.

5.8 Some Applications of Gyroid Morphology

As discussed earlier, blends of diblock or triblock copolymers with homopolymers were used to fine-tune the volume fraction composition of the blocks and study their effect on the morphology. Thomas et al. [81, 82] showed that diblock copolymers of PS-*b*-PI and blends with the corresponding homopolymers that form one-dimensionally (1D) periodic lamellar morphology can function as photonic materials in the visible light range. They used a PS-*b*-PI diblock copolymer having nearly equal molecular weights of the blocks, with the M_w of PS and PI of 194000 and 197000, respectively. The homopolymers (h-PS and h-PI) also had an equal molecular weight of 13000 each. Films (20–50 μm thick) of PS-*b*-PI were made using cumene solutions with equal fractions of the homopolymers added, up to 60% in total homopolymer content. The backscattered electron imaging of a film with 40% homopolymers showed the lamellae seen in Figure 5.25a. In this case, the bright regions are

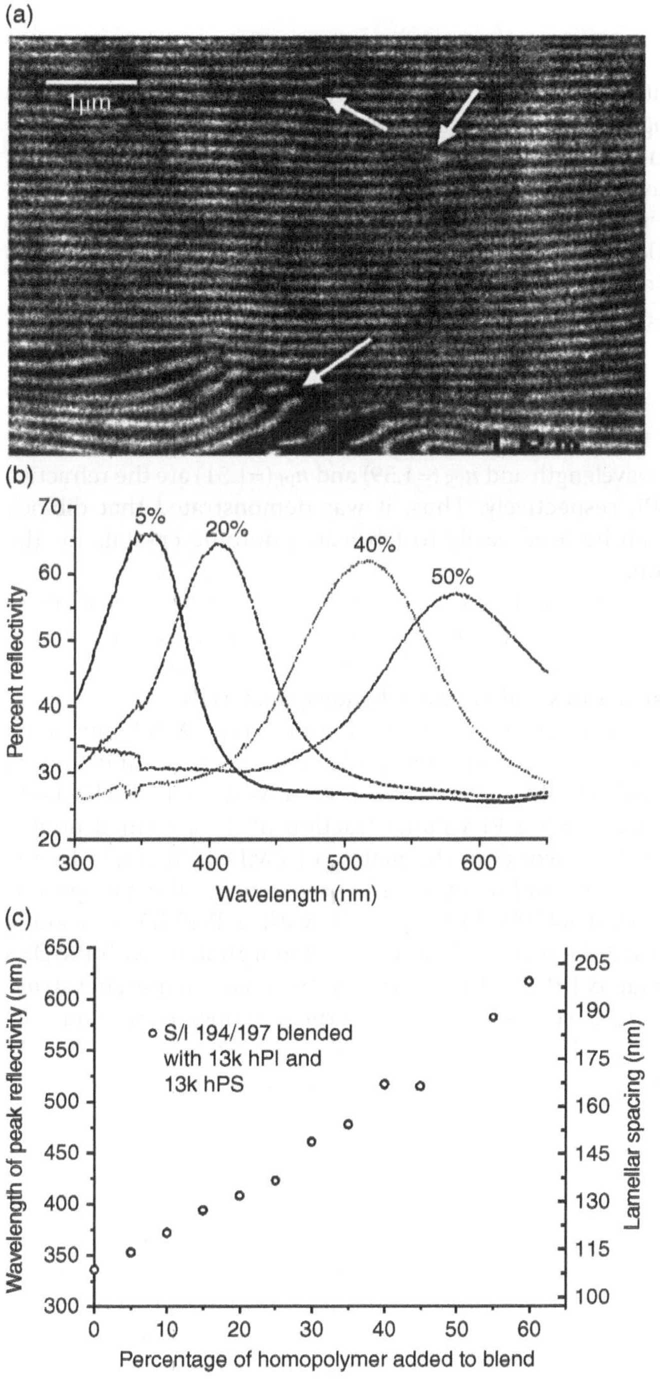

Figure 5.25 (a) Backscattered electron image of a fracture surface of the blend film containing 40% homopolymer. PI stained with OsO_4 appears white and PS is dark. The arrows show defects. (b) Reflectivity of blend films containing 5, 20, 40, and 50% (total) homopolymer. (c) Peak reflectivity wavelength plotted with the homopolymer content in the blend. The lamellar spacings, calculated according to Eq. 5.5, are shown on the right ordinate. (Source: Urbas et al. [81]. Reproduced with permission of Wiley.)

due to PI stained with OsO_4, and PS domains are dark. The reflectivity of the curves corresponding to various blends is shown in Figure 5.25b. The reflectivity peak positions are in the visible wavelengths from 350 to 600 nm and the peaks move to longer wavelengths and become broader as well with increasing homopolymer content. The lamellar spacing increases with the homopolymer content, and the reflectivity peak wavelength also increases almost linearly as seen in Figure 5.25c. The lamellar spacing L was calculated from the peak reflectivity wavelength according to

$$L = \frac{\lambda_p}{n_{PS} + n_{PI}} \tag{5.5}$$

where λ_p is the peak wavelength and n_{PS} (=1.59) and n_{PI} (=1.51) are the refractive indices of PS and PI, respectively. Thus, it was demonstrated that diblock copolymer blends can be used easily to fabricate photonic crystals for the visible spectral region.

It was noted before that Fredrickson [12] proposed that the gyroid phase could be stable in the strong segregation limit, at very high χN. Such gyroid morphology has been targeted for applications such as nanoporous and nanorelief ceramic structures and photonic bandgap materials.

Thomas et al. [30] used a triblock copolymer of the type A_1BA_2, in which A_1 and A_2 were PIs of two different molecular weights, and B was poly(pentamethyl disilylstyrene) (P(PMDSS)). A polymer with PI/PMDSS/PI: 24/100/26 (kg/mol), with a PI volume fraction of 33%, formed double gyroid consisting of PI networks in the matrix of PMDSS (denoted by the authors as P(PMDSS)–DG) and was used as a precursor for the nanoporous structure. Another, with 44/168/112 (kg/mol), with a P(PMDSS) volume fraction of 51%, formed the inverse double-gyroid morphology with PMDSS networks in the PI matrix [PI–DG] and was used to create a nanorelief structure. The double gyroid consists of two interpenetrating, continuous 3D network of the minority block. Samples were film-cast from toluene. The PI was selectively removed with ozone and the P(PMDSS) was converted to silicon oxycarbide. With the P(PMDSS–DG, removal of PI leaves the silicon oxycarbide matrix with interconnected tortuous void network (nanoporous structure) in the ceramic matrix, as seen in the AFM image shown in Figure 5.26. On the other hand, with the PI–DG, upon removal of the PI matrix, network of ceramic struts are obtained (nanorelief structure). Since the P(PMDSS) was converted to a ceramic, both the nanostructures exhibit mechanical robustness, compared to those obtained with the native polymer alone. It was suggested that due to the low dielectric constant and high temperature stability, the ceramic network from the PI–DG could be used in interconnects related to photoresists and the periodic ceramic/void structures could be photonic bandgap materials.

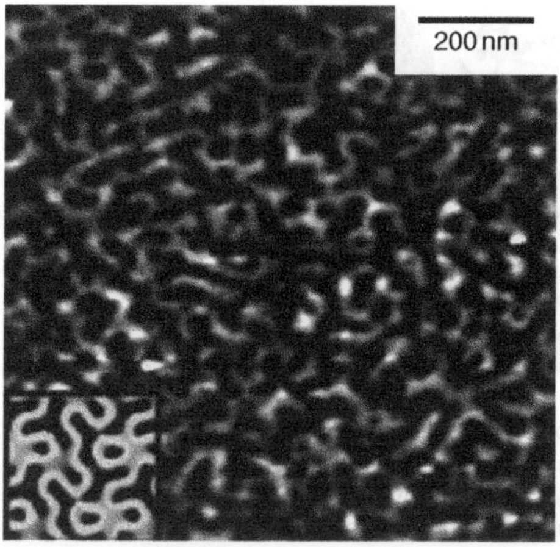

Figure 5.26 An AFM image of the surface morphology of P(PMDSS)–DG after ozonolysis and UV exposure. With the removal of the PI networks, ordered tortuous pathways were formed within a silicon oxycarbide matrix. Bright regions are highest (maximum height on the image is 10 nm) and the dark regions are voids. The inset shows the [012] view at zero height of a volume-rendered surface of the double-gyroid structure with empty strut networks, which appear dark. (Source: Urbas et al. [81]. Reproduced with permission of Wiley.)

In another study from this group, the double-gyroid morphology created by the self-assembly of a PS-*b*-PI diblock copolymer was examined by Thomas et al. for application as photonic crystals [83]. One of the blocks, for example, PI could be etched away, leaving a porous matrix with a tortuous void network. For this case, they used a PS-*b*-PI diblock copolymer with the molecular weights of 300 000 and 450 000 for the PS and PI blocks, respectively. The styrene volume fraction was 38%, and the system exhibited double-gyroid structure. SAXS experiments led to a cubic lattice parameter of $a = 258$ nm and $Ia3d$ symmetry of the lattice. Figure 5.27a shows the TEM image of a thin section of the diblock copolymer which was stained with OsO_4 in which the two styrene networks are seen (these appear light in the image). Upon UV/O_3 exposure of the films to etch out the PI block, free-standing, interpenetrating network of the PS was seen, as in Figure 5.27b. This process yields PS/air photonic crystal. It was found that the lattice parameter of the etched film was comparable to the unetched film which confirmed that the etching process did not affect the network structure of the glassy PS block. Reflectivity measurements showed the main peak reflectivity at 327 nm for the unetched PS-*b*-PI film, with peak width of 70 nm. Additional less-intense peaks were observed at 280 and 235 nm. These were in agreement with the computed positions of the reflective peaks. The PS/air film, after the partial removal of the PI block, showed shift of the peaks to shorter wavelengths due to differences in the average refractive indices. These experiments confirmed that the double-gyroid structure is a photonic crystal in the visible wavelength range.

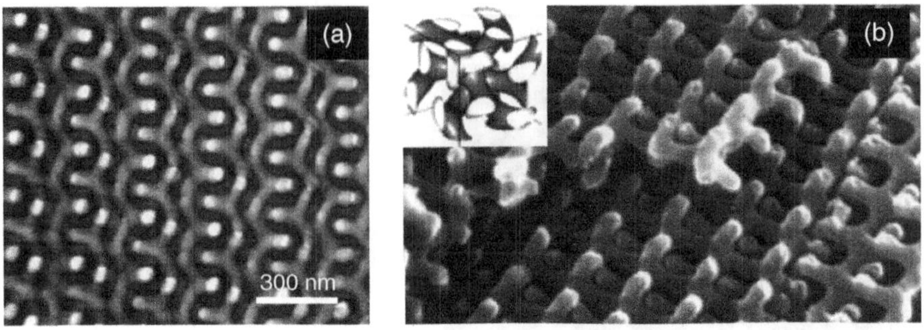

Figure 5.27 (a) A TEM image of a thin section of the PS-*b*-PI diblock copolymer film stained with OsO_4. The two PS networks appear light. (b) A SEM image of the film after the removal of the PI block by etching. The free-standing interpenetrating network of PS is seen. A computed surface of the PS network (with a volume fraction of 34%) is shown in the inset. (Source: Urbas et al. [81]. Reproduced with permission of Wiley.)

A similar approach was used by Crossland et al. [84] to fabricate a hybrid solar cell using the bicontinuous double-gyroid morphology of the diblock copolymer poly(4-fluorostyrene)-*b*-poly(D,L-lactide) (PFSt-*b*-PLA). The PFSt weight fraction was 60%. The BCP film with the gyroid morphology was prepared on a glass substrate with a conductive coating. The PLA block was selectively etched out. The voided network was filled with titanium oxide (Ti(IV) oxide). Annealing the Ti(IV) oxide/PFSt gyroid structure at 500°C under argon and then under oxygen removed the PFSt block, leaving the free-standing inorganic semiconducting gyroid. The array was then backfilled with an organic hole-transporting material and capped with a counter electrode to produce a bicontinuous bulk heterojunction solar cell.

Such a strategy of creating gyroid templates using BCPs and backfilling with inorganic materials was used by Steiner et al. [10] for fabricating metamaterials. The properties of the metamaterials are not usually found in natural or other known materials, and these depend not on the atomic-level arrangements but also on their structural design. They can function as 3D optical and negative index materials as well as exhibiting unusual mechanical properties. There has recently been a summary article on these types of materials in the *Chemical and Engineering News* of the American Chemical Society (see http://cen.acs.org/articles/93/i14/Metamaterials-Unusual-Mechanics.html).

Films of several micrometers thickness of a triblock copolymer of PI-*b*-PS-*b*-PEO were cast from anisole on to a treated glass substrate which led to interpenetrating double gyroid network of polyisoprene and PEO blocks in the matrix of PS. The schematic of the approach followed by Steiner et al. [10] is shown in Figure 5.28. The gyroid structure is shown in Figure 5.28a. The PI segment was removed by UV exposure (Figure 5.28b). Electrodeposition was used to fill the porous network with gold (Figure 5.28c). Plasma etching removed the remaining polymer to leave a 3D continuous gold/air network, as in Figure 5.28d).

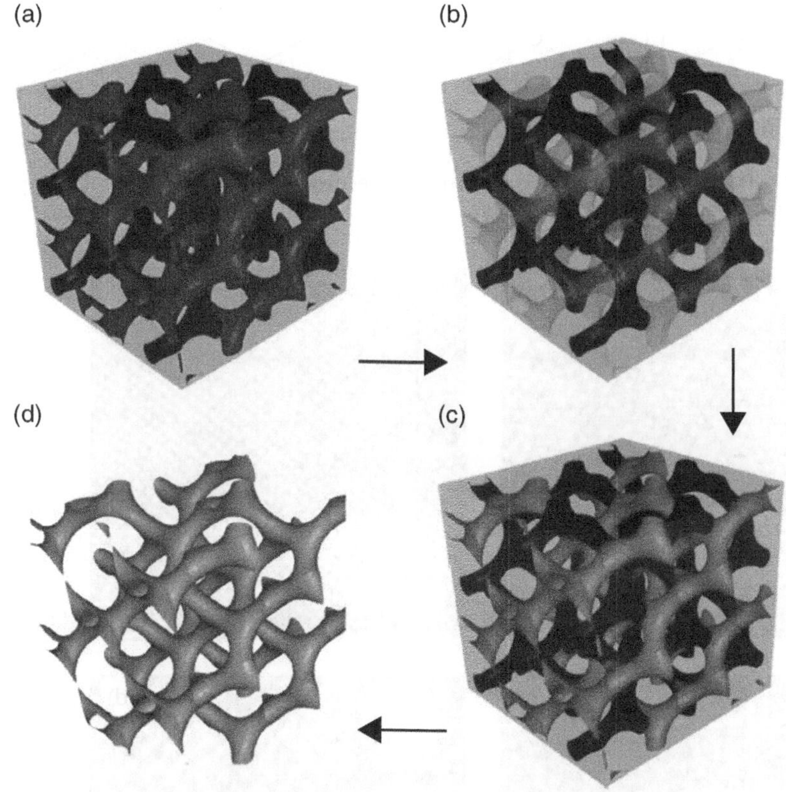

Figure 5.28 Schematic of the steps in metamaterial fabrication. (a) Double-gyroid network of PI and PEO blocks (shown in different shades of grey) in PS matrix; (b) the PI block is removed; (c) the pores are filled with gold; and (d) plasma etching of PS and PEO blocks leaves the 3D continuous gold/air network. See Vignolini et al. [10] for the corresponding color images. (Source: Vignolini et al. [10]. Reproduced with permission of Wiley.)

Wiesner et al. [85–88] used BCPs as structure-directing agents to create ceramic/polymer hybrid morphologies of metal oxide containing mesoporous silica, using a sol–gel "ormocer" process (ormocer is an acronym for organically modified ceramic). They used a PI-*b*-PEO diblock copolymer with a molecular weight of 16 400 and PEO volume fraction of 35%, which showed a melt morphology of hexagonally packed PEO cylinder in the PI matrix. The BCP was mixed with two metal alkoxides shown in Scheme 5.3, namely, (3-glycidyloxypropyl) trimethoxysilane, and aluminum sec-butoxide. The polymer-inorganic composite was then calcinated by step-wise heating up to 600°C, resulting in a mesoporous material. Figure 5.29 shows the TEM images of the original and the calcinated materials in the cubic phase, which confirm the bicontinuous morphology. SAXS analysis indicated $Im\bar{3}m$ symmetry, with lattice lengths of 63.0 and 39.5 nm, for the original and calcinated materials. Although the lattice

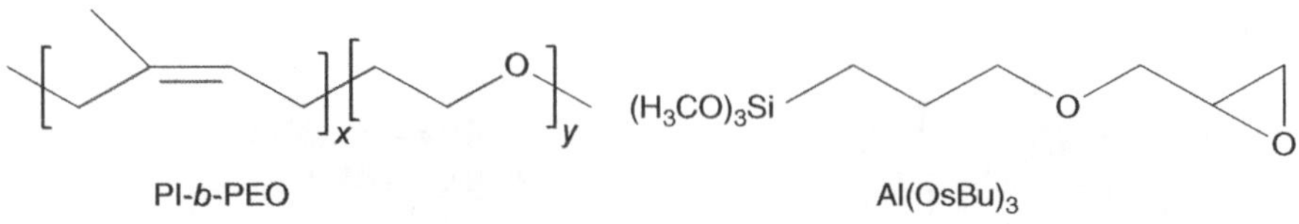

Scheme 5.3 The block copolymer and the ormocer precursors used by Wiesner et al. for preparing mesoporous silica [86].

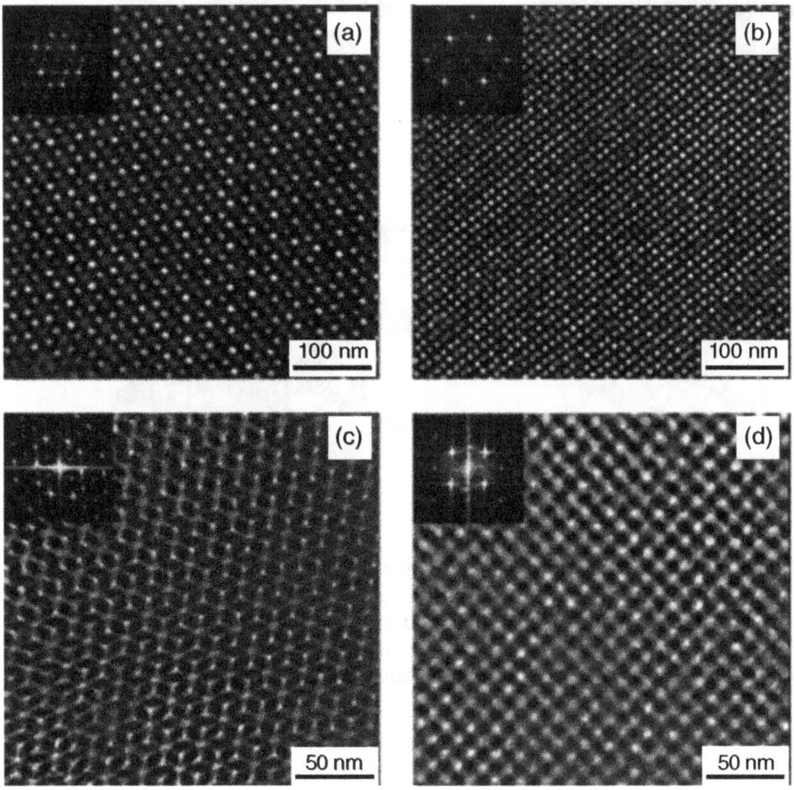

Figure 5.29 Bright-field TEM images from as-made (top) and calcined material (bottom) highlighting the threefold (a, c) and fourfold (b, d) projections of the cubic phase. The calcined material shows greater contrast due to the removal of all organic material, which allows for the convenient identification of the bicontinuous nature of the structure. Insets in each panel show the computed Fourier diffraction patterns. (Source: Finnefrock et al. [86]. Reproduced with permission of Wiley.)

size decreases by 63% upon calcination the symmetry of the structure was preserved indicating the robustness of the BCP-derived ormocer material. The symmetry was consistent with the plumber's nightmare structure, rather than a gyroid, as illustrated in Figure 5.30.

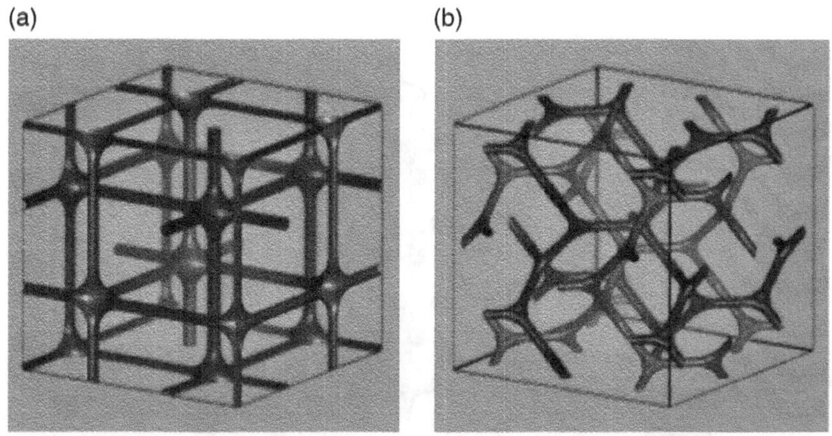

Figure 5.30 (a) Real-space image of a stick structure derived from the plumber's nightmare. For comparison, the real-space image of a gyroid-derived structure is shown in (b). The graphics were generated by using the following website: http://msri.org/publications/sgp/SGP/. The different shades serve to distinguish the two distinct 3D channel systems. (Source: Finnefrock et al. [86]. Reproduced with permission of Wiley.)

Using the ormocer method, Wiesner et al. also fabricated the reverse of the porous materials, namely, ceramic struts which were thick, robust bicontinuous silica-based nanostructures, with $Im\bar{3}m$ symmetry with the plumber's nightmare morphology [85]. These could serve as photonic bandgap materials. A model based on TEM and SAXS is shown in Figure 5.31.

Without the calcination, these authors studied the effect of different amounts of the metal oxides on the morphology of the organic/inorganic hybrids [87]. The 3D network formation of the inorganic framework involves the hydrophilic PEO block. By selectively swelling the (organic) PI block, single hybrid nano-objects with controlled shape can be isolated. By this technique, spheres, micelles, hairy rods, etc., were obtained as nano-objects. A schematic of the various nano-objects which can be fabricated from the designed initial morphologies is shown in Figure 5.32. The "hairy" feature arises from the PEO block.

The use of supramolecular complexation between BCPs and small molecules was mentioned before. The work of Stamm's group on PS-*b*-P4VP complexed with HABA was briefly discussed [60]. Some of the works of ten Brinke on supramolecular assemblies were highlighted in Chapter 4. His group also published a series of papers on modifying the morphologies of BCPs, including creating gyroid phases by complexing BCPs with pentadecylphenol (PDP) [89]. They showed that a core–shell gyroid morphology can be created using the self-assembly of PDP with a triblock copolymer of poly(*tert*-butoxystyrene)-block-polystyrene-block-poly(4-vinylpyridine) (PtBOS-*b*-PS-*b*-P4VP). Figure 5.33 summarizes their work which formed the cover page of *Macromolecular Rapid Communications*, vol. 32, issue 4.

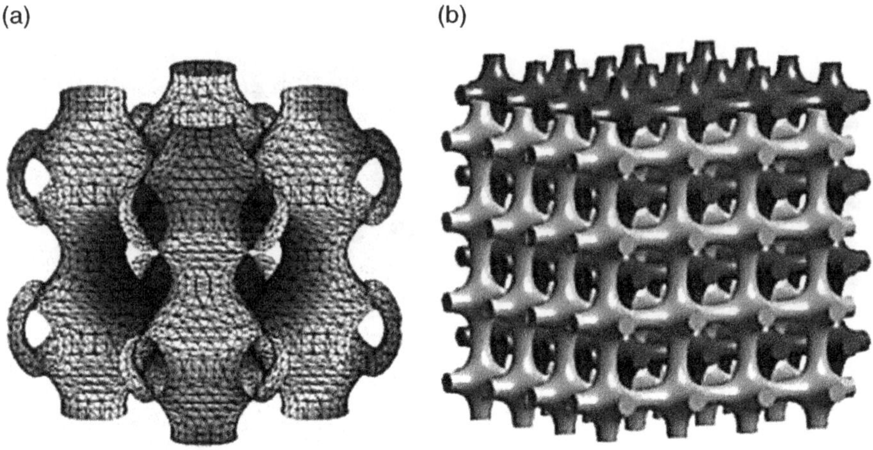

Figure 5.31 (a) The "P" minimal surface and (b) a skeletal structure of the plumber's nightmare with the networks occupying 37% of the volume. (Source: Jain et al. [85]. Reproduced with permission of Wiley.)

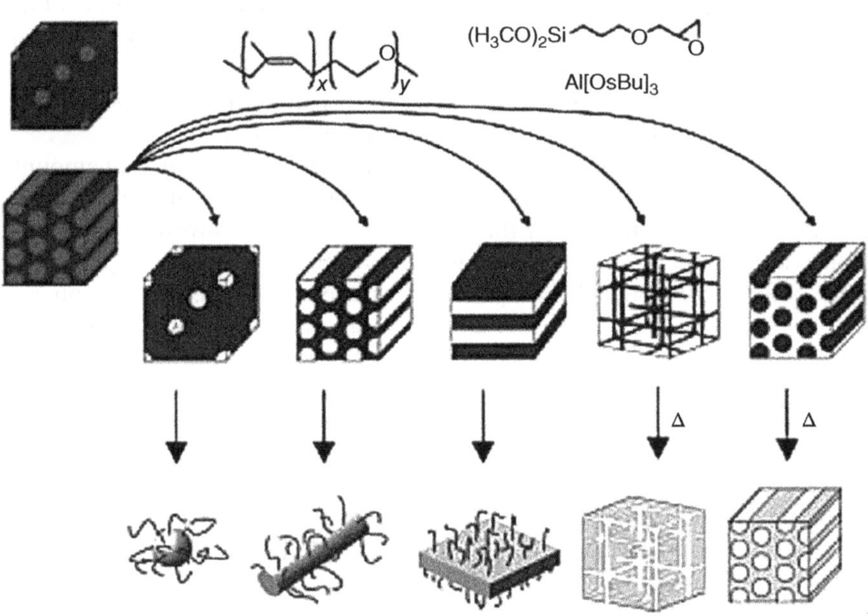

Figure 5.32 An illustration of the approach to create nanoobjects of different shapes using ormocer precursors and PI-*b*-PEO diblock copolymers. (Source: Finnefrock et al. [86]. Reproduced with permission of Wiley.)

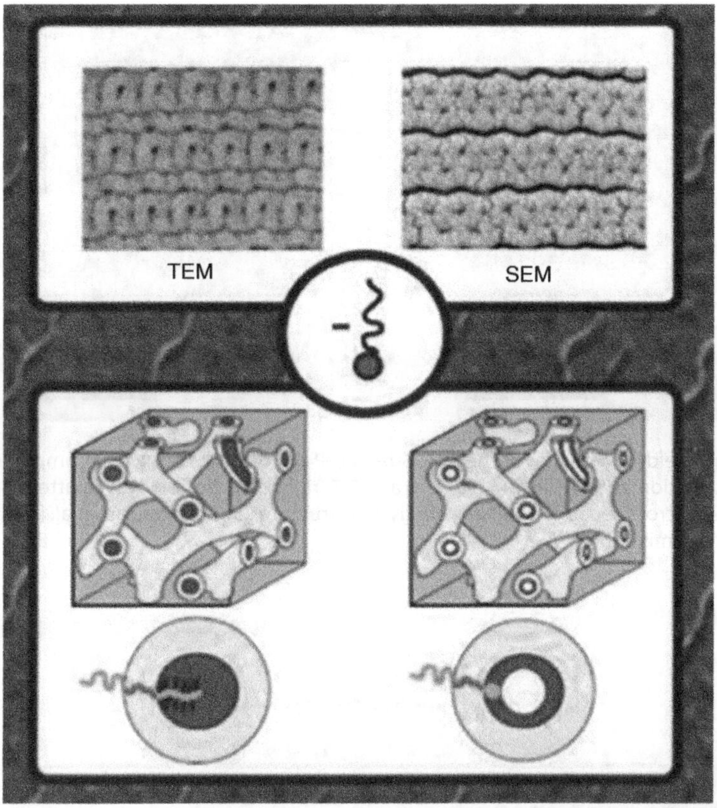

Figure 5.33 Summary of the work on poly(*tert*-butoxystyrene)-block-polystyrene-block-poly(4-vinylpyridine)/PDP complex, which appeared on the cover page of Ref. [89]. Poly(*tert*-butoxystyrene)-block-polystyrene-block-poly(4-vinylpyridine) triblock copolymers and less than stoichiometric amounts of pentadecylphenol (PDP) formed supramolecular complexes and self-assembled into a core–shell gyroid morphology with the core channels formed by the hydrogen bonded P4VP(PDP) complexes. Removal of PDP by washing resulted in well-ordered nanoporous films. These were then used as templates for nickel plating. (Reprinted with permission from Ref. [89]. Copyright (2011) John Wiley & Sons.)

Films were cast from solutions of the triblock copolymer and PDP in chloroform to form the supramolecular complex. Weight fractions of PDP relative to P4VP was varied from $x = 0.3–0.8$. Figure 5.34 shows the bright-field TEM image of the bicontinuous double gyroid morphology of the complex with $x = 0.5$ and 0.7. Iodine was used to stain the P4VP/PDP comb block. The double-wave features represent the projection along the (211) plane of the gyroid unit cell.

Washing the films with ethanol for 3 days at room temperature resulted in the removal of PDP and nanoporous films were formed. Figure 5.35 shows a

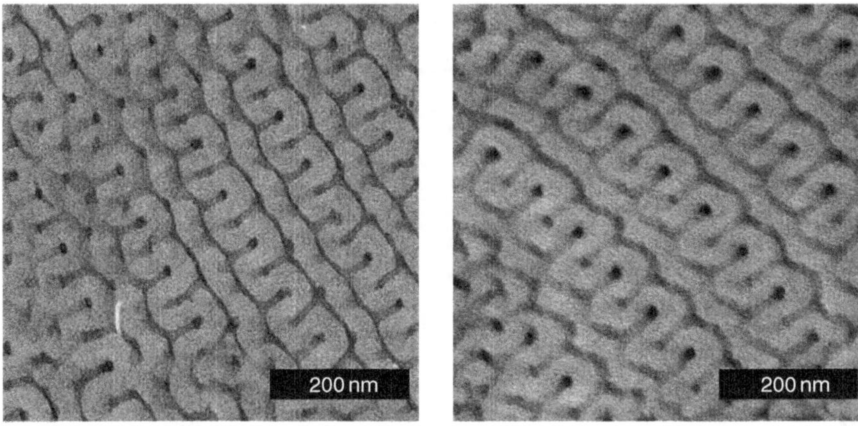

Figure 5.34 Bright-field TEM images of the PtBOS-*b*-PS-*b*-P4VP triblock copolymer complex with PDP weight fraction *x*: (P4VP/PDP) = 0.5 (left) and 0.7 (right). The double-wave pattern arises from the (211) cross section of the double-gyroid structure. (Source: du Sart et al. [89]. Reproduced with permission of Wiley.)

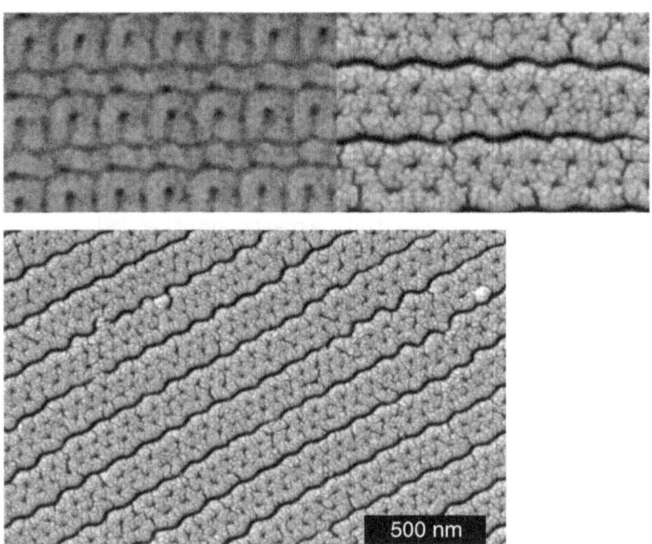

Figure 5.35 TEM (top left) and SEM images of the copolymer/PDP complex with $x = 0.5$ with the same scale. The image on top right corresponds to a washed film, with PDP removed. A large area of the washed film is seen in the SEM image at the bottom. (Source: du Sart et al. [89]. Reproduced with permission of Wiley.)

combined TEM and SEM images of the sample with $x = 0.5$ at the same magnification, but the SEM corresponds to the film that was washed. The nanoporous network formation is evident in the SEM image. The bottom part of Figure 5.35 also confirms the uniformity of the morphology over a large area. Mercury porosimetry measurements indicated a total porosity of 11.54%, a pore volume of 115.4 mm^3/g, an average pore diameter of 12 nm, which were in agreement with calculations based on the morphology. The porous film was used as a template for electroless nickel plating. This is a demonstration of creating inorganic bicontinuous networks using BCP/small-molecule supramolecular complex formation. A further detailed study [90] of such a supramolecular complex with PS-*b*-P4VP and PDP and the double gyroid was later published by these authors.

5.9 Graphoepitaxy

Epitaxy is the growth of crystalline layer of a material over another crystalline surface, often to grow defect-free films. It replicates the surface structure of substrate. The word derives from "Epi" to mean "over" or "upon" in Greek and "Taxis" to mean arrangement. If both the substrate and the deposited material are of the same kind, it is known as homoepitaxy, and if they are different, it is then heteroepitaxy. This technique is used extensively in silicon and related industries and in nanotechnology. The epitaxial growth could be induced by vapor deposition or in the liquid state. Very early studies of Pennings [91] on polyethylene crystallization and of Chanzy [92] on cellulose triacetate (CTA) showed such epitaxial growth in polymer systems. Pennings obtained "shish-kebab" type of morphology of PE by stirring-induced crystallization. With the primary fibers of PE (shish), the stirring-induced crystallization produced chain-folded growth as "kebabs." Chanzy and Roche [92] observed such shish-kebab growth in the case of CTA. They acetylated the polymorph I of cellulose from *Valonia ventricosa*, and the TEM image of the "smooth" primary fibers of CTA I crystalline form is shown in Figure 5.36a. These were then converted to polymorph II of CTA, by autoclaving CTA I in water vapor at 230°C for 15 min. The TEM analysis showed that the original fibrous morphology was preserved, but the extensive dissolution and recrystallization during the polymorphic conversion resulted in the growth of folded-chain lamellae ("kebabs") on the primary extended ("shish") chain fibers (Figure 5.36b). Electron diffraction showed that the directions of the CTA chains in the shish and kebab were the same. Thus, during polymorphic conversion, epitaxial crystallization had occurred.

While the epitaxy described earlier pertains to atomic-scale crystallization, assembly of nano- or microscale objects following a set pattern created on a substrate is called "graphoepitaxy." This is like a brick-layer following the pattern in building a wall. As shown in Figure 5.37, suppose a collection of cubes

Figure 5.36 (a) Fibers of polymorph I of cellulose triacetate (CTA (I)) and (b) epitaxial growth of "kebab" lamellar CTA II crystals on the swollen primary CTA I fibers during polymorphic conversion. (Source: Chanzy and Roche [92]. Reproduced with permission of Wiley.)

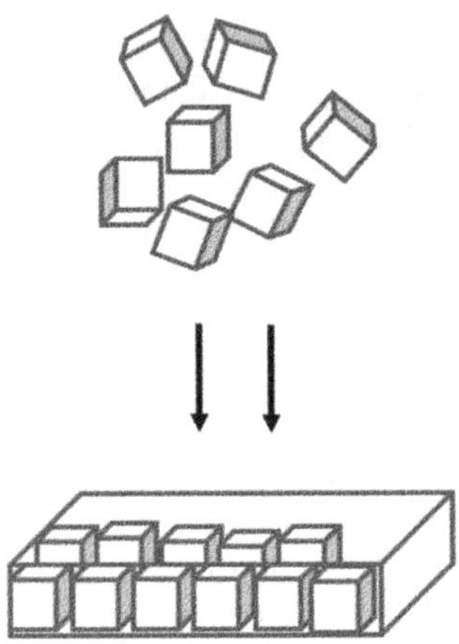

Figure 5.37 An illustration of packing of objects following a pattern.

are dropped on a tray and the tray is shaken to settle the cubes. The cubes will then align themselves along the walls of the tray, similar to the way in which, for example, the balls are collected at the beginning of the game on a pool table. Givargizov [93] applied this technique to grow oriented crystallites on a patterned amorphous substrate. Chou et al. [94] developed a high-throughput method using compression molding (imprint lithography)

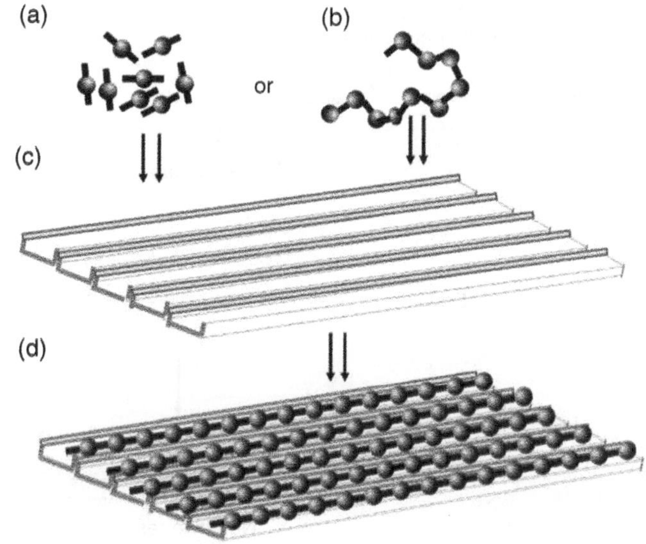

Figure 5.38 A schematic of polymerizing monomers (a) on a patterned substrate (c) or solution casting a polymer (b) on to the substrate (c) to create an aligned pattern (d).

to create a contrast pattern on a substrate and fabricated metal patterns with 25 nm resolution. Schvartzman et al. [95] created a pattern of a catalyst on a sapphire substrate and grew ZnO nanowires with controlled position, orientation, and length.

Figure 5.38 shows a scheme in which a monomer (a) is polymerized in the presence of a (catalyst) patterned substrate and the resultant polymer is aligned along the grooves of the substrate (c). A polymer (b) can also be solution-coated onto the substrate (c) to obtain an aligned pattern as in (d). Although BCPs form phase separated domains in the range of a few tens of nanometers, the uniformity of these domains does not normally extend beyond a submicron grain size. Fabricating films with uniform positional organization of such domains over a large area is required for most applications. Employing predesigned templates and graphoepitaxy to create nanopatterns over a range of several microns has been pursued by a number of authors [96–115]. The basic principle follows that of making a photoresist (a light-sensitive material used in processes such as photolithography and photoengraving to form a patterned coating on a surface) as shown in Figure 5.39 for an SiO_2 layer. A resist layer is deposited on the SiO_2 layer. The mask serves to selectively pattern the resist layer. For a positive resist, the resist layer not covered by the mask is burnt off, whereas it is cross-linked for the negative resist. It is obvious from Figure 5.39 that either the positive or negative resist can serve as template for creating molds. Figure 5.40 shows the scheme in which a photoresist master is used for

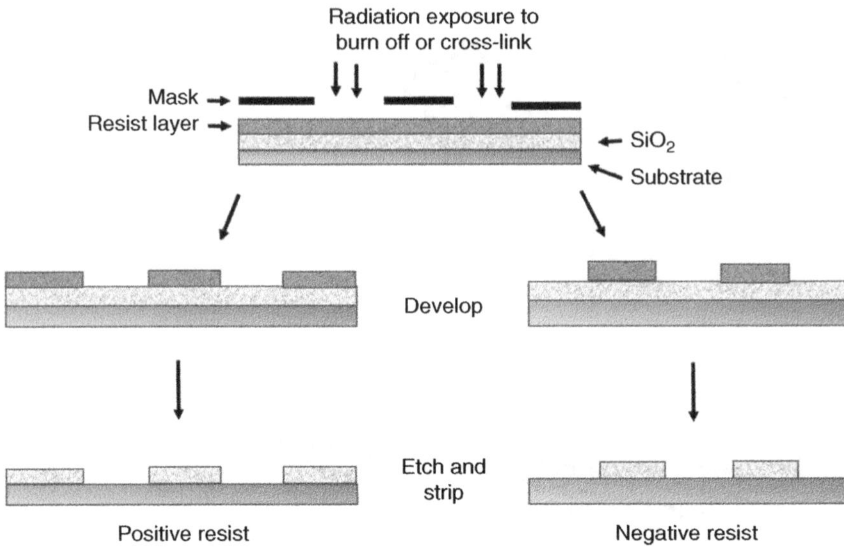

Figure 5.39 An illustration of the steps in making positive and negative photoresists.

soft lithography, using polymers. The master is coated with PDMS; it is then cross-linked and peeled off. This serves as a stamp or a mold. The (soft) mold can then be pressed against a prepolymer to transfer the pattern; the pre-polymer can be cured and removed from the stamp to create a replica.

Del Campo and Arzt [116a] as well as Acikgoz et al. [116b] reviewed the conventional and alternative lithography techniques using polymers on a nanoscale (nano-imprint lithography or NIL). The principle of the methods is similar to macroscale traditional injection molding, that is, a mold is used to shape the melt of a polymer and subsequent cooling of the molded polymeric material and removal from the mold. It is like cookie-cutting or waffle-making. Figure 5.40 shows several variations of the process. Figure 5.40a is just injection molding, by filling a polymer melt in a mold cavity and then separating the polymer impression from the mold. In Figure 5.40b, a mold is pressed on a polymer along with increasing the temperature such that the polymer melts and flows into the mold to get an imprint. The mold is then removed to leave an imprinted polymer. When using a polymer such as PDMS, it can be cross-linked before separating it from the mold. This polymer imprint can be used to create a complementary mold: the hills and valleys would be reversed when using the imprint. Figure 5.40c and d illustrates the use of a polymer precursor (a reactive monomer) to create the imprint and subsequent polymerization using UV exposure or thermal curing.

Once a pattern is created on a substrate by one of the several methods, graphoepitaxy method can be used to organize the BCP domains. Kramer et al.

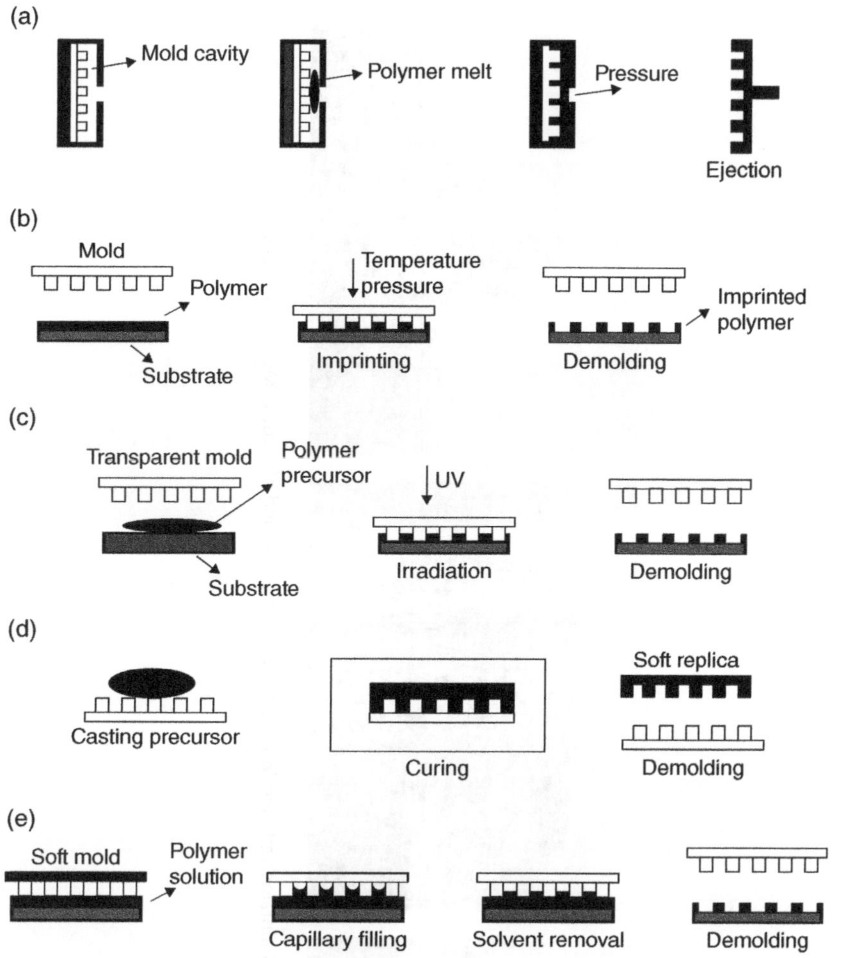

Figure 5.40 Various processes for nano-imprint lithography using a polymer or a prepolymer and a topographically patterned mold. Simple injection molding (a), thermal embossing (b), UV curing (c), thermal curing of patterned prepolymer (d), and solution molding with a soft mold (e) are illustrated. (Source: del Campo and Arzt [116a] and Acikgoz et al. [116b]. Reproduced with permission of American Chemical Society & Elsevier.)

[108] created rectangular mesas with heights of 3–300 nm, and 1–10 μm widths and spacings on a silicon substrate (SiO_2) using photolithography and etching techniques. The surface thus consists of mesas and wells (valleys) between them. (Mesa is a Portuguese and Spanish word for a table. It is also used for geographical formation of a range of large, steep, elevated surfaces surrounded by valleys, as in Figure 5.41. A related formation is a butte, which is a tall single peak.) They used a PS-*b*-P2VP BCP (with a 12% fraction of PVP) which would

Figure 5.41 Photographs of mesas. Glass Mountains (left) and Monument Valley (center) and of a butte (Monument Valley). (Source: From Wikipedia.)

form spherical domains of PVP in the PS matrix. The films of the BCP were spin-cast on a glass slide, floated off in water, and transferred to the silicon substrate. The films were then annealed at 180°C, well above the T_g of PS, to promote well-ordered morphology.

The films were of thickness corresponding to a single layer of the spherical domains. Figure 5.42a shows the film with single-crystal-like order on the mesa of 30 nm height, with grain size corresponding to the width of 4.5 μm. The dark regions on either edge are due to the drop off from the mesa into the well. The fast Fourier transform (FFT) shown in Figure 5.42b contains six diffraction spots, corresponding to hexagonal symmetry. A schematic of the arrangement of the BCP on the mesa and the well is shown in Figure 5.42c. For graphoepitaxy to occur, the mesa height should be at least one layer thickness of the spherical morphology, which was $L_0 = 27$ nm in this case. Successful graphoepitaxy patterning was achieved for mesa heights ranging from L_0 to $10L_0$. The width of the mesa was also found to be important. Since the edges nucleate the epitaxial

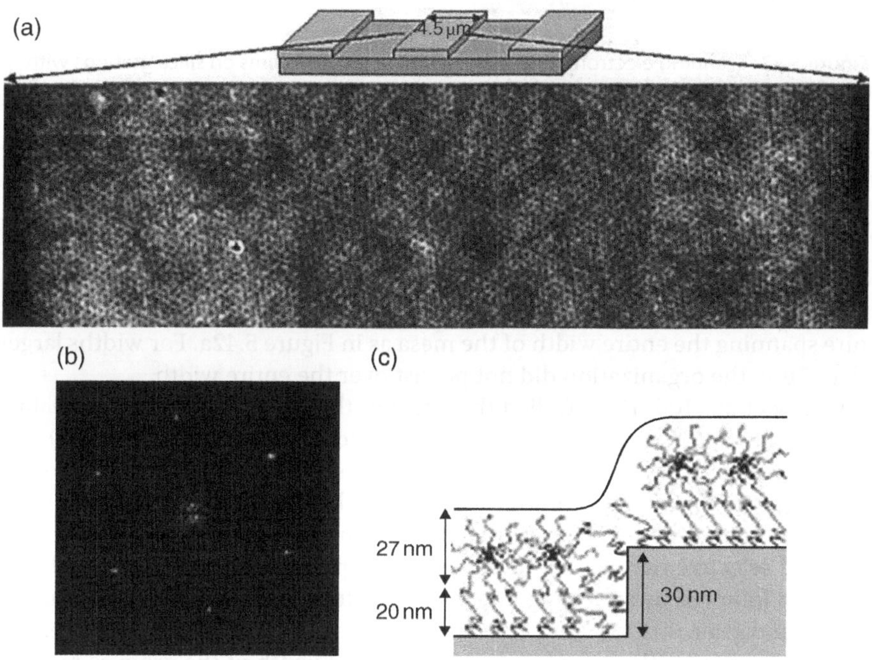

Figure 5.42 (a) SFM of PS–PVP film on top of a mesa of 4.5 m width. When the mesa edges are less than 5 μm apart, a single crystal is formed. (b) The six diffraction peaks in the associated FFT indicating hexagonal symmetry. (c) Schematic of a polyvinylpyridine brush that is observed on all SiO_2 surfaces followed by a layer of spheres encased in a styrene matrix. See Segalman et al. [108] for the corresponding color image. (Source: Segalman et al. [108]. Reproduced with permission of Wiley.)

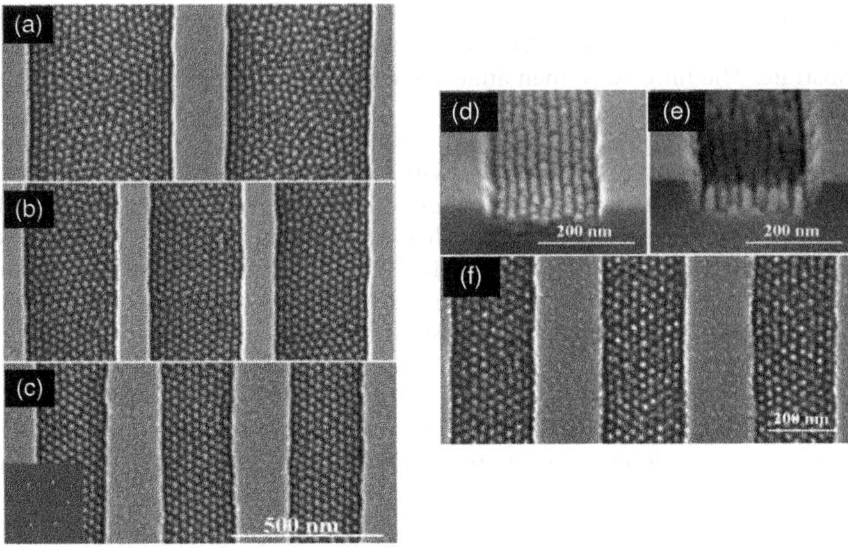

Figure 5.43 Scanning electron micrographs of PS/PFS 50/12 films on silica gratings with (a) 500 nm, (b) 320 nm, and (c) 240-nm wide grooves, respectively. The inset in (c) is the Fourier transform of the plan-view pattern showing sixfold symmetry. The films were spin cast at 3500 rpm, annealed at 140°C for 48 h and plasma-treated to remove PS. (d) Side view of the pattern in (c); (e) and (f) side view and plan view of the patterns following transfer to the underlying silica substrate by ion etching. The silica posts seen in (f) have the same organization corresponding to the PFS in (c). (Source: Cheng et al. [107]. Reproduced with permission of American Institute of Physics.)

growth, if the edges are less than 5 μm apart, the grains join to create a structure spanning the entire width of the mesa as in Figure 5.42a. For widths larger than 5 μm, the organization did not persist over the entire width.

Cheng et al. [106, 107] studied the effect of the groove width and annealing using a PS-*b*-PFS diblock copolymer, with molecular weights of 30 000 and 12 000, respectively, for the PS and PFS blocks. With the volume fraction of 12%, the equilibrium morphology would be such that spherical domains of PFS would be contained in the PS matrix. The use of the organometallic PFS enabled selective removal of PS using oxygen plasma. Spin-cast from toluene solution followed by annealing, the resulting film showed spherical domains of 20 nm and grain size of about 280 nm. Films were then spin-cast on patterned silica substrates with a step height of 50 nm and width of the grooves ranging from 240 to 500 nm. Annealing at 140°C over a long period of 48 h improved the organization compared to the film (not shown here) which was annealed for only 4 h. The PS layer was selectively removed by oxygen plasma treatment to reveal the PFS spheres. Figure 5.43a–c shows the changes in the BCP organization with the width of the grooves. With the 500 nm groove width, three rows of

PFS spheres are aligned parallel to the wall and this extends further in the well with the groove width of 320 nm.

With a further reduction in the width to 240 nm, near-perfect alignment is seen with the grain size extending to the width of the groove. The FFT (inset in Figure 5.43c) shows sixfold symmetry. It was also found that any misalignment and defect that arose due to the edge roughness of the grooves vanished with annealing time. The well-ordered patterns in Figure 5.43c were transferred to the silica substrate underneath, by reactive ion etching using CHF_3 plasma. This led to ordered arrays of silica posts with widths of 20 nm which correspond to the original size of the PFS spheres. Figure 5.43e and f shows the side and plan views of the silica posts, with an aspect ratio of 3. It is seen that the pattern of the posts corresponds to the organization of the PFS spheres in (c). Thus, a pattern transfer from the BCP to the silica substrate was achieved.

Ober et al. [117] showed the feasibility of controlling the self-assembly of a poly(α-methylstyrene)-*b*-poly(4-hydroxystyrene) BCP (PαMS-*b*-PHOST) (PαMS fraction of 33%), which is lithographically patternable by "top-down" (conventional deep UV lithography) and "bottom-up" (self-assembly) methods. Solvent annealing rather than thermal annealing was used to improve the quality of the morphology. Figure 5.44a shows the disordered perpendicular cylindrical domain morphology of films cast on silicon wafers with propylene glycol methyl ether acetate (PGMEA) as the solvent. Upon solvent annealing with THF, which is a nonselective solvent (good for both blocks), the cylindrical domains become parallel to the surface (Figure 5.44b) due to differential surface wetting of the PHOST block. The center-to-center distance of the cylinders was measured to be about 22 nm and the film thickness was 14–16 nm. Annealing in acetone, which is a preferential solvent for PHOST, leads to hexagonal array of dots, seen in Figure 5.44c. Although the figure looks similar to Figure 5.44a with cylindrical morphology, the dots in Figure 5.44c correspond to spherical morphology, as confirmed by grazing incidence SAXS (GISAXS). Preferential swelling of PHOST in acetone- induced OOT from cylinders to spheres, which was kinetically trapped upon drying the film. Figure 5.44d and e shows the SEM images of photolithographic patterns of the BCP, which were produced by exposing the as-cast and THF-annealed films to 248 nm light, in the presence of a photoacid generator (triphenylsulfonium trifluoromethanesulfonate (TPST)) and a cross-linking agent (1,3,4,6-tetrakis(methoxymethyl)glycoluril (TMMGU)), baked at 115°C for 60 s, and developed with a mixed solvent system of cyclohexane and isopropanol. The dark lines in Figure 5.44d and e (with identical resolution) correspond to the remaining cross-linked polymer corresponding to the lithographic pattern after the baking and developing steps.

The graphoepitaxial alignment of PαMS-*b*-PHOST was achieved by depositing the BCP on to SiO_2 substrates fabricated with 30 nm deep topographical patterns created by UV lithography and etching techniques. These substrates were also pretreated by spin coating with PS solution, and baking at 195°C for 2.5 h, in

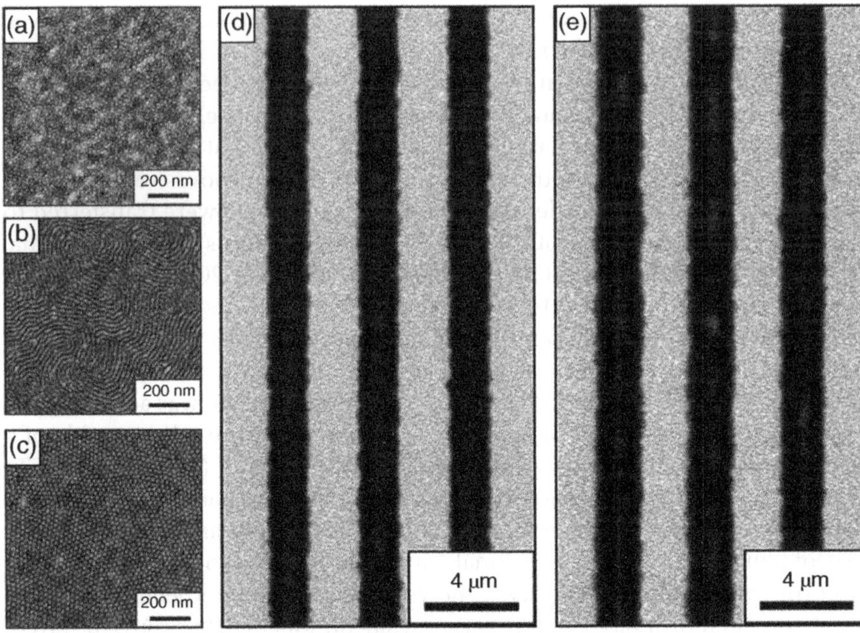

Figure 5.44 Atomic force height micrographs of (a) as-cast PMS-*b*-PHOST film with cylindrical domains oriented perpendicular. (b) Parallel cylinders upon annealing with THF. (c) Acetone annealed film with spherical morphology, showing dot pattern. Scanning electron micrograph of photolithographic patterns of PαMS-*b*-PHOST. (d) and (e) Films containing the photoacid generator and cross-linker were exposed; (d) as-spun and (e) after THF solvent annealing. Dark regions are lines of cross-linked polymer left behind after postexposure bake and development in solvent. (Source: Bosworth et al. [117]. Reproduced with permission of American Chemical Society.)

order that the minority PαMS block would preferentially wet the surface. This is called "PS brush treatment" to avoid surface-pinning effects [118]. Figure 5.45 shows the SEM images of the films after solvent annealing. With BCP films deposited on the patterned substrates with PS brush treatment and annealed with THF, aligned cylindrical domains with parallel orientation are seen in Figure 5.45a. The figure showed that 37 cylinders run defect-free over a distance of 2 μm. Figure 5.45b shows that annealing in THF of films cast on an untreated SiO_2 substrate (no PS brush treatment) produced cylinders without pattern registration. On the other hand, annealing in acetone led to perpendicular cylinders with hexagonal dot pattern, as seen in Figure 5.45c, irrespective of whether the substrate was PS brush treated or not. These authors followed the time evolution of the OOT from cylinders to spherical morphology with acetone annealing. The transition began after 3 h of annealing in the trenches (much earlier in the thinner areas outside the trench), and it was complete after 5 h.

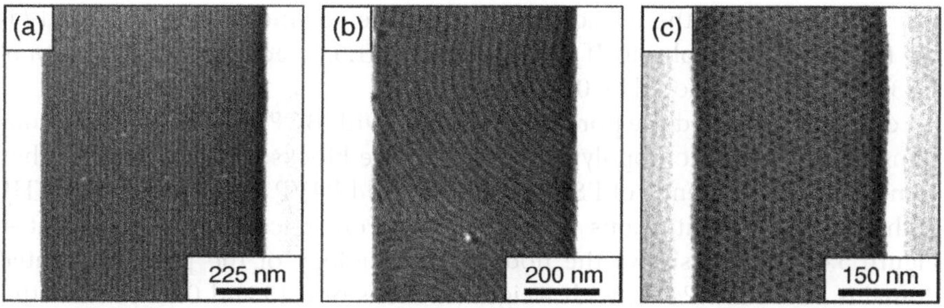

Figure 5.45 Alignment of PαMS-*b*-PHOST on lithographically patterned SiO_2 substrates. (a) Self-aligned cylindrical domains with parallel-orientation on PS brush-treated substrate, after annealing with THF; (b) nonaligned cylinders without pattern registration on substrate without PS treatment, annealed with THF; and (c) vertically aligned hexagonal array of cylinders, with acetone annealing (on substrates with or without PS treatment). Darker regions in the images correspond to PαMS blocks. (Source: Bosworth et al. [117]. Reproduced with permission of American Chemical Society.)

5.10 Porous Structures

A common way to create films with porous structures is to use porogens, which are additives (fillers) used with a matrix material (e.g., a polymer). Once a film is cast, the porogen is selectively washed or etched away to leave pores in the matrix. Other techniques such as nonsolvent addition are also used [119]. Since BCPs form uniform phase separated domains, these are candidates for fabricating nanoporous films. One of the applications explored for such nanoporous films is in antireflective coatings. For an ideal single-layered antireflective coating, the optical thickness of the coating must be one-fourth of the wavelength of the incident light. The refractive index must be such that

$$n_f = (n_0 n_S)^{1/2} \tag{5.6}$$

where n_f, n_0, and n_s are the refractive indices of the film, air, and the substrate, respectively. Since the refractive index of air is 1 and those of glass or plastic substrates are 1.5, the n_f of the antireflective coating must be 1.23 to accomplish zero reflectance at a particular wavelength. If pores smaller than the wavelength of light are homogeneously distributed in a film, it would be possible to meet the aforementioned condition because of the introduction of air. The reflectance also depends on the pore volume. The refractive index and the fractional pore volume are related according to [120]

$$n^2 = n_p^2 (1 - f) + n_a^2 f \tag{5.7}$$

where n_p and n_a are the refractive indices of the polymer film and air, and f is the fractional pore volume. If the value of n_p is 1.5, to achieve an effective n of 1.23, the value of f should be 0.59.

Lee et al. [121] used the porogen technique with BCPs to create thin porous films, by mixing a homopolymer of one of the blocks as the porogen. They spin-cast films of blends of PS-*b*-P4VP BCP and P4VP homopolymer in THF with various concentrations of P4VP on silicon, mica, or glass substrates. Figure 5.46a and b shows the nodular morphology of the phase separated domains of PS and P4VP, with weight fractions of 0.45 and 0.55 of the latter (see Lee et al. [121] for a range of weight fractions of P4VP used). The weight fraction here includes that of the P4VP block in the BCP. Due to different rates of evaporation of THF from the PS and P4VP domains, the morphology shown is not at equilibrium, but kinetically trapped. Figure 5.46c shows a TEM image of a film (0.55 weight fraction of P4VP) that was stained with iodine, which selectively stains the P4VP domains. It shows that the P4VP homopolymer is preferentially associated with the P4VP block of the copolymer. The films were then soaked in ethanol, to selectively remove the P4VP homopolymer and resulting in porous films, as seen in Figure 5.46d and e. An SEM image of the

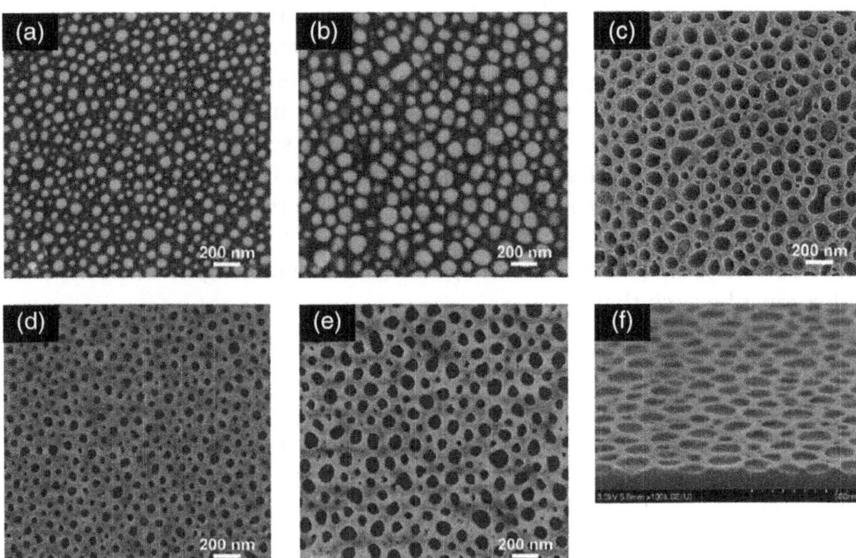

Figure 5.46 (a) and (b) AFM height images of as-spun PS-*b*-P4VP/P4VP blends with 0.45 and 0.55 weight fraction of P4VP. (c) TEM image of iodine-stained PS-*b*-P4VP/P4VP film with 0.55 weight fraction of P4VP. The stained regions correspond to P4VP. (d) and (e) AFM height images after soaking in ethanol, of films with 0.45 and 0.55 P4VP (wt fraction). (f) A tilted SEM image of the film shown in (e). (Source: Lee et al. [121]. Reproduced with permission of Elsevier.)

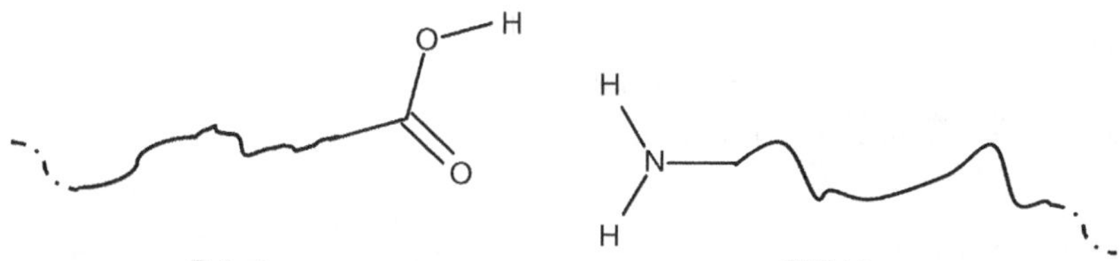

Scheme 5.4 End-functionalized polystyrene and PMMA to form supramolecular block copolymer.

film is shown in Figure 5.46e. With an increase in P4VP weight fraction from 0.03 to 0.55, the average pore diameter increased from 10 to 72 nm. The transmittance over a range of wavelengths was higher than that of bare glass slide.

Gao et al. [120] used a variation of the aforementioned method by using supramolecular assembly. Homopolymers of PS end-functionalized with a COOH group and PMMA terminated with NH_2 group were used as shown in Scheme 5.4, so that a supramolecular BCP could be formed with hydrogen bonding between the terminal groups.

Equimolar ratio of the two polymers were dissolved in THF and stirred at room temperature. The solution was spin-coated on glass slides. By varying the concentration, films with thickness ranging from 50 to 200 nm were made. Although the molecular weights of the two polymers were about the same, lamellar domains were not formed, but nanodomains randomly distributed in the film, without sharp boundaries. This was attributed to the rapid evaporation of THF which prevented the equilibrium morphology to be formed. Porous structure was generated by exposing the films to cyclohexane to remove the PS. The fractional pore volume could be changed by the time of exposure to the solvent. It was found that with a porous film of 116 nm thickness exposed to cyclohexane for 20 min, a high transmittance of 97.93% was achieved for visible light. With thicker films (152 and 203 nm), an inhomogeneous three-layered pore structure was observed upon solvent exposure for 20–40 min, with higher porosity in the top and bottom layers compared to the middle layer. For glass slides coated on both sides with these three-layered films, high transmittance of 98 and 99% was observed in the wavelength regions of 800–1400 nm and 1200–2000 nm, respectively.

As the aforementioned three-layered pore structure enabled high transmission in the broadband region, Han's group [122] designed porous films with gradient refractive index, by surface modification of the substrate and the solvent evaporation. A BCP/homopolymer blend of PS-*b*-PMMA/PMMA was dissolved in a solvent mixture of THF and toluene, and spin-coated on a glass substrate that was treated with octadecyltrichlorosilane. Although both blocks in the BCP were of similar molecular weight, no lamellar (or cylindrical) morphology was observed. Figure 5.47 traces the origin of the gradient refractive index.

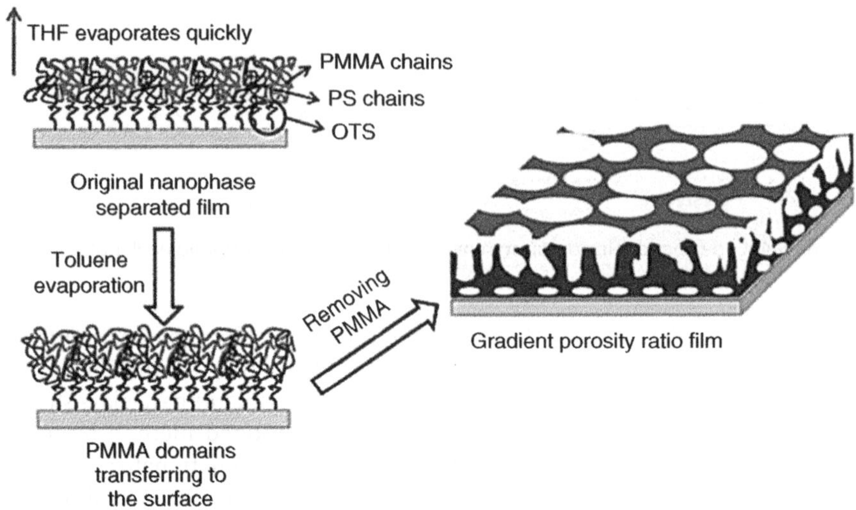

Figure 5.47 Schematic representation of the gradient porosity ratio film formation process. (Source: Li et al. [122]. Reproduced with permission of Wiley.)

With the mixed solvent system used, the THF evaporates fast from the spin-cast film, leaving toluene behind. Because of the surface modification of the substrate and with the help of the remaining toluene, the PMMA block as well as the PMMA homopolymer move vertically toward the surface. Thus, the density of the PMMA domains increases from the glass/film interface to the air surface, creating a gradient. Upon high-intensity UV irradiation of the dry film (365 nm for 3 h), the PMMA chains (both in the BCP and the homopolymer) were degraded and washed away with acetic acid. The purpose of adding the PMMA homopolymer is to vary the porosity and pore size. With such films with gradient refractive index, transmittance of more than 97% was achieved both in the visible and near-infrared region. Further, the film also showed color reproduction response in the visible range.

Using the BCP systems in which the minority component forms cylindrical domains, these can be selectively removed to leave porous channels that run through the thickness of the films. Figure 5.48 illustrates the steps in creating films with well-aligned pores. As the BCP is cast as a film by, for example, spin coating on a substrate, the cylindrical domains will be randomly oriented. These are then aligned to be vertical to the substrate and then removed, leaving pores in the major component of the BCP. As these cylindrical pores are created in the matrix of the other component of the BCP, these can also be used as a nanoreactor to carry out reactions in a confined nanotube. The porous thin films are targeted for applications as photonic bandgap materials, molecular sieves, etc.

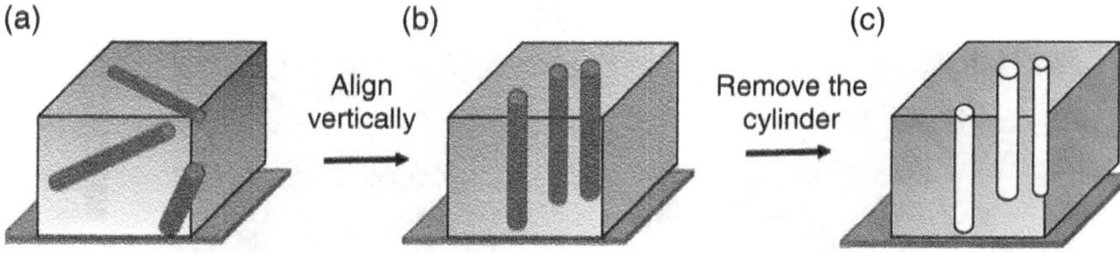

Figure 5.48 Steps involved in creating cylindrical nanocavity structures using block copolymers. (a) Cylindrical domains are initially randomly oriented, (b) cylinders are aligned to be vertical to the substrate surface, and (c) the minor component of the BCP forming the cylindrical phase is removed to leave vertical pores.

Russell et al. [123] used two approaches to align the cylinders vertical to the substrate surface. For thin films of BCP, the surface of the substrate was modified to be neutral, such that the domains of the BCP would align themselves normal to the substrate. Thermal annealing was also required. To this end, Mansky et al. [123] prepared random copolymers of styrene and MMA, which were end-grafted to a silicon surface. Such surfaces were found to be neutral. These authors [124] then spin-coated toluene solutions of styrene–methyl methacrylate BCP (P(S-*b*-MMA)), onto such surface-modified silicon substrate. With the PMMA volume fraction of 30%, the morphology would consist of PMMA cylinders in the PS matrix. The films were then annealed at 170°C for 72 h under vacuum. This temperature is above the glass transition temperatures of both PS (105°C) and PMMA (115°C). Figure 5.49a and b shows the height and phase images obtained from AFM tapping mode, which confirm that the cylinders (14 nm diameter) are normal to the substrate. The films were then exposed to deep UV radiation (254 nm, dose of 25 J/cm^2), which caused the degradation of the PMMA block while cross-linking of the PS matrix. The degraded PMMA was washed away with acetic acid and the film was rinsed with deionized water. This led to an ordered array of nanopores, which were once occupied by the PMMA cylinders, as seen in the AFM images shown in Figure 5.49c and d (height and phase, respectively). Figure 5.50 shows SEM images of the top and cross-sectional images of the film. It is seen from Figure 5.50b that the holes span the entire thickness of the film. These are essentially nanoreactors.

For thicker films (about 1 μm) of the same BCP, Russel et al. used an electric field of 25 V/μm to orient the cylindrical domains. The BCP films were spin-cast from toluene onto a conductive substrate such as silicon, gold-coated silicon, or aluminized Kapton, which is poly(oxydiphenylene pyromellitimide). The film was then annealed at 165°C for 14 h while kept under a direct current electric field of 30–40 V/μm, as shown schematically in Figure 5.51a. This led to the orientation of the PMMA cylinders to orient parallel to the applied field. The field was removed after cooling the film to room temperature.

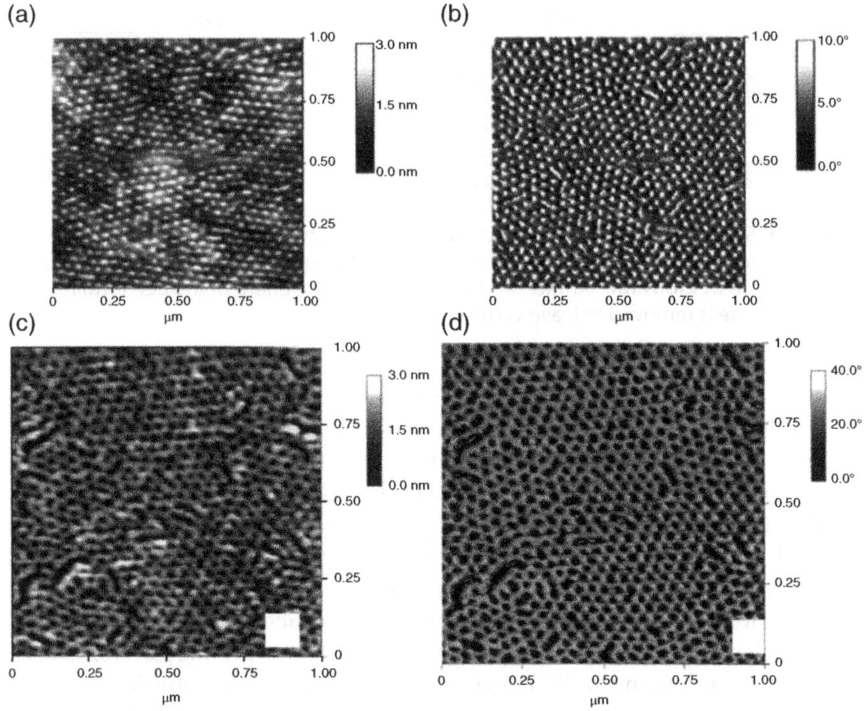

Figure 5.49 AFM image obtained from a thin film (40 nm) of P(S-*b*-MMA) on a neutral substrate after annealing at 170°C in tapping mode; (a) shows the height image and (b) the phase image. After removal of the PMMA material comprising the cylinders, the height (c) and the phase (d) images were obtained. Holes at the original locations of the cylinders can be clearly recognized. (Source: Thurn-Albrecht et al. [124]. Reproduced with permission of Wiley.)

The orientation was confirmed by small angle scattering. As in the previous case, the PMMA was degraded with a deep UV radiation (25 J/cm^2) while simultaneously cross-linking the PS domain. After cleaning the film with acetic acid, optically transparent nanoporous PS film with pores of 14 nm diameter was obtained (Figure 5.51b). As noted earlier, these can be used as nanoreactors. Russell et al. [112] electrodeposited Co or Cu into these pores, to create nanowires (Figure 5.51c).

Ober et al. [125] used a different approach, which they called "additive-driven phase-selective chemistry," in which neither treated surfaces nor an electric field was used. To demonstrate this process, three BCPs were used along with additives as shown in Figure 5.52: (i) Poly(α-methylstyrene-*b*-hydroxy styrene) (P(α-MS-*b*-HOST)); (ii) poly(α-methylstyrene-*b*-isoprene) (P(α-MS-*b*-I)), and (iii) poly(isoprene-*b*-ethylene oxide). The BCPs and the corresponding additives were spin-coated on to silicon substrates, to create films of various

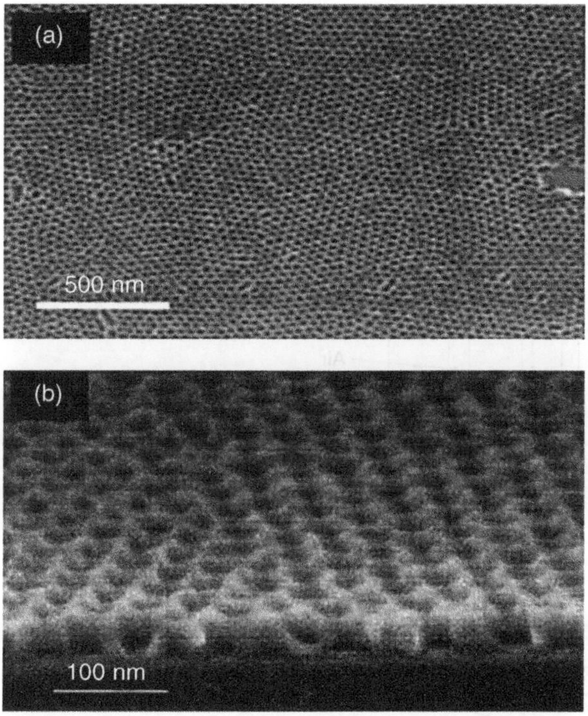

Figure 5.50 FESEM image obtained from a thin film of P(S-*b*-MMA) after removal of the PMMA block inside the cylinders; (a) shows a top view, and (b) a cross-sectional view. (Source: Thurn-Albrecht et al. [124]. Reproduced with permission of Wiley.)

thicknesses, in which the cylindrical domains were vertically aligned upon annealing. In the case of BCP (i), exposure to UV light triggers the photoacid generator to catalyze the reaction of TMMU to cross-link the hydroxystyrene block. Subsequent UV photolysis removes the α-methylstyrene block as the sacrificial block, leaving cylindrical pores. Similarly, with (ii) and (iii), one of the blocks is selectively stabilized and the other removed to leave porous films. These porous films were also used as photoresist templates to create lithographic images.

A different approach using solvent annealing and selective swelling was used by Wang et al. [126, 127]. In this process, there is no elimination of one of the blocks, but only modification of the morphology. Similar to thermal annealing in which a sample is heated in a closed chamber, usually in vacuum, solvent annealing is the process of enclosing the sample in a chamber containing a solvent vapor. The studies on SINC were mentioned earlier. A PS-*b*-P2VP BCP, in which the P2VP was the minor block was spin cast on silicon substrate from chloroform solution [126], varying in thickness from 110 to 600 nm. The films

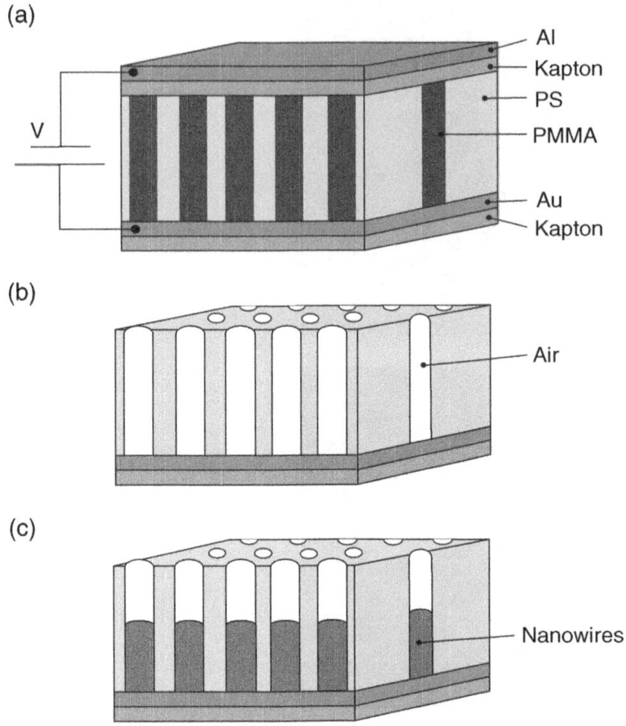

Figure 5.51 A schematic representation of high-density nanowire fabrication in a polymer matrix. (a) An asymmetric diblock copolymer annealed above the glass transition temperature of the copolymer between two electrodes under an applied electric field, forming a hexagonal array of cylinders oriented normal to the film surface; (b) after removal of the minor component, a nanoporous film is formed; and (c) by electrodeposition, nanowires can be grown in the porous template, forming an array of nanowires in a polymer matrix. (Source: Thurn-Albrecht et al. [112]. Reproduced with permission of American Association for Advancement of Science.)

were then solvent-annealed in saturated vapor of chloroform in a closed chamber. It was found that instant evaporation of the solvent after annealing for 40 s at room temperature led to a highly ordered perpendicular orientation of the cylinders. The film was then immersed in hot ethanol (50°C) for 3 h and dried at room temperature. This swelling and drying process created highly uniform, hexagonally organized, monodisperse cylindrical pores of about 12 nm diameter and a spacing of 37 nm between the pores. SEM images of the cross section showed that these pores were straight, through the entire thickness of the film. The initial film can be swollen to different degrees in chloroform at room temperature. Varying the temperature of immersion in ethanol from 40 to 70°C (for 3 h) led to an increase in pore diameter from about 20 to 50 nm and the

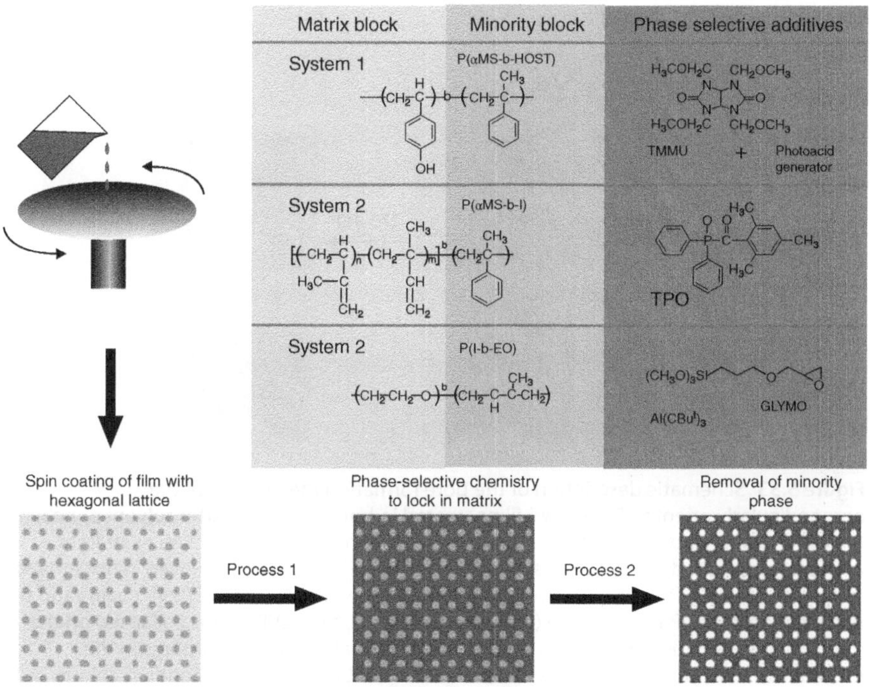

Figure 5.52 Illustration of the steps in fabricating porous films from BCP. The solution of BCP along with the phase-selective additives is spin-coated on to a silicon substrate and annealed, leading to vertically aligned cylindrical domains. Phase selective chemistry locks in the morphology. Removal of the cylindrical minority phase leaves porous films. In Systems 1 and 2, the poly(α-methylstyrene) block is the sacrificial block that gets removed. (Source: Du et al. [125]. Reproduced with permission of Wiley.)

porosity increased from about 3 to 25%. Thus, membranes with various pore diameters can be fabricated. Using the BCP film as templates, atomic layer deposition of aluminum oxide or TiO_2 and subsequent calcination to remove the BCP produced aligned nanotubes, interconnected at one end by the surface layer.

The role of solvent selectivity or neutrality in the aforementioned process was discussed by Yin et al. [126]. Here, selectivity means that the solvent is a good solvent for one of the blocks of the BCP, and it is neutral if the interaction is the same for both blocks. With PS-*b*-P2VP, chloroform is a neutral solvent and annealing causes the solvent to penetrate and increase the mobility of both PS and P2VP blocks in a very short time, leading to perpendicular alignment of the cylindrical domains. When these authors used P4VP as one of the blocks (PS-*b*-P4VP), annealing in chloroform caused the cylindrical domains of the P4VP block to be parallel to the surface of the film. Chloroform is a weaker

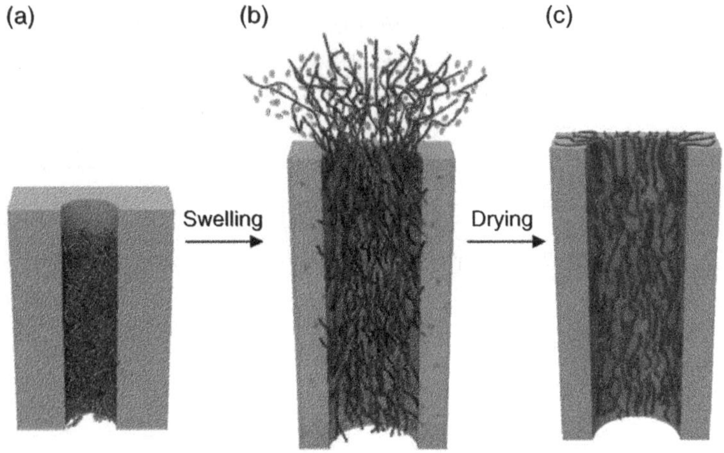

Figure 5.53 Schematic description of the pore formation mechanism of the perpendicularly oriented PS-*b*-P2VP films treated in hot ethanol. (a) After solvent annealing, the P2VP chains (shown as rope-like structures) are condensed and packed as straight cylinders embedded in PS matrix (shown as solid block); (b) when immersed in hot ethanol (ellipsoids), the P2VP chains are swollen, leading to the overflow of the P2VP chains and the deformation of the PS matrix; and (c) after drying in air, the P2VP chains collapsed on the membrane surface and pore walls, leading to the formation of straight pores in the positions of the initial P2VP cylinders. See Yin et al. [126] for the corresponding colour image. (Source: Yin et al. [126]. Reproduced with permission of American Chemical Society.)

solvent for P4VP than for P2VP, and it is a selective solvent for PS-*b*-P4VP, which results in different rates of penetration of the solvent in the blocks. When a mixture of ethanol and chloroform was used to anneal PS-*b*-P4VP, with ethanol being a good solvent for P4VP, highly ordered, perpendicularly oriented cylinders were obtained.

Figure 5.53 illustrates the mechanism of pore formation in the solvent-annealing/swelling process. With solvent annealing in chloroform cylindrical domains of P2VP (rope-like) aligned vertically in the PS matrix (solid block) are obtained (Figure 5.53a). Immersing in hot ethanol (ellipsoids) causes the swelling of the P2VP block, deformation of the PS matrix, and overflow of the P2VP chains (Figure 5.53b). Upon drying in air, the P2VP chains collapse on the pore walls as well as on the surface of the film, creating pores where the cylinders were once present. This model for the mechanism of pore formation would also explain the dependence of pore size and porosity on the temperature of ethanol treatment. Thus, in the solvent-annealing/swelling process, there is no sacrifice of one of the blocks: the chains are just collapsed on the walls of the pores.

Yang et al. [128] used the swelling-induced porous film prepared as before as template to create inorganic antireflective films. As described earlier,

PS-*b*-P2VP BCP was dissolved in chloroform with different concentrations to prepare spin-coated films of various thicknesses. These BCP films formed phase separated, nonporous morphology. By immersing in ethanol at 65°C for 15 h, films with high porous morphology were created. As the solution concentration of the BCP was increased, the thickness of the film increased from 142 nm for a 0.6% concentration to 326 nm with 1.2%. The pores also became more uniform with thickness. Correspondingly, the maximum transmission increased from 97.6% at the wavelength of 612 nm to 99% at 1290 nm. An inorganic replica of the porous BCP was prepared by atomic layer deposition of TiO_2 on the porous membranes and calcination in air to burn off the BCP. This produced highly porous TiO_2 membranes for application as antireflective films.

Cui et al. [129] used solvent annealing and thermal annealing to convert spherical domains to oriented cylindrical domains in a PS-*b*-PMMA BCP. The volume fraction of PS was 25.2%. They used a mixture of dichloromethane, methanol, and acetone to cast thick films in a Petri dish and thin films by spin-casting on silicon wafer. While CH_2Cl_2 evaporated fast, it took several days for the rest of the solvents to do so. The initial morphology consisted of spherical domains of PS in the PMMA matrix, due to the selective solvent used. Upon solvent annealing (in THF for thick films and CH_2Cl_2 for thin films), coalescence of the spheres and their transformation to oriented cylinders were observed. The mechanism for such conversion was rationalized as follows: With the PS volume fraction of 25.2%, cylindrical morphology of the PS in the PMMA matrix would be expected. However, due to the use of selective solvent, nonequilibrium spherical morphology occurs and gets frozen in the drying process. As the solvent evaporates, a chemical potential gradient develops in the film in the direction of evaporation (upward). Upon annealing, the memory of the chemical potential gradient drives the directional coalescence of the spherical domains in the direction perpendicular to the surface of the film (direction of the original solvent evaporation), leading to cylindrical alignment. These authors [130] also used thermal annealing to transform spherical domains to oriented cylindrical alignment in a triblock copolymer of PS-*b*-PEB-*b*-PS. Mixed selective solvent system was used in this case also.

It was discussed earlier that Russell et al. [112] deposited Co or Cu into the cylindrical pores to create nanowires. Others used such vertically oriented cylindrical domains in BCP on, for example, silica substrates, to decorate the cylinders with metal nanoparticle deposits and removed the BCP to create nanodot arrays on substrates. Figure 5.54 shows the process developed by Gowd et al. [131] in which the cylinders formed by the P4VP block in a PS-*b*-P4VP BCP were aligned vertically, and dipped in a dispersion of inorganic nanoparticles. These particles preferentially adsorbed to the P4VP block, forming a hexagonally packed set of nanodots. The polymer was cross-linked

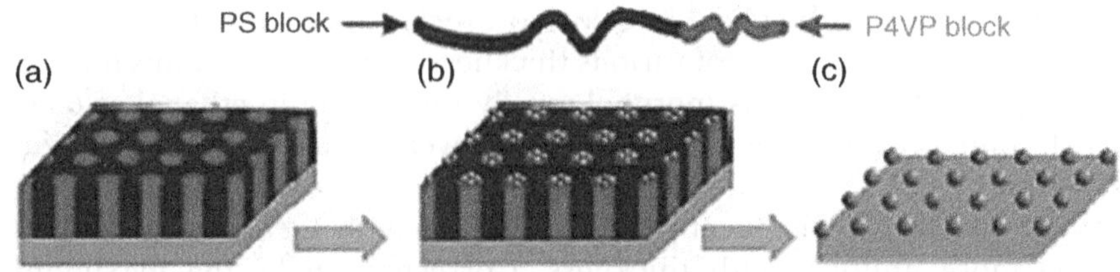

Figure 5.54 Illustration of the route to create nanodots. (a) PS-*b*-P4VP thin film with P4VP cylinders aligned perpendicular to the surface; (b) upon deposition, nanoparticles selectively decorated the swollen P4VP block; and (c) hexagonally close packed inorganic nanodots remain on the silicon substrate after polymer removal. (Source: Gowd et al. [131]. Reproduced with permission of Elsevier.)

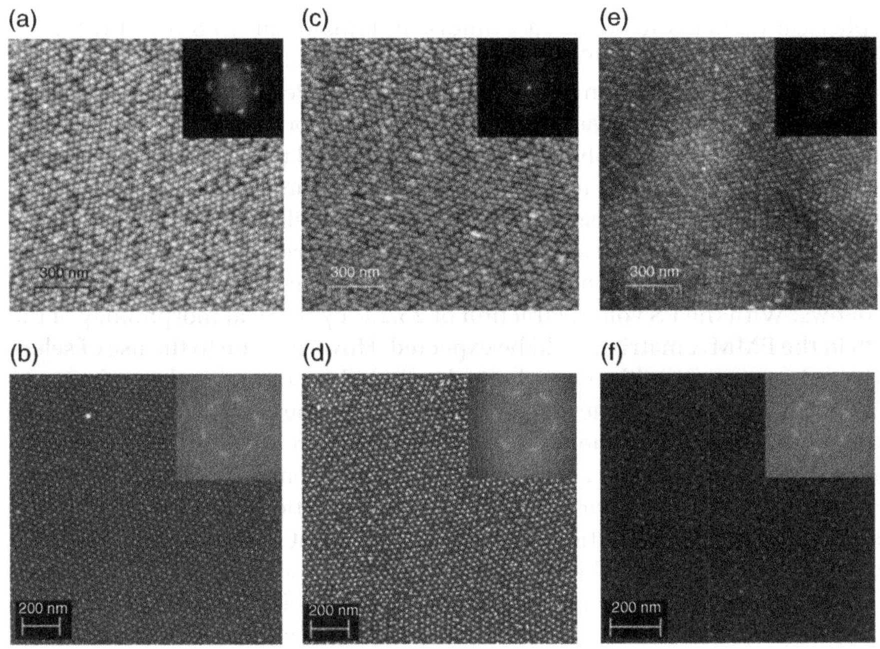

Figure 5.55 SFM (top) and SEM (bottom) images of inorganic nanodots with a center-to-center distance of 29 ± 2 nm on silicon substrates after the removal of polymer thin film. (a, b) Au nanodots; (c, d) Pt nanodots; and (e, f) Pd nanodots. The inset shows an FFT image to show the long-range order. (Source: Gowd et al. [131]. Reproduced with permission of Elsevier.)

and removed, and the nanoparticles remained on the silicon substrate. In these experiments, thin films of PS-*b*-P4VP BCP, with M_n (PS) = 41 500 and M_n (P4VP) = 17 500, were prepared on silicon wafers by dip-coating from chloroform solutions. These were then annealed with 1,4-dioxane vapor for 2 days and dried, which resulted in a film thickness of 28 ± 2 nm, with hexagonally packed cylindrical microdomains of the P4VP block, with an average diameter of 15 ± 1 nm and intercylinder distance of 29 ± 2 nm, as shown in Figure 5.55. The films were dipped in an aqueous dispersion of Au, Pt, or Pd nanoparticles for 2 h. These particles were 3.5 ± 0.5, 2.5 ± 0.5, and 3.8 ± 0.5 nm in diameter, respectively. The incubation for 2 h led to selective deposition of the nanoparticles exclusively on the P4VP domains, with the PS domain unaffected. The polymer film was stabilized by cross-linking with UV radiation and removed by pyrolysis or oxygen plasma, leaving the metallic nanodots on the silicon substrate.

5.11 Crystalline Block Copolymers

If one of the blocks is crystallizable, the nucleation and crystallization of that block would become important and would play a role in phase separation of the domains [132–137]. The work of Bates group [26] was discussed before, on the phase behavior of PE-*b*-PEE diblock copolymers (Scheme 5.1) with the PE volume fraction (f_{PE}) varying from 0.25 to 0.46. The PE block is crystallizable in this case. The BCPs with poly(ferrocenyl silane) block is another example [31, 34]. Müller et al. [138] reviewed the nucleation and crystallization in such BCPs.

5.12 Nanotechnology

The size of the phase separated domains of BCPs provides an opportunity for various applications in the field of nanotechnology. The ability to selectively etch one of the blocks allows for the fabrication of tunable photonic bandgap devices, waveguides, nanoporous polymer membranes or ceramic membranes, antireflective films, and nanoreactors. The morphology of the BCPs can be used as lithographic templates at the nanoscale, and creation of nano-objects. The gyroid morphology, when combined with selective etching, leads to a variety of metamaterials. These applications have been summarized by Park et al. [139]. Chart 5.1 shows the various technologies in which BCPs could have an impact: photonic bandgap materials, nanoporous structures by selective etching, nanolithographic templates, etc., which were illustrated in various sections of this chapter. The subject of nanocomposites (i.e., polymer/clay composites) was not treated here since we focussed mainly on self-assembly. However, molecular composites created by selective association of surfactants with BCPs are discussed in Chapter 8. Amplifying some of the applications listed in Chart 5.1, the various devices that could be fabricated using BCPs are illustrated in Figure 5.56 [139].

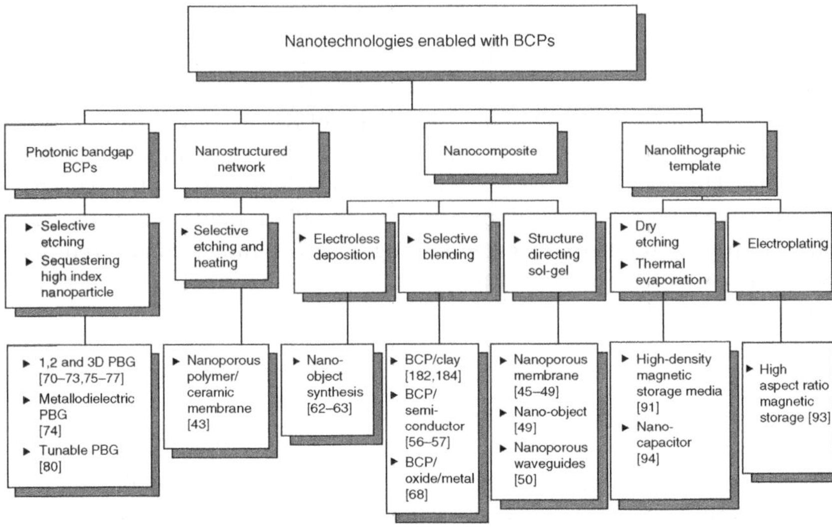

Chart 5.1 A flowchart of the various technologies that are enabled by the unique morphologies of block copolymers. (Source: Park et al. [139]. Reproduced with permission of Elsevier.)

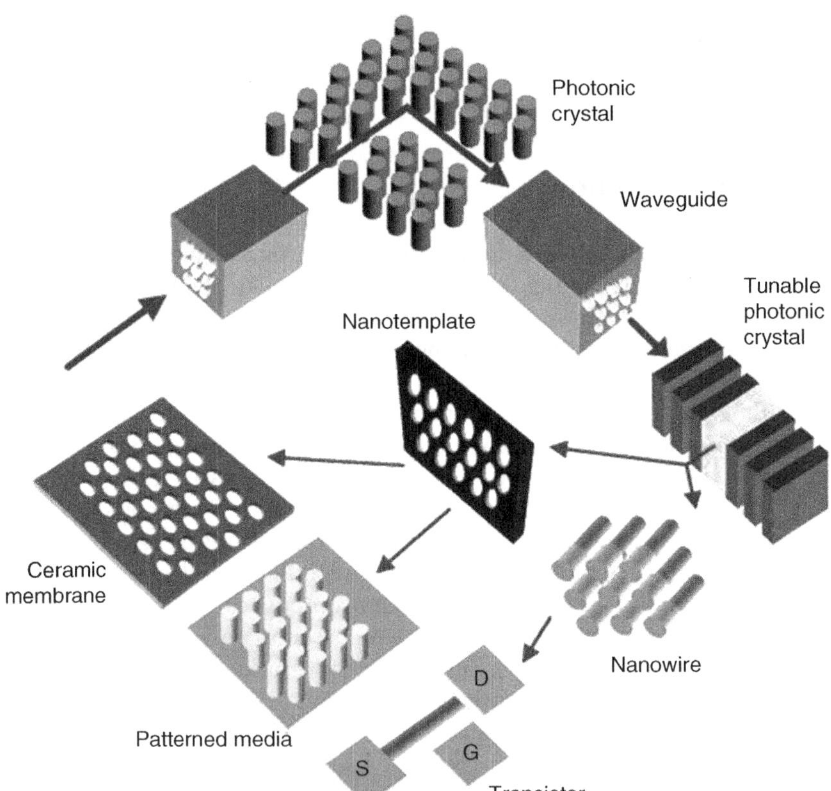

Figure 5.56 An illustration of various devices that could be fabricated by using block copolymer morphologies. (Source: Park et al. [139]. Reproduced with permission of Elsevier.)

References

1 A. S. Dunn and H. W. Melville, *Nature,* **1952**, *169,* 669–700.

2 A. S. Dunn, B. D. Stead, and H. W. Melville, Trans. *Faraday Soc.*, **1954**, *50,* 279–290.

3 H. F. Mark, *Text. Res. J.*, **1953**, *23,* 294–298.

4 F. S. Bates and G. H. Fredrickson, *Phys. Today,* **1999**, *52* (2), 32–38.

5 F. S. Bates and G. H. Fredrickson, *Annu. Rev. Phys. Chem.*, **1990**, *41,* 525–557.

6 B. J. Schmitt, *Angew Chem. Int. Ed. Engl.*, **1979**, *18,* 273–295.

7 L. Leibler, *Macromolecules,* **1980**, *13,* 1602–1617.

8 M. W. Matsen and M. Schick, *Phys. Rev. Lett.*, **1994**, *72,* 2660–2663.

9 M. W. Matsen and F. S. Bates, *Macromolecules,* **1996**, *29,* 1091–1098.

10 S. Vignolini, N. A. Yufa, P. S. Cunha, S. Guldin, I. Rushkin, M. Stefik, K. Hur, U. Wiesner, J. J. Baumberg, and U. Steiner, *Adv. Mater.*, **2012**, *24,* OP23–OP27.

11 A. Schoen, NASA Technical Note TN D-5541, **1970**, Available at: http://ntrs.nasa.gov/search.jsp?R=19700020472 (accessed on July 21, 2016).

12 E. W. Cochran, C. J. Garcia-Cervera, and G. H. Fredrickson, *Macromolecules,* **2006**, *39,* 2449–2451.

13 M. W. Matsen and F. S. Bates, *J. Polym. Sci. B Polym. Phys.*, **1997**, *35,* 945–952.

14 T. Hashimoto, M. Shibayama, and H. Kawai, *Macromolecules,* **1980**, *13,* 1237–1247.

15 H. Hasegawa, H. Tanaka, K. Yamasaki, and T. Hashimoto, *Macromolecules,* **1987**, *20,* 1651–1662.

16 L. Rebenfeld, P. J. Makarewicz, H. D. Weigmann, G. L. Wilkes, *J. Macromol. Sci. Rev. Macromol. Chem.*, **1976**, *C15,* 279.

17 P. J. Makarewicz and G. L. Wilkes, *J. Polym. Sci. Polym. Phys. Ed.*, **1978**, *16,* 1559.

18 N. J. Tyrer, T. L. Bluhm, and P. R. Sundararajan, *Macromolecules,* **1984**, *17,* 2296.

19 A. Perovic and P. R. Sundararajan, *Polym. Bull.*, **1982**, 6, 277–283.

20 M. F. Schulz and F. S. Bates, in *Physical Properties of Polymers Handbook,* **1996**, J. E. Mark, Ed., Chapter 32, Springer.

21 D. A. Hajduk, S. M. Gruner, P. Rangarajan, R. A. Register, L. J. Fetters, C. Honeker, R. J. Albalak, and E. L. Thomas, *Macromolecules,* **1994**, *27,* 490–501.

22 I. W. Hamley, K. A. Koppi, J. H. Rosedale, F. S. Bates, K. Almdal, and K. Mortensen, *Macromolecules,* **1993**, *26,* 5959–5970.

23 D. A. Hajduk, P. E. Harper, S. M. Gruner, C. C. Honeker, G. Kim, E. L. Thomas, and L. J. Fetters, *Macromolecules,* **1994**, *27,* 4063–4075.

24 T. Hashimoto, Y. Ijichi, and L. J. Fetters, *J. Chem. Phys.*, **1988**, *89,* 2463–2472.

25 R. L. Lescanec and M. Muthukumar, *Macromolecules,* **1993**, *26,* 3908–3916.

26 J. Zhao, B. Majumdar, M. F. Schulz, F. S. Bates, K. Almdal, K. Mortensen, D. A. Hajduk, and S. M. Gruner, *Macromolecules,* **1996**, *29,* 1204–1215.

27 S. Foerster, A. K. Khandpur, J. Zhao, F. S. Bates, I. W. Hamley, A. J. Ryan, and W. Bras, *Macromolecules*, **1994**, *27*, 6922–6935.
28 P. M. Lipic, F. S. Bates, and M. W. Matsen, *J. Polym. Sci. B Polym. Phys.*, **1999**, *37*, 2229–2238.
29 D. A. Hajduk, H. Takenouchi, M. A. Hillmyer, F. S. Bates, M. E. Vigild, and K. Almdal, *Macromolecules*, **1997**, *30*, 3788–3795.
30 V. Z.-H. Chan, J. Hoffman, V. Y. Lee, H. Iatrou, A. Avgeropoulos, N. Hadjichristidis, R. D. Miller, and E. L. Thomas, *Science*, **1999**, *286*, 1716–1719.
31 R. G. H. Lammertink, M. A. Hempenius, E. L. Thomas, and G. J. Vancso, *J. Polym. Sci. B Polym. Phys.*, **1999**, *37*, 1009–1021.
32 J. D. Vavasour and M. D. Whitmore, *Macromolecules*, **1993**, *26*, 7070.
33 M. W. Matsen and M. Schick, *Macromolecules*, **1994**, *27*, 4014.
34 C. Kloninger, D. Knecht, and M. Rehahn, *Polymer*, **2004**, *45*, 8323–8332.
35 D. A. Rider, K. A. Cavicchi, K. N. Power-Billard, T. P. Russel, and I. Manners, *Macromolecules*, **2005**, *38*, 6931–6938.
36 J.-C. Eloi, D. A. Rider, J.-Y. Wang, T. P. Russell, and I. Manners, *Macromolecules*, **2008**, *41*, 9474–9479.
37 J. Gwyther and I. Manners, *Polymer*, **2009**, *50*, 5384–5389.
38 J. Gwyther, G. Lotze, I. Hamley, I. Manners, *Macromol. Chem. Phys.*, **2011**, *212*, 198–201.
39 T. Hashimoto, Y. Nishikawa, K. Tsutsumi, *Macromolecules*, **2007**, *40*, 1066.
40 T. Hashimoto, K. Yamasaki, S. Koizumi, and H. Hasegawa, *Macromolecules*, **1993**, *26*, 2895–2904.
41 T. Hashimoto, S. Koizumi, and H. Hasegawa, *Macromolecules*, **1994**, *27*, 1562–1570.
42 F. Court and T. Hashimoto, *Macromolecules*, **2001**, *34*, 2536–2545.
43 F. Court and T. Hashimoto, *Macromolecules*, **2002**, *35*, 2566–2575.
44 A.-C. Shi and J. Noolandi, *Macromolecules*, **1994**, *27*, 2936.
45 A.-C. Shi, J. Noolandi, and H. Hoffmann, *Macromolecules*, **1994**, *27*, 6661.
46 A.-C. Shi and J. Noolandi, *Macromolecules*, **1995**, *28*, 3103.
47 M. M. Disko, K. S. Liang, S. K. Behal, R. J. Roe, and K. J. Jeon, *Macromolecules*, **1993**, *26*, 2983–2986.
48 R. J. Spontak, S. D. Smith, and A. Ashraf, *Macromolecules*, **1993**, *26*, 956–962.
49 T. Hashimoto, K. Tsutsumi, and Y. Funaki, *Langmuir*, **1997**, *13*, 6869–6872.
50 K. I. Winey, E. L. Thomas, and L. J. Fetters, *J. Chem. Phys.*, **1991**, *95*, 9367.
51 K. I. Winey, E. L. Thomas, and L. J. Fetters, *Macromolecules*, **1991**, *24*, 6182.
52 K. I. Winey, E. L. Thomas, and L. J. Fetters, *Macromolecules*, **1992**, *25*, 422.
53 K. I. Winey, E. L. Thomas, and L. J. Fetters, *Macromolecules*, **1992**, *25*, 2645.
54 C. Kloninger and M. Rehahn, *Macromolecules*, **2004**, *37*, 8319–8324.
55 A. Sidorenko, I. Tokarev, S. Minko, and M. Stamm, *J. Am. Chem. Soc.*, **2003**, *125*, 12211–12216.
56 I. Tokarev, R. Krenek, Y. Brukov, D. Schmeisser, A. Sidorenko, S. Minko, and M. Stamm, *Macromolecules*, **2005**, *38*, 507–516.

57 D. Zschech, A. P. Milenin, R. Scholz, R. Hillebrand, Y. Sun, P. Uhlmann, M. Stamm, M. Steinhart, and U. Gosel, *Macromolecules*, **2007**, *40*, 7752–7754.
58 O. Seifarth, R. Krenek, I. Tokarev, Y. Burkov, A. Sidorenko, S. Minko, M. Stamm, and D. Schmeisser, *Thin Solid Films*, **2007**, *515*, 6552–6556.
59 V. Luchnikov, A. Kondyurin, P. Formanek, H. Lichte, and M. Stamm, *Nano Lett.*, **2007**, *7*, 3628–3632.
60 B. Nandan, M. K. Vyas, M. Böhme, and M. Stamm, *Macromolecules*, **2010**, *43*, 2463–2473.
61 O. Ikkala and G. ten Brinke, *Science*, **2002**, *295*, 2407–2409.
62 O. Ikkala and G. ten Brinke, *Chem. Commun.*, **2004**, 2131–2137.
63 W. van Zoelen, T. Asumaa, J. Ruokolainen, O. Ikkala, and G. ten Brinke, *Macromolecules*, **2008**, *41*, 3199–3208.
64 W. van Zoelen, E. Polushkin, and G. ten Brinke, *Macromolecules*, **2008**, *41*, 8807–8814.
65 S.-H. Tung, N. C. Kalarickal, J. W. Mays, and T. Xu, *Macromolecules*, **2008**, *41*, 6453–6462.
66 S.-H. Tung and T. Xu, *Macromolecules*, **2009**, *42*, 5761–5765.
67 M. Morton, *J. Polym. Sci. Polym. Symp.*, **1977**, *60*, 1.
68 T. Uchida, T. Soen, T. Inoue, and H. Kawai, *J. Polym. Sci. A-2*, **1972**, *10*, 101–121.
69 J. F. Beecher, L. Marker, R. D. Bradford, and S. L. Aggarwal, *J. Polym. Sci. C*, **1969**, *26*, 117–134.
70 W. Zheng and Z.-G. Wang, *Macromolecules*, **1995**, *28*, 7215–7223.
71 R. Stadler, C. Auschra, J. Beckmann, U. Krappe, I. Voight-Martin, and L. Leibler, *Macromolecules*, **1995**, *28*, 3080–3097.
72 E. Helfand and A. M. Sapse, *J. Chem. Phys.*, **1975**, *62*, 1327–1331.
73 Y. Mogi, H. Kotsuji, Y. Kaneko, K. Mori, Y. Matsushita, and I. Noda, *Macromolecules*, **1992**, *25*, 5408–5411.
74 Y. Mogi, K. Mori, Y. Matsushita, and I. Noda, *Macromolecules*, **1992**, *25*, 5412–5415.
75 S. P. Gido, D. W. Schwark, and E. L. Thomas, *Macromolecules*, **1993**, *26*, 2636–2640.
76 C. Auschra and R. Stadler, *Macromolecules*, **1993**, *26*, 2171–2174.
77 U. Krappe, R. Stadler, and I.-G. Voigt-Martin, *Macromolecules*, **1995**, *28*, 4458.
78 K. Jung, V. Abetz, and R. Stadler, *Macromolecules*, **1996**, *29*, 1076.
79 U. Breiner, U. Krappe, V. Abertz, and R. Stadler, *Macromol. Chem. Phys.*, **1997**, *198*, 1051–1083.
80 J. Suzuki, M. Seki, and Y. Matsushita, *J. Chem. Phys.*, **2000**, *112*, 4862–4868.
81 A. Urbas, R. Sharp, Y. Fink, E. L. Thomas, M. Xenidou, and L. J. Fetters, *Adv. Mater.*, **2000**, *12*, 812–814.
82 A. Urbas, Y. Fink, and E. L. Thomas, *Macromolecules*, **1999**, *32*, 4748–4750.
83 A. M. Urbas, M. Maldovan, P. DeRege, and E. L. Thomas, *Adv. Mater.*, **2002**, *14*, 1850–1853.

84 E. J. W. Crossland, M. Kamperman, M. Nedelcu, C. Ducati, U. Wiesner, D.-M. Smilgies, G. E. S. Toombes, M. A. Hillmyer, S. Ludwigs, U. Steiner, and H. J. Snaith, *Nano Lett.*, **2009**, *9*, 2807–2812.

85 A. Jain, G. E. S. Toombes, L. M. Hall, S. Mahajan, C. B. W. Garcia, W. Probst, S. M. Gruner, and U. Wiesner, *Angew. Chem. Int. Ed.*, **2005**, *44*, 1226–1229.

86 A. C. Finnefrock, R. Ulrich, A. Du Chesne, C. C. Honeker, K. Schumacher, K. K. Unger, S. M. Gruner, and U. Wiesner, *Angew. Chem. Int. Ed.*, **2001**, *40*, 1208–1211.

87 R. Ulrich, A. Du Chesne, M. Templin, and U. Wiesner, *Adv. Mater.*, **1999**, *11*, 141–146.

88 M. Templin, A. Franck, A. Du Chesne, H. Leist, Y. Zhang, R. Ulrich, V. Schädler, and U. Wiesner, *Science*, **1997**, *278*, 1795–1798.

89 G. G. du Sart, I. Vukovic, Z. Vukovic, E. Polushkin, P. Hiekkataipale, J. Ruokolainen, K. Loos, and G. ten Brinke, *Macromol. Rapid Commun.*, **2011**, *32*, 366–370.

90 I. Vukovic, T. P. Voortman, D. H. Merino, G. Portale, P. Hiekkataipale, J. Ruokolainen, G. ten Brinke, and K. Loos, *Macromolecules*, **2012**, *45*, 3503–3512.

91 A. J. Pennings, J. M. A. A. van der Mark, and H. C. Booji, *Kolloid Z. Z. Polym.*, **1970**, *236*, 99.

92 H. D. Chanzy and E. J. Roche, *J. Polym. Sci. Polym. Phys. Ed.*, **1974**, *12*, 2583–2586.

93 E. I. Givargizov, *J. Cryst. Growth*, **2008**, *310*, 1686–1690.

94 S. Y. Chou, P. R. Krauss, and P. J. Renstrom, *Science*, **1996**, *272*, 85–87.

95 M. Schvartzman, D. Tsivion, D. Mahalu, O. Raslin, and E. Joselevich, *Proc. Natl. Acad. Sci.*, **2013**, *110*, 15195–15200.

96 J. Y. Cheng, C. A. Ross, H. I. Smith, and E. L. Thomas, *Adv. Mater.*, **2006**, *18*, 2505–2521.

97 S. Xiao, X. Yang, E. W. Edwards, Y.-H. La, and P. F. Nealey, *Nanotechnology*, **2005**, *16*, S324–S329.

98 S. H. Kim, M. J. Misner, T. Xu, M. Kimura, and T. P. Russell, *Adv. Mater.*, **2004**, *16*, 226–231.

99 P. Du, M. Li, K. Douki, X. Li, C. B. W. Garcia, A. Jain, U. Wiesner, and C. K. Ober, *Adv. Mater.*, **2004**, *16*, 953–957.

100 M. Li, K. Douki, K. Goto, X. Li, C. Coenjarts, D. M. Smilgies, and C. K. Ober, *Chem. Mater.*, **2004**, *16*, 3800–3808.

101 C. T. Black and O. Bezencenet, *IEEE Trans. Nanotechnol.*, **2004**, *3*, 412–415.

102 J. Y. Cheng, A. M. Mayes, and C. A. Ross, *Nat. Mater.*, **2004**, *3*, 823–828.

103 H. Li and W. T. S. Huck, *Nano Lett.*, **2004**, *4*, 1633–1636.

104 M. Lazzari and M. A. Lopez-Quintela, *Adv. Mater.*, **2003**, *15*, 1583–1594.

105 I. W. Hamley, *Angew. Chem. Int. Ed.*, **2003**, *42*, 1692–1712.

106 J. Y. Cheng, C. A. Ross, E. L. Thomas, H. I. Smith, and G. J. Vancso, *Adv. Mater.*, **2003**, *15*, 1599–1602.

107 J. Y. Cheng, C. A. Ross, E. L. Thomas, H. I. Smith, and G. J. Vancso, *Appl. Phys. Lett.*, **2002**, *81*, 3657–3659.

108 R. A. Segalman, H. Yokoyama, and E. J. Kramer, *Adv. Mater.*, **2001**, *13*, 1152–1155.

109 L. Rockford, S. G. J. Mochrie, and T. P. Russel, *Macromolecules*, **2001**, *34*, 1487–1492.

110 W. A. Lopes and H. M. Jaeger, *Nature*, **2001**, *414*, 735–738.

111 K. Fukunaga, H. Elbs, R. Magerle, and G. Krausch, *Macromolecules*, **2000**, *33*, 947–953.

112 T. Thurn-Albrecht, J. Schotter, G. A. Kästale, N. Emley, T. Shibauchi, L. Krusin-Elbaum, K. Guarini, C. T. Black, M. T. Tuominen, and T. P. Russell, *Science*, **2000**, *290*, 2126–2129.

113 X. M. Yang, R. D. Peters, P. F. Nealey, H. H. Solak, and F. Cerrina, *Macromolecules*, **2000**, *33*, 9575–9582.

114 S. Förster and M. Antonietti, *Adv. Mater.*, **1998**, *10*, 195–217.

115 Z.-R. Chen, J. A. Kornfield, S. D. Smith, J. T. Grothaus, and M. M. Satkowski, *Science*, **1997**, *277*, 1248–1253.

116 (a)A. del Campo and E. Arzt, *Chem. Rev.*, **2008**, *108*, 911–945;(b)C. Acikgoz, M. A. Hempenius, J. Huskens, and G. J. Vancso, *Eur. Polym. J.*, **2011**, *47*, 2033–2052.

117 J. K. Bosworth, M. Y. Paik, R. Ruiz, E. L. Schwartz, J. Q. Huang, A. W. Ko, D.-M. Smilgies, C. T. Black, and C. K. Ober, *ACS Nano*, **2008**, *2*, 1396–1402.

118 C. Harrison, P. M. Chaikin, D. A. Huse, R. A. Register, D. H. Adamson, A. Daniel, E. Huang, P. Mansky, T. P. Russell, C. J. Hawker, D. A. Egolf, I. V. Melnikov, and E. Bodenschatz, *Macromolecules*, **1999**, *33*, 857–865.

119 M. Vaysse, M. K. Khan, and P. Sundararajan, *Langmuir*, **2009**, *25*, 7042–7049.

120 J. Gao, X. Li, B. LI, and Y. Han, *Polymer*, **2010**, *51*, 2683–2689.

121 W. Lee, X. Zhang, and R. M. Briber, *Polymer*, **2010**, *51*, 2376–2382.

122 X. Li, J. Gao, L. Xue, and Y. Han, *Adv. Funct. Mater.*, **2010**, *20*, 259–265.

123 P. Mansky, Y. Liu, E. Huang, T. P. Russell, and C. Hawker, *Science*, **1997**, *275*, 1458–1460.

124 T. Thurn-Albrecht, R. Steiner, J. DeRouchey, C. M. Stafford, E. Huang, M. Bal, M. Tuominen, C. J. Hawker, and T. P. Russell, *Adv. Mater.*, **2000**, *12*, 787–791.

125 P. Du, M. Li, K. Douki, X. Li, C. B. W. Garcia, A. Jain, D. M. Smilgies, L. J. Fetters, S. M. Gruner, U. Wiesner, and C. K. Ober, *Adv. Mater.*, **2004**, *16*, 953–957.

126 J. Yin, X. Yao, J.-Y. Liou, W. Sun, Y.-S. Sun, and Y. Wang, *ACS Nano*, **2013**, *7*, 9961–9974.

127 Y. Wang and F. Li, *Adv. Mater.*, **2011**, *23*, 2134–2148.

128 J. Yang, L. Tong, Y. Yang, X. Chen, J. Huang, R. Chen, and Y. Wang, *J. Mater. Chem. C*, **2013**, *1*, 5133–5141.

129 G. Cui, M. Fujikawa, S. Nagano, M. Sano, H. Takase, T. Miyazaki, S. Sakurai, and K. Yamamoto, *Polymer*, **2014**, *55*, 1601–1608.
130 S. Sakurai, H. Bando, H. Yoshida, R. Fukuoka, M. Mouri, K. Yamamoto, and S. Okamoto, *Macromolecules*, **2009**, *42*, 2115–2121.
131 E. B. Gowd, B. Nandan, N. C. Bigall, A. Eychmüller, P. Formanek, and M. Stamm, *Polymer*, **2010**, *51*, 2661–2667.
132 M. Baiardo and G. C. Alfonso, *Macromol. Chem. Phys.*, **2001**, *202*, 2509–2517.
133 V. Balsarnoa, Y. Paolini, G. Ronca, and A. J. Müller, *Macromol. Chem. Phys.*, **2000**, *201*, 2711–2720.
134 V. Balsarnoa, A. J. Müller, F. von Gyldenfeldt, and R. Stadler, *Macromol. Chem. Phys.*, **1998**, *199*, 1063–1070.
135 M. A. Hillmyer and F. S. Bates, *Macromol. Symp.*, **1997**, *117*, 121–130.
136 V. Balsarnoa and R. Stadler, *Macromol. Symp.*, **1997**, *117*, 153–165.
137 V. Balsarnoa, F. von Gyldenfeldt, and R. Stadler, *Macromol. Chem. Phys.*, **1996**, *197*, 3317–3341.
138 A. J. Müller, V. Balsamo, and M. L. Arnal, *Adv. Polym. Sci.*, **2005**, *190*, 1–63.
139 C. Park, J. Yoon, and E. L. Thomas, *Polymer*, **2003**, *44*, 6725–6760.

Further Reading

V. Abetz and P. F. W. Simon, *Adv. Polym. Sci.*, **2005**, *189*, 125–212.
S. B. Darling, *Prog. Polym. Sci.*, **2007**, *32*, 1152–1204.
J.-F. Gohy, *Adv. Polym. Sci.*, **2005**, *190*, 65–136.
N. Hadjichristidis, S. Pispas, G. Floudas, *Block Copolymers: Synthetic Strategies, Physical Properties, and Applications*, **2003**, Wiley-Interscience.
N. Hadjichristidis, H. Iatrou, M. Pitsikalis, S. Pispas, and A. Avgeropoulos, *Prog. Polym. Sci.*, **2005**, *30*, 725–782.
I. W. Hamley, *Prog. Polym. Sci.*, **2009**, *34*, 1161–1210.
M. A. Hillmyer, *Adv. Polym. Sci.*, **2005**, *190*, 137–181.
J. K. Kim, S. Y. Yang, Y. Lee, and Y. Kim, *Prog. Polym. Sci.*, **2010**, *35*, 1325–1349.
M. Li, C. A. Coenjarts, and C. K. Ober, *Adv. Polym. Sci.*, **2005**, *190*, 183–226.
B. Nandan, B. K. Kuila, and M. Stamm, *Eur. Polym. J.*, **2011**, *47*, 584–599.
A. Nunns, J. Gwyther, and I. Manners, *Polymer*, **2013**, *54*, 1269–1284.

6

Rotaxanes and Polyrotaxanes

The word "rotaxane" derives from Latin to mean wheel and axle. The forerunner is the catenane, which again stems from the word "catena" in Latin, to mean a chain (like the Olympic rings). It appears [1] that the idea of covalently trapping one molecule into another to create interlocked molecules (catenane) was proposed by Willstätter around 1906. Of course, that required that both molecules be cyclizable. Note that the Olympic ring was originally designed in 1912. In their papers published in 1952 and 1953, Mark [2] and Frisch et al. [3] discussed the possibility of cyclics while investigating the molecular weight of poly(dimethyl siloxane) (PDMS) chains. Because of the large difference in the skeletal bond angles Si—O—Si (about 140°) and O—Si—O (110°), a PDMS chain in the *trans* conformation would close upon itself after 12 skeletal bonds ($n = 360/30$). Mark et al. [2, 3] concluded that "the solubility characteristics suggest that the structure is a chain of rings, each of which has a weight of about 5000." This perhaps is the first observation of interconnected rings. Since then, there have been a number of publications [4–7] on macrocyclization equilibrium of PDMS as well as cyclic PDMS.

6.1 Definitions and Early Work

In a 1961 paper entitled "Chemical Topology," Frisch and Wasserman [8] presented several scenarios of "topological isomerism," which perhaps was visionary. They described several possibilities, such as knotted and unknotted loops, interlocked and noninterlocked rings, knots and chains, and Möbius strips. When a linear molecule threads through a cyclic without covalent bonds between them, it is termed "pseudorotaxane," as shown in Figure 6.1a. When blocking groups are attached to the ends to prevent the dethreading of the linear molecule, it becomes a rotaxane (Figure 6.1b). If the linear molecule itself cyclizes and forms an interlocked loop, it is a catenane (Figure 6.1d). Polyrotaxane results when the rotaxane is polymerized (Figure 6.1c). The International Union of Pure and Applied Chemistry (IUPAC) has developed

Physical Aspects of Polymer Self-Assembly, First Edition. P. R. Sundararajan.

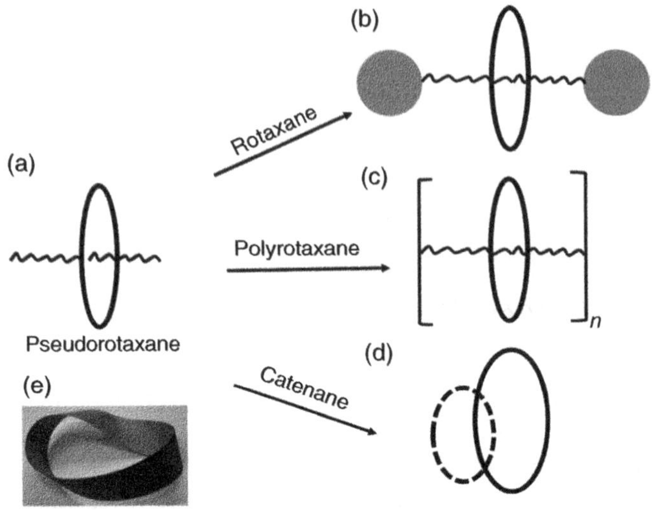

Figure 6.1 (a)–(e) Illustration of several possible assemblies of linear and cyclic molecules.

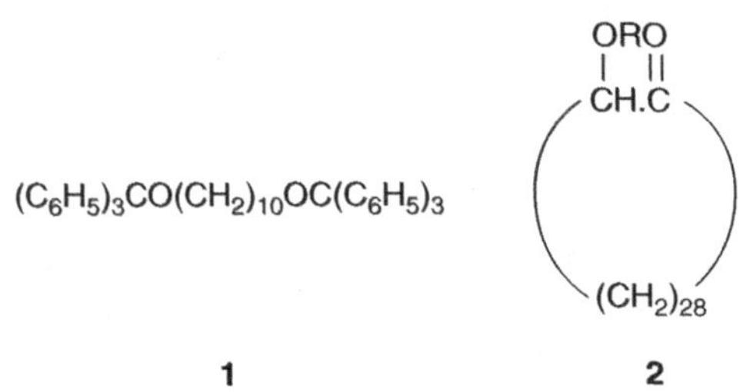

Scheme 6.1 The linear molecule and the macrocycle used by Harrison and Harrison [10] to build a rotaxane.

appropriate nomenclatures for rotaxanes and related structures [9]. Frisch and Wasserman also introduced the isomer for Möbius strip, as in Figure 6.1e. They illustrated how this could be converted to a catenane. The rotaxanes and catenanes are prime examples of supramolecular chemistry.

Following the suggestion of Frisch and Wasserman in the aforementioned paper on the possibility of threading a linear molecule into a macrocycle and preventing its dethreading by a bulky group, Harrison and Harrison [10] synthesized a rotaxane, with the linear chain **1** in Scheme 6.1 and a macrocycle **2**. The triphenyl end groups prevent dethreading.

Further classification leads to (a) a homorotaxane, on the one hand, in which the cyclic and the linear molecules bear chemical resemblance. The linear and the cyclic pair shown in Scheme 6.1 and a polyrotaxane constructed with poly(ethylene oxide) (PEO) and a crown ether are examples of such a system.

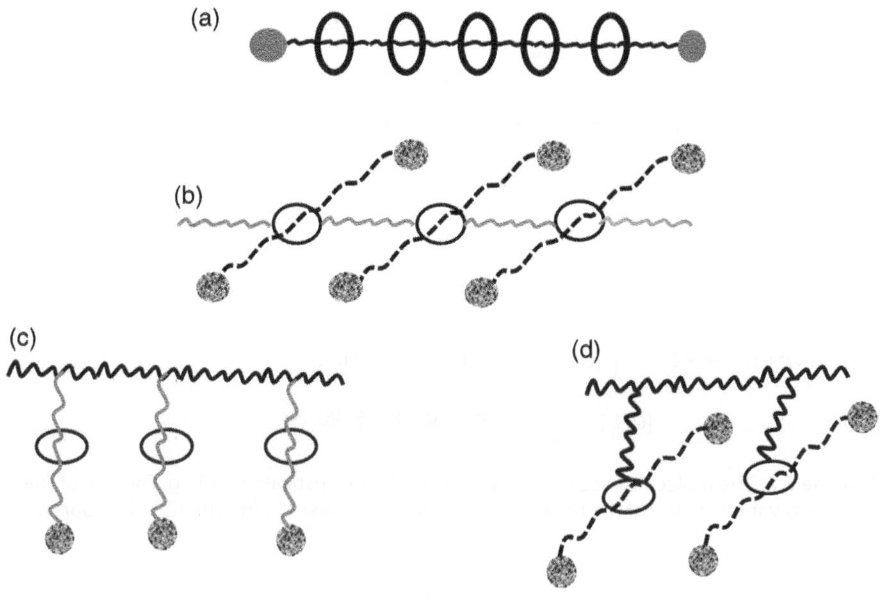

Figure 6.2 Schematics of main chain (a, b) and side chain (c, d) rotaxanes. (Source: Gibson and Marand [1]. Reproduced with permission of Wiley.)

(b) A heterorotaxane, on the other hand, involves a linear and a cyclic species of different chemical structures. When a linear chain of PEO or a polyester is threaded through a cyclodextrin, a heterorotaxane is formed. Thus, these classifications apply to rotaxanes as well as polyrotaxanes. Naturally, the cyclic must have a large enough cavity for the polymer chain to thread through and the chain should be end-capped to prevent slippage. The rotaxane could also be of the main chain or side chain threaded structure, as illustrated in Figure 6.2.

Various methods have been used over the decades to synthesize rotaxanes and polyrotaxanes. In the statistical approach (shake or stir), the linear and the cyclic molecules are mixed together and the threading would be entropically driven with no specific interaction between them. The advantage is that as long as the cyclics are large enough, the choice between the linear and the cyclic species is wide. However, a large excess quantity of the cyclic might be needed to achieve significant threading. In the template method, the threading is enthalpically directed, with attractive interactions between the cyclic and the linear species. The noncovalent forces, such as hydrogen bonding, metal complexation, charge transfer, and Coulombic interactions, that enable self-assembly could be used to drive the rotaxane formation. An example of the statistical threading in a homorotaxane is the work of Schill et al. [11] in which they used cyclic alkanes of different size with a linear alkane (Scheme 6.2a) and a cyclic alkane with linear alkanes of various lengths (Scheme 6.2b) to study the threading

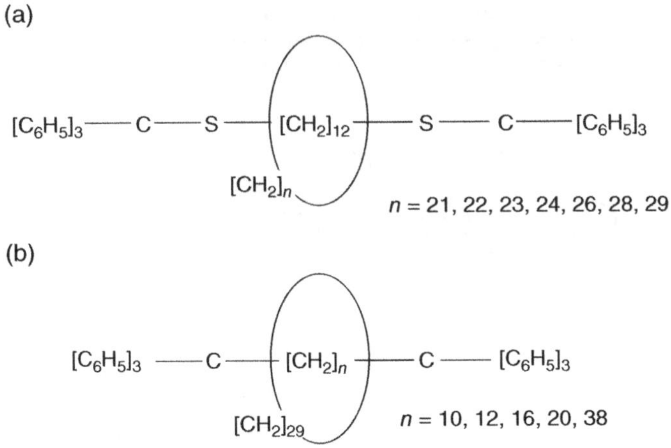

Scheme 6.2 The molecules used by Schill et al. [11] for statistical threading. The size of the cyclic was varied in (a) and the length of the linear component in (b). The $(C_6H_5)_3$ group was used as the stopper.

efficiency. They reported yields of 0.35–6.8% in the case of (**a**) and 4.5–11% for (**b**) depending on the ring size and the length of the linear chain.

Apart from mixing the cyclic and linear species, polyrotaxanes could be formed by (i) cyclization in the presence of linear polymers, (ii) polymerize monomers to form linear chains in the presence of cyclics, and (iii) polymerize monomeric rotaxanes. Agam et al. [12] used statistical threading as well as polymerization of monomers, using the components shown in Scheme 6.3. A crown ether (dibenzo-30-crown-10: Scheme 6.3b) was threaded on to poly(ethylene glycol) (PEG) of molecular weights ranging from 400 to 1000 by mixing them (equimolar) and heating at 120°C for 30 min. The threading was arrested by adding naphthalene-1,5-diisocyanate (Scheme 6.3c) to the mixture and heating for another 30 min, forming high-molecular-weight polyurethanes (Scheme 6.3d). They also used dibenzo-44.1-crown-14.7 and dibenzo-58.2-crown-19.4. With the 30-atom crown, the threading was determined to be about 11–12%. When the ring size was increased to 44-atom crown, the threading was 34 and 41% for PEG 400 and PEG 600, respectively, but decreased to 30% with PEG 1000. Similarly, with the 58-atom crown, the threading was 52 and 63% for PEG 400 and PEG 600, respectively, and in this case also, it decreased to 48% with PEG 1000. The cyclics have to find the chain ends to thread and it appears that as the chain length is increased, this process is slowed. Since the large 58-atom crown and PEG showed the maximum threading in the earlier experiments, Agam et al. [12] also prepared rotaxanes using PEG 600 and dibenzo-58.2-crown-19.4, and using triphenyl group as the stopper. They also polymerized ethylene oxide in the presence of the crown ether and terminated the polymerization after a week by adding triphenylchloromethane, which also served as the stopper. Techniques

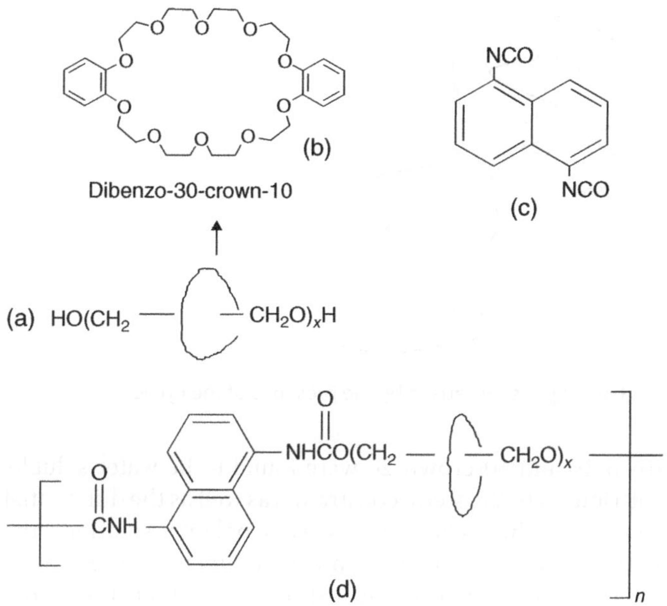

Scheme 6.3 The procedure used by Agam et al. [12]. A linear PEG chain (a) and the crown ether (b) form a rotaxane statistically. The diisocyanate (c) was reacted with the rotaxane to form a polyurethane polyrotaxane (d).

such as infrared (IR), ultraviolet (UV), nuclear magnetic spectroscopy (NMR), and chromatography were used to establish the formation of the rotaxanes.

Following this work, a number of papers were published on the synthesis of rotaxanes and polyrotaxanes [13, 14]. For example, the Gibson group [15] reported on the synthesis of polyrotaxanes with polyurethane chains and crown ether cyclics. They used crown ethers with ring sizes from 36 to 60 and formed polyrotaxanes by synthesizing polyurethanes in the presence of the macrocycles. Since the melting temperatures of these cyclics are in the range of 30–50°C, these were used as the solvent medium to react tetrakis (ethylene glycol) with bis(*p*-isocyanatophenyl) methane to form the polyurethane. It was expected that threading occurred prior to polymerization. The threading efficiency was measured as x/n, which was the mole ratio of the cyclics per repeat unit of the polymer. For a constant crown ether/glycol molar feed ratio of 1.5, the threading efficiency increased with the ring size, from about 0.19 for 36-crown to 0.9 for 60-crown. The polyrotaxane formation was confirmed by ^{1}H-NMR and gel permeation chromatography analysis. A consistent observation was that the solubilities of the polyrotaxanes markedly differed from those of the parent polymers. The polyrotaxanes described in this work were more soluble in common solvents than the parent polymer and the solubility increased with the weight percentage of the crown ethers in the rotaxane. The polyurethane

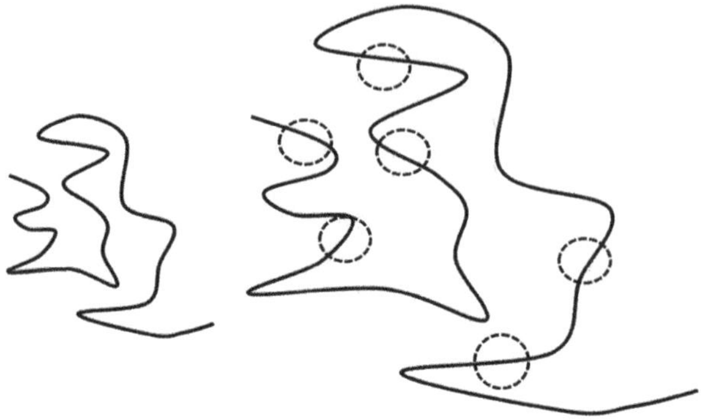

Figure 6.3 Illustration of chain expansion caused by the presence of the cyclics.

rotaxane with 48-crown-16 and 60-crown-20 were found to be water soluble. Such a behavior was attributed to the steric constraints as well as the differential solubilities of the polymer and the cyclics. In the case of polymer solutions, the polymer–solvent interactions and the polymer–polymer interactions (excluded volume) determine the chain conformation and distinguish between a good and a poor solvent. The balance between these also determines the theta condition. The theoretical treatment of excluded volume interactions in the case of cyclic polymers has been presented by Benmouna and Maschke [16]. With the cyclics threaded, the steric constraints would expand the polymer chain as illustrated in Figure 6.3. Thus, although a solvent might be poor for the parent polymer, the effect of the cyclics would be to expand the polymer causing a change in the polymer/solvent interaction. The cyclic would then be acting as a compatibilizer for the polymer/solvent pair. The end-to-end distance of the polymer and the flexibility of the system would be influenced by the presence of the cyclics. It appears that there has been no report on the determination of theta temperatures for polyrotaxanes. In the work of the Gibson group discussed here, the glass transition temperature (T_g) of the parent polyurethane was 51°C, and it decreased linearly to about 15°C with 36-crown and to −40°C with 60-crown.

As another approach, Gibson et al. [17] prepared polyrotaxanes by simply mixing crown ethers with preformed polyurethane, as shown in Scheme 6.4. The polymer was mixed with the molten crown ether (42-crown-14 or 30-crown-10) as the solvent. Hydrogen bonding between the N—H group of the polymer and the oxygen of the crown ether led to as much as 22% by mass of the 42-crown-14 incorporated in the polyrotaxane. Lower extent of threading was found with 30-crown-10 due to fewer oxygen atoms to hydrogen bond. The intrinsic viscosity (in tetrahydrofuran or THF) of the polyrotaxane was higher than that of the parent polymer. Due to the "locking in" of the cyclic to the polymer via hydrogen bonding, the poly(urethane rotaxane)s behaved as a single-phase material. A single glass transition temperature was observed, which depended on the number of threaded cyclics.

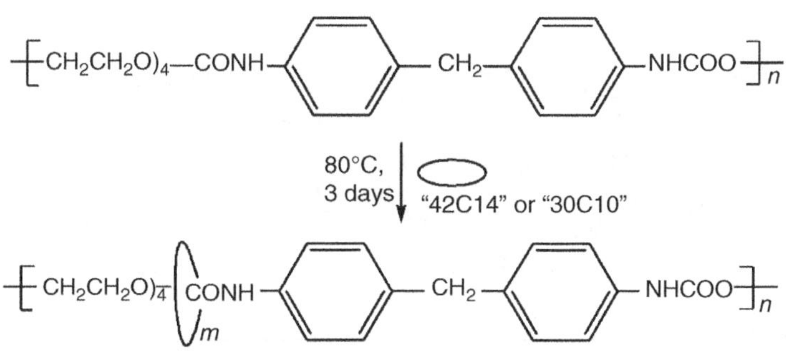

Scheme 6.4 Polyrotaxane synthesis by mixing preformed polyurethane with molten crown ethers.

6.2 Cyclodextrins for Inclusion

Cyclodextrins (CDs) have also been extensively used to form rotaxanes. These are cyclic α-1, 4 linked D-glucose molecules, which are produced by enzymatic degradation of starch. Also known as Schardinger dextrins, these are obtained from the action of an amylase from *Bacillus macerans* on starch. Such naturally produced cyclic amyloses consist of only six, seven, or eight D-glucose units in their structures. These are known as α, β, and γ CDs (cyclohexaamylose, cycloheptaamylose, and cyclooctaamylose, respectively). From energy calculations, Sundararajan and Rao [18] showed that CDs with less than five or greater than eight are not sterically favored when the dihedral angles at each of the α-1,4 linkages were kept the same. Figure 6.4 shows the molecular conformations and the cone-like shape of the molecules. The crystal structure of α-CD potassium acetate complex was first published by Hybl et al. [20]. As seen in Figure 6.4a, the periphery is characterized by hydrogen bonds between successive O-2 and O-3 hydroxyl groups. The interior is predominantly hydrophobic. The size of the interior cavities are 0.57, 0.78, and 0.95 nm, respectively, for α-, β-, and γ-CD, and this enables complexation with small molecules as well as threading of polymers. The history of CDs has been reviewed by Crini [21]. The encapsulation capability allows for several biological applications [22]. A recent review [23] cites that CDs can be found in at least 35 pharmaceutical products and that the CD-based inclusion complexes are of great interest in improving drug bioavailability. The applications of CDs for building supramolecular structures have been reviewed [24, 25].

As the CDs can form complexes with small molecules as well as polymers, a few authors have used the term "inclusion complex" when a polymer was threaded through the CDs. Polymers threading through crystalline occlusions have also been called "inclusion complexes" [19], as shown in Figure 6.5 for a poly(phenylenevinylene) threaded through the channels of mesoporous silica [26, 27]. A summary of the features of host–guest inclusion compounds was presented by Nassimbeni [28]. Generally, when a polymer is threaded through

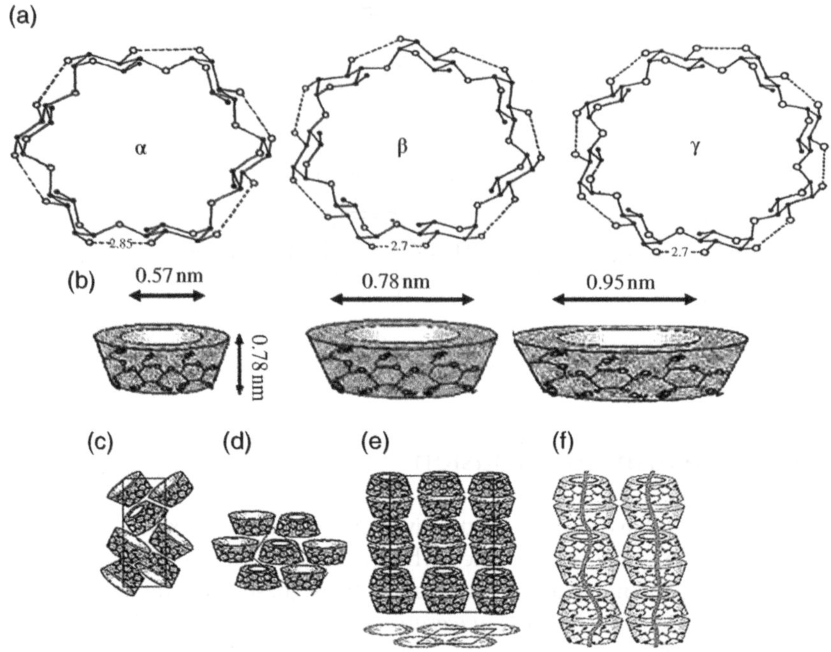

Figure 6.4 (a) Molecular conformations of α, β, and γ CDs, with dotted lines showing the hydrogen bond between the OH-2 and OH-3 hydroxyl groups. (b) Schematics of the cone shapes of the CDs. The molecular depth and the interior diameters are marked. Various modes of stacking are shown: (c) cage, (d) layer, (e) head-to-tail channel, and (f) an inclusion compound with polymer guests. (Source: Hernández et al. [19]. Reproduced with permission of American Chemical Society.)

a cyclic such as a CD, it is then an inclusion complex or a pseudo-polyrotaxane and when stoppers are attached to this complex, it becomes a polyrotaxane.

There have been several studies on CD-based polyrotaxanes. A few will be discussed here. Harada and Kamachi [29, 30] first showed that α-CD can complex with PEG. When saturated aqueous solutions of α-CD were mixed with aqueous solutions of PEG, the precipitates were found to be the inclusion compounds of the two. The rate of complex formation and precipitation depended on the PEG chain length, with a rapid rate for a molecular weight of 1000, and decreasing with higher molecular weights. As the chain length increased, the relative number of chain ends (for the cyclic to thread through) decreased, resulting in the maximum rate for $M_w = 1000$. From X-ray diffraction data, it was concluded that the crystalline structure corresponded to the channel-type organization as in Figure 6.4e and f. They also observed that PEG did not form complex with β-CD due to the large cavity size and that poly(propylene glycol) did not form complex with α-CD due to the small cavity size of the latter. Harada et al. [30] also prepared PEGs $R(CH_2CH_2O)_nCH_2CH_2R'$

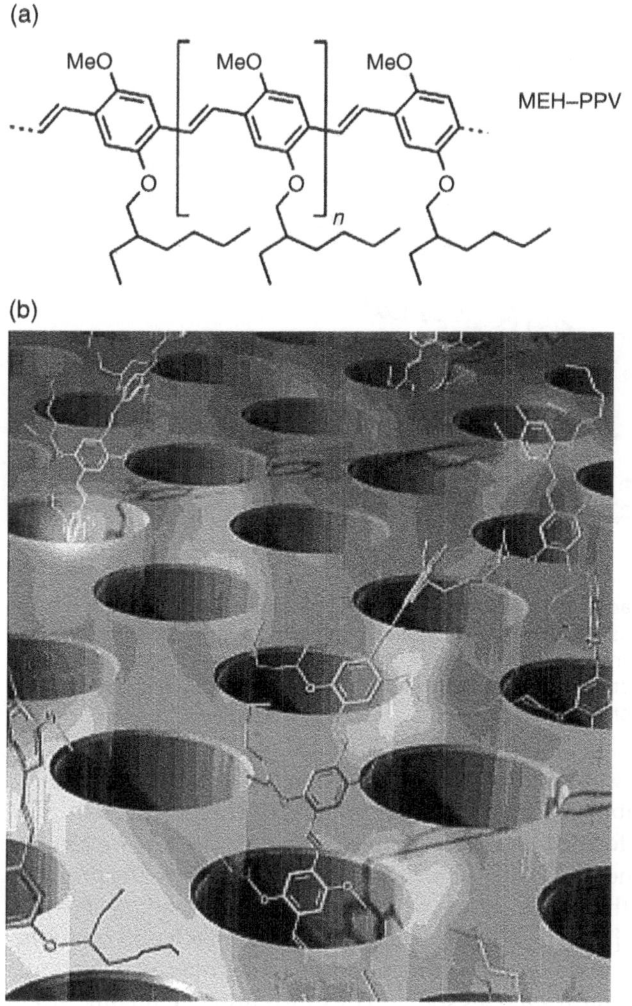

Figure 6.5 (a) Schematic representation of a poly(phenylenevinylene) chain threaded into the channels of mesoporous silica. (b) Each pore, with a diameter of 2.2 nm, can accommodate only one polymer chain. (Source: Nguyen et al. [26]. Reproduced with permission of Wiley.)

with R and R′ corresponding to various small end groups, such as —OH, $—NH_2$, and $—OCH_3$, with M_w ranging from 500 to 2000. The yields in these cases were somewhat higher than those for unmodified PEG. In a similar study, Harada et al. [31] used polyisoprene (PIP) as the axle, with α-, β-, and γ-CD. The PIP molecular weight ranged from 615 to 10 000, with predominantly cis-1,4 configuration (about 70%). When an aqueous solution of γ-CD was added to PIP, it formed inclusion complex, with the yield increasing from 6.3 to 19% as the molecular weight was increased from 615 to 3000. As in the case of PEG

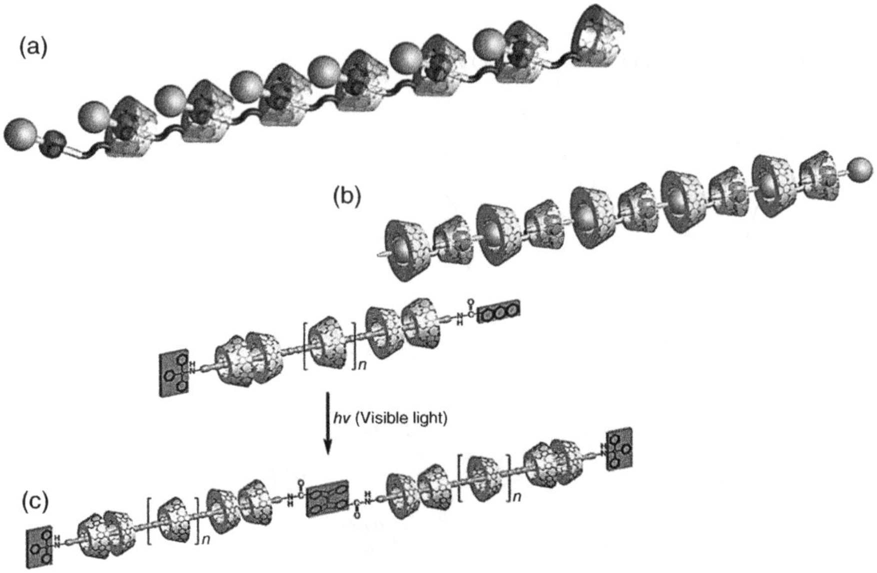

Figure 6.6 Models of supramolecular polyrotaxanes: (a) Polyrotaxane with cinnamic-substituted α-CD in aqueous solution, (b) polyrotaxane formed via heterodimer formation: AdCx-occluded β-CD and cinnamoyl-occluded α-CD with trinitrobenzene and β-CD as stoppers, and (c) photoreaction-induced polyrotaxane based on β-CD. (Source: Harada [32]. Reproduced with permission of Wiley.)

discussed earlier, there was no threading with the long PIP chain of $M_w = 10000$ due to the fewer number of chain ends for the CD to thread through. The β-cyclodextrin (β-CD) formed inclusion complex only with the low-molecular-weight (615) PIP, and the α-CD did not form complex at all, due to the presence of the methyl group in the PIP structure, which impedes the threading of α-CD.

In a later publication [32], Harada constructed polyrotaxanes based on supramolecular polymers. Figure 6.6a shows the model of a polyrotaxane which was formed from a [2] rotaxane in aqueous solution by substituting a cinnamic group at the secondary hydroxyl of the α-CD via an amide linkage (3-CiNH-α-CD). Trinitrobenzene sulfonic acid sodium salt was used as the stopper. In another case, cinnamoyl substituted β-CD was found to be insoluble in water.

When adamantane carboxylic acid (AdCx) was added to the aqueous suspension of cinnamoyl-β-CD, a clear solution resulted, indicating that AdCx was included in the cavity of β-CD, and the cinnamoyl group was expelled. Adding α-CD to this solution/dispersion caused the cinnamoyl group to be occluded in the α-CD cavity and a supramolecular heterodimer was formed. Adding a stopper led to a [2] rotaxane, which formed supramolecular polyrotaxanes in water (Figure 6.6b). In yet another construction using supramolecular chemistry, Okada and Harada [32, 33] prepared poly(propylene glycol) functionalized with triphenylmethyl group on one end and a 2-anthryl group at the other end. This axle was added to a saturated solution of β-CD and stirred for 3 days to form a turbid solution containing a pseudorotaxane (Figure 6.6c). The solution was diluted and was exposed to visible light ($\lambda > 340$ nm) for 5 days under vigorous stirring. This led to dimerization of the 2-anthryl groups leading to a polyrotaxane. It was determined from ^{1}H NMR spectroscopy that about 10 β-CD were interlocked in the polyrotaxane shown in Figure 6.6c.

As for the earlier competition between the cinnamoyl and AdCx substitutions, Inoue et al. [34] functionalized β-CD with PEG of various lengths and a hydrocinnamoyl spacer, for example, mono-6-*O*-[4-poly(ethylene glycol) 10 000 acid-hydrocinnamoyl]-β-CD (6-PEG10000acid-HyCiO-β-CD) (Figure 6.7a). Here, 10 000 refers to the M_n of PEG. In aqueous solutions of 6-PEG10000 acid-HyCiO-β-CD, NMR spectra showed that the cinnamoyl aromatic part was occluded in the cavity of the β-CD. Upon addition of equimolar AdCx, the cinnamoyl part was excluded from the cavity by the inclusion of AdCx. With half-molar inclusion of AdCx, a threading–dethreading process was seen. The conformational exchange rate constant decreased exponentially as the number of PEG units in the substitution increased.

Arai et al. [35] prepared polyrotaxanes by a statistical approach, using amine-terminated poly(tetrahydrofuran) (ATPT) or PEG (ATPEG) as axle, with α-CD or permethylated α-CD (PMeCD) as the cyclics, as shown in Schemes 6.5 and 6.6.

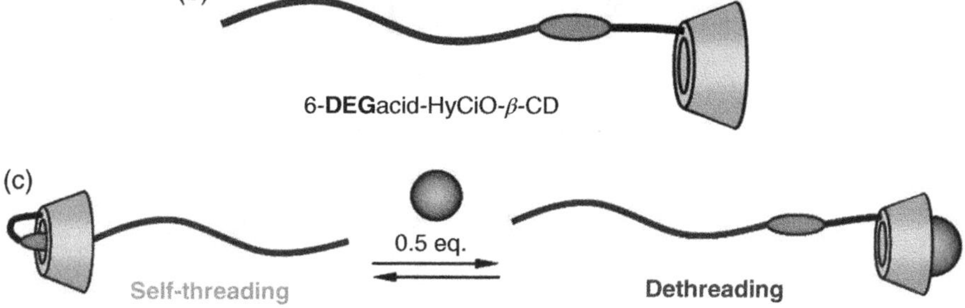

Figure 6.7 (a) β-CD functionalized with PEG of various lengths and a hydrocinnamoyl spacer; (b) schematic of the structure in (a) with the hydrocinnamoyl spacer represented by the ellipse; (c) self-threading is shown on the left with cinnamoyl (ellipse) included-PEG, and de-threading on the right. The sphere represents the adamantane carboxylic acid. (Source: Inoue et al. [34]. Reproduced with permission of American Chemical Society.)

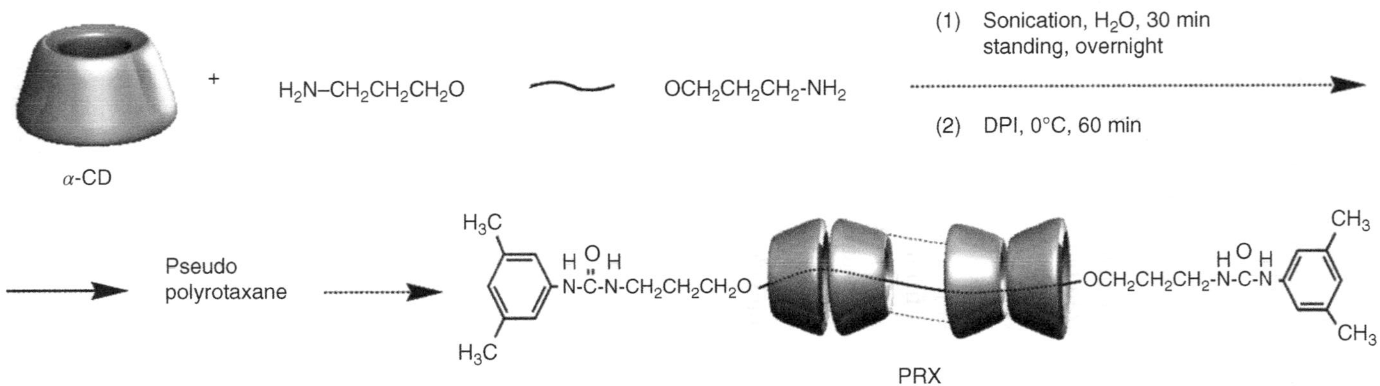

Scheme 6.5 Synthesis of polyrotaxane with α-CD and amine-terminated linear polymer with 3,5-dimethylphenyl isocyanate end-caps [35].

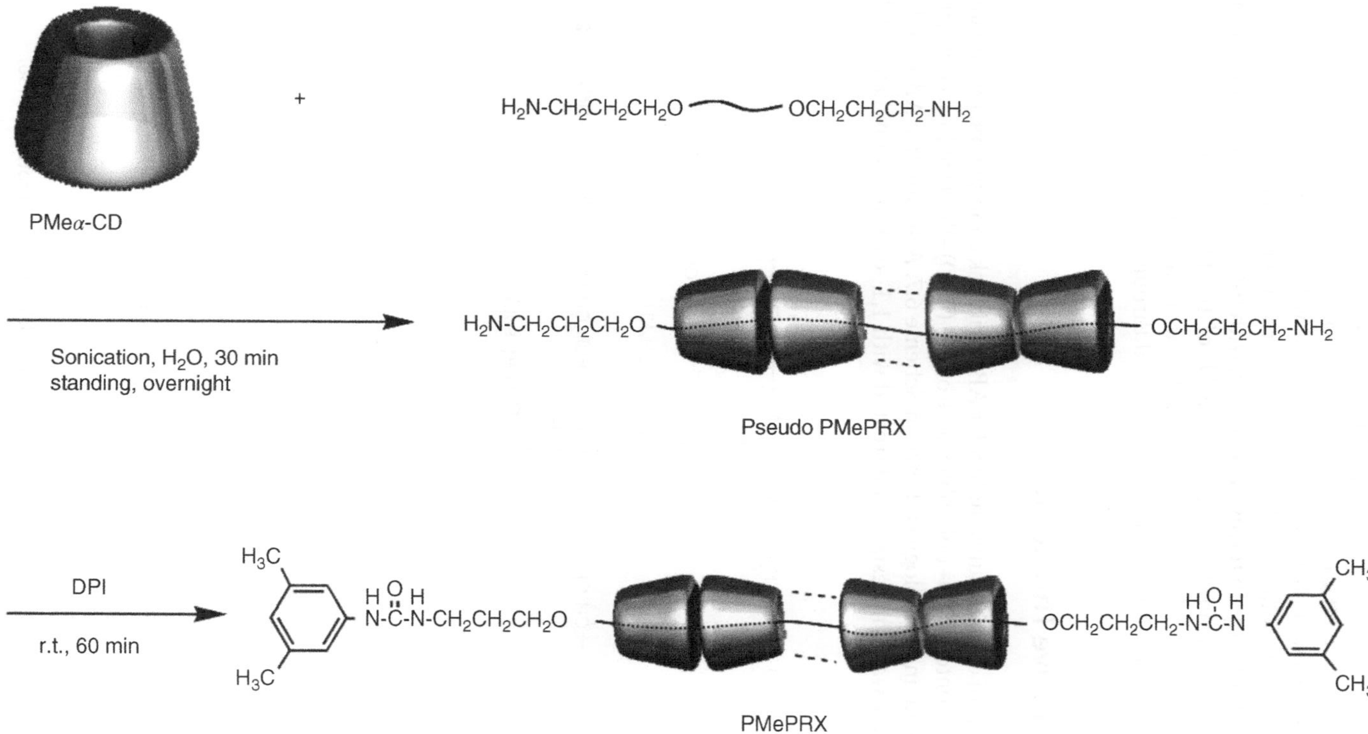

Scheme 6.6 Synthesis of polyrotaxane with permethylated α-CD and amine-terminated linear polymer, with 3,5-dimethylphenyl isocyanate end-caps [35].

In a simple procedure, α-CD or permethylated α-CD (PMeCD) and the linear polymer were sonicated in water for 30 min or kept standing overnight to form pseudo-polyrotaxane. It was then end-capped with 3,5-dimethylphenyl isocyanate to form polyrotaxane. The rotaxane formation was confirmed by NMR and chromatography. In terms of coverage, with α-CD, the ratio of the axle and the cyclic ranged from 85 to 96% with a M_n of 1000 to 1800, but decreased to 54% when a longer axle with a M_n of 7700 was used. It was as if the cyclic had to find the ends of the longer chain to thread through.

6.3 Selective Threading

Yang et al. [36] created supramolecular ABA triblock copolymers, by threading α- and γ-CD onto a PEO–poly(phenylene oxide) (PPO)–PEO triblock copolymer. The synthetic methodology is shown in Scheme 6.7. A symmetric ABA triblock copolymer 1 was synthesized and treated with 1,1′-carbonyldiimidazole (CDI)

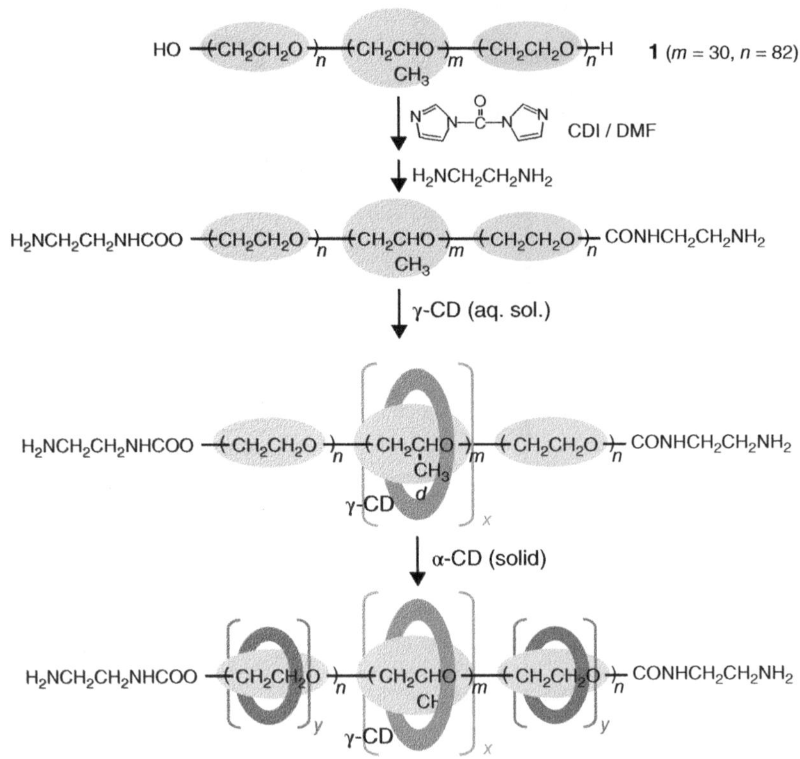

Scheme 6.7 Synthesis of supramolecular poly(triblock/cyclodextrin rotaxane) [36].

followed by reaction with ethylenediamine to replace the hydroxyl groups and yield an amino-terminated triblock copolymer <u>2</u>. The polymer was then added to an aqueous solution of γ-CD, resulting in an inclusion complex of γ-CD with the PPO segment (<u>3</u>). The poly(pseudorotaxane) was then mixed with α-CD and the mixture was kept at 80°C in vacuum for a week, to let the α-CD thread into the PEO blocks to form the poly(pseudorotaxane) (<u>4</u>). It was then capped with 2,4,6-trinitrophenyl stoppers. From ^{1}H NMR analysis of the composition, it was determined that in the final polyrotaxane, the central block had 11 γ-CD units (x in Scheme 6.7) and there were nine α-CD units on the PEO blocks. Thus, it is an ABA-type triblock polyrotaxane with the CDs.

In a further study, Li et al. [37] synthesized cationic polyrotaxanes for application to gene delivery, since cationic polymers have been investigated as promising nonviral gene delivery vectors. The synthetic procedure is shown in Scheme 6.8a and b. A (Pluronic L64) PEO-*b*-PPO-*b*-PEO triblock copolymer

(a)

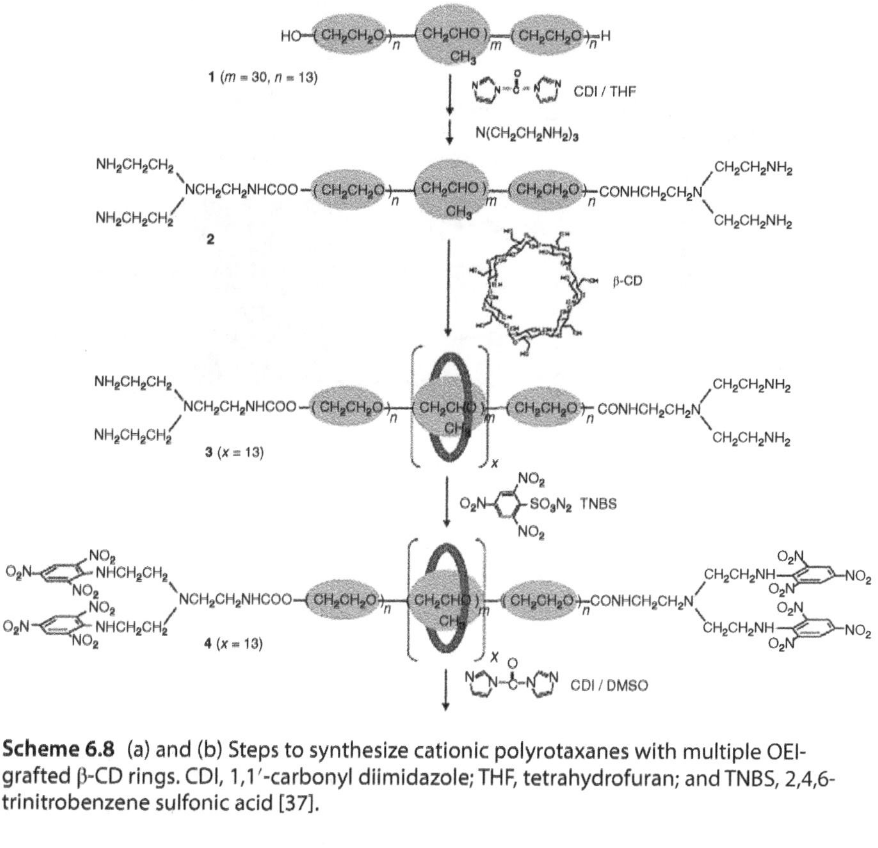

Scheme 6.8 (a) and (b) Steps to synthesize cationic polyrotaxanes with multiple OEI-grafted β-CD rings. CDI, 1,1′-carbonyl diimidazole; THF, tetrahydrofuran; and TNBS, 2,4,6-trinitrobenzene sulfonic acid [37].

(b)

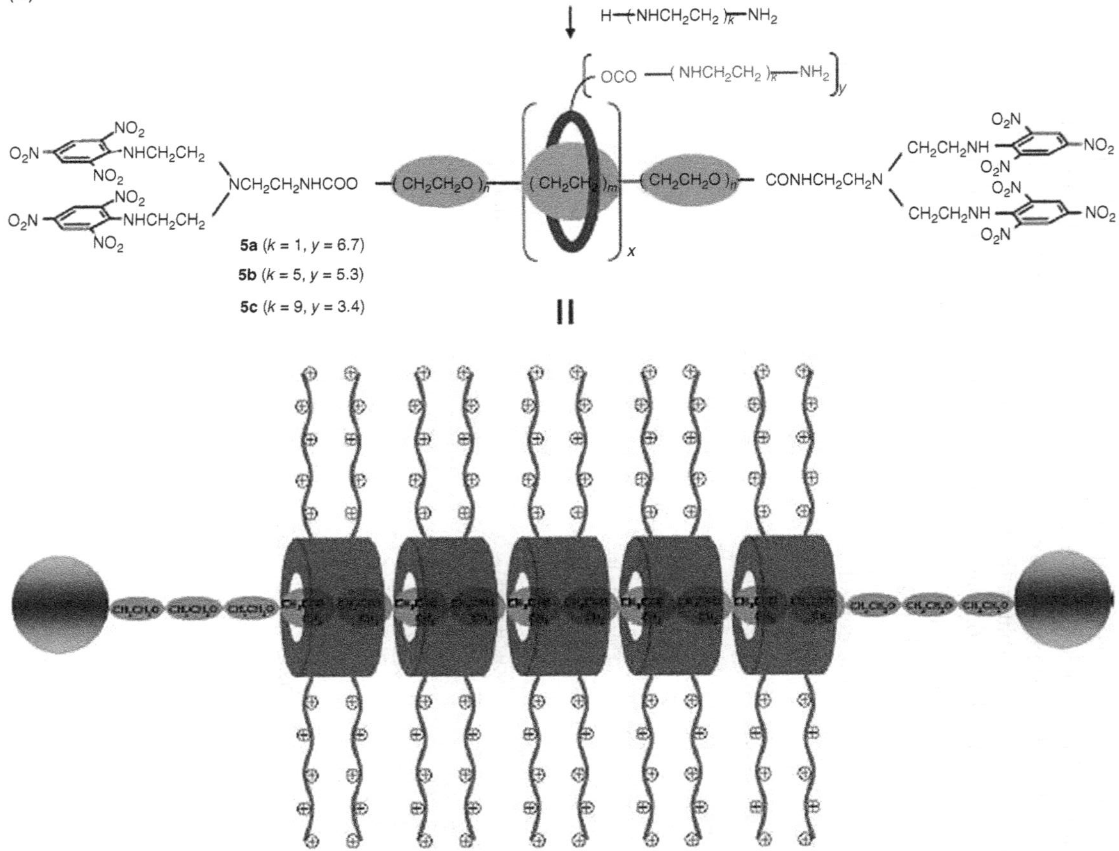

Scheme 6.8 (Continued)

(1 in Scheme 6.8a with 30 propylene oxide at the center and 13 ethylene oxide units on either end: $EO_{13}PO_{30}EO_{13}$) was treated with CDI and then reacted in DMF with tris(2-aminoethyl)amine [$N(CH_2CH_2NH_2)_3$] to form the tetraamine derivative of the triblock copolymer 2 in Scheme 6.8a. This was then added to an aqueous solution of β-CD and end-blocked with sodium salt of 2,4,6-trinitrobenzene sulfonic acid. NMR analysis showed that 13 β-CD were included in the polyrotaxane (4). Oligoethylenimine [(ethylenimine)$_k$, $k = 1–9$] was then grafted to the CD, to make OEI-grafted polyrotaxanes 5a, 5b, and 5c in Scheme 6.8b. A model of the cationic polyrotaxane is shown in Scheme 6.8b. It was confirmed that these cationic polyrotaxanes condense plasmid DNA (p-DNA) into particulate structures and that the condensation capability was similar to or slightly better than polyethylenimine.

A variation of such a triblock polyrotaxane system was reported by Zhang et al. [38]. They created an amphiphilic ABA triblock copolymer as shown in Scheme 6.9, with β-CD threaded through the central block and a hydrophilic polymer was used as the stopper. Starting from the commercial Pluronic F-127 triblock copolymer, which consists of poly(propylene oxide) (65 units) and PEO (100 units), the hydroxyl end groups were replaced to convert Pluronic

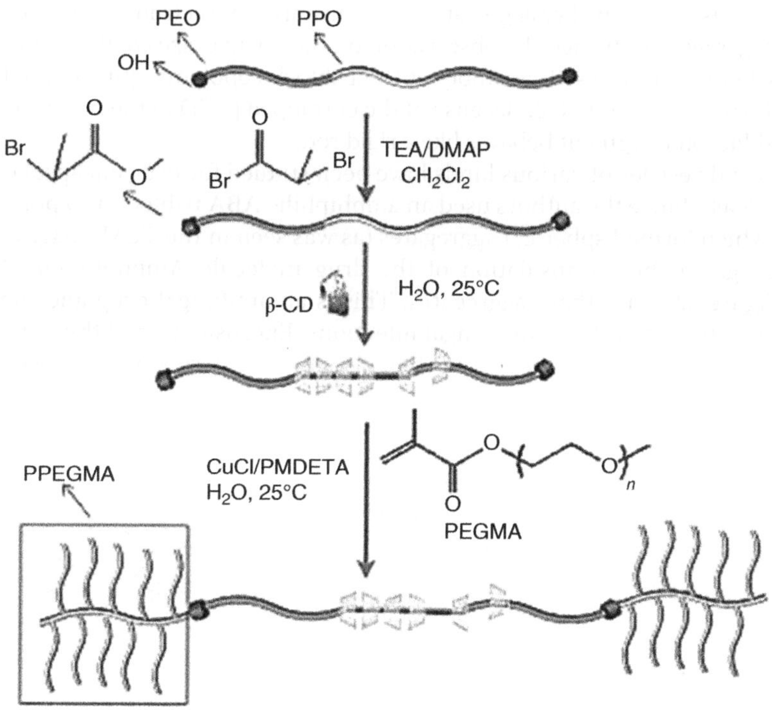

Scheme 6.9 Synthetic route to polyrotaxane-based amphiphilic triblock copolymers [38].

F-127 to atom transfer radical polymerization (ATRP) macro-initiator 2-bromopropionyl (BrP)-terminated Pluronic F-127. It was then treated with β-CD in aqueous solution at 25°C. In Scheme 6.9, TEA is trimethylamine, DMAP is 4-dimethylaminopyridine, and PMDETA is *N,N,N′,N″,N″*-pentamethyldiethylenetriamine. Polyethylene glycol methyl ether methacrylate (PEGMA) was thus coupled to the Pluronic F-127 as end groups. There appeared to be some competition between PEGMA and Pluronic F-127 to form inclusion complex with β-CD. It was found that when the feed molar ratio of PEGMA/BrP-F127-BrP was varied from 40 to 100, keeping the β-CD/BrP-F127-BrP at 30, the degree of polymerization (DP) of PEGMA oligomers ranged from about 51 to 87. When the feed ratio β-CD/BrP-F127-BrP was changed from 20 to 40 keeping the PEGMA/BrP-F127-BrP ratio at 60, the number of β-CD included in the F127 chain increased from about 15 to 37. The self-assembly of the poly(triblock/CD rotaxane) and poly(triblock rotaxane) (i.e., without the CD) in aqueous media was also studied. These authors identified these samples as follows: the rotaxane with no CD and a molar ratio of PEGMA/F127 60 was denoted as F-0-β-CD-60. With a CD/F127 molar ratio of 30 and PEGMA/F127 60, the sample was denoted as F-30 β-CD-60. Measurement of the radius of gyration by light scattering showed that the R_g with β-CD (79 nm) was much smaller than that without it (188 nm), that is, much smaller aggregate size. Transmission electron microscopy (TEM) images also confirmed the observation on the relative sizes of the aggregates. A schematic of the self-assembly of the triblock copolymer/β-CD-based rotaxane is shown in Figure 6.8. Because of the entrapped β-CD in the rotaxane, the central Pluronic segment behaves like a rigid rod.

Micelles and vesicles of various kinds have been studied for drug encapsulation and release. Since the authors used an amphiphilic ABA triblock/CD polyrotaxane which formed spherical aggregates (as was seen in the TEM images), they investigated the encapsulation of the drug molecule Amphotericin B (AmB) (Figure 6.9) into these aggregates. This is an antifungal drug and the only effective treatment for some fungal infections. Encapsulation of this drug molecule in PEO-*b*-poly(L-amino acid) block copolymer derivatives has been

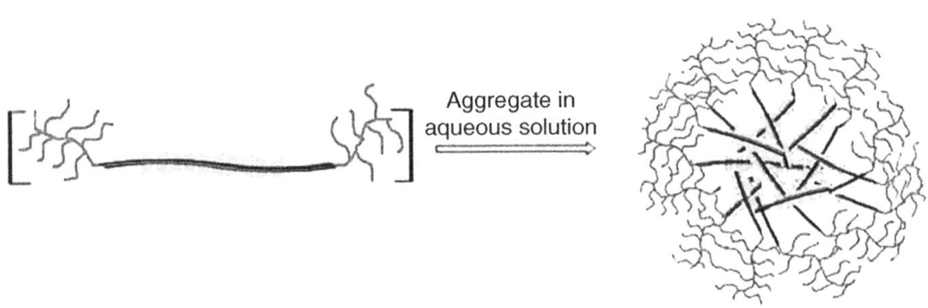

Figure 6.8 Illustration of self-assembly of poly(triblock/CD rotaxane) in water. (Source: Zhang et al. [38]. Reproduced with permission of Elsevier.)

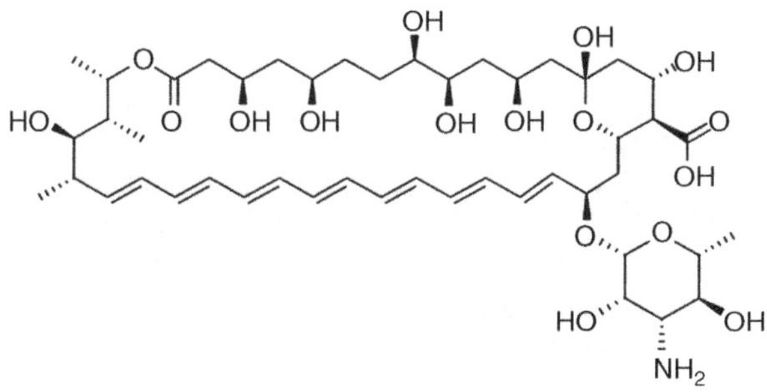

Figure 6.9 Molecular structure of Amphotericin B.

studied by Adams et al. [39] Dialysis was used to load the AmB to the polyrotaxane micelles. The drug loading content (DLC) and the drug loading efficiency (DLE) were measured according to

$$\mathrm{DLC}(\%) = \frac{\text{Amount of AmB in the micelles}}{\text{Amount of AmB loaded micelles}} \times 100$$

$$\mathrm{DLE}(\%) = \frac{\text{Amount of AmB in the micelles}}{\text{Amount of AmB used for the preparation of the loaded micelles}} \times 100$$

It was found that DLC and DLE were higher with the β-CD including polyrotaxane micelles than the ones without the CD.

6.4 Micelles of Double-Hydrophilic Block Copolymers via Rotaxane Formation

In the case of amphiphilic block copolymers, the hydrophobic block would self-assemble to form a core and the hydrophilic block would form the corona to be in contact with the aqueous medium. Such an assembly would result even if both segments of a block copolymer are hydrophobic but solvent-selective. Huang et al. [40] discussed the case of the block copolymer PEO-*b*-poly((dimethyl amino)ethyl methacrylate) (PEO-*b*-PDMA) in which both segments are hydrophilic and would form a homogeneous solution in water. When α-CD was introduced, the cyclics threaded selectively onto the PEO segments, which led to aggregates of the PEO/CD which behaved as rod-like segments (see Figure 6.10a for PEO-*b*-PAA block copolymer). This transformation leads to selective organization, similar to those in block-selective solvents. Both blocks of the copolymer being hydrophilic and biocompatible is an advantage for

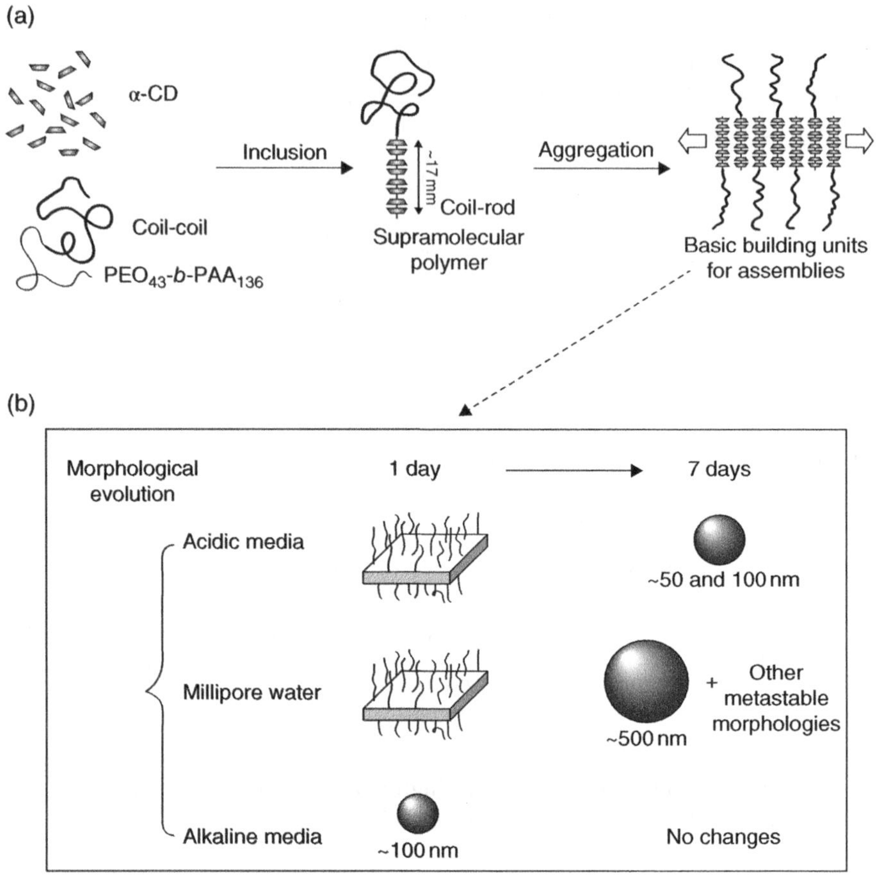

Figure 6.10 (a) A model of the self-assembly of the PEO-*b*-PAA block copolymer induced by the inclusion of CD. (b) Evolution of spherical morphology with time. (Source: Huang et al. [41]. Reproduced with permission of Wiley.)

biomedical applications. X-ray diffraction confirmed that the PEO/α-CD formed channel-type organization (as in Figure 6.4e). While the block copolymer formed a homogeneous solution in water, selective aggregation occurred upon the formation of the PEO/CD pseudo-polyrotaxane, which would lead to micellar morphology. In this two-stage process, supramolecular diblock copolymer is formed upon threading of α-CD on to the PEO block, which then aggregates into a channel-type structure. Micellar organization then occurred, with the rod-like pseudo-polyrotaxane forming the core and the PDMA block as the corona. Thus, it is an example of a CD-induced self-assembly of a block copolymer. These authors [41] also studied the polyrotaxane formation of α-CD with PEO-*b*-poly(acrylic acid) (PEO_{43}-*b*-PAA_{136}) and the dependence of

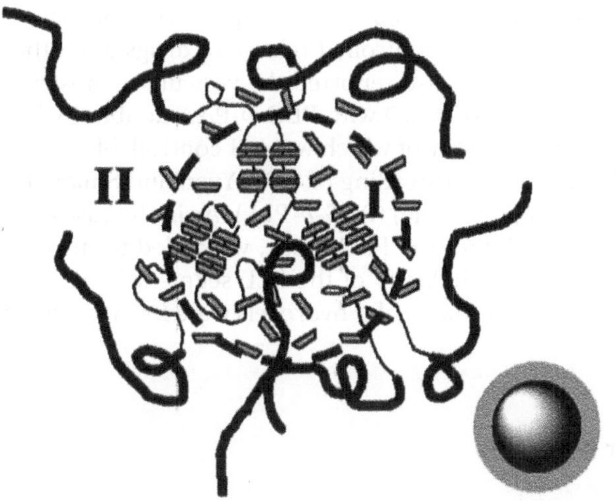

Figure 6.11 A model for the formation of particles upon complexation of PEO-*b*-PDMA/cyclodextrin in water. Region I shows the spherical core and the corona (gray color) is marked by II. (Source: Huang et al. [40]. Reproduced with permission of Wiley.)

morphology on the pH of the medium. The DP of PEO and PAA were 43 and 136, respectively. For the purpose of studying the self-assembly in aqueous media with various pH, stock solutions of the block copolymer and the α-CD were prepared, using HCl aqueous solutions of pH = 2, Millipore water (pH = 7), and NaOH aqueous solution. The pH values of the mixtures of these solutions were 2.5, 3.1, and 6.5, respectively. The hydrodynamic radii were determined using dynamic light scattering, and TEM was used to study the morphology. As in the case of PEO-*b*-PDMA, the α-CD selectively threaded through the PEO block, as illustrated in Figure 6.10a. In the Millipore water medium, the TEM images showing square-shaped particles were seen after 1 day. Various other morphologies such as hollow spheres and doughnut shapes were seen after 7 days, as shown in Figure 6.10b. The authors changed the self-assembly pattern by changing the ionization of the COOH group of PAA by varying the pH of the solutions. Nonionized COOH can form hydrogen bonds with the hydroxyl groups of the α-CD. With a solution of pH = 2.5, square-shaped particles with a side of about 350 nm were seen in the TEM after mixing for a day. After 7 days of mixing, these particles changed to spherical micelles of two distinct dimensions of 50 and 150 nm in diameter. When the mixing was performed in a solution of pH = 6.5, with the hydrogen bonds between the COOH and the α-CD eliminated, uniform spherical micelles of 100 nm diameter were seen and did not change with time. In all these cases, the uncovered PAA block forms the corona. A model for the formation of spherical morphology is shown in Figure 6.11. The ability to change the morphology with pH in aqueous media and the fact that both blocks are biocompatible promises potential application of such a system for encapsulation and controlled release of drugs and other biomedical applications. Liu et al. [42] also reported the vesicular nanostructures formed with the complexation of α-CD with PEO-*b*-PAA double-hydrophilic block copolymers, the morphology of which was pH controllable.

In a similar study on block-selective threading of CDs, Yuan and Shuai [43] used a poly(ethylene glycol)-*b*-poly(L-lysine) (PEG-*b*-PLL). As in the cases discussed earlier, the α-CD threaded onto the PEG block, which led to micellar formation. By controlling the hydrophobicity of the PLL segment by varying the pH of the aqueous medium, supramolecular hydrogels were formed, which were reversible.

6.5 Homopolymer Micelles

In the cases discussed before, block copolymers were used, and selective threading of the CDs onto one of the blocks occurred, leading to aggregation of the rotaxanes and micellar structures. Dong et al. [44] used a homopolymer, poly(ε-caprolactone), with α-CD to create supramolecular polymer micelles (SMPMs). The route that they followed to create the SMPMs is shown in Figure 6.12. An amphiphilic complex of PCL/α-CD formed in THF/water mixtures, in which only a part of PCL threaded through the CDs. With the removal of THF and

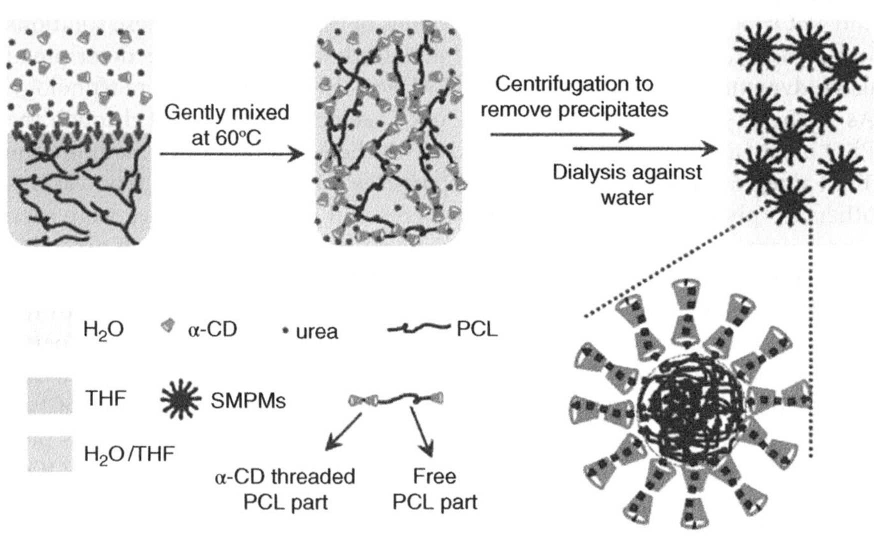

Figure 6.12 Illustration of the steps in the formation of supramolecular polymer micelles consisting of PCL and α-CD. (Source: Dong et al. [44]. Reproduced with permission of Wiley.)

dialysis with water, the system goes through a secondary association process to form SMPMs. To sustain the micelle formation and prevent the precipitation of the inclusion complex as a crystalline powder, they used urea in the solution, a maleic anhydride modified α-CD, as well as a high-molecular-weight PCL (M_n = 37 000). As seen in Figure 6.12, with the partial rotaxane formation, the hydrophobic, non-CD-included segment of the polymer forms the core and the α-CD-included part of PCL forms the corona. The TEM images clearly showed the core–shell morphology, with the shell thickness of about 6 nm. To examine these SMPMs as candidates for drug delivery applications, prednisone acetate, an anti-inflammatory drug with poor solubility in water was included in the micelles and their release rate was studied. The SMPMs showed a high drug loading of 39.5% and a slow release over a period of 700 h. Since the PCL and α-CD are the US Food and Drug Administration-approved materials for human use, the aforementioned SMPMs have potential applications in various therapies.

As mentioned in the first paragraph of this chapter, the geometry of the PDMS repeat unit and the low energy barrier between bond rotations (rotational isomeric states) enable macrocyclization. Brown and Slusarczuk [45] have shown that under certain equilibration conditions, cyclics with as many as a few hundred skeletal bonds can be isolated. A number of studies have been reported on linear and cyclic PDMS as well as their topological blends in which cyclics were entrapped with the linear chains. The groups of James Mark, John Semlyen, and their associates made some of the pioneering contributions to this area. Clarson et al. [46] compared the glass transition and crystallization behavior of cyclic and linear PDMS with up to about 650 skeletal bonds. While the T_g of the linear PDMS increased with the length of the chain, it decreased in the case of cyclic PDMS. Using a cryo-SEM and PDMS samples of M_w = 38 900 and 182 000 (with polydispersity of 2.8 and 1.7, respectively), Sundararajan [47] showed that the polymer crystallized at about −70°C, with spherulitic morphology (about 100 μm in size). The cyclics formed discrete single crystals as overgrowth on the polymer spherulites. Even a silicon wax lubricant showed single crystals at −103°C.

6.6 Linear and Cyclic PDMS

PDMS is one of the most widely used polymers, and it is often cross-linked into a network structure; its elastomeric properties have been studied by several authors. The Semlyen group [48, 49] has investigated the threading of linear PDMS chains into the cyclics. Typically, a certain amount of cyclics (e.g., 15 wt%) with well-defined ring size was mixed with hydroxyl-terminated linear chains. The end-linking of the PDMS was carried out with tetraethylorthosilicate and curing to produce a tetrafunctional network that could be as ideal as possible. A comparable network without the cyclics was also produced as a reference. The untrapped cyclics and noncrosslinked linear chains were

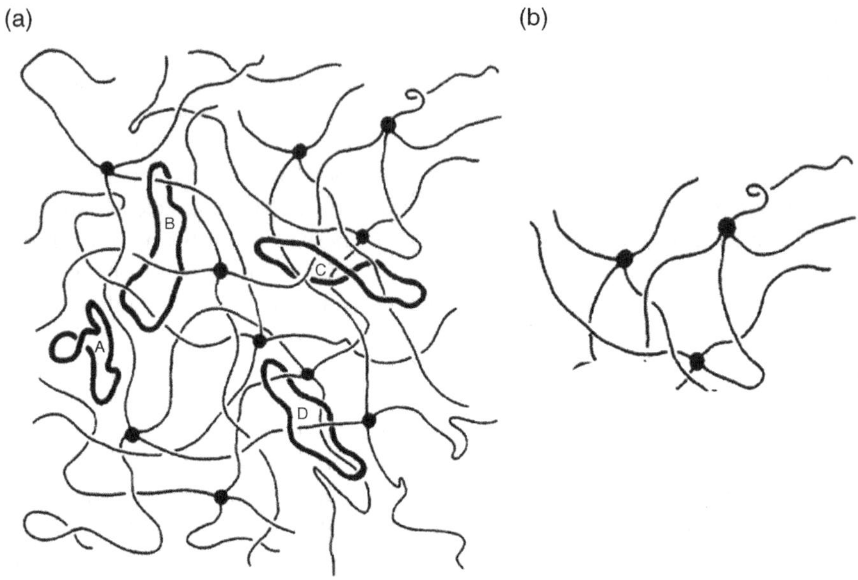

Figure 6.13 Schematic of (a) trapping of cyclic PDMS in tetrafunctional network and (b) intermolecular network (entanglement) without cyclics. (Source: DeBolt and Mark [50]. Reproduced with permission of American Chemical Society.)

extracted, to determine the network efficiency and the amount of trapped cyclics. It was found that cyclics with less than 38 skeletal bonds (n) were not trapped. With $38 \leq n \leq 517$, the percentage of cyclics that were trapped increased with n, reaching 94% with $n = 517$. A schematic of such a topologically cyclic-trapped tetrafunctional network [50] is shown in Figure 6.13a. In the figure, the cyclics A, B, C, and D are shown to represent entrapment with multiple segments as well as a single segment of the network chains. The entangled network without the cyclics is illustrated in Figure 6.13b.

Thus, the rotaxane formation with large cyclics could occur involving multiple chains. Such multiple-segment binding, in addition to the chain-end cross-linking, would have a profound effect on the mechanical properties of these networks. Mark's group used molecular simulations [50, 51] to develop a model for such topological trapping of cyclics PDMS. Figure 6.13b illustrates the chemical cross-linking of entangled chains without the cyclics.

6.7 Abrasion Resistance

In a typical xerographic copy machine, making a copy of an original involves five basic steps. The heart of the machine is a photoreceptor which creates an electrostatic image of the original. The photoreceptor consists of a charge generating layer (CGL) and a charge transport layer (CTL) coated one over the other on to a Mylar® substrate. The CGL consists of pigments such as phthalocyanines, perylene derivatives, and azo pigments in a polymer binder. The CTL consists of a polycarbonate layer, with a dispersion of molecules, such as *N,N′*-diphenyl-*N,N′*-bis(3-methylphenyl)-(1,1′-biphenyl)-4,4′-diamine (known as TPD) that are capable of transporting electrons or holes through the organic layer. In the first step, the photoreceptor (a belt in the case of large volume machines or a drum in small volume copiers) is negatively charged with a corona. When the original is exposed to light, the unprinted areas reflect the light on to the photoreceptor, discharging the corresponding areas, and the charge remains on the image of the printed parts of the document. Thus, an electrostatic image of the original document is obtained. Positively charged toner particles are then deposited which adhere to the negative charge on the image areas. The toner layer is then transferred to paper by contact transfer. The residual toner, if any, is removed by cleaning photoreceptor with a blade or a brush (see Dan A. Hays, *Scientific American, Ask the Experts,* March 31, 2003 "How does a photocopier Work?"). During the last step of cleaning the photoreceptor, abrasion can occur. Although the glass transition temperature (T_g) of the polycarbonate is typically about 150°C, dispersing the charge transport molecule (CTM) (with a high loading of close to 50 wt%) would reduce the T_g significantly. Abraded areas can act as nucleating sites for the CTM to aggregate and crystallize. This would also result in the deterioration of the performance of the photoreceptor.

It has been shown that abrasion is related to cohesive energy. Giltrow [52] measured the abrasion of various types of polymers and showed that the wear decreases linearly with an increase in the cohesive energy of the polymers. The latter is simply the energy required to remove one polymer chain from the bulk. When the author of this book was employed at Xerox, it was thought that similar to the case of cyclic-loaded PDMS system, if a polycarbonate rotaxane is used for the CTL, with the linear chains tangled with the cyclics, the cohesive energy would be increased and the abrasion would be reduced. It was at that time that Brunelle [53–55] had developed routes to synthesize cyclic polycarbonates, with the intent of using these cyclics to synthesize super-high-molecular-weight polycarbonates.

Sundararajan et al. [56] used Brunelle's approach to make cyclic polycarbonates, and mixed them with linear chains in dichloromethane for fabricating the CTL and to build a lab-scale device. The procedure was described in detail in the patent. The abrasion resistance was evaluated using the Taber abrader according to the test method T-476 of Technical Association of the Pulp and Paper Industry (TAPPI). The weight loss due to abrasion after 10 000 cycles was measured as the rate of abrasion in μg/cycle. It was found that without the presence of the cyclics, the rate of abrasion was 1.5 μg/cycle, and it reduced to 1.2, 1.05, and 0.86 μg/cycle with cyclic contents of 5, 10, and 20 wt%, respectively. Thus, with 20 wt% of the macrocyclic polycarbonates with the linear chains in the CTL, the abrasion was reduced by almost 50%, from 1.5 to 0.86 μg/cycle. It was also shown that the devices based on the polyrotaxane exhibited excellent xerographic properties when compared with that based on the conventional CTL and there was no performance penalty due to the incorporation of the cyclics.

6.8 Beyond Linear Polymers and α-, β-, and γ-CDs

While the studies described earlier dealt with linear polymers and cyclic species, such as crown ethers and CDs, a family of rotaxanes using single-wall carbon nanotubes (SWNTs) has emerged. Apart from the α-, β-, and γ-CDs that are commonly used, a paper published in mid-1960s by French et al. [57] described the isolation of higher CDs, namely, δ, ε, ζ, and η CDs, with 9, 10, 11, and 12 glucose units, respectively, in milligram quantities. They even obtained a theta-fraction. While much later, Frömming and Szejtli [58] noted that "preparation of large-ring CDs would seem to be an unrewarding challenge," Ueda et al. [59] had at that time isolated δ-CD (cyclomaltononaose), containing nine α-1,4′ linked glucose units. They also reported the isolation and purification of η-cyclodextrin (cyclomaltododecaose) with 12 glucose units. The availability of these cyclics with a large internal cavity affords the possibility of threading linear species such as SWNTs, which are bulkier than single polymer chains.

de Juan and Pérez [60] have given an overview of the encapsulation of various species such as fullerenes (commonly known as "peapod") and various organic molecules as well into the cavities of SWNT. In the early years of carbon nanotube (CNT) research, several approaches have been explored to solubilize CNTs, such as functionalization, use of surfactants, and complexation with synthetic polymers. The Stoddart group [61] used the solubility of the amylose component of starch to create a supramolecular assembly of amylose/SWNT. To trigger the helix formation of amylose, an aqueous solution of starch–iodine complex was formed (the first step in Figure 6.14c), which is known to be blue in color. Amylose itself forms a left-handed helix, as shown in Figure 6.14a. SWNT was dispersed in the blue solution with agitation followed by mild sonication. Amorphous carbon was removed from the solution of starch/SWNT by centrifugation. As illustrated in Figure 6.14c, the SWNT displaces the iodine in a "pea shooting" fashion, and forms amylose-wrapped SWNT complex. Figure 6.14a shows the conformation of the D-glucopyranose unit in the amylose helix. Figure 6.14b shows the results of the molecular dynamics simulation with a short SWNT and maltooctaose (an amylose fragment with eight repeat units). The simulation showed that the CH hydrogens at C-3 and C-5 positions (Figure 6.14a) interact with the hydrophobic surface of the SWNT, while the —OH groups attached to C-2, C-3, and C-6 atoms of the pyranose ring project

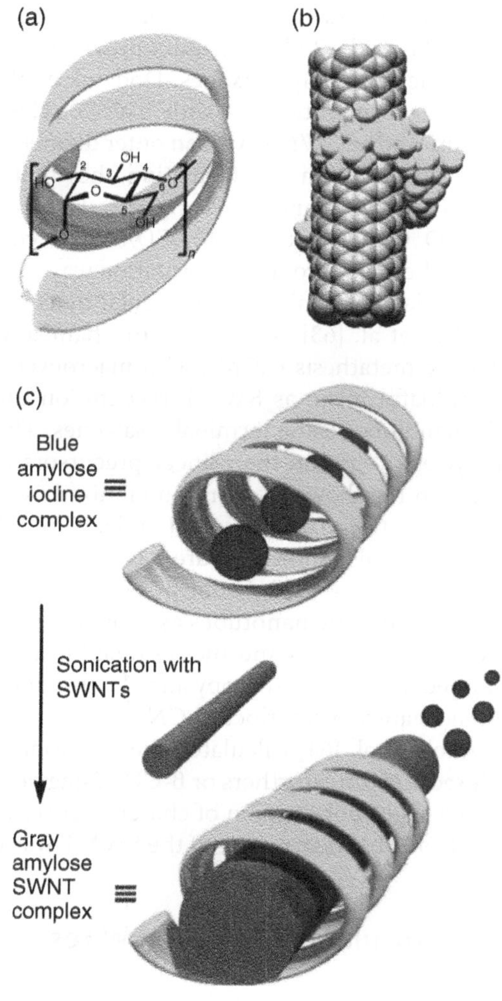

Figure 6.14 (a) Schematic representation of the complex between the left-handed helix of amylose and a small molecule. One α-1,4-linked D-glucopyranose residue is shown inside the helix. (b) Computer simulated model of the wrapping of a maltooctaose over a short (6,6)-single-wall nanotube. (c) Schematic representation of the carbon nanotubes displacing iodine molecules from the amylose helix by the "pea-shooting" type of mechanism. (Source: Star et al. [61]. Reproduced with permission of Wiley.)

outward to form hydrogen bonds with water. UV–Vis spectra were used to confirm the release of iodine from amylose–iodine complex during the formation of amylase–SWNT complex. It was found that similar to iodine, the helical conformation of amylose is also promoted by n-butanol and enabled amylase–SWNT complex. Enzymatic hydrolysis of the starch complex led to the precipitation of SWNT, which upon filtration, resulted in "bucky paper." Thus, in this case, the rotaxane formation involves the whole helical segment of amylose and the SWNT, which serves to solubilize the CNT, as well as to purify it from the other carbonaceous material. Pure SWNT is obtained with the removal of starch by enzymatic hydrolysis.

As mentioned earlier, the availability of larger CDs, δ, ε, ζ, and η CD, with 9, 10, 11, and 12 glucose units provides more options for the fabrication of rotaxanes and polyrotaxanes. Dodziuk et al. [62] used η-CD with 12 glucose units to form rotaxanes with SWNT. The internal diameter of this CD is 1.8 nm, and an SWNT with an outer diameter of about 1.2 nm can be threaded. To enable the rotaxane formation, they first cut the SWNT by grinding along with β- or γ-CD, and then sonicated the mixture in a 2.6 mM solution of η-CD in D_2O. Clear solutions of η-CD/SWNT complex were obtained upon filtration. The pseudorotaxane formation was confirmed by NMR spectra as well as molecular mechanics and molecular dynamics simulations.

Pérez et al. [63] synthesized mechanically interlocked SWNT, using ring-closure metathesis (RCM), with macrocycle precursors with two π-extended tetrathiafulvalene as SWNT recognition units and closed the rings around the nanotubes using terminal bisalkenes. The chemical structures of the macrocycles (**1**–**3**) and their linear precursors (**4**–**6**) are shown in Scheme 6.10. Figure 6.15a shows an energy-minimized molecular model of a pseudorotaxane formed with the macrocycle **1** and a (7, 6) SWNT. The diameters of **1** and the SWNT are also shown. Figure 6.15b shows the reaction scheme using the linear precursors **4**–**6** and Grubbs' second-generation catalyst. The macrocycles form around the nanotubes as shown in the figure. The nonthreaded macrocycles and SWNTs and the residual linear precursors were then washed and filtered. NMR spectroscopy and electron microscopy confirmed the formation of mechanically interlocked CNT.

Akola et al. [64] calculated the electronic properties of SWNT when complexed with crown ethers or β-CD. Using density functional theory, they found that a slight polarization of charge occurred with rotaxane formation, but the electronic band structure of the SWNT was not affected significantly.

6.9 Insulated Molecular Wires

In addition to the potential applications of rotaxanes and polyrotaxanes for drug encapsulations and delivery and the others mentioned earlier, these supramolecular assemblies also allow control of electronic properties that could be of immense interest in areas, such as organic light-emitting diodes (OLEDs). When conjugated polymers, which are "molecular wires," are used as axle with cyclics, such as CDs, "insulated molecular wires" (IMWs) are achieved. Frampton and Anderson [27] presented a review of such IMW, with rotaxane architectures as well as "polymer-wrapped" configurations. The encapsulation of poly(phenylenevinylene) (PPV) in the channels of mesoporous silica (Figure 6.5) is an example of isolating individual chain, so that they are insulated from intermolecular associations of their own kind (see also Figure 6.4). Frampton and Anderson described rotaxane and polyrotaxane systems involving metal-directed

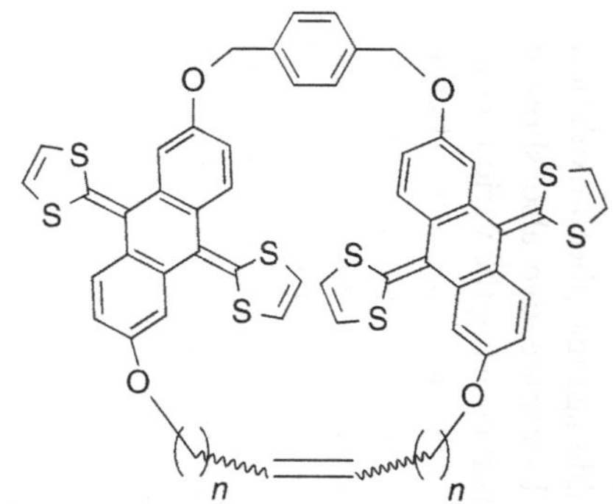

1, $n = 6$
2, $n = 8$
3, $n = 9$

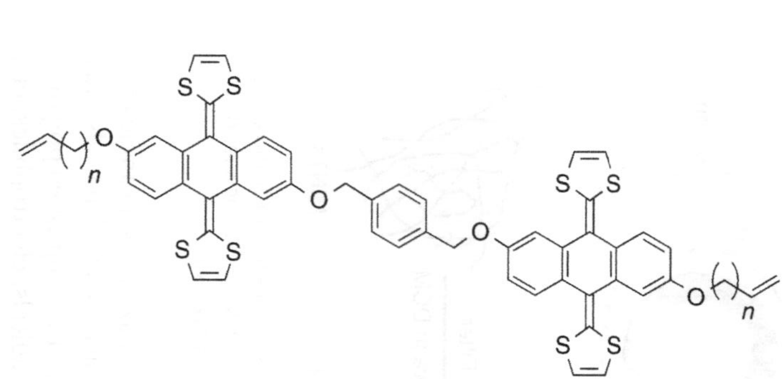

4, $n = 6$
5, $n = 8$
6, $n = 9$

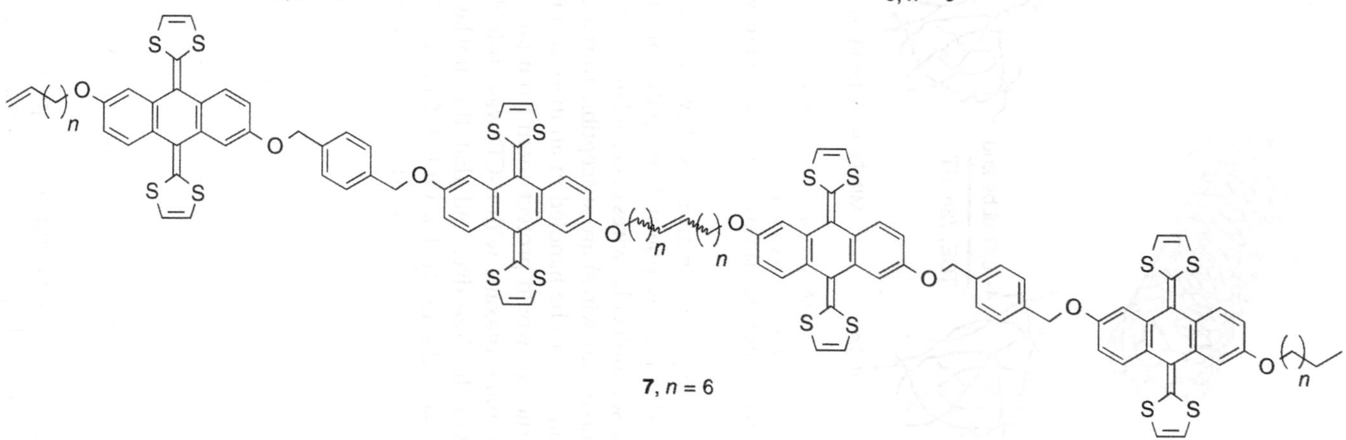

7, $n = 6$

Scheme 6.10 Macrocycles consisting of tetrathiafulvalene (**1–3**) and their linear precursors (**4–6**). A linear oligomer is also shown here [63].

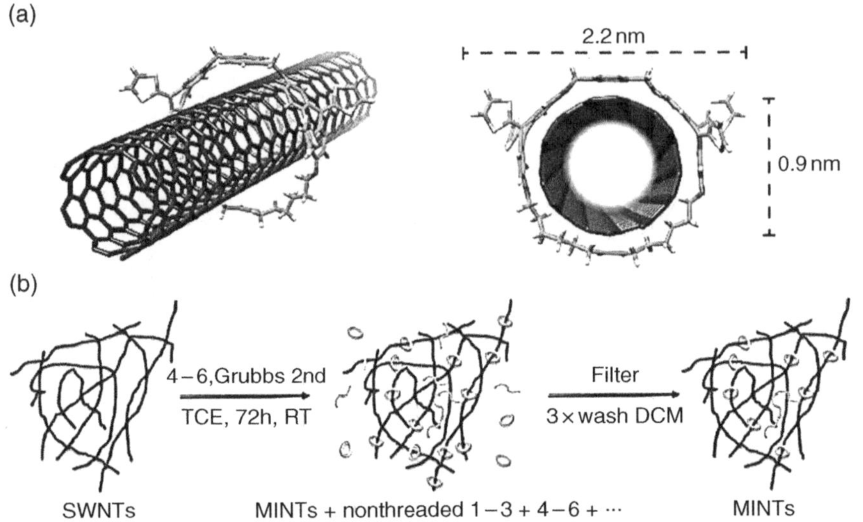

Figure 6.15 (a) Molecular models of pseudorotaxanes with 1 (see Scheme 6.10) and a (7,6) SWNT. Their diameters are also shown. (b) The reaction scheme to wrap macrocycles on to SWNT. (Source: de Juan et al. [63]. Reproduced with permission of Wiley.)

threading, cyclodextrin-based systems, and so on, with azo dyes, cyanine dyes, polythiophenes, polyanilines, polyazomethines, polysilanes, etc. They also discussed molecular wires encapsulated by amylose. Since the conjugated polymers (molecular wires) are susceptible to reactions with radicals, electrophiles, etc., and have to be handled in an inert atmosphere, insulating them in a rotaxane structure would protect them from any such reactions. It was found that the cyanine rotaxane with α-CD was 40-fold more stable against photo-oxidation than the free dye, and that the amylase-wrapped cyanine dye also showed enhanced stability. The OLEDs fabricated from rotaxanes using the various conjugated polymers showed that they preserved their semiconducting properties in the presence of the CDs. Thus, apart from the CDs, polymers such as amylose which form helices with large internal cavity are also candidates for components of insulated molecular wires. The work of Stoddart et al. that was described earlier on the CNT–starch complex was prompted to enhance the solubility of CNTs. Likewise, the use of amylose to wrap the conjugated polymers would enable solubility of these otherwise mostly insoluble polymers.

6.10 Molecular Switches and Machines

The ability to manipulate the positions of the cyclics with respect to the axle has opened up a whole new area of studies aimed at using them as molecular shuttles or molecular switches. The rotaxanes based on β-CD and the triblock

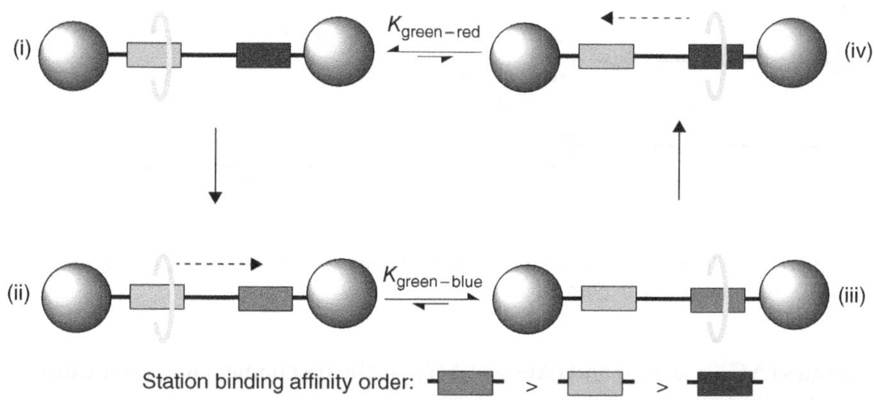

Figure 6.16 An illustration of stimuli responsive molecular shuttle. See Ref. [67] for the color version of this figure. (Source: Altieri et al. [67]. Reproduced with permission of American Chemical Society.)

copolymer (polyethylene glycol-*b*-polypropylene glycol-*b*-polyethylene glycol) (PEG–PPG–PEG, Pluronic® polymer) were discussed earlier. Yui et al. [65, 66] analyzed the temperature dependence of the location of the β-CD along the triblock copolymer. With the polyrotaxanes end-capped with fluorescein-4-isothiocyanate (FITC), there was interaction detected between the β-CD and the end-caps at 10°C. With an increase in temperature, the majority of the CDs moved toward the PPG block. In their polyrotaxane systems, it was estimated that seven β-CD rings were threaded into the polymer with the FITC end-cap. ^{1}H NMR was used to estimate the location of the cyclics along the chain. At low temperatures, the cyclics were distributed evenly and became more localized with the PPG segment with increasing temperature. Thus, this is a simple case of moving the population of the cyclics to the central segment of the triblock copolymer.

Most of the research reported to date on the molecular switches involves catenanes or short rotaxanes. In the case of the latter, the process requires that the cyclic be moved along the axle from one position to another by stimuli involving means such as photochemical, electrochemical, electrostatic, thermal, pH, solvent medium, or metal complexation/decomplexation. A schematic of one such process described by Altieri et al. [67] is illustrated in Figure 6.16, for a hydrogen bonded molecular shuttle that is electrochemically switchable. In Figure 6.16, the cyclic is initially in its preferred site (i) (rectangle of the left), which has a higher binding affinity than the site (iv) (rectangle on the right). A reaction is induced to change the site (iv) to the characteristic of (iii) (rectangle on the right), which has a higher affinity for the cyclic than the site (i). The cyclic now moves to station (iii). If the reaction is reversed so that the affinity of the station (iii) is reduced to that of (iv), the cyclic would move back to its original station (ii). The sum of the differences in the binding

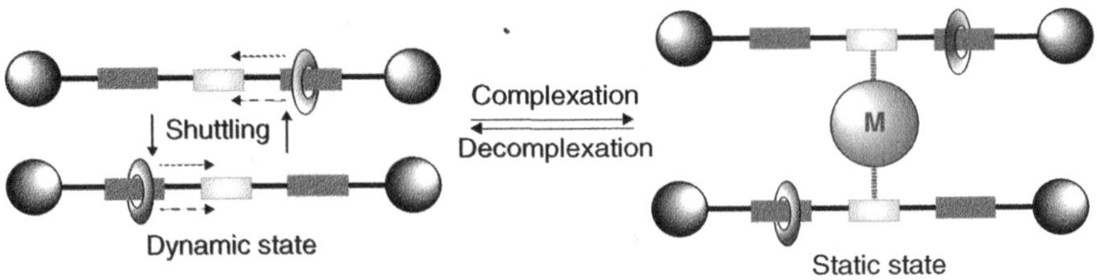

Figure 6.17 An illustration of dynamic and static binary shuttling modes. (Source: Jiang et al. [70]. Reproduced with permission of Wiley.)

energies ($\Delta G_{(i)} - \Delta G_{(iv)}$) and ($\Delta G_{(iii)} - \Delta G_{(i)}$) is the mechanical work done during this shuttling process.

In the aforementioned work, the authors used an axle containing succinamide (SI) and naphthalimide (NI) hydrogen bonding stations and a benzylic amide macrocycle. In the neutral state, the macrocycle positions itself with the SI since the NI is a poorer hydrogen bond acceptor. Upon reduction, $NI^{-\cdot}$ becomes a better H-bond acceptor and the macrocycle moves to associate with NI. Upon reoxidation, the cyclic shuttles back to its original position.

In this context, it is useful to recall that Leigh et al. [68, 69] set some definitions in their extensive reviews. Note that their 2007 review (Ref. [68]) cited 641 references and the one in Ref. [69] contained 1949 literature references. Leigh et al. distinguished between terms such as motors, machines, and switches. For example, they note that in a rotaxane "switch," the position of the macrocycle on the axle changes and that influences its function. Return of the macrocycle to its original position in such a switch reverses the work done, so that there is no net "work done." A "motor," on the other hand, performs work when taken through a cycle continuously, and there is no loss of work done unless it is reversed in its trajectory. Note that the work is taken here as a vectorial quantity, with magnitude and direction.

In the case of the example cited with respect to Figure 6.16, the system is a nondegenerate molecular shuttle in that the two stations are nonequivalent and could be a binary logic system. A degenerate system, on the other hand, would allow dynamic shuttling, as described by Otera et al. [70]. This is described in Figure 6.17. A nonpreferential shuttling between the two pink states occurs in the dynamic situation, which gets arrested in one of the configurations to a static state upon, for example, a metal complexation. Decomplexation returns the system to the dynamic state.

The molecular system investigated by Otera et al. is shown schematically in Figure 6.18a and b. By heating a solution of macrocycle **3** and the thread **2** with a central 2,2′-bypyridine in acetonitrile at 55°C for 6 days, the rotaxane **1** (with substitutions at the 4,4′ positions in **1a**) was obtained (Figure 6.18a). ^{1}H NMR spectrum at room temperature showed that the system was centrosymmetric,

(a)

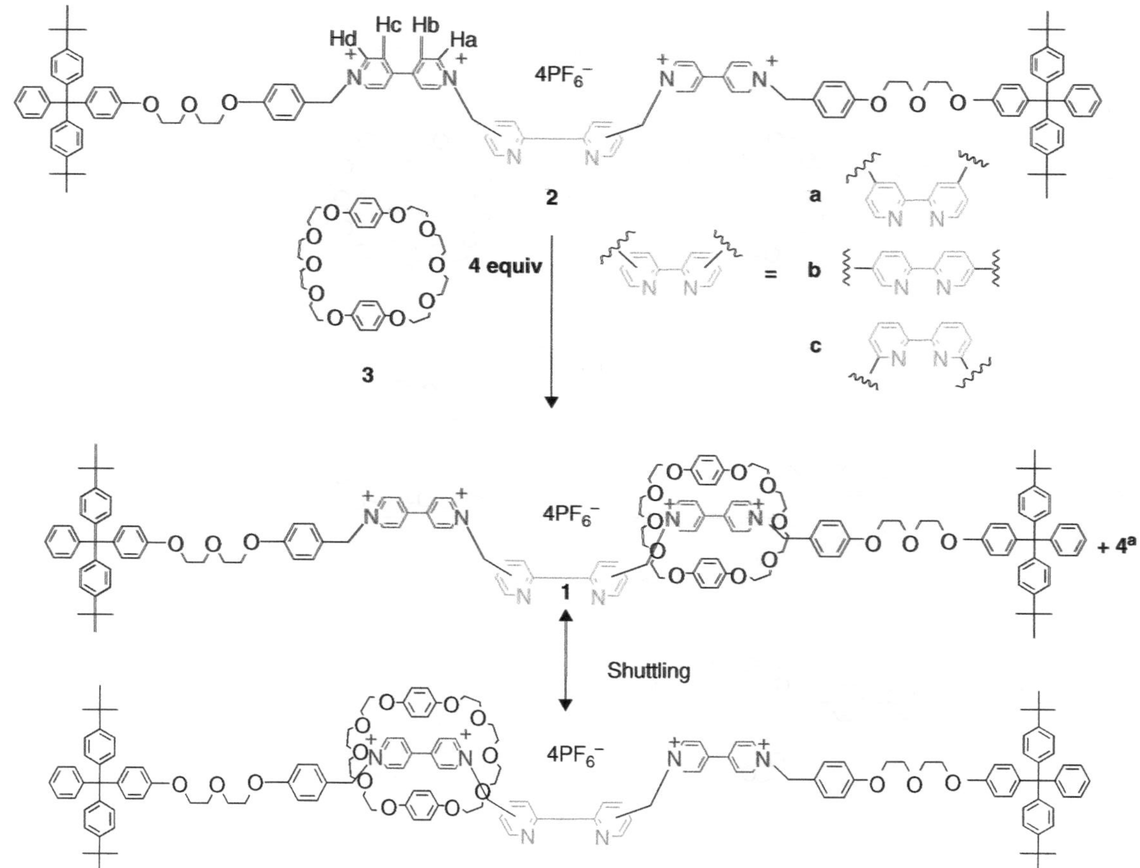

Figure 6.18 (a) Schematic of the design of a degenerate rotaxane system showing shuttling between the stations at the NMR time scale. (b) Schematic of the rotaxane **1a** and **4a** brought to a static state **5a** or **5b** by metal complexation. (Source: Jiang et al. [70]. Reproduced with permission of Wiley.)

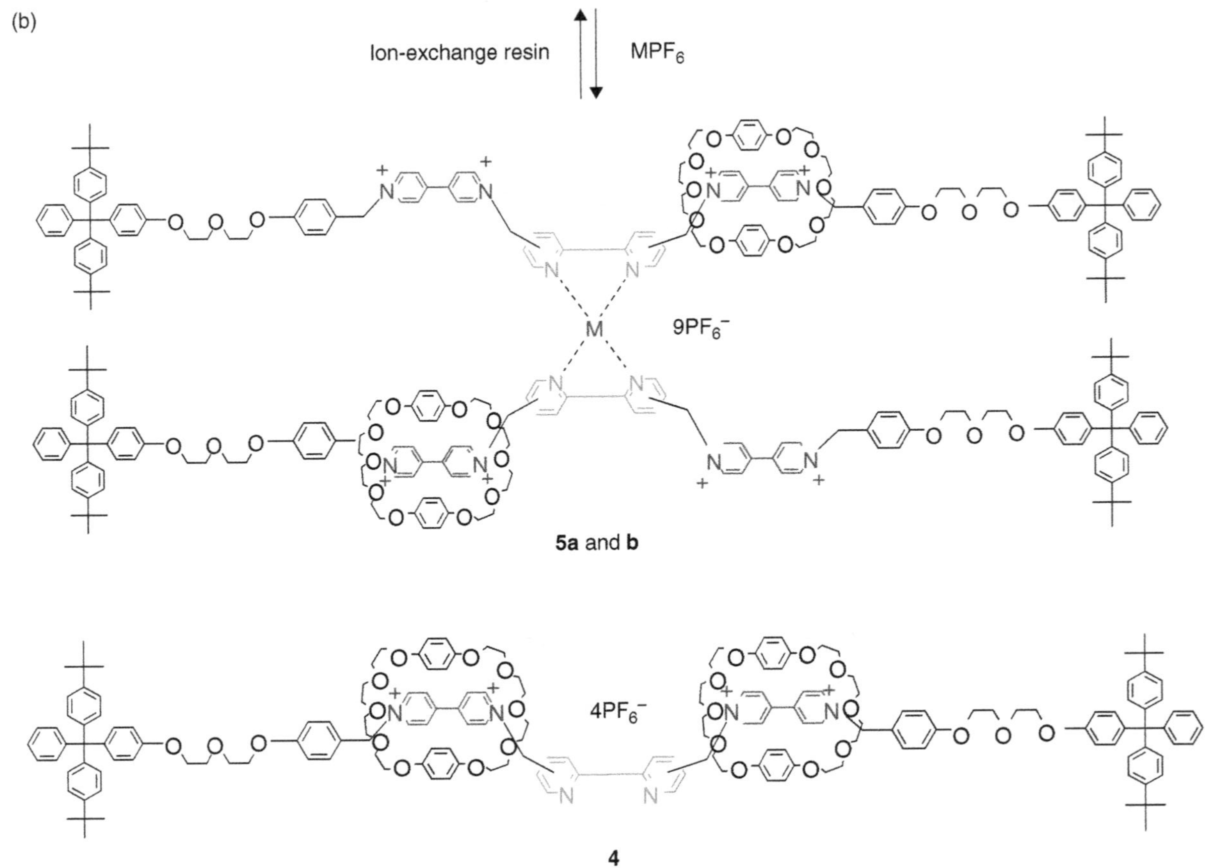

Figure 6.18 (Continued)

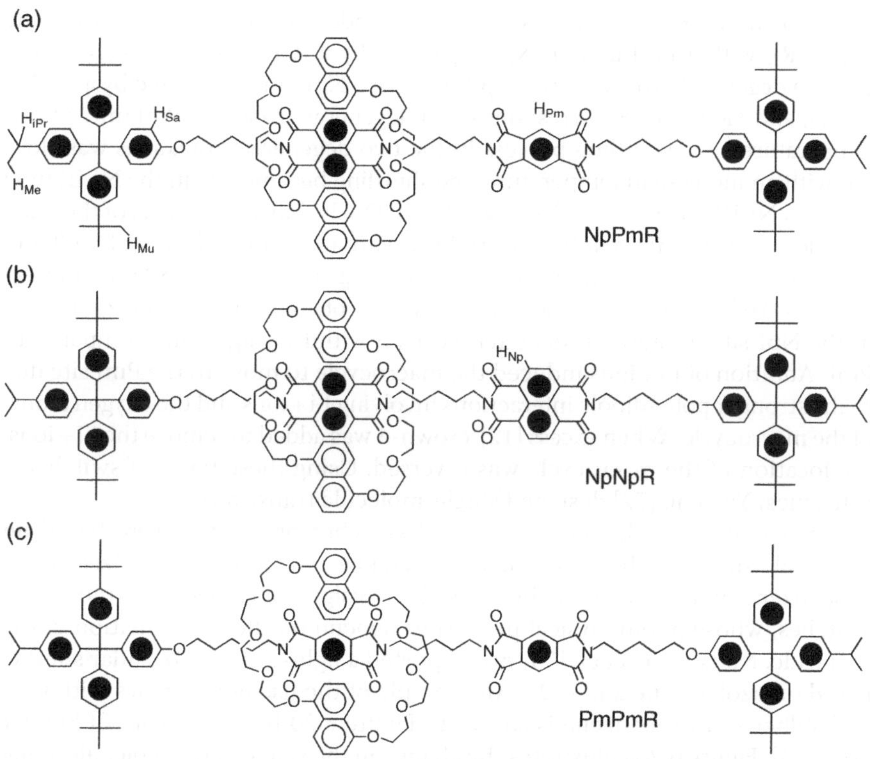

Figure 6.19 Structures of the (a) bistable [2] rotaxane NpPmR and the degenerate [2] rotaxanes, (b) NpNpR, and (c) PmPmR.

which indicated that the crown ether shuttled back and forth between the two equivalent bipyridinium sites (as in **4a**) at a rapid rate that could not be measured on the NMR time scale. When a solution of $[Cu(CH_3CN)_4]PF_6$ in acetone was added to a solution of **1a** (also in acetone), the color of the solution turned from orange to dark red. Upon evaporation of the solvent, a 2 : 1 complex of **5a** was obtained. The chelation of the Cu^I ion at the central ligands as shown in Figure 6.18b blocks the movement of the macrocycle and hampers the shuttling. Adding a suspension of an ion-exchange resin in acetone with pH = 5.5 led to decomplexation of the metal within 15 min and the original color and the shuttling motion were restored. The same dynamic/static behavior (with Cu^I complexation/decomplexation) was also observed with **1b**, but not in the case of **1c**. With the latter, the steric crowding prevented the complexation.

Iijima et al. [71] investigated the systems shown in Figure 6.19 with two electron-accepting recognition sites pyromellitic diimide (PmI) and 1,4,5,8-naphthalene-tetracarboxylic diimides (NpI) and the macrocycle 1,5-di-naphtho

[38] crown-10. Figure 6.19a shows the nondegenerate (bistable) rotaxane (NpPmR), with a PmI and an NpI separated by a $(CH_2)_5$ spacer. Figure 6.19b shows a degenerate rotaxane (NpNpR) with two NpI units separated by a $(CH_2)_5$ spacer, and Figure 6.19c corresponds to a degenerate rotaxane PmPmR. At low temperatures, the shuttling process in the two degenerate rotaxanes was slow, but with an increase in temperature, the shuttling became fast in the NMR time scale. ^{1}H NMR spectra recorded at 277 K in CD_2Cl_2 showed shuttling of the macrocycle occurred in NpNpR and PmPmR with activation barriers of 14.0 and 10.9 kcal/mol, respectively. This indicates a stronger donor–acceptor interaction in the case of NpNpR. In the case of the bistable rotaxane, the macrocycle resides in the NpI site because of its better electron withdrawing ability compared to PmI. Addition of Li + ions induced the macrocycle to move to the PmI site due to the stronger pole–dipole interactions involving Li + ions and the oxygen atoms of the macrocycle. When excess [12]-crown-4 was added to remove the Li + ions, the location of the macrocycle was reversed. Using these types of switchable rotaxanes, Yu et al. [72] designed single-molecule transistors.

Whereas most of the aforementioned switches or shuttles were based on chemical stimuli, photo-activated, reversible *cis–trans* isomerization of azobenzene and stilbene has been used to design rotaxane-based molecular switches, whose photophysical properties depend on the conformation of the molecule. Nakashima et al. [73] first reported a light-driven molecular shuttle, based on azobenzene and CDs. An example of the use of rotaxanes with CDs and stilbene or azobenzene is shown in Figure 6.20 from the work of Stanier et al. [74]. Figure 6.20a illustrates the change in the shape of the molecule along with the cyclodextrin as the photo-isomerization changes the *trans* conformation (E-isomer) to *cis* (Z-isomer). In this case, the cyclodextrin moved from the azo site to the biphenyl segment. However, the size of the CD and the size of the blocking group determine the steric hindrance involved in the conformational change. These authors studied the photo-isomerization of the rotaxanes as well as the corresponding dumbbells alone (axle, without the presence of the CD). In the case of the azobenzene rotaxane shown in Figure 6.20b, for the dumbbell alone irradiation at 361 nm induced E/Z isomerism, with the absorption at 255 nm. Irradiation at 255 nm led to reversibility to the E-isomer. However, when the α-CD was included, no isomerization occurred, due to the steric crowding that prevented the CD from moving. Note that although the axles in Figure 6.20a and b contain azobenzene, the groups surrounding them are different. Compound **2** with β-CD (Figure 6.20c) and compound 3 with α-CD (with stilbene axle) as well as the corresponding axles underwent reversible E/Z isomerization upon irradiation at 340 nm and then at 265 nm. Compound **2** with α-CD did not show any photo-isomerization due to steric crowding, similar to the case of the rotaxane shown in Figure 6.20b. The fluorescence quantum yield (Φ) depended on the kinetics of the photo-isomerization, which was dictated by the location of the CD on the axle.

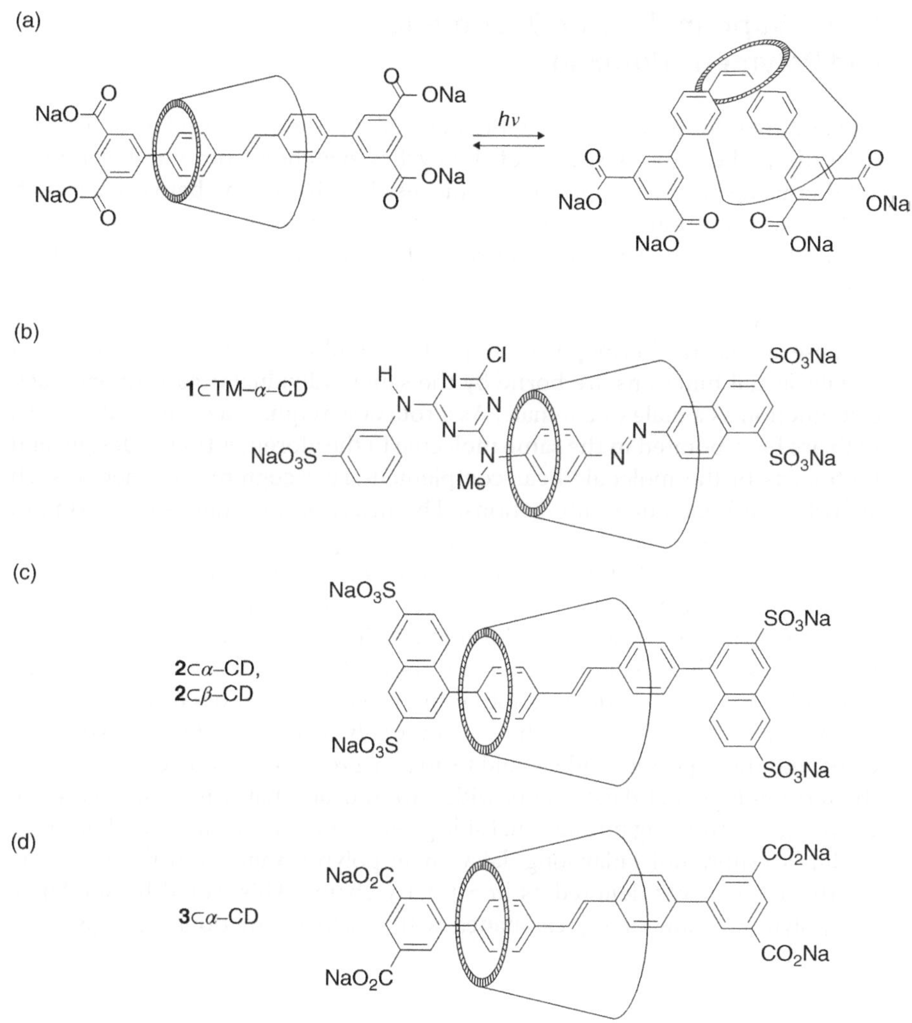

Figure 6.20 (a)–(d) Structures of rotaxanes based on azobenzene and stilbene axles. TM-α-CD is hexakis (2,3,6-tri-*O*-methyl)-α-cyclodextrin. Compounds **1**, **2**, and **3** are the free dumbbells. (Source: Stanier et al. [74]. Reproduced with permission of Wiley.)

Tian et al. [75] investigated systems with both azobenzene and stilbene in one axle, and two CDs, one on each of these sites. A number of systems containing CDs and azobenzene or stilbene units have been studied with respect to photo-isomerization as a means to achieve a switch with the binary state of the systems ("0" or "1") interconvertible with light.

6.11 Supramolecular Oligomeric and Polymeric Rotaxanes

The examples for molecular switches described earlier are based on short-axle rotaxanes, where the positions of the cyclics switched from one location to another using chemical, optical, etc., means. The distinction between a switch and a motor was pointed out before [68, 69]. A strategy of supramolecular recognition and assembly was developed by the Stoddart group [76, 77], in which a molecule was designed to have both the ring and a tail in its structure, as shown in Figure 6.21a.

These are like the hermaphrodite species in biology, in which both male and female sexual functions are borne by the same individual organism and each can function as a male or a female. As a rotaxane requires a cyclic and an axle, both are incorporated in the same molecule in the aforementioned design, and both parts of the molecule bear complementary recognition elements, such as hydrogen bonds or π-interactions. The insertion of a tail into the ring of another molecule leads to the dimeric form (Figure 6.21b), and propagation of such assembly would lead to an oligomeric rotaxane (Figure 6.21c). A cationic dibenzo[24]crown-8 with a secondary dialkylammonium side chain (Figure 6.22a) was chosen by Ashton et al. [76] to study such a system. This molecule assembles as shown schematically in Figure 6.22a, with the tail of one molecule forming the axle for another molecule. The authors termed this as a "daisy chain." Note that in **A**, the tails mutually form the axles for two molecules, and the supermolecule would form a macrocycle. Ashton et al. [76] used the terminology [*c*2] daisy chain, with *c* to indicate that it is cyclic, with two components. By using bifunctional side groups, polymerization could be done, leading to supramolecular long daisy chain polyrotaxane (**B** in Figure 6.22a). (In this case, it was denoted as [*a* ∞] daisy chain.) This would be similar to the polymerization of [2] catenanes with reactive functional groups from

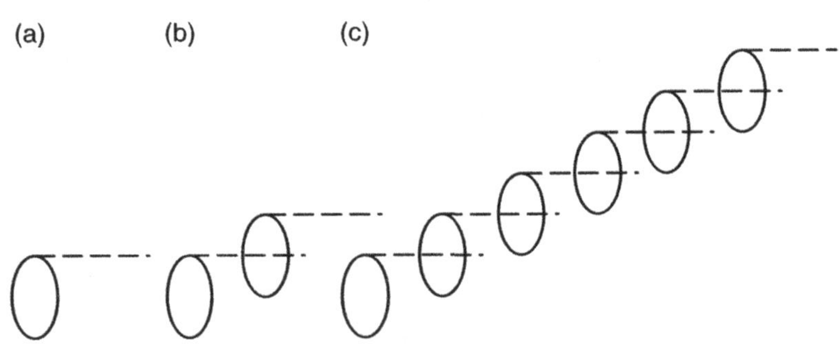

Figure 6.21 Schematic of a "hermaphrodite" molecule (a) with both the cyclic and the axle. The dimeric and oligomeric supramolecular assembly is shown in (b) and (c), respectively.

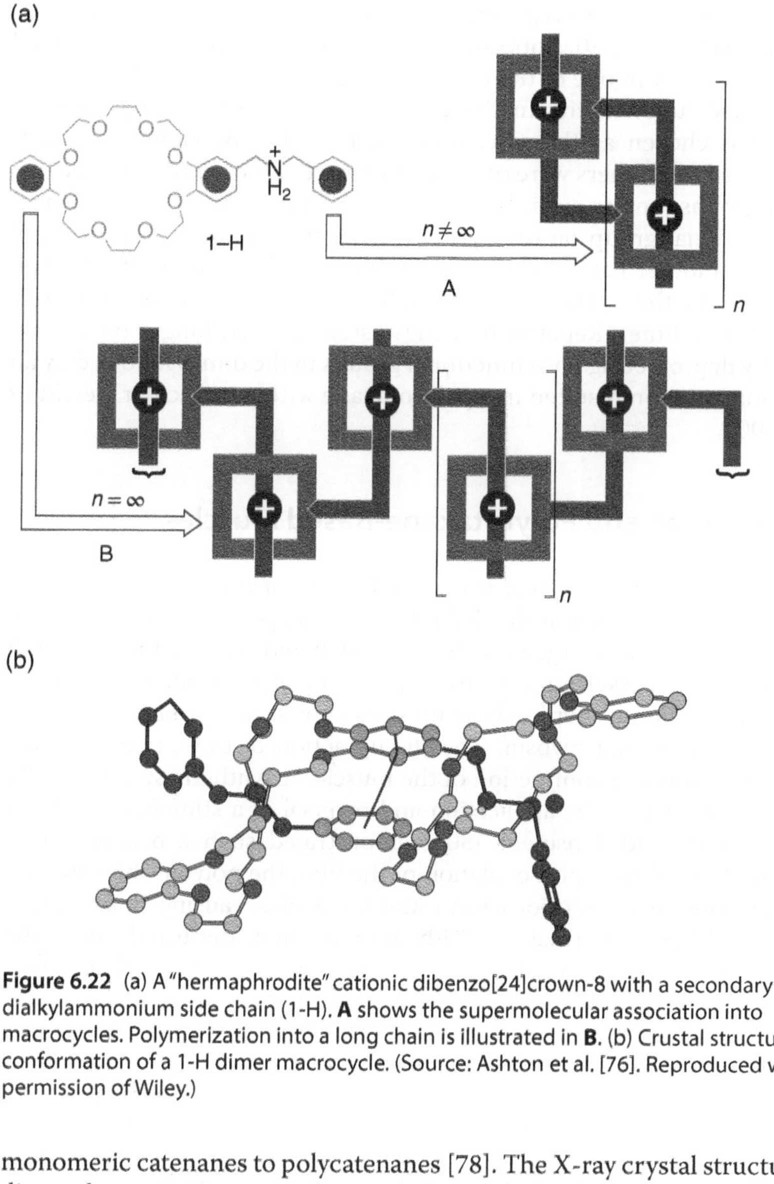

Figure 6.22 (a) A "hermaphrodite" cationic dibenzo[24]crown-8 with a secondary dialkylammonium side chain (1-H). **A** shows the supermolecular association into macrocycles. Polymerization into a long chain is illustrated in **B**. (b) Crustal structure conformation of a 1-H dimer macrocycle. (Source: Ashton et al. [76]. Reproduced with permission of Wiley.)

monomeric catenanes to polycatenanes [78]. The X-ray crystal structure of the dimer shown in Figure 6.22b revealed π–π stacking with a distance of 3.36 Å between the catechol rings bearing the NH_2^+ group and hydrogen bonds involving one of the hydrogen atoms on each NH_2^+ group and a polyether oxygen atom from each crown component, with distances of 2.91 and 2.02 Å between $N^+\cdots O$ and $H\cdots O$, respectively, and an almost linear hydrogen bond angle ($N^+—H\cdots O$) of 170°.

A stepwise approach to preparing mechanically linked polyrotaxane was devised by Werts et al. [79]. Rotaxanes were synthesized with silyl-protected phenol functionality in one of the two stopper groups and an allyl-protected carboxylic acid functionality in the cyclic part, which was a cyclophane. Polyether was chosen as the axle. Individual batches of monofunctional/monoprotected monomers were obtained by selectively deprotecting the two functional groups (on the cyclic or the stopper), that is, in one batch, deprotect the functional group on the cyclic, leaving the one of the stopper still protected. In another batch, deprotect the functional group on the stopper, with the one on the cyclic intact. Coupling reaction of these individual batches led to a dimer. Repetition of these steps yielded longer oligomers. Sequentially deprotecting both functional groups in the dimer followed by an esterification reaction resulted in a polyrotaxane with a molecular weight of about 400 000.

6.12 Rotaxane and Polyrotaxane-Based Muscles

Muscle is a part of the body that is essential to perform tasks by humans or animals. The muscle system is divided into three categories: skeletal, smooth, and cardiac. The internal organs such as stomach and bladder function with smooth muscles. The skeletal muscles that are part of, for example, arms and legs, are of the voluntary type. These muscles consist of myofibril filaments, which contain actin and myosin. The sliding action of these two filaments results in extension and contraction of the muscle. A synthetic system should expand and contract by the application and reversal of a stimulus and do so repeatedly. Otero and Sansiñena [80] demonstrated such a behavior with polypyrrole (PPy) films. Upon oxidation of the film, the polymer film swelled to accommodate the counterions and water molecules, causing chain expansion and a rod-like conformation. With the reduction, the expulsion of the counterions and water molecules results in contraction of the polymer to a random coil conformation, causing the film to shrink. These authors called this phenomenon the "electrochemomechanical" property.

Since the muscle consists of two filaments which slide past each other to expand and contract, a pair of hermaphrodite molecules which interpenetrate each other (homodimeric pair) could serve as a model. Figure 6.23a shows a schematic of such a system [81–83] studied by the Sauvage group. With close proximity of the cyclics, the end-to-end distance would be large (left side of Figure 6.23a). When the cyclics slide apart, the system would contract (right side of Figure 6.23a). The precursor molecule [**1**] designed by the Sauvage [81–83] group is shown in Figure 6.23b. It consists of a bidentate chelate (phenanthroline) in the ring as well as the filament (axle). By mixing **1** with Cu^{I} stoichiometrically, a mutually threaded, hermaphrodite dimer was formed as

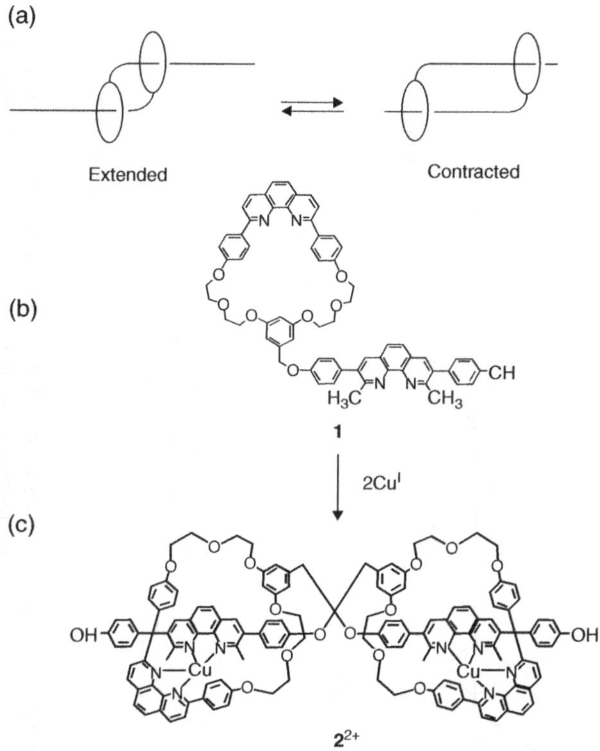

Figure 6.23 (a) Schematic of a pseudorotaxane dimer which could contract and extend by mutually sliding motion, (b) precursor molecule, and (c) the dimer formed with Cu complex. (Source: Jiménez et al. [81]. Reproduced with permission of Wiley.)

shown in Figure 6.23c, with the Cu complex. From the X-ray crystal structure of the complex, it was found that the length of the dimer was about 36 Å. To add muscle-like functionality to **1**, another ligand had to be attached, so that the changes in the metal coordination (ligand exchange) would lead to a change in the length of the dimer. A stopper had to be added as well, to prevent slippage. A terdentate (2,2′,6′,2″ terpyridine) was attached to the filament and a tetraphenyl group was used as the stopper. The modified precursor formed mutually threaded dimer, as shown in Figure 6.24a, with Cu coordination with the phenanthroline unit (black sphere in Figure 6.24). Treatment with KCN dislodged the Cu and addition of Zn^{2+} led to a five-coordinate complex with the terpyridine unit. The length of the extended dimer in Figure 6.24a was 85 Å and upon contraction (Figure 6.24c), it reduced to 65 Å. The extended dimer was regained with metal exchange, by adding excess Cu^{I} to the Zn complex.

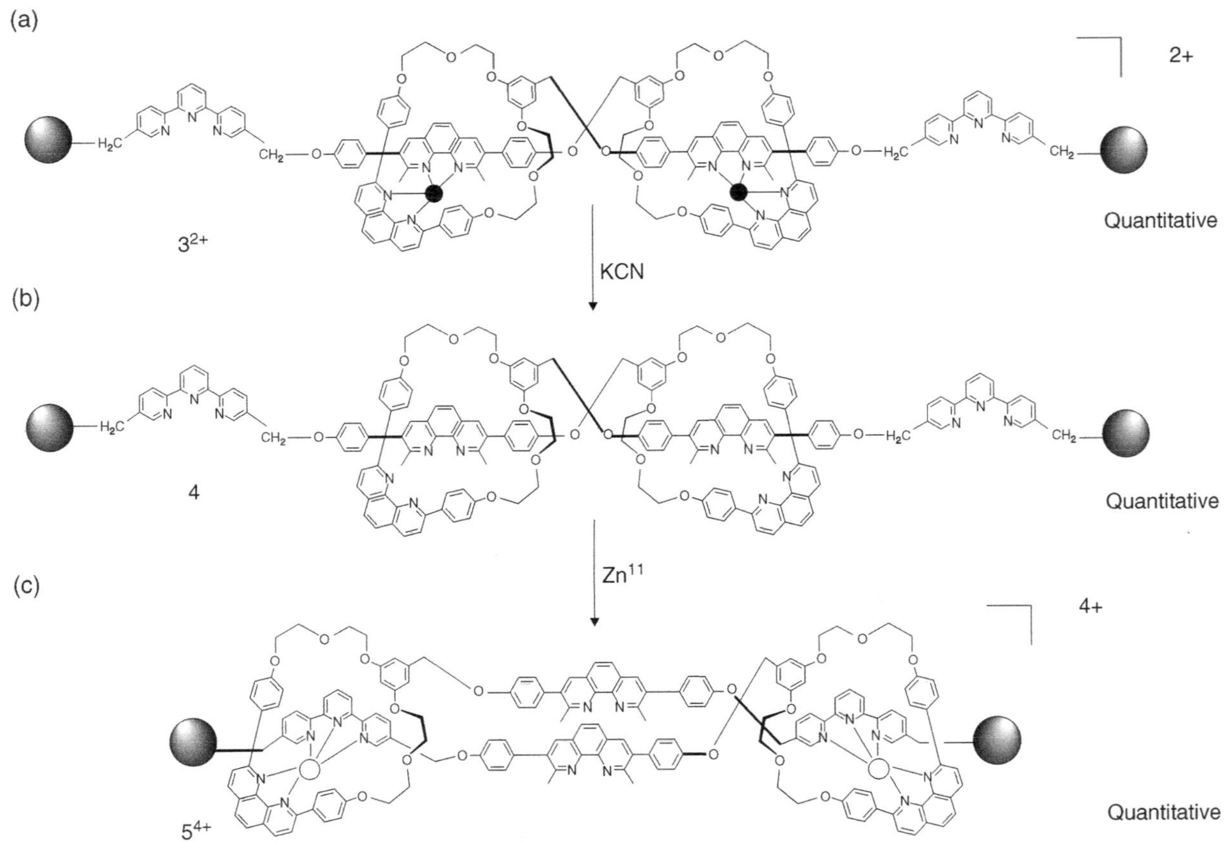

Figure 6.24 (a) Schematic of the dimer formation of the four coordinate Cu complex in Figure 6.23c, with the added 2,2′,6′,2″ terpyridine and tetraphenyl stopper (black spheres); (b) KCN displaces the copper; and (c) addition of Zn^{2+} leads to a five-coordinate complex with terpyridine segment, leading to contraction of the dimer. (Source: Jiménez et al. [81]. Reproduced with permission of Wiley.)

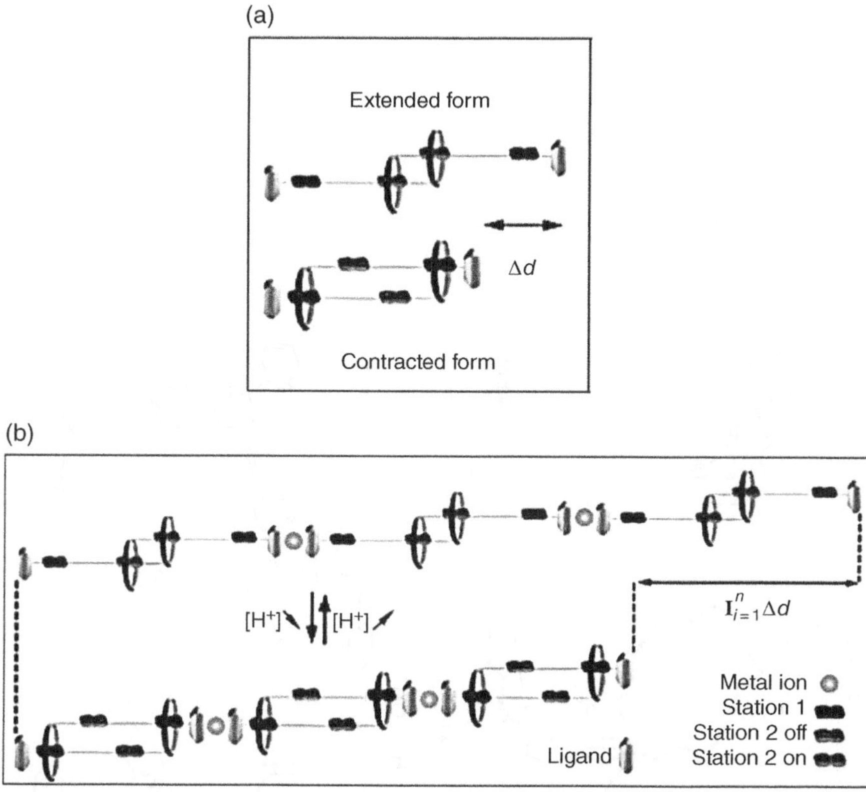

Figure 6.25 The principle of building a polymeric daisy chain. (a) Homodimeric [c2] daisy chain, with a ligand as a stopper. The contraction is denoted by Δd. (b) Supramolecular polymerization using metal coordination by the stoppers. The total contraction is the sum $\sum_{i=1}^{n} \Delta d$. (Source: Du et al. [84]. Reproduced with permission of Wiley.)

Thus, it was shown to be a reversible extension–contraction process, as shown schematically in Figure 6.23a.

While the earlier example was based on a homodimeric pair, building polymeric daisy chains by supramolecular assembly was reported by Du et al. [84]. The principle of their approach is shown in Figure 6.25a, in which a rotaxane with a homodimeric pair ([*c*2] daisy chain) was used as the monomer. The difference in length between the extended and contracted states was denoted as Δd. The cyclic was a [24] crown-8 and terpyridine was used as the blocking group which would also serve as a ligand, as illustrated in Figure 6.26. The axle contained a secondary ammonium and the triazolium ions. In the protonated state, the crown ether binds the ammonium ion

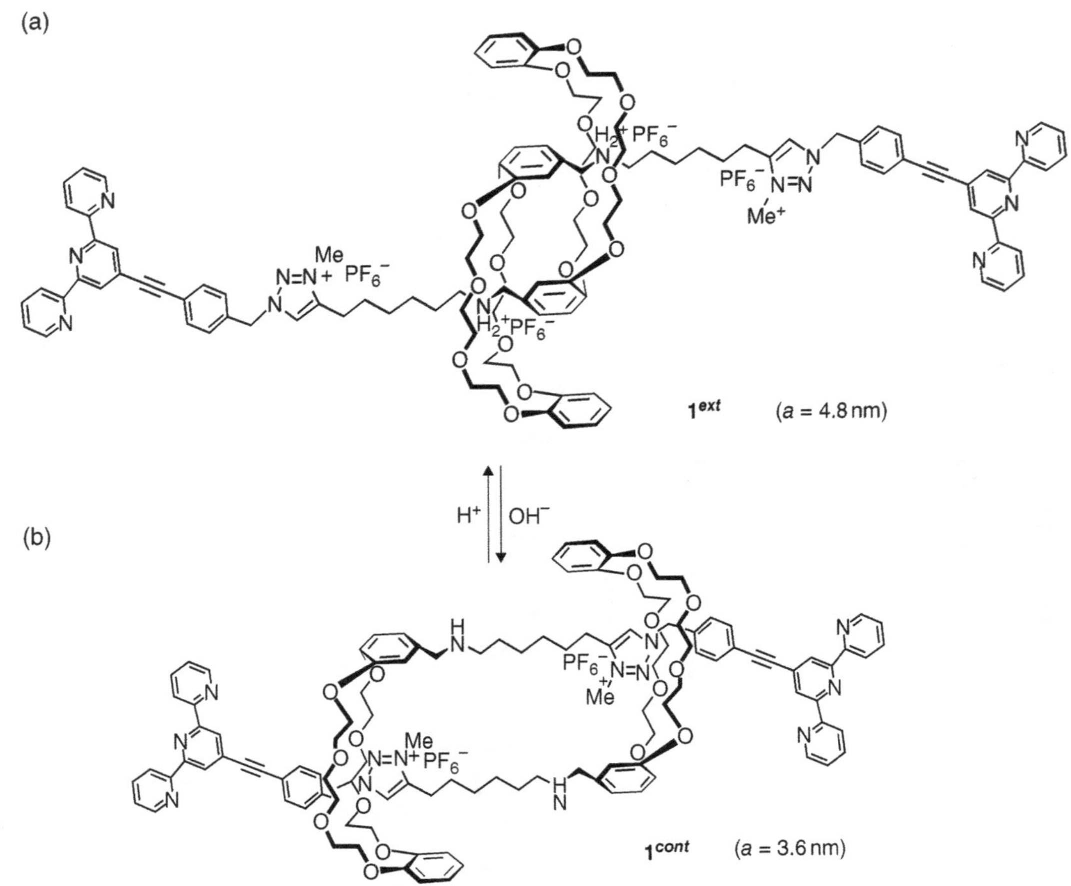

Figure 6.26 (a) and (b) Illustration of the extended and contracted conformations of the homodimeric pair ([c2] daisy chain) designed by Du et al. [84]. (Source: Du et al. [84]. Reproduced with permission of Wiley.)

which leads to the extended conformation (1^{ext} in Figure 6.26a), and upon deprotonation, it moves to the triazolium site, causing a contraction of the system (1^{cont} Figure 6.26b). The experimental procedure for obtaining the daisy chains 1^{ext} and 1^{cont} in Figure 6.26 is given in the supplementary information section of Ref. [84]. Using molecular modeling, the lengths of the molecule in the extended and contracted states were calculated to be 48 and 36 Å, respectively ($\Delta d = 12$ Å).

The route to build a supramolecular polymer is shown schematically in Figure 6.25b. The stoppers bind the metal ions and form the coordination polymer. Supramolecular polymerization of 1^{ext} and 1^{cont} was carried out with one equivalent of $Zn(OTf)_2$ or $FeCl_2$ in a 1 : 1 solution of $CDCl_3/CD_3CN$ with a 10 mM concentration of **1**. UV/Vis spectra confirmed the bisterpyridine complex formation. Light scattering and small-angle neutron scattering (SANS) were used to determine the conformational properties, such as the persistent length. The DP of the Zn complex of 1^{cont} was 1967, and it was 2937 for the Fe coordination. With the Fe complex of 1^{ext} and 1^{cont}, the persistent lengths (L_p) were 17.6 and 12.8 nm, the radii of gyration (R_g) were 200 and 305 nm, and the hydrodynamic radii (R_H) were 167 and 189 nm, respectively. The polymers possess a wormlike chain behavior. Thus, coordination polymerization of the homodimeric molecular machine using Zn or Fe ions led to muscle-like supramolecular chains at the mesoscale.

Harada et al. [85] described a supramolecular polymeric actuator system based on β-CD. Although it is not exactly a polyrotaxane, it involves an inclusion complex of ferrocene (Fc). Hydrogels of random copolymers of poly(acrylamide) with β-CD and Fc functionalization were used, with different ratios of the monomers (x, y, z) (Figure 6.27a). Redox-driven inclusion complexation and decomplexation of Fe with β-CD led to contraction and expansion of the gels, as shown in Figure 6.27. Ceric ammonium nitrate (CAN) was used as an oxidant. To study the size variation of the gels, the β-CD–Fc gel (3,3,1) (i.e., $x/y/z = 3,3,1$) was immersed in Tris/HCl (0.1 M) buffer. It was then kept in the buffer with CAN (50 mM) for an hour. The oxidized gel was then placed in the Tris/HCl buffer. Figure 6.27b shows the photographs of the gel upon oxidation and reduction, in which the diagonal length of the gel is indicated. An illustration of the contraction and expansion with the formation and release of the β-CD–Fc inclusion is given in Figure 6.27c. In another experiment, these authors recorded the elongation and contraction of a rectangular piece of the gel to which a load was attached (291 mg). Using this data, the work done was calculated to be 2.0 μJ. Similar experiments were performed with different ratios of x, y, and z. The strength of the gels was found to depend on the relative concentrations of β-CD and Fc in the polymer.

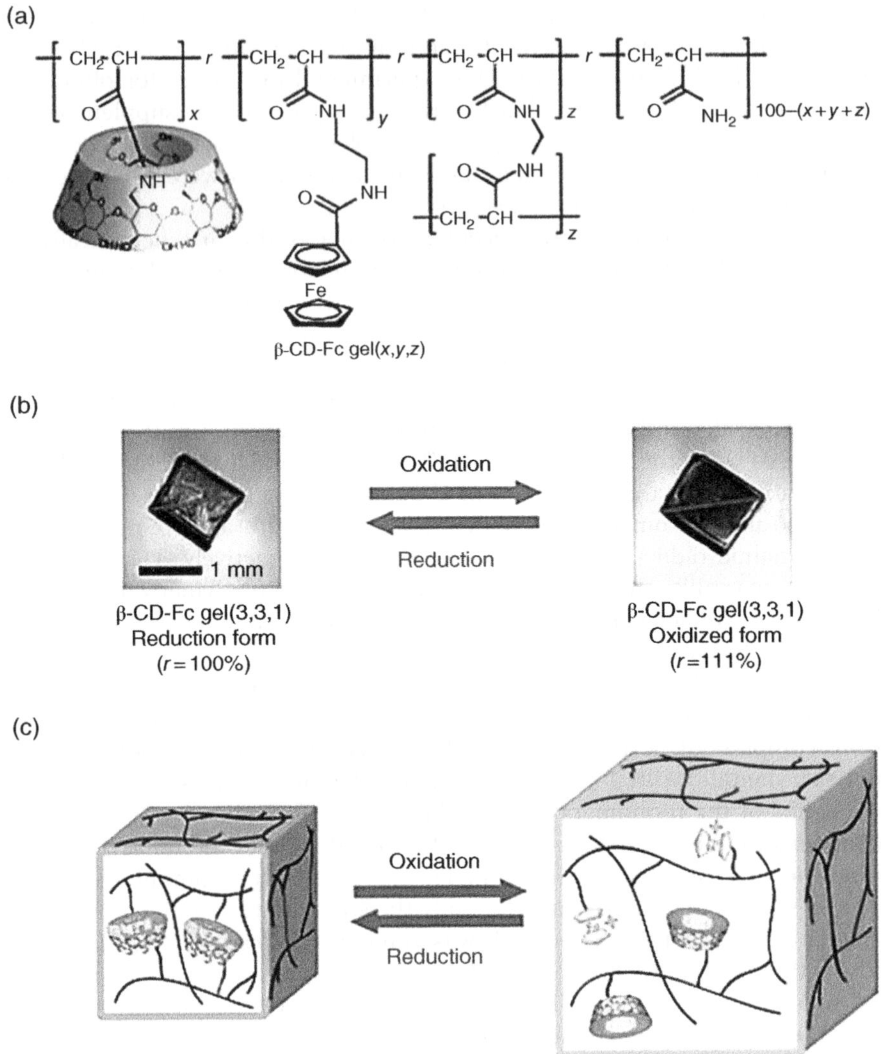

Figure 6.27 (a) The random copolymer with β-CD and Fc-functionalized monomers; (b) photograph of the hydrogel of the (3,3,1) polymer showing the change in size with oxidation and reduction; and (c) illustration of contracted structure of the gel with β-CD–Fc inclusion and expansion with decomplexation of Fc from β-CD. (Source: Nakahata et al. [85]. Reproduced with permission of Wiley.)

References

1 H. W. Gibson and H. Marand, *Adv. Mater.*, **1993**, *5*, 11–20.
2 H. F. Mark, *Monatsh. Chem.*, **1952**, *83*, 545.
3 H. Frisch, I. Martin, and H. Mark, *Monatsh. Chem.*, **1953**, *84*, 250–256.
4 P. J. Flory, *Statistical Mechanics of Chain Molecules*, **1969**, John Wiley & Sons, Inc., New York.
5 H. Jacobson and W. H. Stockmayer, *J. Chem. Phys.*, **1950**, *18*, 1600–1606.
6 J. A. Semlyen, Ed., *Cyclic Polymers*, **1986**, Elsevier (2nd Ed., 2013, Springer).
7 S. J. Clarson, in *Cyclic Polymers*, 2nd Ed., **2003**, J. A. Semlyen, Ed., Chapter 5, Springer.
8 H. L. Frisch and E. Wasserman, *J. Am. Chem. Soc.*, **1961**, *83*, 3789–3795.
9 J. Vohlídal, E. S. Wilks, A. Yerin, A. Fradet, K.-H. Hellwich, P. Hodge, J. Kahovec, W. Mormann, and R. F. T. Stepto, *Pure Appl. Chem.*, **2012**, *84*, 2135–2165.
10 I. T. Harrison and S. Harrison, *J. Am. Chem. Soc.*, **1967**, *89*, 5723–5724.
11 G. Schill, W. Beckmann, N. Schweickert, and H. Fritz, *Chem. Ber.*, **1986**, *119*, 2647–2655.
12 G. Agam, D. Graiver, and A. Zilkha, *J. Am. Chem. Soc.*, **1976**, *98*, 5206–5214.
13 H. W. Gibson and E. J. Mahan, in *Cyclic Polymers*, 2nd Ed., **2003**, J. A. Semlyen, Ed., Chapter 12, Springer.
14 C. J. Bruns and J. F. Stoddart, *Acc. Chem. Res.*, **2014**, *47*, 2186–2199.
15 Y. X. Shen, D. Xie, and H. W. Gibson, *J. Am. Chem. Soc.*, **1994**, *116*, 537–548.
16 M. Benmouna and U. Maschke, in *Cyclic Polymers*, 2nd Ed., **2003**, J. A. Semlyen, Ed., Chapter 16, Springer.
17 C. Gong, Q. Ji, C. Subramaniam, and H. W. Gibson, *Macromolecules*, **1998**, *31*, 1814–1818.
18 P. R. Sundararajan and V. S. R. Rao, *Carbohydr. Res.*, **1970**, *13*, 351–358.
19 R. Hernández, M. Rusa, C. C. Rusa, D. López, C. Mijangos, and A. E. Tonelli, *Macromolecules*, **2004**, *37*, 9620–9625.
20 A. Hybl, R. E. Rundle, and D. E. Williams, *J. Am. Chem. Soc.*, **1965**, *87*, 2779–2788.
21 G. Crini, *Chem. Rev.*, **2014**, *114*, 10940–10975.
22 X. Ma and Y. Zhao, *Chem. Rev.*, **2015**, *115*, 7794–7839.
23 S. V. Kurkov and T. Loftsson, *Int. J. Pharm.*, **2013**, *453*, 167–180.
24 G. Wenz, *Angew. Chem. Int. Ed.*, **1994**, *33*, 803–822.
25 A. Harada, Y. Takashima, and M. Nakahata, *Acc. Chem. Res.*, **2014**, *47*, 2128–2140.
26 T.-Q. Nguyen, J. Wu, S. H. Tolbert, and B. J. Schwartz, *Adv. Mater.*, **2001**, *13*, 609–611.
27 M. J. Frampton and H. L. Anderson, *Angew. Chem. Int. Ed.*, **2007**, *46*, 1028–1064.

28 L. Nassimbeni, *Acc. Chem. Res.*, **2003**, *36*, 631–637.
29 A. Harada and M. Kamachi, *Macromolecules*, **1990**, *23*, 2821–2823.
30 A. Harada, J. Li, and M. Kamachi, *Macromolecules*, **1993**, *26*, 5698–5703.
31 T. Michishita, Y. Takashima, and A. Harada, *Macromol. Rapid Commun.*, **2004**, *25*, 1159–1162.
32 A. Harada, *J. Polym. Sci. A Polym. Chem.*, **2006**, *44*, 5113–5119.
33 M. Okada and A. Harada, *Org. Lett.*, **2004**, *6*, 361.
34 Y. Inoue, M. Miyauchi, H. Nakajima, Y. Takashima, H. Yamaguchi, and A. Harada, *Macromolecules*, **2007**, *40*, 3256–3262.
35 T. Arai, M. Hayashi, N. Takagi, and T. Takata, *Macromolecules*, **2009**, *42*, 1881–1887.
36 C. Yang, J. Yang, X. Ni, and J. Li, *Macromolecules*, **2009**, *42*, 3856–3859.
37 J. Li, C. Yang, H. Li, X. Wang, S. H. Goh, J. L. Ding, D. Y. Wang, and K. W. Leong, *Adv. Mater.*, **2006**, *18*, 2969–2974.
38 X. Zhang, X. Zhu, F. Ke, L. Ye, E.-Q. Chen, A.-Y. Zhang, and Z.-G. Feng, *Polymer*, **2009**, *50*, 4343–4351.
39 M. L. Adams, D. R. Andes, and G. S. Kwon, *Biomacromolecules*, **2003**, *4*, 750–757.
40 J. Huang, L. Ren, H. Zhu, and Y. Chen, *Macromol. Chem. Phys.*, **2006**, *207*, 1764–1772.
41 J. Huang, L. Ren, and Y. Chen, *Polym. Int.*, **2008**, *57*, 714–721.
42 J. Liu, H. R. Sondjaja, and K. C. Tam, *Langmuir*, **2007**, *23*, 5106–5109.
43 R. Yuan and X. Shuai, *J. Polym. Sci. B Polym. Phys.*, **2008**, *46*, 782–790.
44 H. Dong, Y. Li, S. Cai, R. Zhuo, X. Zhang, and L. Liu, *Angew. Chem. Int. Ed.*, **2008**, *47*, 5573–5576.
45 J. F. Brown and B. M. Slusarczuk, *J. Am. Chem. Soc.*, **1965**, *87*, 931.
46 S. J. Clarson, K. Dodgson, and J. A. Semlyen, *Polymer*, **1985**, *26*, 930–934.
47 P. R. Sundararajan, *Polymer*, **2002**, *43*, 1691–1693.
48 S. J. Clarson, J. E. Mark, and J. A. Semlyen, *Polym. Commun.*, **1986**, *27*, 244.
49 S. J. Clarson, J. E. Mark, and J. A. Semlyen, *Polym. Commun.*, **1987**, *28*, 151.
50 L. C. DeBolt and J. E. Mark, *Macromolecules*, **1987**, *20*, 2369–2374.
51 S. Z. Wu and J. E. Mark, *Polym. Rev.*, **2007**, *47*, 463–485.
52 J. P. Giltrow, *Wear*, **1970**, *15*, 71.
53 D. J. Brunelle, T. L. Evans, T. G. Shannon, E. P. Boden, K. R. Stewart, L. P. Fontana, and D. K. Bonauto, *Polym. Prepr. (Am. Chem. Soc., Div. Polym. Chem.)*, **1989**, *30*(2), 569–570.
54 D. J. Brunelle and E. P. Boden, *Macromol. Symp.*, **1992**, *54/55*, 397–412.
55 D. J. Brulelle, "Macrocyclic Polymers," in *Encyclopedia of Polymer Science and Technology*, 4th Ed., **2014**, H. F. Mark, Series Ed., Vol. *7*, John Wiley & Sons, Inc., 674–692.
56 P. R. Sundararajan, D. K. Murti, P. G. Odell, and G. K. Hamer, Photoreceptor Including Rotaxanes, US Patent 6,096,464, **August 1, 2000**.
57 D. French, A. O. Pulley, J. A. Effenberger, M. A. Rougvie, and M. Abdullah, *Arch. Biochem. Biophys.*, **1965**, *111*, 153–160.

58 K. H. Frömming and J. Szejtli, *Cyclodextrins in Pharmacy*, **1994**, Kluwer Academic, Dordrecht, pp. 1–4.

59 T. Endo, H. Ueda, S. Kobayashi, and T. Nagai, *Carbohydr. Res.*, **1995**, *269*, 369–373.

60 A. de Juan and E. M. Pérez, *Nanoscale*, **2013**, *5*, 7141–7148.

61 A. Star, D. W. Steuerman, J. R. Heath, and J. F. Stoddart, *Angew. Chem. Int. Ed.*, **2002**, *41*, 2508–2512.

62 H. Dodziuk, A. Ejchart, W. Anczewski, H. Ueda, E. Krinichnaya, G. Dolgonos, and W. Kutner, *Chem. Commun.*, **2003**, 986–987.

63 A. de Juan, Y. Pouillon, L. Ruiz-González, A. Torres-Pardo, S. Casado, N. Martín, A. Rubio, and E. M. Pérez, *Angew. Chem. Int. Ed.*, **2014**, *53*, 5394–5400.

64 J. Akola, K. Rytkonen, and M. Manninen, *J. Phys. Chem. B*, **2006**, *110*, 5186–5190.

65 H. Fujita, T. Ooya, and N. Yui, *Macromol. Chem. Phys.*, **1999**, *200*, 706–713.

66 T. Ikeda, N. Watabe, T. Ooya, and N. Yui, *Macromol. Chem. Phys.*, **2001**, *202*, 1338–1344.

67 A. Altieri, F. G. Gatti, E. R. Kay, D. A. Leigh, D. Martel, F. Paolucci, A. M. Z. Slawin, and J. K. Y. Wong, *J. Am. Chem. Soc.*, **2003**, *125*, 8644–8654.

68 E. R. Kay, D. A. Leigh, and F. Zerbetto, *Angew. Chem. Int. Ed.*, **2007**, *46*, 72–191.

69 S. Erbas-Cakmak, D. A. Leigh, C. T. McTernan, and A. L. Nussbaumer, *Chem. Rev.*, **2015**, *115*, 10081–10206.

70 L. Jiang, J. Okano, A. Orita, and J. Otera, *Angew. Chem. Int. Ed.*, **2004**, *43*, 2121–2124.

71 T. Iijima, S. A. Vignon, H.-R. Tseng, T. Jarrosson, J. K. M. Sanders, F. Marchioni, M. Venturi, E. Apostoli, V. Balzani, and J. F. Stoddart, *Chem. Eur. J.*, **2004**, *10*, 6375–6392.

72 H. Yu, Y. Luo, K. Beverley, J. F. Stoddart, H.-R. Tseng, and J. R. Heath, *Angew. Chem. Int. Ed.*, **2003**, *42*, 5706–5711.

73 H. Murakami, A. Kawabuchi, K. Kotoo, M. Kunitake, and N. Nakashima, *J. Am. Chem. Soc.*, **1997**, *119*, 7605–7606.

74 C. A. Stanier, S. J. Alderman, T. D. W. Claridge, and H. L. Anderson, *Angew. Chem. Int. Ed.*, **2002**, *41*, 1769–1772.

75 D.-H. Qu, Q.-C. Wang, X. Ma, and H. Tian, *Chem. Eur. J.*, **2005**, *11*, 5929–5937.

76 P. R. Ashton, I. Baxter, S. J. Cantrill, M. C. T. Fyfe, P. T. Glink, J. Stoddart, A. J. P. White, and D. J. Williams, *Angew. Chem. Int. Ed.*, **1998**, *37*, 1294–1297.

77 P. R. Ashton, I. W. Parsons, F. M. Raymo, J. F. Stoddart, A. J. P. White, D. J. Williams, and R. Wolf, *Angew. Chem. Int. Ed.*, **1998**, *37*, 1913–1916.

78 S. Menzer, A. J. P. White, D. J. Williams, M. Bělohradský, C. Hamers, F. M. Raymo, A. N. Shipway, and J. F. Stoddart, *Macromolecules*, **1998**, *31*, 295–307.

79 M. P. L. Werts, M. van den Boogaard, G. M. Tsivgoulis, and G. Hadziioannou, *Macromolecules*, **2003**, *36*, 7004–7013.

80 T. F. Otero and J. M. Sansiñena, *Adv. Mater.*, **1998**, *10*, 491–494.
81 M. C. Jiménez, C. Dietrich-Buchecker, and J.-P. Sauvage, *Angew. Chem. Int. Ed.*, **2000**, *39*, 3284–3287.
82 M. C. Jiménez, C. Dietrich-Buchecker, J.-P. Sauvage, and A. De Cian, *Angew. Chem. Int. Ed.*, **2000**, *39*, 1295–1298.
83 M. C. Jimenez-Molero, C. Dietrich-Buchecker, and J.-P. Sauvage, *Chem. Comm.*, **2003**, 1613–1616.
84 G. Du, E. Moulin, N. Jouault, E. Buhler, and N. Giuseppone, *Angew. Chem. Int. Ed.*, **2012**, *51*, 12504–12508.
85 M. Nakahata, Y. Takashima, A. Hashidzume, and A. Harada, *Angew. Chem. Int. Ed.*, **2013**, *52*, 5731–5735.

Further Reading

V. Balzani, A. Credi, F. M. Raymo, and J. F. Stoddart, *Angew. Chem. Int. Ed.*, **2000**, *39*, 3348–3391.
B. L. Feringa, Ed., *Molecular Switches*, **2001**, Wiley-VCH Verlag GmbH, Weinheim.
N. Hadjichristidis, A. Hirao, Y. Tezuka, and F. Du Prez, Ed. *Complex Macromolecular Architectures, Synthesis, Characterization and Self-assembly*, **2011**, John Wiley & Sons, Inc., Hoboken, NJ.
J. E. Mark and B. Erman, *Rubberlike Elasticity: A Molecular Primer*, 2nd Ed., **2007**, Cambridge University Press (1988, John Wiley & Sons, Inc., New York, 196 pp.).
F. M. Raymo, *Adv. Mater.*, **2002**, *14*, 401–414.
H. Tian and Q.-C. Wang, *Chem. Soc. Rev.*, **2006**, *35*, 361–374.
X. Zhang, X. Zhu, X. Tong, L. Ye, A.-Y. Zhang, and Z.-G. Feng, *J. Polym. Sci. A Polym. Chem.*, **2008**, *46*, 5283–5293.

7

Polymer Gels

7.1 One-Dimensional Growth

The three states of matter, with molecules in the form of gas, liquid, and solid are well known. With the first two, the molecules do not necessarily have any order. The solid too can be amorphous without any three-dimensional (3D) order or be crystalline. The fourth state of matter, the liquid crystal which is neither a liquid nor a 3D crystal, exhibits 1D or 2D order. Yet another state is the commonly known "gel." In the context of self-assembly, gelation is the hierarchical formation of 3D network superstructures from self-assembled nanostructures. This system is based on the self-assembly of the solute molecules (which are known as the gelators) into ordered aggregates trapping the solvent. The morphology of self-assembled gels may be either discrete structures, such as empty vesicles, or fibrillar networks that form due to predominantly unidirectional extended assemblies. These in turn trap the solvent in a 3D network. Figure 7.1 shows the schematic of a monocarbamate molecule which is a model compound for polyurethanes (PUs). It has one N—H···C=O hydrogen bonding pair and short alkyl side chains on either side. Because of the hydrogen bonding and van der Waals interactions along one direction, the supramolecular association leads to predominant unidirectional growth in the presence of the solvent and the morphology of the gel consists of fibers of several hundred microns in length [1]. (The xerogel refers to the gel that has been dried. Although there might be a shrinkage, the overall morphology is maintained.) Crystallization from the melt was also quite rapid in this case that it was not possible to determine the rate of growth using conventional optical microscopy. Figure 7.2 shows the schematic of a biscarbamate molecule, with two hydrogen bonding motifs separated by a $(CH_2)_6$ spacer and alkyl side chains on either side [2]. The figure shows the directions of the hydrogen bonds and van der Waals interactions. In this case again, fibers are seen in the morphology of the xerogels. A feature of these fibers is that they are not flat but are folded into an eaves trough morphology, which is a precursor to hollow

Physical Aspects of Polymer Self-Assembly, First Edition. P. R. Sundararajan.

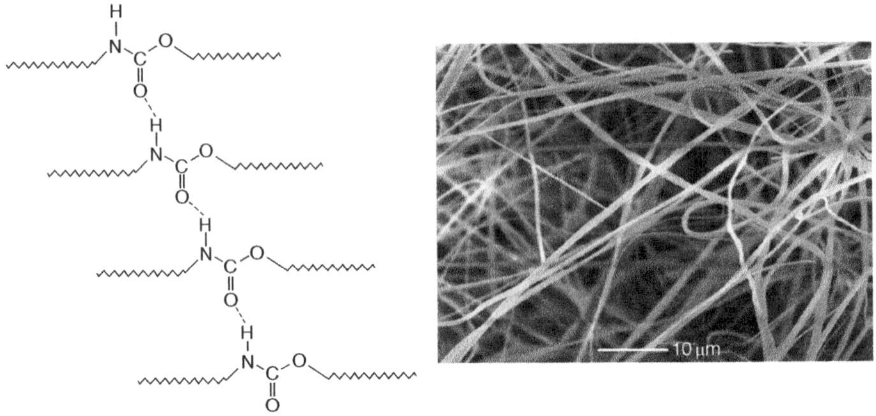

Figure 7.1 Schematic of a monocarbamate molecule and the morphology of the gel fibers. (Source: Moniruzzaman and Sundararajan [1]. Reproduced with permission of American Chemical Society.)

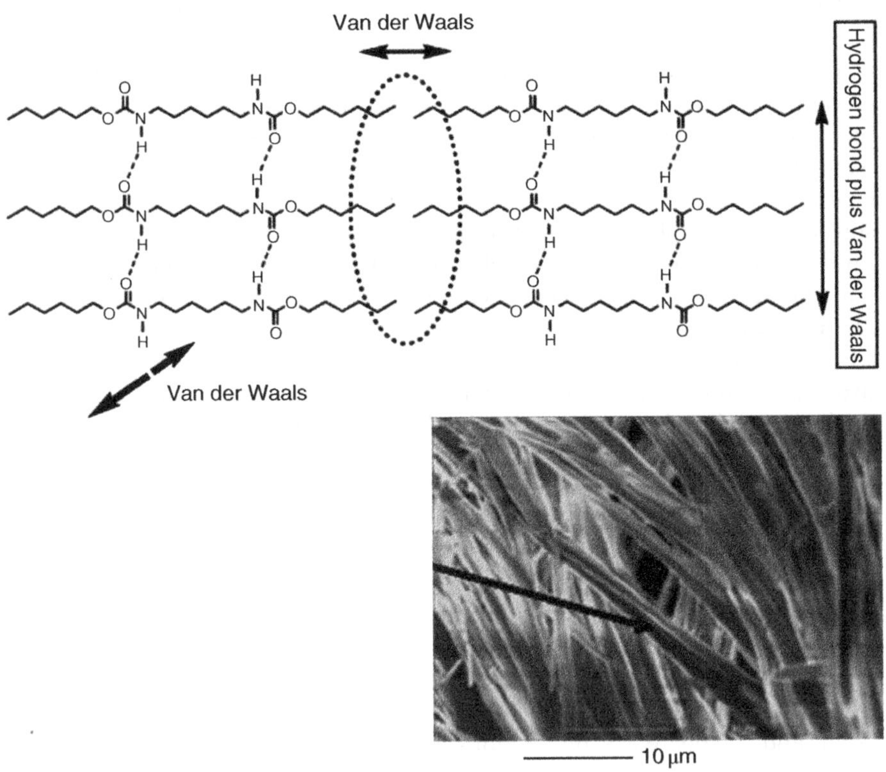

Figure 7.2 Schematic of a biscarbamate molecule and the morphology of the gel fibers. The arrow in the SEM micrograph points to eaves trough type of folded morphology of the fibers. (Source: Khanna et al. [2]. Reproduced with permission of American Chemical Society.)

tube formation. The gelation caused by the self-assembly of urea and bis-urea (cousins of carbamates) has been studied [3–8]. Thus, the predominant directional growth resulting in fibers and the trapping of the solvent by the physical network formation of these fibers characterize the gels of the organic small molecules. The commonality for designing a gelator is that the formation of gels indicates a hierarchical process ranging from individual dissolved molecules to 3D entangled superstructures. Therefore, by rational design of molecular structures and precise control of their aggregation by altering various factors, such as the solvent, temperature, and concentration, it is possible to direct the self-assembly pathway toward the formation of gels. The choice of the gelator–solvent pair is still a matter of trial and error, although approaches based on solubility parameters, etc., have been proposed to enhance the chance of success [9]. The whole process of making a gel is simple: choose a solvent in which the gelator does not dissolve at ambient conditions. Heat the mixture to make a homogeneous solution at a higher temperature. This would allow the solute to interact with the solvent and in the case of polymers, chain expansion would occur due to the solvent becoming a good solvent. Upon cooling, the solvent would become progressively poor promoting the self-association of the gelator and in the case of polymers the chain entanglements would occur, trapping the solvent.

7.2 Definitions and Classifications

With regard to polymers, much effort has been devoted to define a "gel." In a classic paper, Flory [10] summarized the types of polymeric gels conceptually and derived the mathematical conditions for gelation through the formation of infinite networks mentioned earlier. To define a gel itself, Flory cited Herman's definitions that these "(i) are coherent colloid systems of at least two components, (ii) exhibit mechanical properties similar to those of solids, and (iii) both the dispersed component and the dispersion medium extend themselves continuously throughout the system." The criterion (i) includes those in which a single component is swollen by the second component.

Based on this definition, a gel must develop a 3D network that permeates its volume and remains stable within specific ranges of concentration and temperature. Flory [10] classified gels into four types, based on the structural features: (i) Well-ordered lamellar structures, including gel mesophases. (ii) Covalently connected polymer networks, which are completely disordered. These would be chemically cross-linked networks. (iii) Polymer networks formed through physical aggregation. These would include those with crystalline aggregates formed and would have local order, but otherwise disordered. (iv) Particulate, disordered structures. This would be a case discussed in a

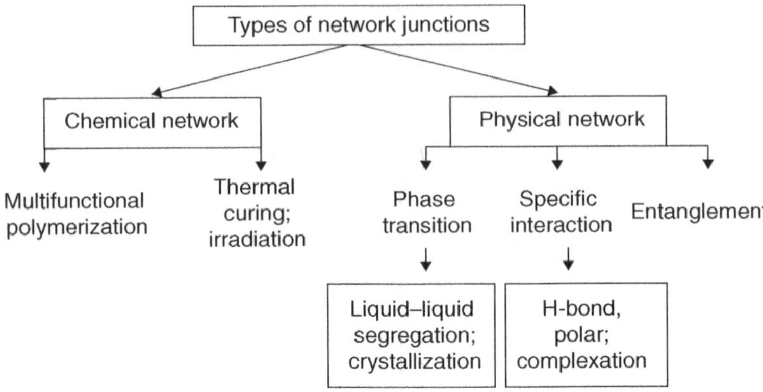

Figure 7.3 Classification of different methods of creating network structures in polymers for enabling gel formation. (Source: Keller [12]. Reproduced with permission of Royal Society of Chemistry.)

later section that relates to the morphology of gels using atactic polystyrene (a-PS) and PDMS, where glassy, polymer-rich spheres are connected by isolated solvated chains. In proposing phenomenological definitions for a "gel," Almdal et al. [11] added that the storage modulus G' should show a pronounced plateau over times of the order of seconds and that the loss modulus G'' should be smaller than G' in the plateau region.

In a seminal paper that was presented at a Faraday Discussion meeting on gels in 1995, Andrew Keller [12] discussed various structural aspects of polymer gels in detail. In parallel with Flory's classifications, Keller distinguished between the chemical and physical gels (cases (ii) and (iii) of the Flory classification). These are shown as a flow chart in Figure 7.3.

In the case of chemical gels, the polymer chains are cross-linked using methods such as thermal curing or (UV) radiation curing. The network thus created is swollen by the solvent creating the gel. The physical gels, which are often thermo-reversible, result upon phase transitions and entanglements. Figure 7.4 illustrates four different cases, of which Figure 7.4a is that of small molecules mentioned earlier with regard to Figures 7.1 and 7.2. The cross-linked network that is used for chemical gel is shown schematically in Figure 7.4b. The network that arises simply out of entanglements between the chains is shown in Figure 7.4c. In the schematic shown in Figure 7.4d, the entanglements or chemical cross-links are replaced by local crystalline domains in which intramolecular chain folding and the tie molecules bridging the lamellae constitute the network. This would correspond to cases (i) and (iii) of the Flory classification.

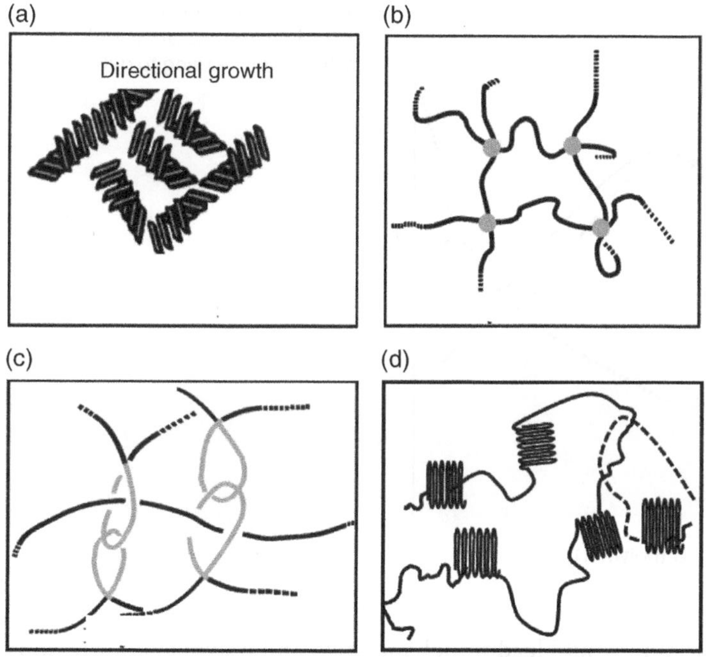

Figure 7.4 Illustration of network formation: (a) small-molecule directional aggregation, (b) chemically cross-linked network, (c) entanglement causing physical network, and (d) connectivity through crystalline lamellae in crystallizable polymers.

7.3 Gels with Noncrystallizable Polymers

a-PS is an example of such a polymer, which does not have groups that give rise to specific interactions such as hydrogen bonds or Coulombic interactions. Gels with a-PS have been studied using solvents such as carbon disulfide, decalin, and cyclohexanol. The gelation in this case has been attributed to liquid–liquid (L–L) phase separation with vitrification [13, 14]. This concept is illustrated in Figure 7.5, with polymer concentration on the abscissa and the temperature on the left ordinate. The solid curve in the figure shows the phase boundary and the dashed curve corresponds to the spinodal. A case of liquid–liquid transition without macroscopic phase separation has been discussed by Murata and Tanaka [15]. It is known that the glass transition (T_g) temperature of a polymer would decrease, in the presence of a solvent, the extent of the reduction depending on the solvent volume/weight fraction. This is schematically shown by the dotted curve in Figure 7.5, corresponding to the right ordinate. From the single phase region, as the temperature is decreased, the

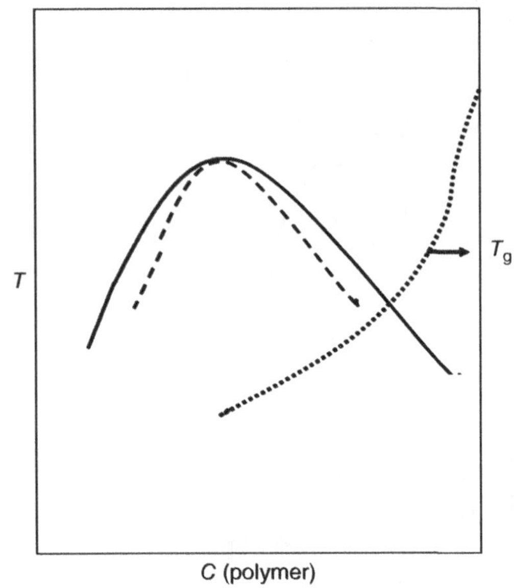

Figure 7.5 An illustration of liquid–liquid phase separation and vitrification, leading to gels.

L–L phase separation would lead to polymer-rich and polymer-poor regions. At a temperature where the glass transition is reached, vitrification would set in (at the intersection of the solid and dotted curves in Figure 7.5), arresting the solvated morphology. The morphology of the gels was found to depend on polymer concentration. Keller classified them into polymer-rich, polymer-poor, and intermediate states. Based on the scanning electron microscopy (SEM) images, these are sketched in Figure 7.6. With low polymer concentration, the solute formed glassy beads, connected by strands of polymer chains, as in Figure 7.6a. With an increase in polymer concentration, a bicontinuous structure (Figure 7.6b) occurred. Solvent droplets in the polymer matrix, as in Figure 7.6c, were seen with further increase of polymer concentration. Coalescence of polymer beads was also seen, as in Figure 7.6d. The SEM images corresponding to low and high polymer concentrations [13] are shown in Figure 7.7a and b. With the 1% gel sample that was quenched prior to freeze-drying, beads of aggregated polymer are seen, with strands of polymer connecting these beads. This would correspond to the model shown in Figure 7.6a. The morphology seen in Figure 7.7b for the 57% gel would correspond to Figure 7.6c with the solvent forming droplets in the glassy polymer matrix. The "dots" seen in Figure 7.7b are the holes left in the matrix following the solvent evaporation.

Another system that we discuss in this category is poly(dimethyl siloxane) (PDMS), although PDMS can be crystallized at low temperatures. Sundararajan [16] discussed the crystalline spherulitic morphology of PDMS upon lowering

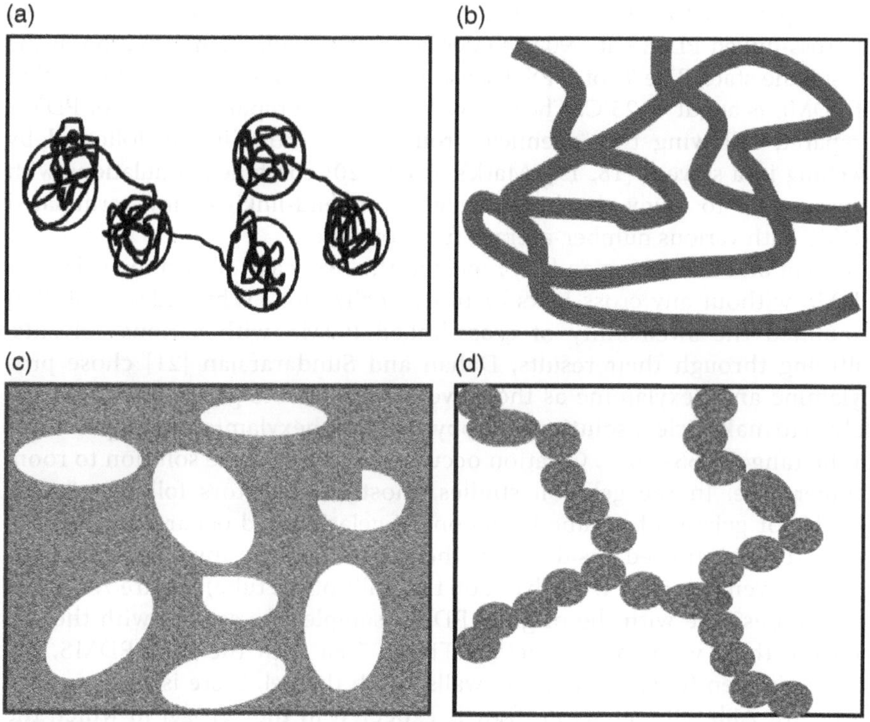

Figure 7.6 Schematic of the change in morphologies with an increase in polymer concentration from (a) to (d). Filled areas denote the polymer and the blank regions correspond to the solvent. (Source: Keller [12]. Reproduced with permission of Royal Society of Chemistry.)

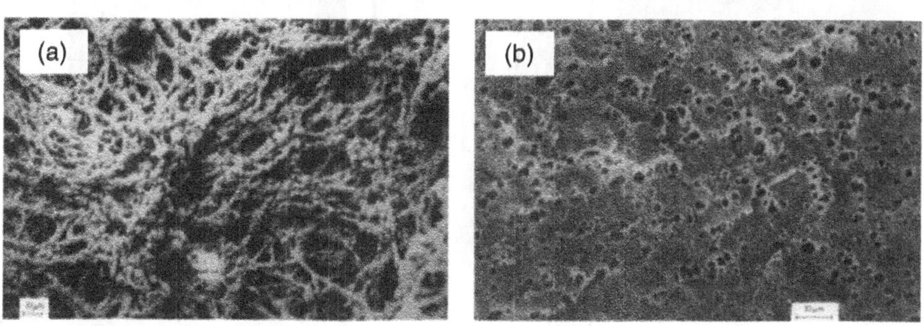

Figure 7.7 SEM images of freeze-dried gel samples of a-PS: (a) 1% gel sample and (b) 57% gel sample quenched prior to freeze-drying. (Source: Hikmet et al. [13]. Reproduced with permission of Elsevier.)

the temperature to −70°C. Damaschun [17] recorded X-ray diffraction pattern of cross-linked PDMS at −90°C to derive the molecular conformation in the crystalline state. The T_g of a-PS discussed earlier is about 100°C, whereas that of PDMS is about −125°C. There have been several reports on gels of PDMS prepared following the "chemical" route, with cross-linking followed by swelling in a solvent [18, 19]. Mark's group [20] combined simulations with experiments to study the hydrosilylation on end-linking vinyl-terminated PDMS with various number-average functionalities.

Dahan and Sundararajan [21] reported the first case of forming gels with PDMS without any cross-links or functionalization. Whitesides et al. [22] examined the swellability of cross-linked PDMS with various solvents. Filtering through their results, Dahan and Sundararajan [21] chose propylamine and hexylamine as the solvents to promote gelation. PDMS dissolved to make a clear solution in propylamine or hexylamine at temperatures in the range of 65–90°C. Gelation occurred by cooling the solution to room temperature. In the gelation studies, most investigators follow a "quick check" for gelation by "tube inversion." If gelation had occurred, when the vessel (tube) is turned upside down, there should not be any flow of the sample (not even oozing of the solvent on the walls of the tube). Figure 7.8 shows three tubes, one with the original PDMS sample, the second with the gel, and the third with the xerogel. In Figure 7.8a with the bulk PDMS, the sample is seen flowing along the walls. With the gel, there is no flow seen in Figure 7.8b. Of course, no flow is expected in the xerogel in which the solvent had evaporated.

Figure 7.8 Tube inversion method to check for gelation. (a) Bulk PDMS flowing along the walls of the tube, (b) gel with no flow, and (c) xerogel.

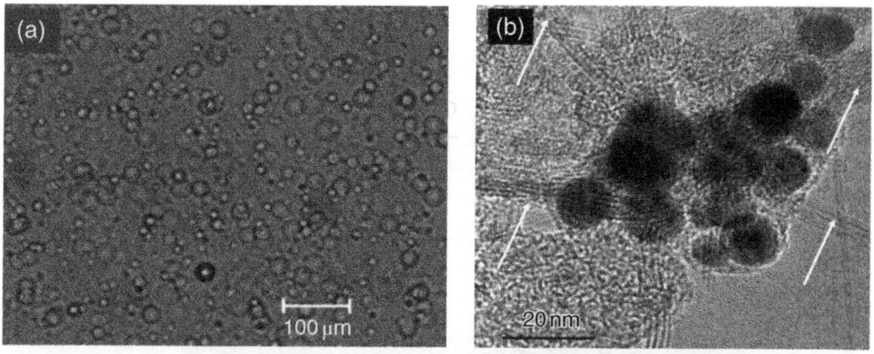

Figure 7.9 (a) Optical micrograph of the PDMS/propylamine gel and (b) TEM image of the xerogel. The arrows indicate the strands of the polymer connecting the spherical domains. (Source: Dahan and Sundararajan [21]. Reproduced with permission of American Chemical Society.)

The morphologies of the gel and the xerogel are shown in Figure 7.9a and b, respectively, in the optical and TEM micrographs. Spherical features are seen in Figure 7.9a, and the TEM shows these spherical domains connected by strands of the polymer (indicated by arrows).

The spherical domains seen in the figure is similar to the schematic shown in Figure 7.6a. Apart from the TEM image (Figure 7.9b), fibrils connecting the domains were also seen in the fracture surface using an environmental SEM (E-SEM) (Figure 7.10). The model for the gel formation is shown in Figure 7.10a. The chain conformation of PDMS differs from other polymers such as PS. The bond angles at O—Si—O (110°) and Si—O—Si (145°) differ significantly, and as a consequence, the all-*trans* conformation of the chain would lead to chain closure with about eleven repeat units (360/35 = 10.2). Macrocyclization of PDMS is well known [23–26]. The lengths of the Si—O and Si—C bonds are 1.64 and 1.90 Å, respectively, and the energy difference [24, 26] between the *trans* and *gauche* conformations is only of the order of RT. Due to these conformational features, entanglement and spherical domain formation during gelation is facilitated. With the average diameter of 10–15 nm for the spherical domains, the radius of gyration (R_g) is about 3.9–5.8 nm. These values are similar to the R_g measured from neutron scattering experiments. Beltzung et al. [27] reported a value of 4 nm for the root-mean-square *unperturbed* R_g of PDMS. For the PDMS samples used for membranes, R_g ranging from 6 to 13 nm was reported by Jadav et al. [28].

Specific interactions between PDMS and propylamine could not be detected in the nuclear magnetic resonance (NMR) spectra. The Fourier transform infrared (FTIR) spectra of the gels showed broadening in the Si—O—Si region, compared to the xerogel and the bulk samples. However, dipole interactions

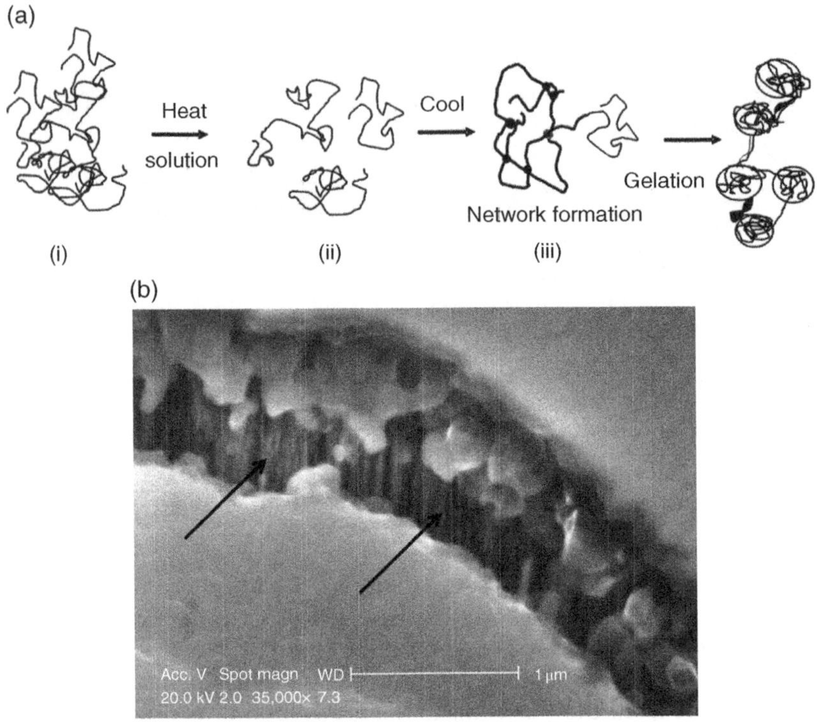

Figure 7.10 (a) A model for the formation of network and spherical domains during gelation with PDMS. (b) E-SEM image (recorded at −140°C) of a fracture domain showing the fibrils (shown by arrows) between domains. (Source: Dahan and Sundararajan [21]. Reproduced with permission of American Chemical Society.)

could not be ruled out, since the dipole moment (μ) of propylamine is 1.4 D and the μ of Si—O and O—Si bonds is 0.60 D as determined by Mark [29].

The physical gels are often thermo-reversible, that is, upon heating, the gel–sol transition would occur and cooling would restore the gel again (sol–gel transition). Such thermo-reversibility could be confirmed using methods other than visual. We will discuss microscopy, differential scanning calorimetry (DSC), and rheology techniques that were used to confirm thermo-reversibility in the case of PDMS gels. Figure 7.11 shows a series of hot-stage optical microscope images at different temperatures. Starting from the gel at room temperature, the spherical domains gradually vanish upon heating to 65°C (Figure 7.11c) (gel–sol). Upon lowering the temperature, the spherical domains form again from about 45°C (Figure 7.11d) and the sol–gel conversion is complete at 25°C (Figure 7.11e).

DSC can be used to determine the gel–sol temperature when heating the gel and the sol–gel transition when cooling. The DSC curves for the PDMS/propylamine

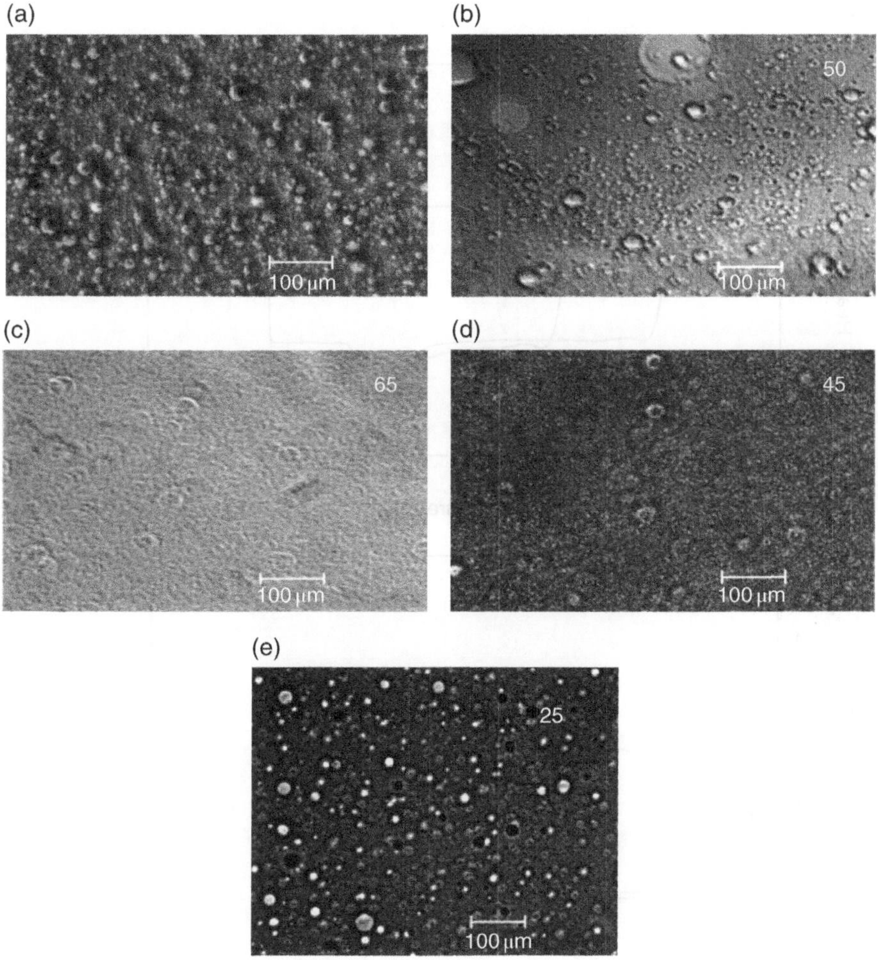

Figure 7.11 Hot-stage optical micrographs to show thermo-reversibility: (a) gel at room temperature, (b) at 50°C, (c) at 65°C showing the transition to sol, (d) at 45°C, and (e) at 25°C with the spherical domains reformed. (Source: Dahan and Sundararajan [21]. Reproduced with permission of American Chemical Society.)

gel system are shown in Figure 7.12. With the first heating cycle, an endotherm is seen at 65°C in Figure 7.12a, corresponding to the gel–sol transition. An exothermic peak for the sol–gel phase change is seen at about 45°C. An endotherm is seen again in the second heating cycle, confirming the thermo-reversibility. Note that the three peaks referred to are well defined. Often, in such DSC studies of gels, the endotherm for the second heating cycle would be very shallow. We overcame this as follows: During the cooling cycle, we arrested the DSC just above 45°C for

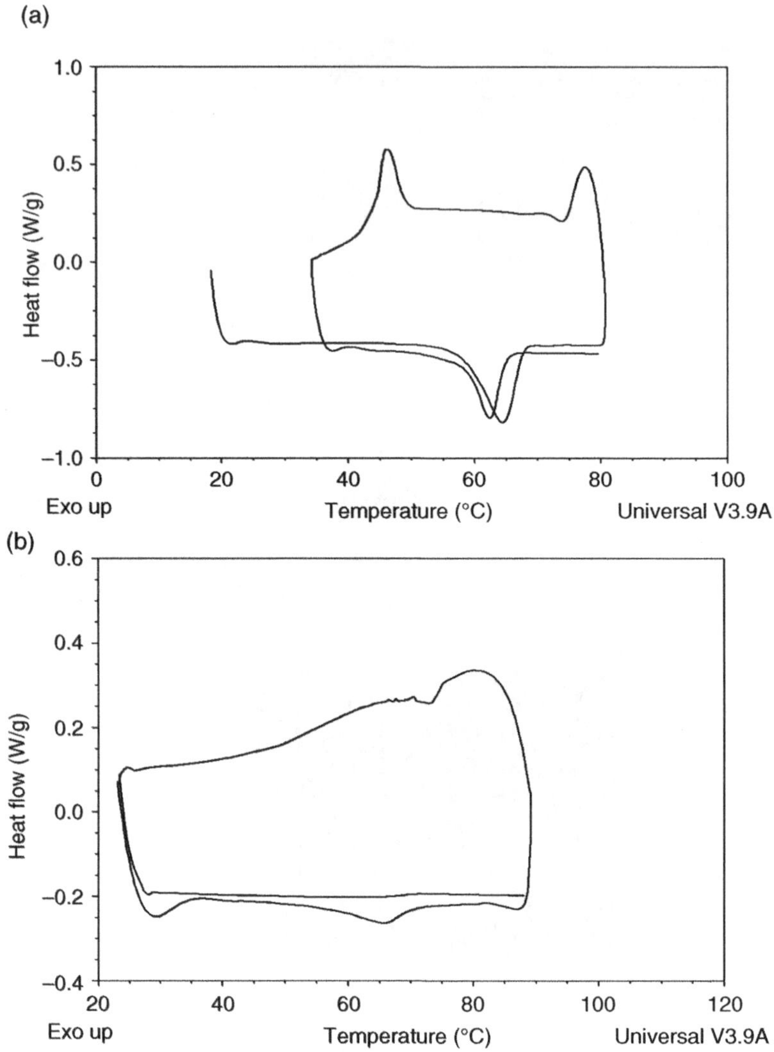

Figure 7.12 DSC traces for (a) the PDMS/propylamine gel showing the first heating/cooling/and second heating cycles and (b) the xerogel. (Source: Dahan and Sundararajan [21]. Reproduced with permission of American Chemical Society.)

30 min, to enable the formation of the gel and then continued with the cooling and second heating cycles. This resulted in a well-formed endotherm during the second cycle. Otherwise, with a scan rate of 5 or 10 per minute, the sample does not have sufficient time to form a gel. Figure 7.12b shows the DSC scan for the xerogel, with a weak endotherm.

Figure 7.13a shows the variation of the storage (G') and loss (G'') moduli with angular frequency for the bulk PDMS, the gel, and the xerogel. For the gel and the xerogel, the G' is higher than G'' and both are almost invariant with frequency, confirming that these are gels, not paste. For the bulk PDMS, both moduli increase with frequency. Figure 7.13b shows the variation of G' of the gel with temperature, during the heating and cooling cycles. It is seen that the mechanical property of the gel is fully reversible, confirming that it is a thermo-reversible gel. Thus, the thermo-reversibility is demonstrated using three different techniques.

Another noncrystallizable polymer that has been studied with respect to gelation is atactic poly(methyl methacrylate) (PMMA). Creation of micro-porous membranes via liquid–liquid demixing/vitrification was reported by Vandeweerdt et al. [30] using atactic PMMA with 1-butanol and cyclohexanol. Porous membrane-like morphologies were seen, with the initial concentration and the cooling determining the porosity. Membrane formation via gelation by the addition of a nonsolvent was discussed by Lin et al. [31]. Vaysse et al. [32] studied the nonsolvent-triggered porous morphology of PMMA gels and PMMA/carbon nanotube (CNT) composite gels. They prepared four types of gels as follows: (i) PMMA was dissolved in acetone by raising the temperature to 55°C and the solution was quenched in dry ice/acetone bath to form a gel. (ii) The same method was used along with CNT dispersed in acetone to make a PMMA/CNT composite gel. (iii) After dissolution of PMMA in acetone, water was added dropwise inducing precipitation of the polymer. The precipitate dissolved quickly when the solution was heated briefly to 90°C. It was then quenched in ice to yield a porous gel. (iv) CNT was dispersed in water by sonication and the dispersion was added dropwise to PMMA/acetone solution. Upon quenching, porous PMMA/CNT nanocomposite gel formed.

While the PMMA gel by itself was a solid and was not porous, the addition of the nonsolvent created the porosity. SEM analysis showed that the addition of CNT reduced the number of wide pores, the pore size and distribution. The same trend was seen with the volume fraction of added water. Other nonsolvents such as methanol, propanol, and cyclohexane also caused porous gels, but the size of the pores and the distribution was related to the solubility parameter. The solubility parameters (δ) of PMMA, acetone, water, methanol, propanol, and cyclohexane are 19, 19.7, 48, 29, 24.9, and 16.8 MPa$^{1/2}$, respectively. The δ for acetone/methanol (50/50), acetone/propanol (25/75), and acetone/cyclohexane (50/50) are 24.4, 22.9, and 18.3 MPa$^{1/2}$, respectively. It is seen in Figure 7.14 that the pores are smaller and more uniform as the δ of the mixture approaches that of PMMA. The CNT was located on the interior of the pores and the elongation of the gels increased with CNT content. For example, upon incorporating CNT in the porous PMMA gels prepared with acetone/water (80/20 v/v), the elongation increased from 6.2 (no CNT) to 11.3 with 0.75% CNT.

(a)

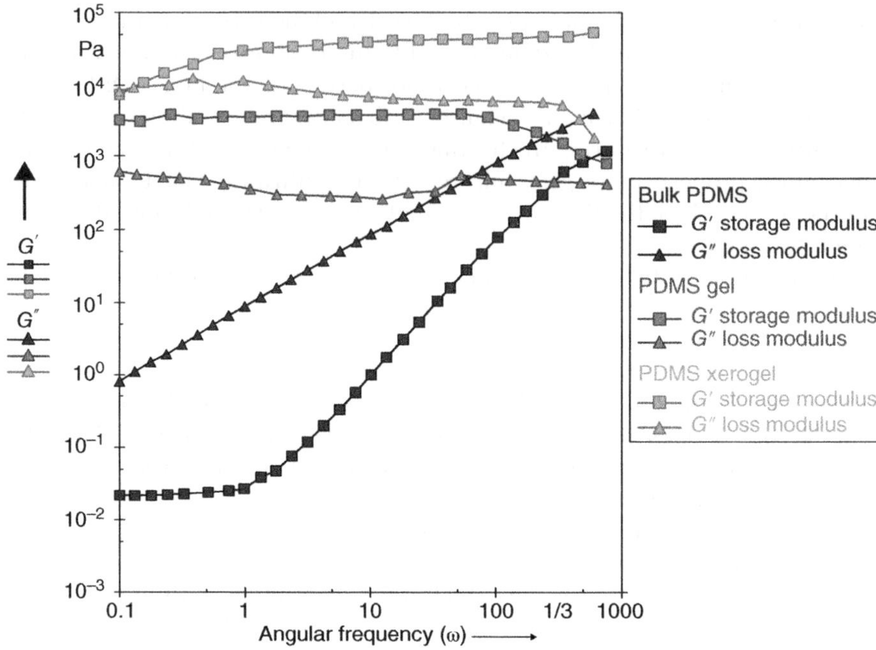

(b)

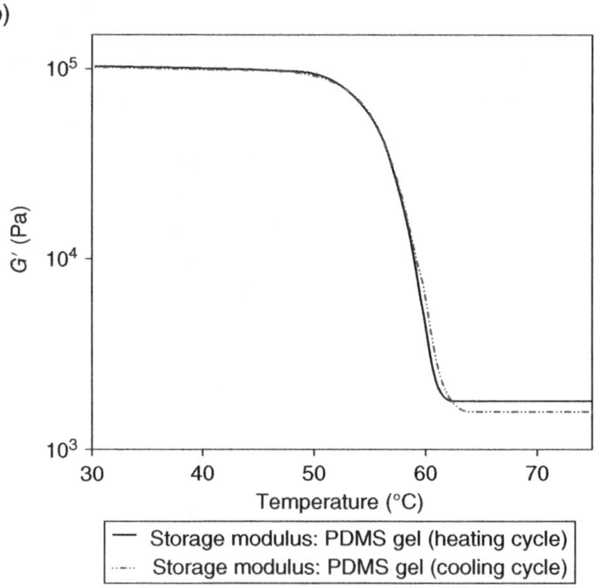

Figure 7.13 (a) Rheological data for the bulk PDMS, the gel, and the xerogel. (b) Variation of G' with temperature during the heating (solid curve) and cooling cycles (dash-dot curve). (Source: Dahan and Sundararajan [21]. Reproduced with permission of American Chemical Society.)

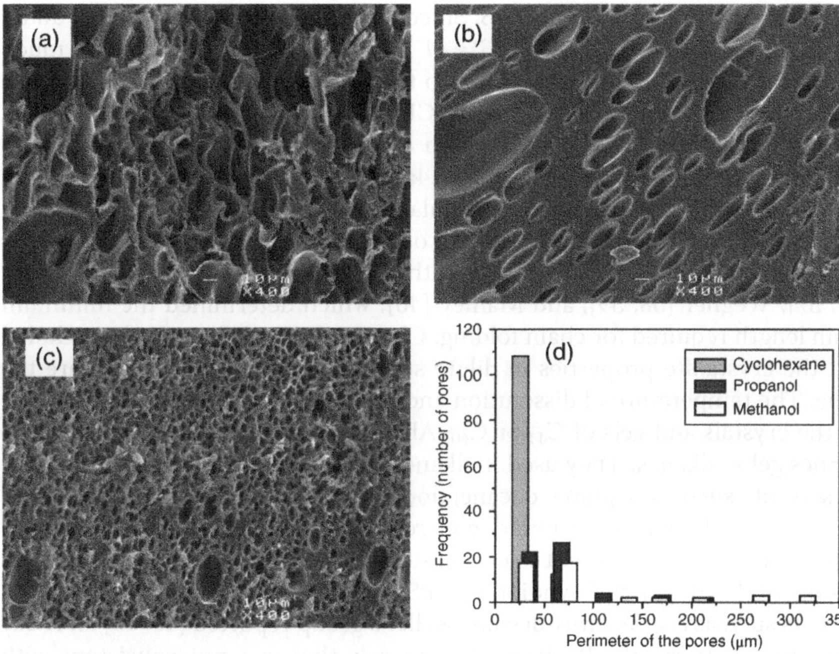

Figure 7.14 SEM images of PMMA gels with/acetone/nonsolvent (v/v) mixtures (a) acetone/methanol: 50/50, (b) acetone/propanol: 25/75, and (c) acetone/cyclohexane: 50/50. The pore size distribution is shown in (d). (Source: Vaysse et al. [32]. Reproduced with permission of American Chemical Society.)

7.4 Gels with Crystallizable Polymers

As is well known, one of the structurally simple polymers is polyethylene (PE), which would crystallize readily as well. Thermo-reversible gelation of linear PE with a range of molecular weights as well as ethylene copolymers with different branch contents was discussed by Mandelkern [33, 34]. Toluene was used as the solvent except that *p*-xylene or decalin was used for very high-molecular-weight polymers. With the high-molecular-weight samples, gelation involves lamellar formation, with a chain traversing more than one lamella as a condition for forming the gel. This is similar to the illustration in Figure 7.4d. The minimum concentration of the polymer required for gel formation depended on the molecular weight and temperature. The temperature dependence was more significant with lower molecular weight samples. At a given temperature, for M_w greater than 3×10^3, the relationship follows

$$C^* = kM^{-a} \tag{7.1}$$

with a varying between 0.2 and 0.3. Since the concentration needed for dilute solution overlap is expressed as $C^* \approx M^{-0.5}$ the C^* for gel formation is much higher. Chan and Mandelkern [34] also found that low-molecular-weight PE samples with $M_n = 574$, 785, and 910 ($(CH_2)_n$, with $n = 41$, 56, and 65) formed extended chain crystals during gelation and not folded lamellae. This would resemble the model shown in Figure 7.4a. It was discussed in Chapter 3 that according to molecular dynamics simulations, the PE chain must be at least 150 units in length for stable folding to occur [35] (to stabilize the stem–stem interaction). This is in accordance with the experimental results of Keller [36, 37], Wegner [38, 39], and Manley [40], which determined the minimum chain length required for chain folding. Chan and Mandelkern [34] concluded that the crystallite properties in dilute solution crystals and the gels are the same. The temperature of dissolution and heat of fusion were nearly the same for the crystals and gels of C_{44} or C_{60}. Abdallah and Weiss [41] found that *n*-alkanes gel *n*-alkanes. They used *n*-alkanes ranging from C_{24} to C_{36}, and these gel solvents such as heptane, decane, dodecane, hexadecane, and a few other liquids. The gel stability was found to increase from C_{24} to C_{36}. Thus the van der Waals interaction is sufficient to enable supramolecular assembly and gelation.

Isotactic PS (i-PS) and syndiotactic PS (s-PS), respectively, form gels with solvents such as *cis*- or *trans*-decalin. Keller's group reported [42, 43] an X-ray diffraction pattern of i-PS/*trans* decalin gels that was not consistent with the usual threefold helical conformation and suggested an extended helical structure with 12 monomers per turn and a pitch of 3.06 nm. Sundararajan et al. [44, 45] proposed an extended chain conformation, with the above helix parameters, in which the chain would adopt a nonstaggered *trans–trans* conformation of the contiguous bonds. The phenyl side chains would interact between themselves, whereas in the conventional threefold helix, the phenyl groups would be accessible to the solvent molecules. Ten different solvents were used by Sundararajan et al. [45], and X-ray diffraction corresponding to the extended conformation was seen with bulky solvents such as hexahydroindan, cyclodecane, and cyclooctane, whereas diffractions attributable to both the threefold helix and the extended helix were seen with aromatic solvents such as ethylbenzene, *tert*-butylbenzene, *n*-butylbenzene, and benzaldehyde. Since the gel corresponding to the 12_1 conformation melted at a lower temperature, it was suggested that the solvent molecules intercalate with these chains in the extended conformation and stabilize the structure. Concurrently, Sundararajan [46, 47] applied the solvent-induced crystallization (SINC) method to i-PS bulk films using solvents of different size, and again, the X-ray diffraction pointed to the extended helix conformation when bulky solvents were used. They were able to get a compound X-ray pattern corresponding to both the threefold helical and extended structures in the same film with some solvents. Exposure of films already containing the threefold helix conformation to the bulky solvents promoted the appearance of the crystalline regions with extended helix conformation, in addition to the threefold helix structure. Partial melting experiments showed that the crystalline regions of the threefold helix were bridged by the chains in the extended conformation. Conformational energy calculations [48] had showed that the *trans–trans* conformation of i-PS is equally favored as the *trans*-gauche.

Guenet et al. [49, 50] disagreed with the proposal of an extended chain conformation for i-PS gels. Based on X-ray and neutron scattering experiments, he proposed that the unusual X-ray pattern seen in the case of i-PS/decalin gels was due to intercalation of the solvent molecules with the chains in the threefold helical conformation. It was argued that since the distance between the phenyl groups on the adjacent turns (along the pitch) is 6.6 Å, there was sufficient room for such intercalation. Klein and Guenet [50] developed relationships between the molecular volume of various solvents and the occlusion of these in the cavities formed between the side groups in the case of isotactic and syndiotactic PMMA and i-PS and a-PS. Keller or Sundararajan did not pursue this work further, albeit it remains that the SINC work of Sundararajan showed the presence of the solvent to be a requirement even in the bulk.

s-PS also forms gels with decalin and other solvents. A series of publications [51–58] showed that s-PS can exist in various polymorphs depending on how the sample was crystallized. Daniel et al. [59–62] discussed the gel and clathrate phases formed by s-PS and 1,2-dichloroethane (DCE) or 1-chloropropane (CP). Chatani et al. [63] determined the crystal structure of the molecular complex of s-PS with toluene. A clathrate delta form of s-PS/DCE was reported by De Rosa et al. [64]. The role of solvents on the conformation on the chains in the gels was discussed by Kobayashi [65]. Two types of helical conformation of the chain have been identified. In the α- and β-crystalline forms, the chain was found to be in an all-*trans* (*tttt*) conformation, where as a 2_1 helical conformation (with *ttgg* sequence of bonds) occurred in the γ, δ, and ε polymorphs. The 2_1 helical conformation can accommodate guest molecules. Thus, s-PS offers a rich variety of structural polymorphs that can be manipulated by different means. As for gelation and crystallization, Berghmans et al. [58] found a competition between these events with the s-PS/*cis*-decalin system. Figure 7.15 shows the phase diagram of the structures obtained with various annealing temperatures, from gels at lower temperatures to other polymorphic forms at higher temperatures.

Itagaki et al. [66] examined the effect of the solvent size on the physical gelation of s-PS. They prepared high-temperature solutions of s-PS in *n*-alkylbenzenes (NBs), with the alkyl side chain length ranging from 1 to 10 [C_6H_5-(C_nH_{2n+1}), $n = 1–10$]. When the solutions were cooled, with $n = 1–3$, gels with a network of fibers were obtained. A paste, with the s-PS chains in the γ-crystalline phase was seen with $n = 4–6$. With longer alkyl chains ($n = 7–10$), a sherbet-like morphology was seen, with the chains in the *trans*-zigzag

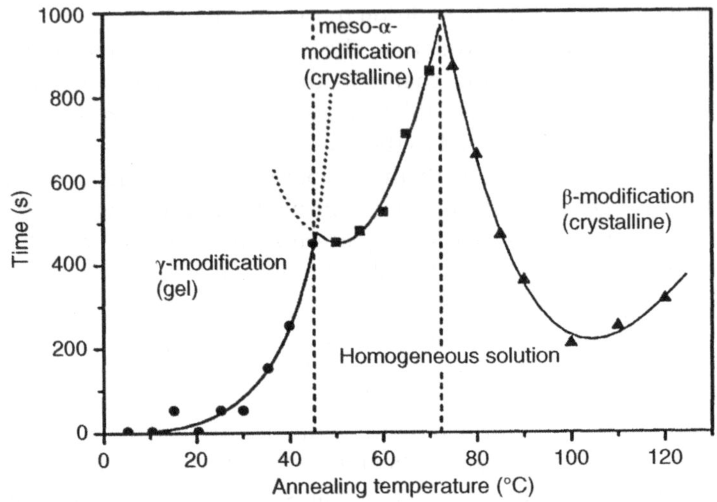

Figure 7.15 Phase diagram of the different structures obtained with s-PS/*cis*-decalin systems at different temperatures. (Source: De Rudder et al. [58]. Reproduced with permission of Wiley.)

conformation in the β-crystalline polymorph. Thus the size of the solvent molecule influences the resulting polymorph when the hot solutions are cooled to room temperature. In a further study, these authors [67] used different fragrant molecules as solvents with the premise that if the fragrance molecules are intercalated with the 2_1 helical conformation in the gels, s-PS can be used for fragrance release. Eight different fragrance molecules were used and the results were rationalized in terms of the molecular volume of the solvents and their solubility parameters (δ). For s-PS, they calculated $\delta = 10.4\,(\text{cal/cm}^3)^{1/2}$. A solvent molecule, whose δ is close to that of s-PS and the molecular volume V is less than about 160 Å^3 can interact with the side group of s-PS to promote the 2_1 helical conformation. Among the solvents used, (−) β-pinene ($V = 155$ Å^3; $\delta = 9.2\,(\text{cal/cm}^3)^{1/2}$) and *cis*-jasmone ($V = 178$ Å^3; $\delta = 9.4\,(\text{cal/cm}^3)^{1/2}$) meet these conditions. With 5 wt% in the fragrant molecules, the s-PS was found in the 2_1 helical conformation with benzaldehyde, 2-methoxynaphthalene, 4-methylanisole, *cis*-jasmone, and (−) β-pinene. The *trans*-zigzag conformation resulted with citral, D-limonene, and L-menthol. Figure 7.16 shows the SEM images of the morphology in three different solvents. In the case of s-PS with benzaldehyde, 2-methoxynaphthalene, and 4-methyl anisole, fibrillar morphology of s-PS was observed as in Figure 7.16a. Lamellar morphology as in Figure 7.16c was observed with citral, D-limonene, and L-menthol. Spherulitic morphology was seen with *cis*-jasmone and (−) β-pinene (Figure 7.16b). The correlation between the molecular volume, solubility parameters, and the crystalline form

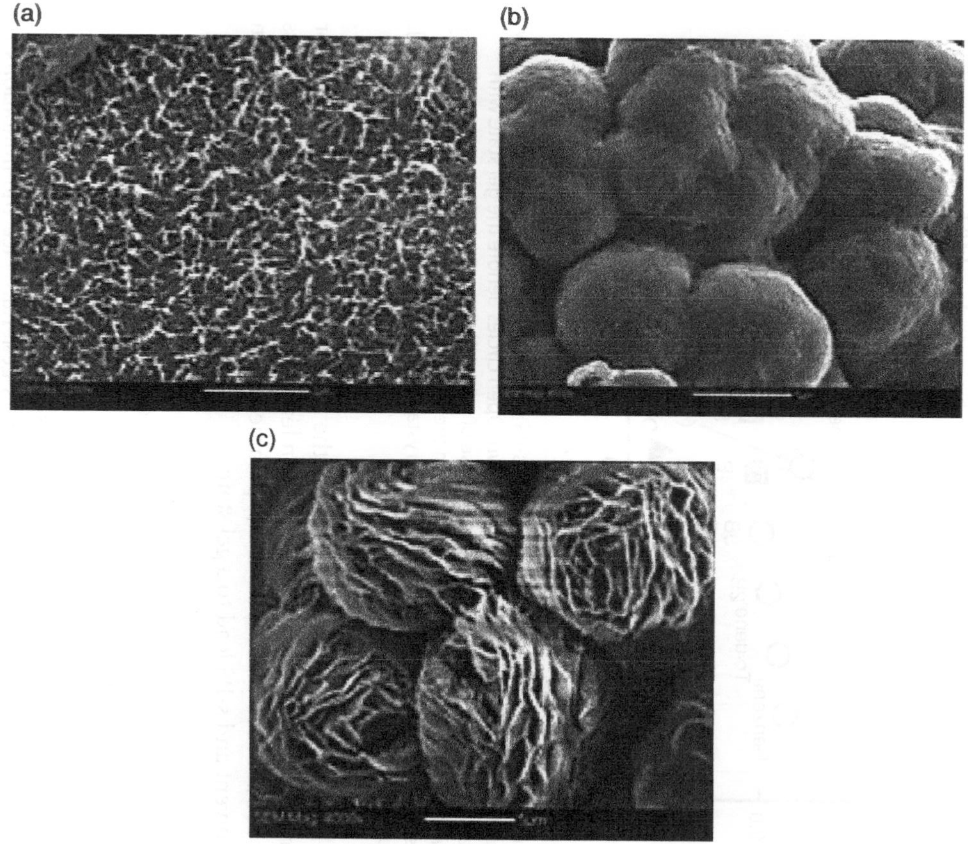

Figure 7.16 SEM images of s-PS/fragrant solvents (with 10 wt% of the polymer) dried under vacuum. (a) s-PS/benzaldehyde, (b) s-PS/*cis*-jasmone, and (c) s-PS/L-menthol. The scale bar is 5 μm. (Source: Mochizuki et al. [67]. Reproduced with permission of Wiley.)

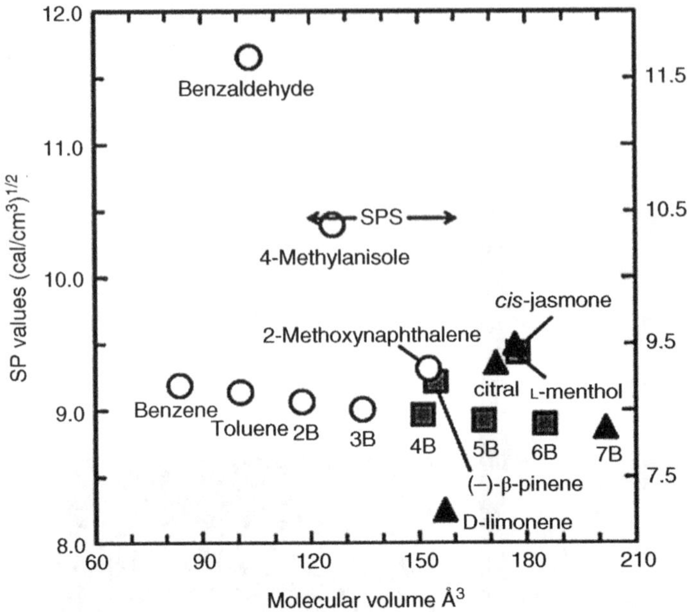

Figure 7.17 A plot of the crystalline phase of s-PS with the molecular volume and solubility parameters of the fragrance molecules is shown. Open circles denote the gel state with fibrillar morphology and the δ-crystalline phase of s-PS; filled squares represent paste-like materials with nonlamellar spherulitic phase in the γ-crystalline phase; Filled triangles correspond to sherbet-like structure in the β-crystalline phase. (Source: Mochizuki et al. [67]. Reproduced with permission of Wiley.)

of s-PS is shown in Figure 7.17. Here, the 2B, 3B, etc., refer to NBs, with, for example, 2B denoting an alkyl side chain length of $n = 2$. It is seen that V of less than about 160 Å^3 and the solubility parameter difference of less than 1 between the solvent and s-PS lead to a gel state.

7.5 Add a Sergeant to the Soldiers to Cause Gelation

Although several stereo-regular polymers such i-PS and s-PS, PMMA, etc., lead to gelation, organogelation of polymers generally face the limitations as in the case of small molecules in terms of selecting the appropriate polymer/solvent pair. For example, in the case of PDMS, it was only recently thermo-reversible physical gels were reported by Dahan and Sundararajan [21] using propylamine and hexylamine as solvents. Prior to their publication, only chemical gels were known, which were, for example, prepared with cross-linking [18] and a swelling solvent [19].

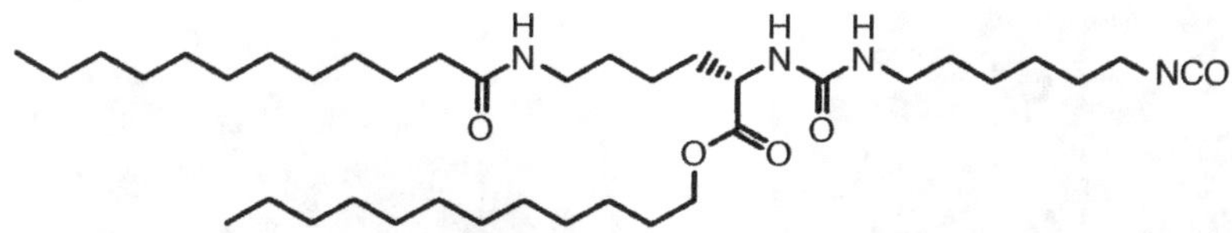

A: Dodecyl ester
B: Ethyl ester

Scheme 7.1 Lysine derivatives used by Suzuki et al. [68, 69] for directing polymer gelation.

Since not all polymer structures have the groups conducive to hydrogen bonding or π-stacking interactions to cause organogelation, a possible approach could be to functionalize the polymers with a small molecule that would provide the supramolecular assembly capability. In an attempt to create polymer gels using such an approach, with a small-molecule gelator to direct the gelation, Suzuki et al. [68, 69] attached L-lysine derivatives shown in Scheme 7.1 to PDMS. They synthesized *N*α-(6-isocyanatehexylaminocarbonyl)-*N*ε-lauroyl-L-lysine dodecyl ester (**A**) and ethyl ester (**B**). These were then attached to two samples of bis(3-aminopropyl)-poly(dimethylsiloxane) with $M_w = 900$–1000 (**1**) and ≈3000 (**2**) and a third sample of poly[dimethyl siloxane-*co*-(3-aminopropyl) methylsiloxane] (**3**) ($M_w = 4000$–5000). The L-lysine derivatives **A** and **B** formed organogels with cyclohexane, aromatic solvents, triolein, and myristic acid isopropyl ester. The derivative **A** with the dodecyl ester could also form gels with oils. Directed by the attached lysine derivatives, the **A**- and **B**-functionalized PDMS polymers **1**, **2**, and **3** also formed gels with several solvents. FTIR analysis showed that the gelation mechanism involves hydrogen bonding between the lysine segments and that the interactions between the polymer/polymer and polymer/lysine segments were weak. The supramolecular assembly of the lysine segment was mainly responsible for the gelation. This is similar to the sergeant–soldier phenomenon discussed in Chapter 3, with the lysine segment acting as the sergeant. The SEM images of the dried gels shown in Figure 7.18 exhibit fibrillar morphology that is commonly seen with small-molecule gelators, with nanofibers of 50–150 nm diameter. Note that the PDMS polymers that were used were of low molecular weight.

This work on derivatization of polymers with L-lysine segments was extended by these authors [70] to other polymers such as polycarbonate and polycaprolactone. These prepolymers and the L-lysine-derived polymers are shown in Scheme 7.2. The dodecyl ester derivative **A** shown in Scheme 7.1 was used in this study. As mentioned before, the low-molecular-weight organogelator, the L-lysine-derivative (**A** in Scheme 7.1), formed gels in several solvents such as alkanes, aromatics, and oils. The L-lysine-functionalized polymers **1A–5A** also formed gels with several organic solvents, mineral oil, and vegetable oil. It was also noted that apart from some differences in the

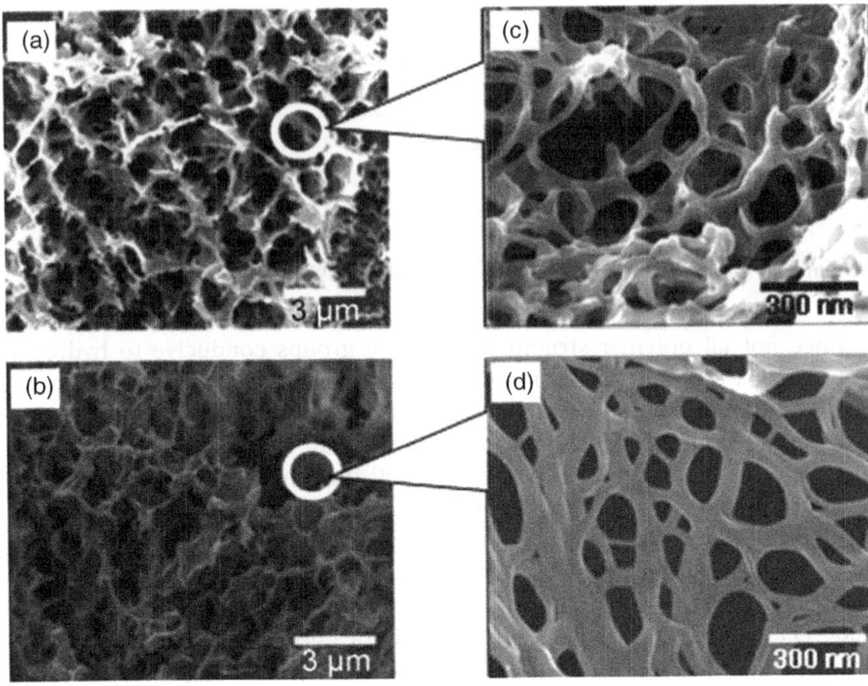

Figure 7.18 Field emission SEM images of dried gels. (a) and (c) dioxane/**1**A; (b) and (d) dioxane/**2**A. (1A and 2A: PDMS polymers **1** and **2**, respectively, functionalized with dodecyl ester A). Note that the scale bar is 3 μm on the left and 300 nm on the right. c and d are the magnified images of the circled areas of a and b, respectively. (Source: Suzuki et al. [68]. Reproduced with permission of Wiley.)

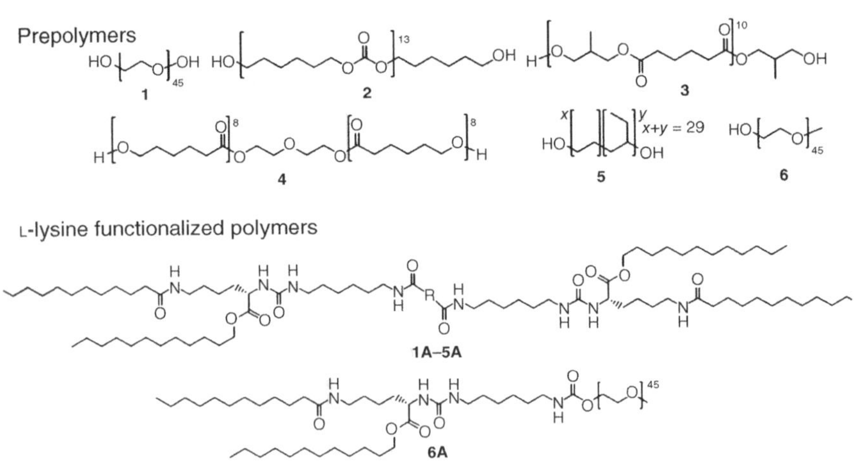

Scheme 7.2 The prepolymers and the polymers functionalized with L-lysine dodecyl derivative (**A** in Scheme 7.1) studied by Suzuki et al. [70]. The **R** in **1A–5A** denotes the prepolymers **1–5**.

solvent selectivity for gelation, the organogelation properties of **1A–5A** were similar, despite the differences in the molecular structures. The SEM images of the dried gels prepared using 1,4-dioxane are shown in Figure 7.19 for all the polymers **1A–6A**. As characteristic of organogels, fibrous or sheet-like morphology is seen. It was also observed that the dimensions of the nanofibers were similar to that formed by the L-lysine derivative **A** by itself in its gels and were independent of the polymer backbone. As for the case of L-lysine-functionalized PDMS discussed earlier, detailed FTIR studies confirmed that the supramolecular assembly was facilitated by the self-assembly of L-lysine derivative (**A**) through hydrogen bonds involving the urea, urethane, and amide groups in **A** as well as the van der Waals interactions of the alkyl chains with the polymer backbones contributing very little. However, the polymers had a role to play in the thermal and mechanical strength of the gels. The gel–sol temperature of **A** by itself was 70°C with dimethyl sulfoxide (DMSO) or 1,4-dioxane, and it decreased to 50–60°C in the case of polymer gels **1A–5A**. Further, the gel strength (or gel hardness measured by the force required to sink a cylindrical bar of 10 mm diameter to a depth of 4 mm in the gels) showed significant increase in the case of polymers. For the L-lysine derivative **A** by itself, the gel strengths were 1.75 and 1.31 kPa for the DMSO and 1,4-dioxane gels. These increased to 22.2 and 7.26 kPa, respectively, when the polymer **3A** was the gelator.

Another case of a small-molecule organic gelator leading the polymer gelation was reported by Khan et al. [71]. As mentioned previously, these authors have studied the crystallization and gelation characteristics of biscarbamates, with various alkyl side chain lengths [2, 72]. These are model compounds for PUs. In an effort to design biodegradable segmented PUs, a few studies have been reported on the use of poly(ε-caprolactone) (PCL) as the soft segment. van Bogart et al. [73] used PCL of $M_w = 830$ or 2000 with aromatic or aliphatic type of hard segments and found that there was compatibility between the soft and hard segments when the molecular weight of PCL was low. Microphase separation occurred when the molecular weight was increased to 2000. Wang et al. [74] also used PCL of $M_w = 530$ or 2000 with the hard segment shown in Scheme 7.3 and came to similar conclusions with regard to phase separation between the soft and hard segments.

Since the structure of biscarbamates resemble that of the hard segment of aliphatic PU, Khan et al. [71] blended biscarbamates with various alkyl side chain lengths as a minor component with PCL ($M_w = 65\,000$) (Scheme 7.4) and studied the phase separated crystallization. In the studies of Wang et al. [74] and van Bogart et al. [73] on microphase morphology of PU, the PCL was part of the chain. In the work of Khan et al., PCL was a component in the blend and the molecular weight (65 000) was much higher than those used in PU. Interestingly, it was found that the biscarbamates and PCL are immiscible and that the microphase morphology depended on whether the biscarbamate/PCL

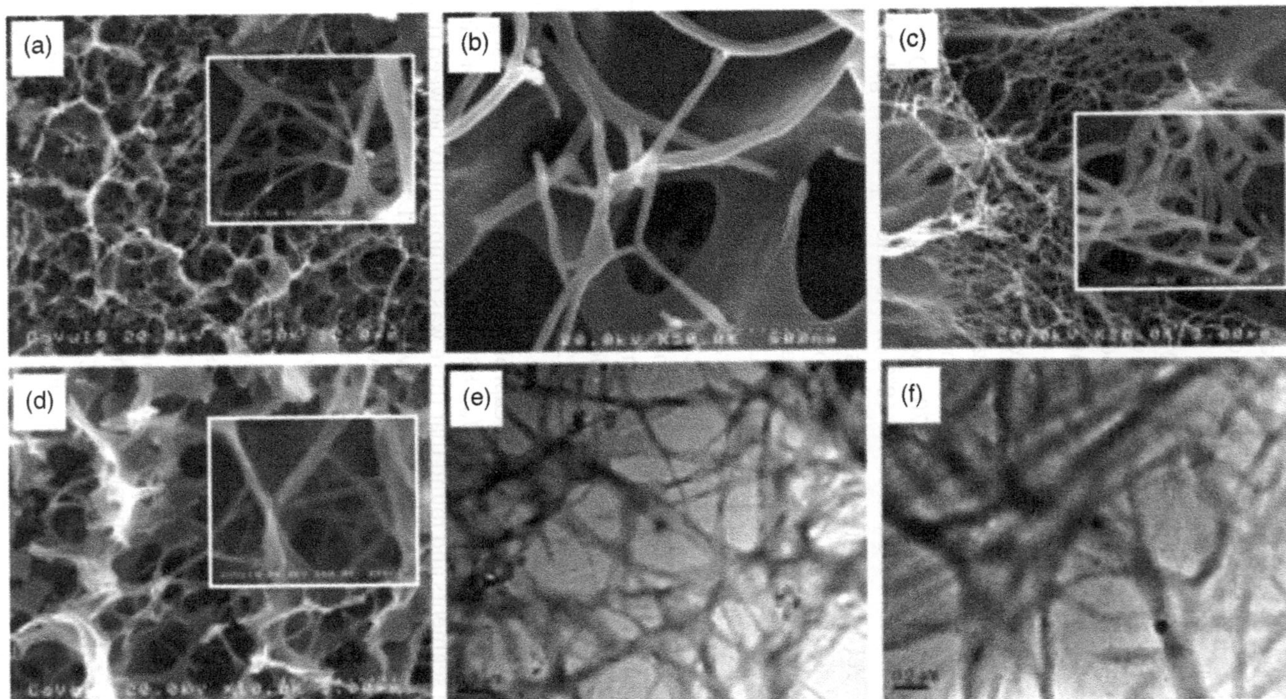

Figure 7.19 SEM (a)–(d) and TEM (e) and (f) images of xerogels with 1,4-dioxane. (a) **1A**, (b) **4A**, (c) **5A**, (d) **6A**, (e) **2A**, and (f) **3A**. (Source: Suzuki et al. [70]. Reproduced with permission of Wiley.)

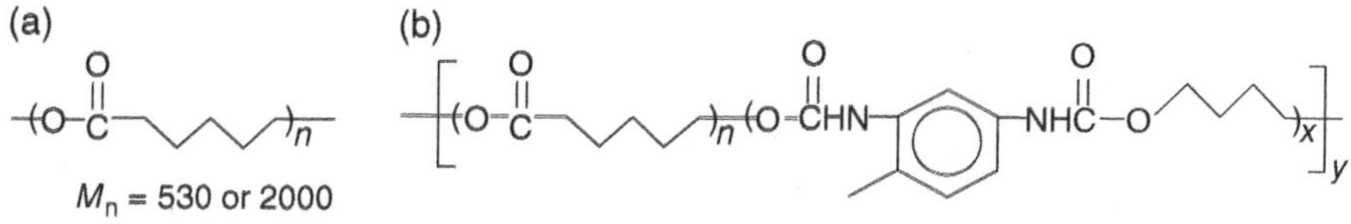

Scheme 7.3 (a) The PCL soft segment and (b) the hard segment used by Wang et al. [74].

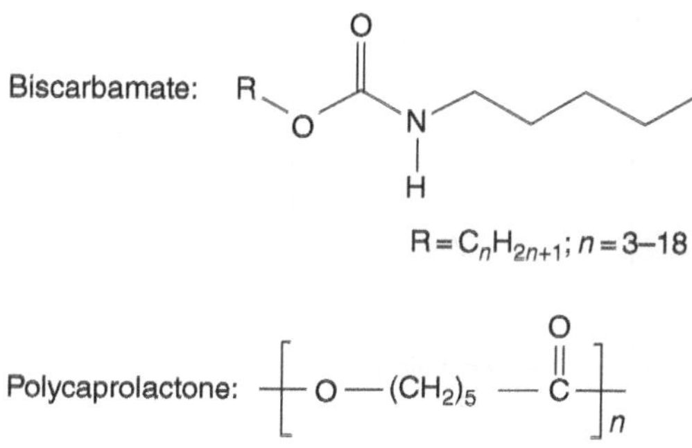

Scheme 7.4 Structures of biscarbamates used and polycaprolactone.

blend films were solvent cast from chloroform or from the melt (bulk). In the solvent-cast films, the biscarbamates formed aggregated crystals in the PCL matrix. However, in the melt-cast films, PCL (although it was the major component) formed droplets in the matrix of spherulites of the biscarbamates. This behavior was attributed to the faster rate of crystallization of the biscarbamate melt relative to that of PCL. Studies of Khanna et al. [2] showed that biscarbamates form gels with several solvents, but not chloroform. PCL, by itself, has not been found to form gels with any solvent. Both biscarbamates and PCL are soluble in chloroform. Upon heating mixtures of biscarbamates and PCL to 60°C and cooling (either slowly or by quenching), gelation was found to occur only in a select range of alkyl side chain lengths of C_{11}–C_{13} as shown in Figure 7.20. With alkyl chains shorter than C_{11}, the mixture stays in solution, and longer than C_{13}, precipitation occurred. Note that in Figure 7.20, the tube in frame b is inverted to demonstrate the gelation.

FTIR spectra showed no mutual noncovalent interactions between the biscarbamates and PCL in the gels. Thus gelation was driven by mutual phase separation (self-sorting). The morphology of the xerogels of the composites is shown in Figure 7.21. The biscarbamates are the minor components. The SEM images show that the PCL forms a matrix in which the network of gel fibers of biscarbamates is dispersed. Since PCL melts at approximately 60°C, and the melting temperature of biscarbamates is above 100°C, heating the samples to partially melt the PCL confirmed the network of fibers of biscarbamates in the

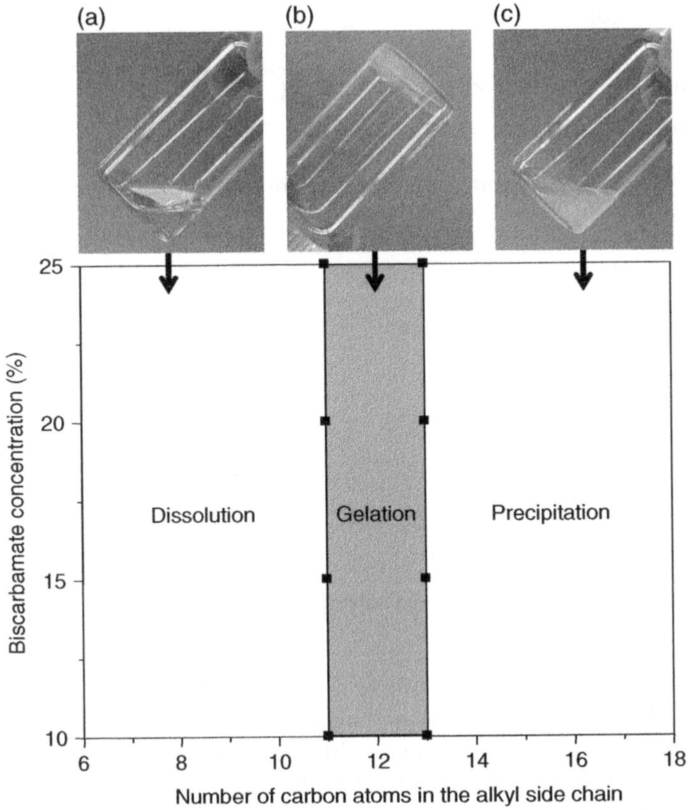

Figure 7.20 (a)–(c) Phase diagram of the gelation/precipitation with PCL/biscarbamate mixtures. Note that the tube in frame b is inverted to show gel formation. (Source: Khan et al. [71]. Reproduced with permission of Royal Society of Chemistry.)

PCL matrix. Further, the biscarbamate fibers exhibited the same eaves trough tubular morphology in the PCL matrix (shown by arrows in Figure 7.21). Impregnation of biscarbamate gel fibers with such a morphology in the PCL matrix would obviously lead to tubular pores in the latter. Such porous structure was confirmed by the SEM images of the fracture surfaces of these xerogels, as shown in Figure 7.22. The pore morphology depends on the alkyl side chain length of the biscarbamates. With C_{11} biscarbamate/PCL gel (Figure 7.22a), pores of 5–10 μm in diameter are fairly uniformly distributed. Larger, more irregular pores are seen in Figure 7.22b for the C_{12}/PCL xerogel. Large macrovoids are seen with C13/PCL xerogel. An enlarged view of one of these voids is shown in Figure 7.22d. Thus, the self-sorting of the biscarbamates and PCL leads to gelation and the tubular morphology of the small molecule leads to a porous composite with PCL as the matrix.

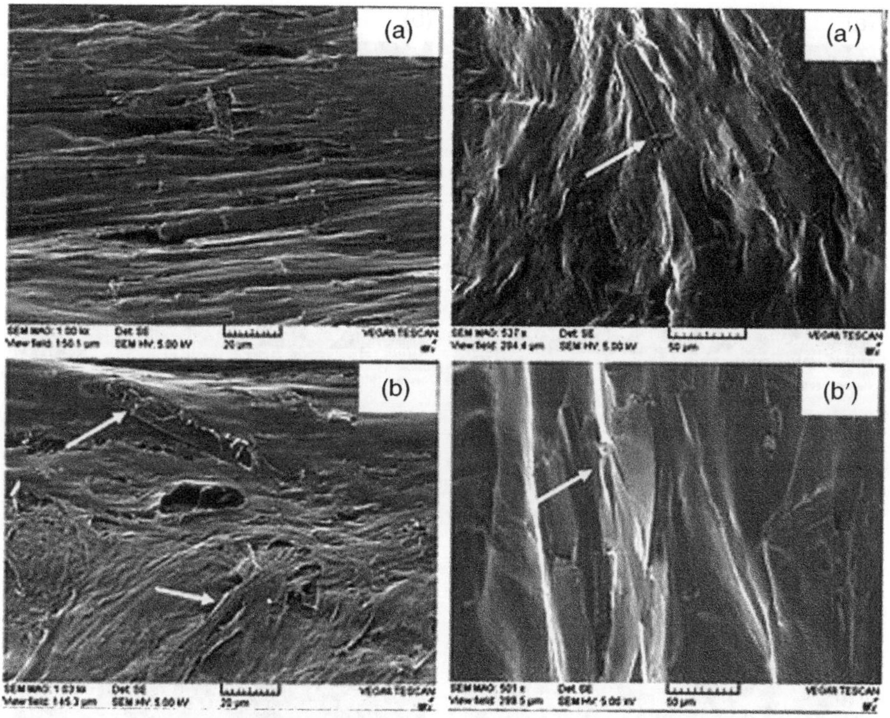

Figure 7.21 SEM micrographs of xerogels of PCL/biscarbamate: 90/10 wt% blends. (a) and (a′) PCL/C_{11}; (b) and (b′) PCL/C_{12}. The left column represents the micrographs of quenched samples and the right column corresponds to the slow-cooled samples. The arrows show the eaves trough morphology of the fibers. (Source: Khan et al. [71]. Reproduced with permission of Royal Society of Chemistry.)

While in the earlier case, there was no specific interaction between the biscarbamates and PCL, supramolecular organogelation of a pseudo-polyrotaxane with an alkyne-functionalized PCL threading through β-cyclodextrin was reported by Jazkewitsch and Ritter [75]. The pseudo-polyrotaxane was formed in dimethylformamide (DMF). Similar gelation with supramolecular rotaxane assembly of functionalized PCL was discussed by Ge et al. [76]. As shown schematically in Figure 7.23, these authors created a four-armed star PCL with dibenzo[24]crown-8 (DB24C8) and a two-armed PCL with dibenzyl ammonium salt (DBAS). When the four-armed PCL–OH and two-armed PCL–OH were mixed in chloroform, there was no association that would indicate supramolecular assembly, as seen from the invariance of reduced viscosity with concentration seen in Figure 7.24. With four-armed PCL functionalized with DB24C8 and two-armed PCL with DBAS, the supramolecular assembly created pseudo-rotaxanes and the reduced viscosity increased exponentially

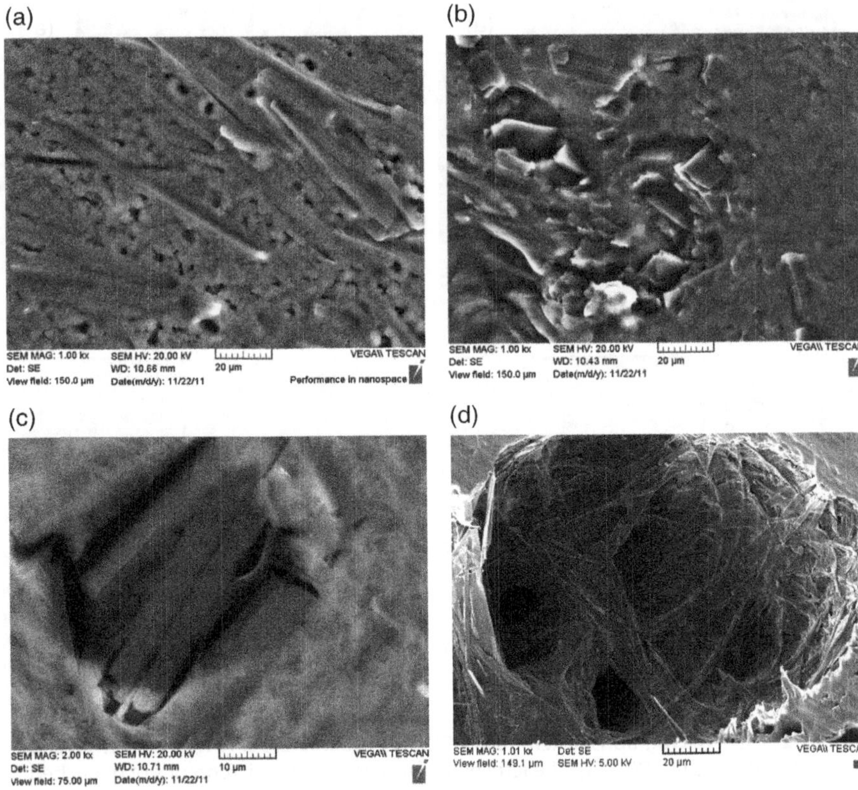

Figure 7.22 SEM images of cross sections of PCL/biscarbamate (85/15 wt%) gels with different number of carbon atoms in the alkyl side chains of the biscarbamate molecules: (a) C_{11}, (b) C_{12}, (c) C_{13}, and (d) the magnified view of a void. (Source: Khan et al. [71]. Reproduced with permission of Royal Society of Chemistry.)

as seen in Figure 7.24. With higher concentrations in chloroform, upon heating the solution and cooling to 20°C resulted in a gel as seen from the inverted tube shown in Figure 7.25a. The gelation was thermo-reversible or pH-reversible, as seen in Figure 7.25b and c.

7.6 π-Interaction-Mediated Gelation

The π-interaction-based gels are attractive due to their optical response. The absorption and emission depend on the stacking of the aromatic segments. The Ajayaghosh group [77, 78] first reported the organogels based on oligo-phenylenevinylenes (OPVs). Since then, a number of studies have been

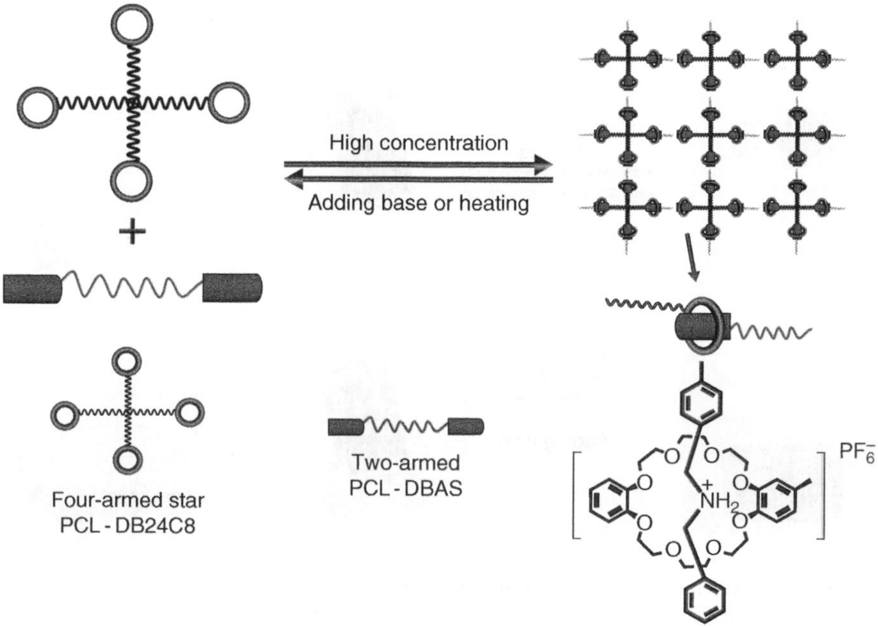

Figure 7.23 Schematic representation for the fabrication of responsive supramolecular networks from four-armed star PCL–DB24C8 and two-armed PCL–DBAS as a result of molecular recognition between dibenzo[24]crown-8 (DB24C8) and dibenzylammonium salt (DBAS) moieties. (Source: Ge et al. [76]. Reproduced with permission of Wiley.)

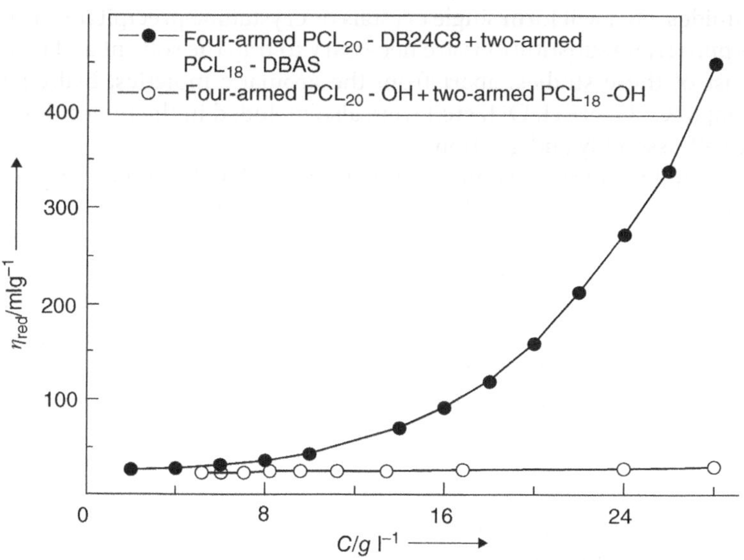

Figure 7.24 Variation of reduced viscosity with concentration for the crown- and DBAS-functionalized PCL and the nonfunctionalized. (Source: Ge et al. [76]. Reproduced with permission of Wiley.)

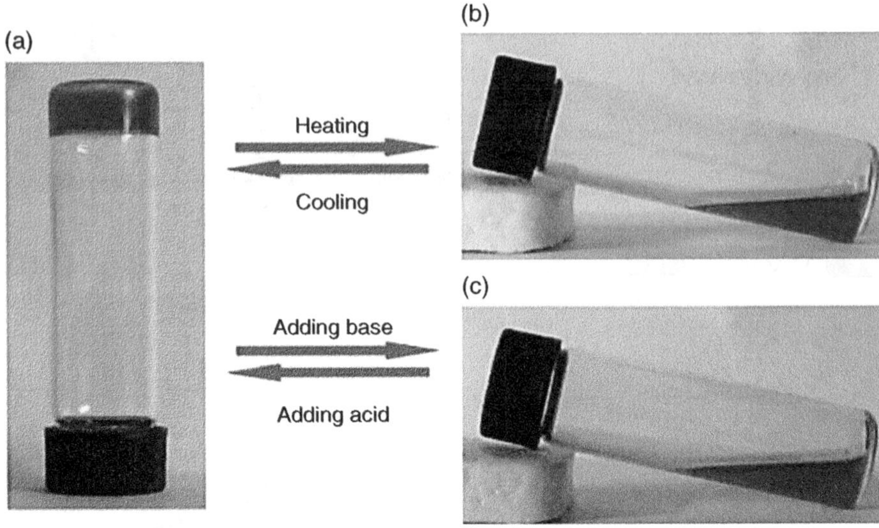

Figure 7.25 (a) Supramolecular gel formed with the association of four-armed PCL–DB24C8 and two-armed PCL–DBAS; (b) gel–sol transition upon heating (a) to 60°C; and (c) gel–sol transition upon treating (a) with 1 : 1 triethylamine. (Source: Ge et al. [76]. Reproduced with permission of Wiley.)

published on gels involving phthalocyanines and perylenes functionalized with various types of substituents. By themselves phthalocyanines, perylenes, perylene diimides, etc., will form single crystals or crystalline precipitates from solutions. Appropriate substitutions are necessary to trap the solvent and form a gel. In most of these studies, apart from the aromatic moieties, hydrogen bonding groups (C═O, N—H, OH, etc.) were also included in the molecules to assist in the self-assembly and gelation.

The self-assembly of perylene diimides functionalized with *commodity polymers* such as PDMS, PEO-*co*-PPO (Jeffamine®), and oligostyrene was discussed in Chapter 4. These molecules do not have a hydrogen bonding group of the types mentioned earlier in their structures. Dahan and Sundararajan discussed the gelation of these perylene-based molecules, using the same solvents as those in the gelation of the homopolymers of PDMS and PS. It was discussed earlier that PDMS gelled propylamine [21]. Both propylamine and hexane/water or diisopropylamine/water were used with Mono-PDMS and Di-PDMS to achieve gelation [79]. Note that Mono-PDMS is PDMS–PDI and Di-PDMS is PDMS–PTCDI–PDMS (see Figure 4.21). Upon heating propylamine/Mono-PDMS or propylamine/Di-PDMS mixture to about 50–80°C to form a homogeneous solution and cooling to room temperature resulted in gelation. Hexane dissolves these solutes and would not normally qualify as a candidate for gelation studies. It was serendipitously found that upon heating hexane solution, phase separation occurred at about 70°C. At that stage, adding water dropwise caused gelation. If the solution was cooled without the addition of water, precipitation occurred. A solution of PDMS polymer by itself in hexane does not phase separate upon heating to 70°C. Hence, the cloudiness that developed with the hexane solutions of Mono- and Di-PDMS was rationalized on the basis of aggregation of the perylene segment at that temperature. IR spectroscopy showed hydrogen bond between the N–H of propylamine and the Si—O—Si, and a weak C—H···N interaction. In the case of hexane-/water-based gels, hydrogen bonds between the water molecules and Si—O—Si were indicated. The SEM and TEM images of the morphologies of the Di-PDMS-based xerogels with propylamine and hexane/water are shown in Figure 7.26. As mentioned previously, the Di-PDMS is a nonaqueous, *macromolecular Gemini surfactant*. The SEM images (Figure 7.26a–d) show that the fibers are several microns in length. As shown by the arrows, these fibers are not flat but curved, as a precursor to hollow-tube formation. As the growth proceeds predominantly in one direction, there is an asymmetry in the interactions between the molecules in the three directions, which causes the fibers to fold [2]. The TEM images (Figure 7.26e and f) show the fine structure of these fibers and the primary fibers are nanowires of 1–1.2 nm thickness. The optical microscope images of the wet gels can be seen in Dahan and Sundararajan [79].

The Mono-PDMS is an inverse macromolecular surfactant: the PDMS tail is soluble in nonaqueous organic solvents and the head group (PDI) is not. As seen in Figure 7.27, the morphology of the gels with propylamine consisted of vesicles of 2–4 μm diameter, with a wall thickness of about 50 nm. Figure 7.27b shows even smaller vesicles. The TEM image in Figure 7.27c corresponds to a vesicle with about 10 nm diameter and a wall thickness of 2 nm. A trilayer vesicular morphology of Mono-PDMS drop cast from solutions was seen before [80]. The H-stacking of the perylene segment was indicated in both cases in ultraviolet–visible (UV–vis) and fluorescent spectra.

In the case of two-component gels that were discussed before, the two participating molecules were of different chemical structures, although in most cases each of them bore hydrogen bonding groups that enabled mutual association between them. In the case of Mono- and Di-PDMS, the molecules are homologous, in that one bears a PDMS chain on one side, and the other on both sides. Dahan and Sundararajan [79] examined if these two molecules would intercalate and form a composite gel, since both of them gel propylamine with about the same sol–gel and gel–sol temperatures. Gels prepared with various ratios of the two showed that Mono- and Di-PDMS *self-sorted* and formed their own morphologies, that is, vesicular and fibrous, respectively. Figure 7.28a shows predominantly spherical morphology with the major component being Mono-PDMS. Similar is the case with the SEM image in Figure 7.28d. With the 50/50 composition, the OM in Figure 7.28b shows more density of fibers coating the spheres. The corresponding SEM image

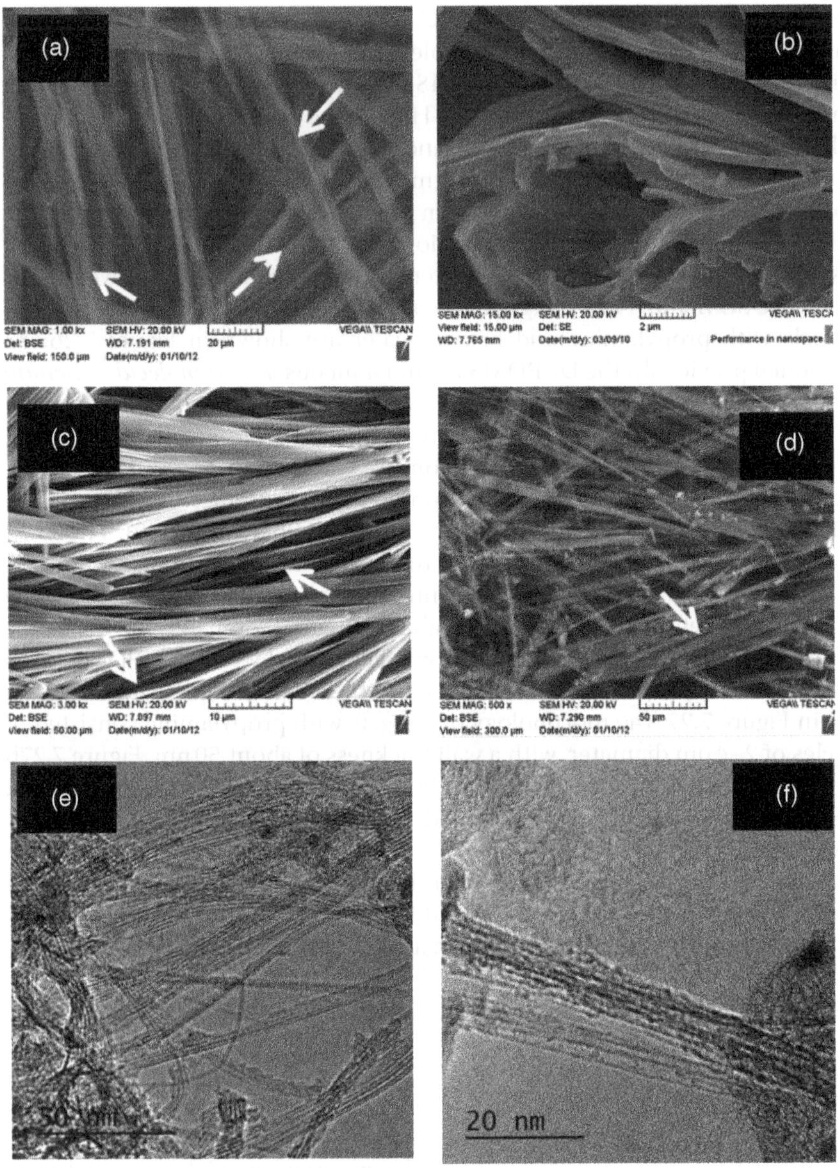

Figure 7.26 (a) SEM image of Di-PDMS/propylamine xerogel; (b) cross section of the sample in (a); (c) xerogel with hexane/water (90/10); (d) xerogel with hexane/water (85/15); TEM image of (e) Di-PDMS/propylamine xerogel; and (f) xerogel with hexane/water (95/5). The arrows in (a), (c), and (d) point to the tubular morphology. (Source: Dahan and Sundararajan [79]. Reproduced with permission of Royal Society of Chemistry.)

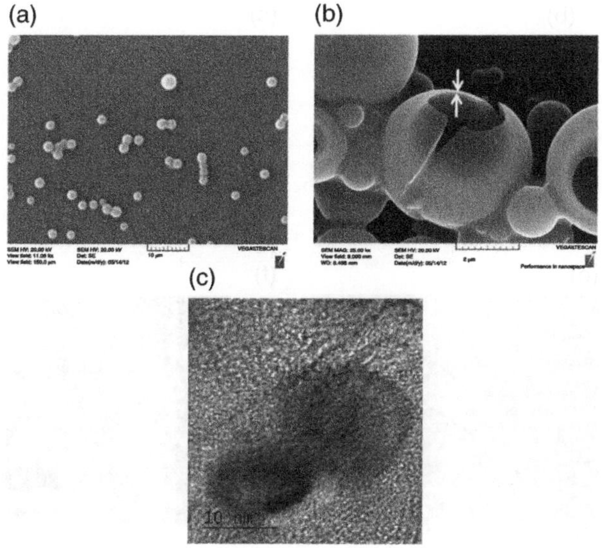

Figure 7.27 (a) SEM image of Mono-PDMS/propylamine xerogel, (b) SEM image of a fractured vesicle showing the wall thickness, and (c) a TEM image of the xerogel. (Source: Dahan and Sundararajan [79]. Reproduced with permission of Royal Society of Chemistry.)

(Figure 7.28e) shows that both fibers and the spheres are present. The enlarged views shown in Figure 7.28g and h would indicate that a significant number of the Mono-PDMS spheres are in fact encapsulated in the fibers of Di-PDMS. Figure 7.28c and f shows that with Di-PDMS as the major component, fibers dominate on the surface of the Mono-PDMS spheres.

It is known that Gemini surfactants can self-assemble in a number ways [81, 82], as micelles, vesicles, and rod-like aggregates, similar to the phospholipid molecules. Such widely different assemblies were seen in the case of the Gemini molecule PS–PTCDI–PS bearing perylene diimides spacer and oligostyrene tails (we use PS for the oligo-styrene also here). In addition to *cis*- and *trans*-decalin which are the traditional solvents that had been used for gelation with PS, Dahan and Sundararajan [83] used cyclohexane and toluene also. The temperature of dissolution varied from 79°C for cyclohexane to 89°C for *cis*-decalin. Although gelation was achieved with all four solvents, the temperature gaps between the dissolution and onset of gelation were 25, 30, 42, and 50°C for *trans*-decalin, cyclohexane, toluene, and *cis*-decalin, respectively. Thus, during the cooling, the onset of gelation occurred at a significantly higher temperature (smaller gap) with *trans*-decalin than with *cis*-decalin. Although IR spectra showed some frequency shifts that could be associated with interactions between C=O and C–H bonds, there were no distinguishing features in the spectra that could account for the differences in gelation using

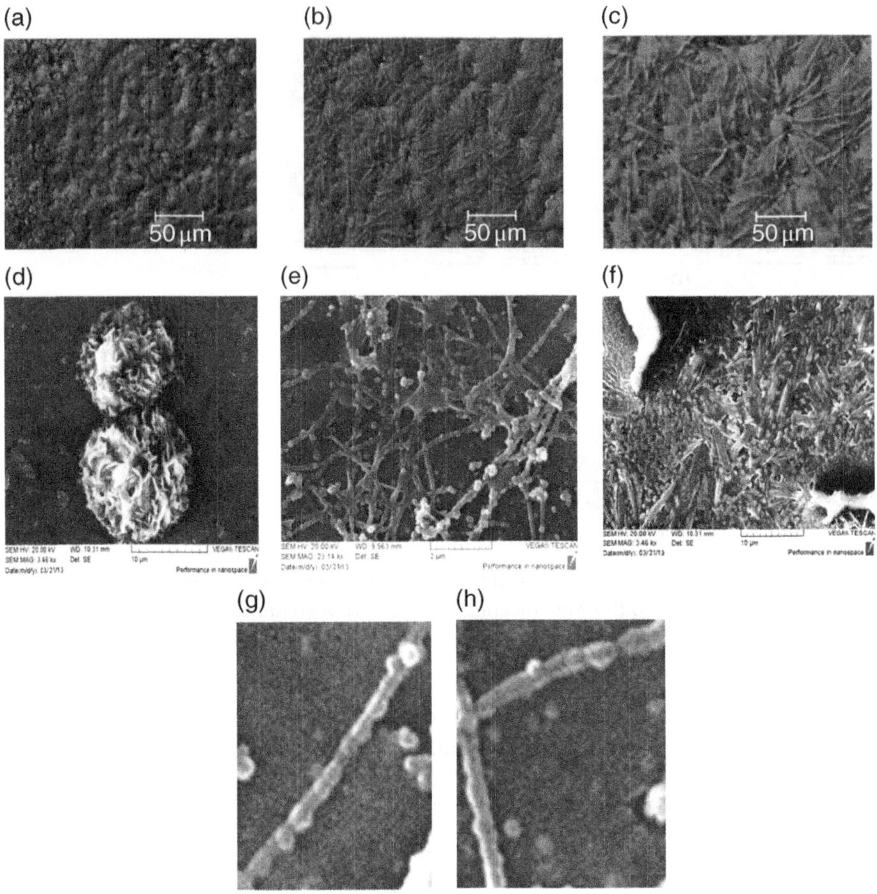

Figure 7.28 Self-sorting in gels of Mono- and Di-PDMS blends. Top row: OM of gels of Mono-/Di-PDMS with propylamine: (a) 80/20, (b) 50/50, and (c) 20/80. Middle row: SEM images of xerogels with Mono-/Di-PDMS: (d) 80/20, (e) 50/50, and (f) 20/80; (g) and (h) are enlarged views of the fibers in (e). (Source: Dahan and Sundararajan [79]. Reproduced with permission of Royal Society of Chemistry.)

these solvents. However, X-ray diffraction from the xerogels of *trans*-decalin was different from those of the other three solvents. With *trans*-decalin, diffraction peaks were seen that would correspond to lamellar organization of the PTCDI segment and intermolecular side-to-side packing distances. These features were not seen with the xerogels of other solvents. In addition, UV–vis and fluorescent spectra also showed differences. The morphology of the xerogels is shown in Figure 7.29. While the gels from the other solvents showed vesicular morphology, fibers of over 100 μm in length are seen in the case of the *trans*-decalin gel. Note that in this case also, the fibers are curved as an eaves

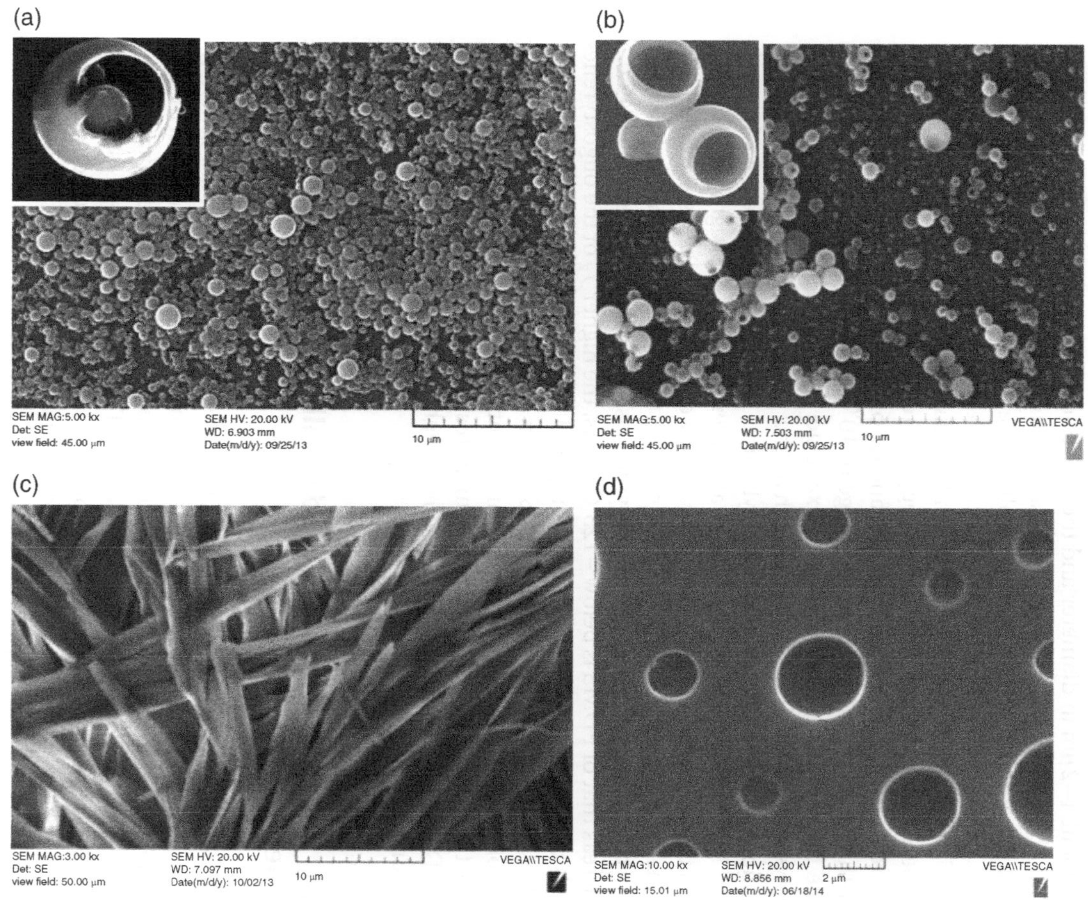

Figure 7.29 SEM of Xerogels from (a) cyclohexane, (b) toluene, (c) *trans*-decalin, and (d) *cis*-decalin. The concentration was 8 mM. (Source: Dahan and Sundararajan [83]. Reproduced with permission of Royal Society of Chemistry.)

trough, similar to the cases discussed earlier. When the self-assembly and growth happen primarily in one direction due to the predominant molecular interactions along a direction (π-interaction in this case), the resultant fibers tend to fold [2]. As for the other solvents, the insets in Figure 7.29a and b show the fracture morphology and that these are vesicular structures. The vesicles are about 1–2 μm in diameter and the wall thickness varies from 40 to 120 nm. With the oligostyrene segments attached to PTCDI, the molecular length would be about 3.5 nm. Hence, the walls of the vesicles would consist of several layers of the molecule. Models of packing of the molecules in the vesicles and the fibers were proposed by Dahan and Sundararajan [83].

These differences in the morphology are reflected in the rheological data. Figure 7.30 shows the variation of G' and G'' of the gels with frequency. The G' for the *trans*-decalin gel is in the range of 10^9 Pa, which is an order of magnitude higher than that of the cyclohexane gel (10^8 Pa). The toluene-based gel shows even a smaller G' ($<10^8$ Pa) than the cyclohexane gel. In the case of Mono- and Di-PDMS gels with propylamine, the G' values were 8.2×10^4 and 9.9×10^4 Pa, respectively [79]. These are several orders of magnitude smaller than those of gels of PS–PTCDI–PS with any of the four solvents. Although all of them have a π-stacking PTCDI segment in their structures, the glass transition temperature (T_g) of the oligostyrene used in this work is about 45°C, whereas that of PDMS is below −125°C. Thus, the higher T_g of the oligostyrene leads to a higher G'. Further, when compared with PDMS gel (Figure 7.13), the values of G' of Mono- and Di-PDMS are an order of magnitude higher than that of PDMS, although the molecular weight of the latter was 182 600. In this case, the presence of the perylene segment results in a higher G' for Mono- and Di-PDMS gels. Note that the molecular weight of PDMS attached to perylene was only 3035.

7.7 Polymer Compatibilized Small Molecule/Polymer Gels

The polymer composite films made with perylene or PTCDI would generally show phase separated crystals of the small molecule and not a uniform dispersion (see Chapter 8). Dahan and Sundararajan [84] showed that short-chain-functionalized PTCDI, such as PS–PTCDI–PS, when dispersed in the corresponding polymer, formed uniform self-assembled aggregate dispersions in the solvent cast films. In the earlier sections, the gelation of PS as well as PS–PTCDI–PS was discussed. The latter bears a short PS chain. Based on the observations on the uniform morphology of the polymer-functionalized PTCDI in the corresponding polymer matrix, Dahan and Sundararajan [84] prepared composite two-component gels of PS–PTCDI–PS with PS. Both of them showed fibrous morphology with *trans*-decalin. The composite gel also

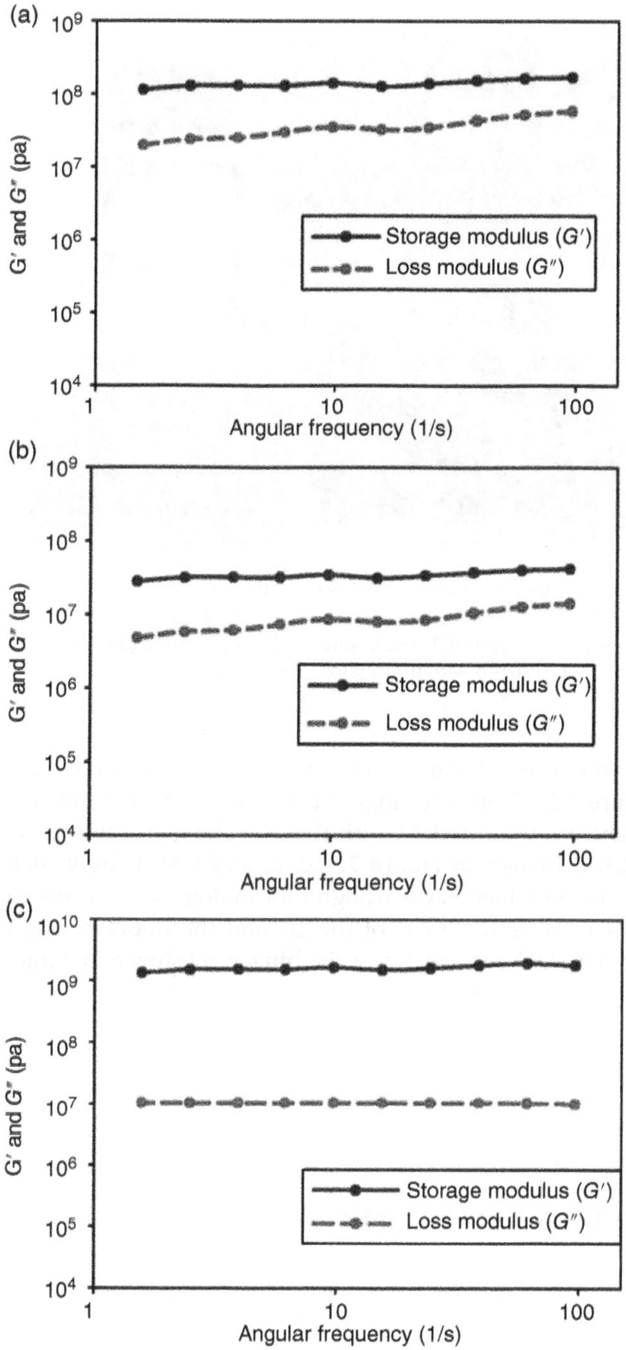

Figure 7.30 Comparison of the rheological parameters G' and G'' of the PS–PTCDI–PS-based gels with (a) cyclohexane, (b) toluene, and (c) *trans*-decalin. (Source: Dahan and Sundararajan [83]. Reproduced with permission of Royal Society of Chemistry.)

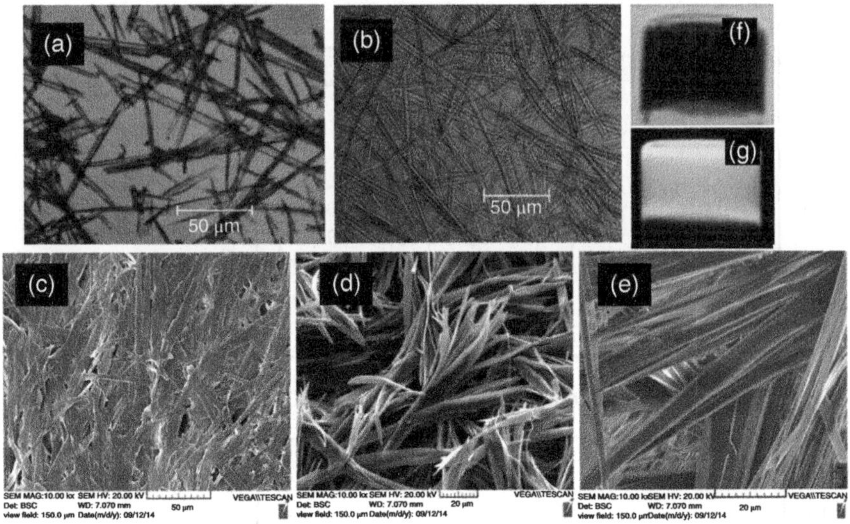

Figure 7.31 (a) and (b) OM of gels with 5 and 10% of PS–PTCDI–PS in PS; (c–e) SEM of xerogels with 2, 5, and 10% PS–PTCDI–PS in PS; and (f) and (g) show the color of the gel and the emission color, respectively (λ_{ex} = 503 nm). These colors can be seen in Ref. [84]. (Source: Dahan and Sundararajan [84]. Reproduced with permission of Elsevier.)

formed fibers of several microns in length and no crystals of PTCDI were seen. The OM images in Figure 7.31 show birefringent fibers, which would indicate that the PTCDI gel fibers are intermeshed with those of the high-molecular-weight polymer. The SEM images in Figure 7.31c, d, and e also show such uniform fibers. Of note is the folded eaves trough morphology seen clearly in Figure 7.31e. Figure 7.31f shows the color of the gel and the fluorescent gel is seen in Figure 7.31g. The UV–vis spectra of the binary gel showed characteristics similar to those recorded for the PS–PTCDI–PS gel.

Similar behaviour was seen [84] with the two component gels made with PDI–PDMS/PDMS as well as PDMS–PTCDI–PDMS/PDMS. Thus, the short polymer chains attached to PTCDI act as compatibilizers when the pigment is dispersed in the corresponding high-molecular-weight polymers.

7.8 Monomer Self-Assembly and Polymer Gels

Cyanate ester resins [poly(cyanurates)] are candidates for applications in structural, aerospace, microelectronics, etc., due to their high thermal stability, low moisture absorption, desirable dielectric properties (2.7–3.2 at 1 MHz), etc. Many of their properties are comparable to or superior to those of epoxy

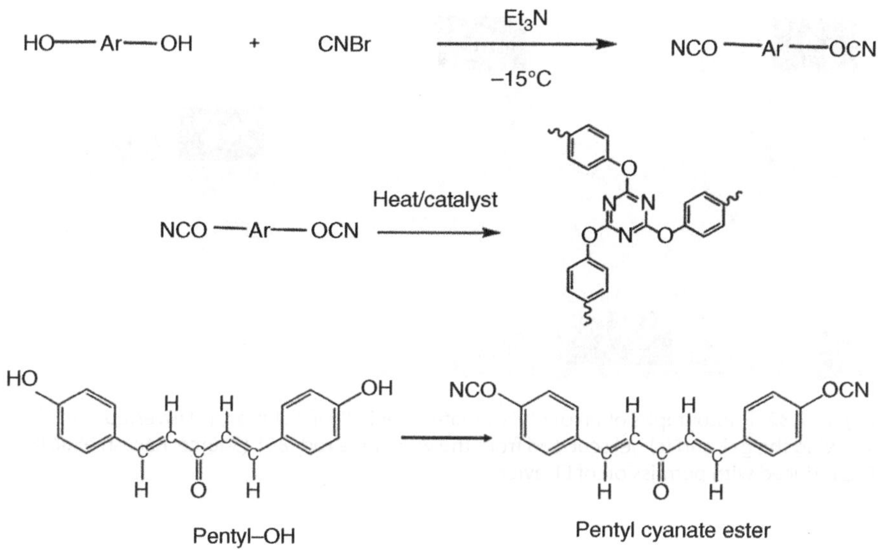

Scheme 7.5 Illustration of the synthesis route to cyanate esters [85].

or phenolic resins and polyimides. The usual route to their synthesis is shown in Scheme 7.5. Treating a hydroxyl substituted aromatic monomer with cyanogen bromide in the presence of triethylamine leads to a cyanate ester. Curing it with thermal energy or photo curing with a catalyst results in cyclotrimerization to form triazine rings and a cross-linked network. Sundararajan et al. [85] used three different monomers, one of which is shown (as pentyl-OH and pentyl cyanate ester) in Scheme 7.5. The cyanate ester monomer can be packed in a mold and cured for preparing the resin. Multistep curing is often used, for example, at temperatures T_1, T_2, and T_3 for times t_1, t_2, and t_3 respectively. Such a procedure was developed for what are known as polymerizable monomer reactant (PMR) resins. When the monomer shown in Scheme 7.5 was cured in the bulk, a solid resin was obtained.

Since the monomers present the prospect of self-assembly via aromatic interactions as well as weak C—H···O=C type of hydrogen bonds, an attempt was made to prepare organogels of the monomers and subsequently polymerize them, with a view to obtain porous poly(cyanurate). The gels were prepared using DMF, and the photographs of the solution and gel are shown in Figure 7.32. The solution in DMF at 60°C is shown in Figure 7.32a. The inverted vial showing the gel formation is seen in Figure 7.32b. The gel could be scooped from the vial, keeping its shape, using a glass rod (Figure 7.32c). The OM of the gel is shown in Figure 7.33b and c. The gel fibers are several hundred μm in

Figure 7.32 Photographs of (a) pentyl cyanate ester in DMF solution, (b) inverted vial showing the gel, and (c) gel scooped from the vial with a rod [85]. (Source: Rakesh et al. [85]. Reproduced with permission of Elsevier.)

length. Upon thermally curing the xerogel, a porous poly(cyanurate) resin was obtained. Figure 7.33e shows the cross section of the porous resin. This may be compared with the cross section of the conventionally cured resin shown in Figure 7.33d, which shows a solid bulk morphology. The monomer gels were also prepared with the addition of CNT (1–5 wt%), and the SEM image of the xerogel of monomer/CNT is shown in Figure 7.33f. The SEM of a cross section of the xerogel-polymerized resin containing CNT is shown in Figure 7.33g, and an enlarged view of a section of this image (Figure 7.33h) shows the presence of CNT entangled with the polymer fibers. Differences in thermal behavior such as curing time/temperatures, decomposition, etc., were found between the bulk curing and gel curing to obtain the poly(cyanurate). Since the onset of decomposition occurs above 400°C, these porous poly(cyanurates) could be used, for example, high temperature filtration and membrane applications. Their compressive strength would also be different from that of bulk-polymerized resins.

In the fabrication of porous poly(cyanurate) described earlier, the gels of cyanate esters were prepared and were thermally cured. The authors did not explore photo curing. Although organogels have been used for preparing porous polymers, in those cases, gels of organic small molecules were made with solvents such as acrylates or styrene in the presence of a cross-linking agent [86]. The medium (acrylate or styrene) was then polymerized in the presence of the gelator. Since the gelator formed network of fibers in the matrix polymer, removal of the organogelator resulted in a porous polymer matrix. Thus, the organogelator was used as a porogen. With the poly(cyanurate) prepared with the gelation of cyanate ester monomers, there was no sacrificial component. Addition of CNT to the gel also enhanced the properties.

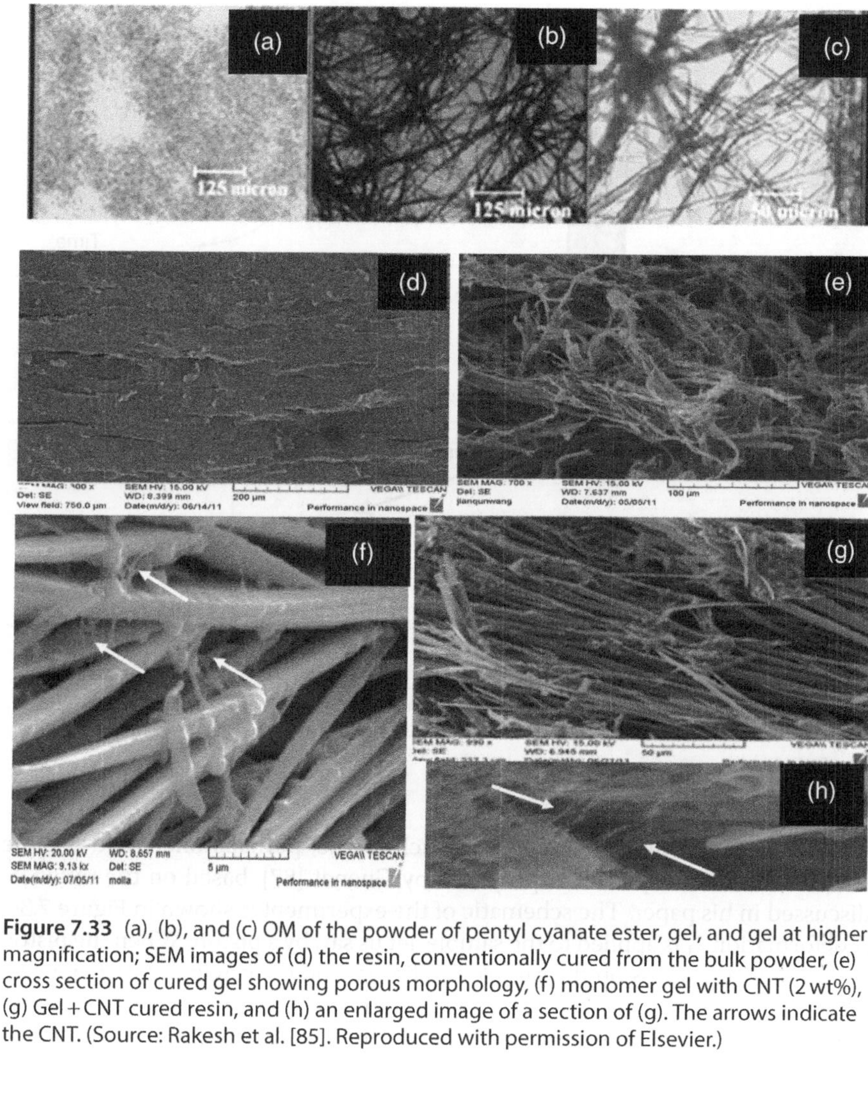

Figure 7.33 (a), (b), and (c) OM of the powder of pentyl cyanate ester, gel, and gel at higher magnification; SEM images of (d) the resin, conventionally cured from the bulk powder, (e) cross section of cured gel showing porous morphology, (f) monomer gel with CNT (2 wt%), (g) Gel + CNT cured resin, and (h) an enlarged image of a section of (g). The arrows indicate the CNT. (Source: Rakesh et al. [85]. Reproduced with permission of Elsevier.)

7.9 Poor Man's Rheology

It was mentioned earlier that for a material state to be qualified as a gel, the storage modulus G' should be larger than G'' and that it should be invariant with frequency in an oscillatory shear experiment. Tube inversion is used by several authors to confirm gelation. Referees of manuscripts, who demand more rigor, would insist on mechanical measurements to distinguish between a gel and a paste, with the gel satisfying the above G' behavior. Not all laboratories have

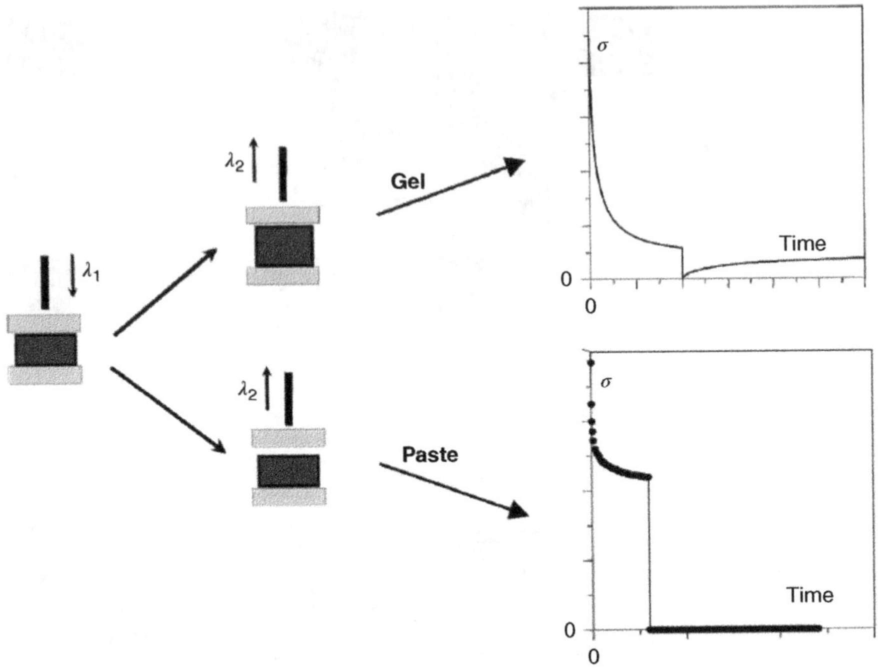

Figure 7.34 Schematic of a relaxation experiment to distinguish between a gel and a paste. (Source: Guenet [87]. Reproduced with permission of Wiley.)

the luxury of having a rheometer. In such a case, perhaps one could use the simple relaxation experiment proposed by Guenet [87], based on the rationale discussed in his paper. The schematic of the experiment is shown in Figure 7.34. A deformation λ_1 is applied to the sample, let us say, by a piston. When the piston is pulled up, the stress on the sample is relieved, and the original state reappears in the case of the gel and not for the paste. This was used by Guenet to illustrate that i-PS/*trans*-decalin system was a gel and that sPS/*trans*-decalin corresponded to a paste consisting of spherulites.

Sundararajan et al. used these quick experiments to confirm the gel state, in addition to the visual tube inversion method in the case of PDMS/propylamine as well as PS–PTCDI–PS gels with *trans*-decalin, cyclohexane, and toluene. The sample was inserted into a syringe, and compressed. The piston was then withdrawn. As seen in Figure 7.35a, following the retraction of the piston, the bulk PDMS did not recover from the stressed state, but the gel did. Likewise, in the case of PS–PTCDI–PS/*trans*-decalin gel, after compression and relaxation, the gel recovered to its original form. This poor man's rheology experiment could augment tube inversion.

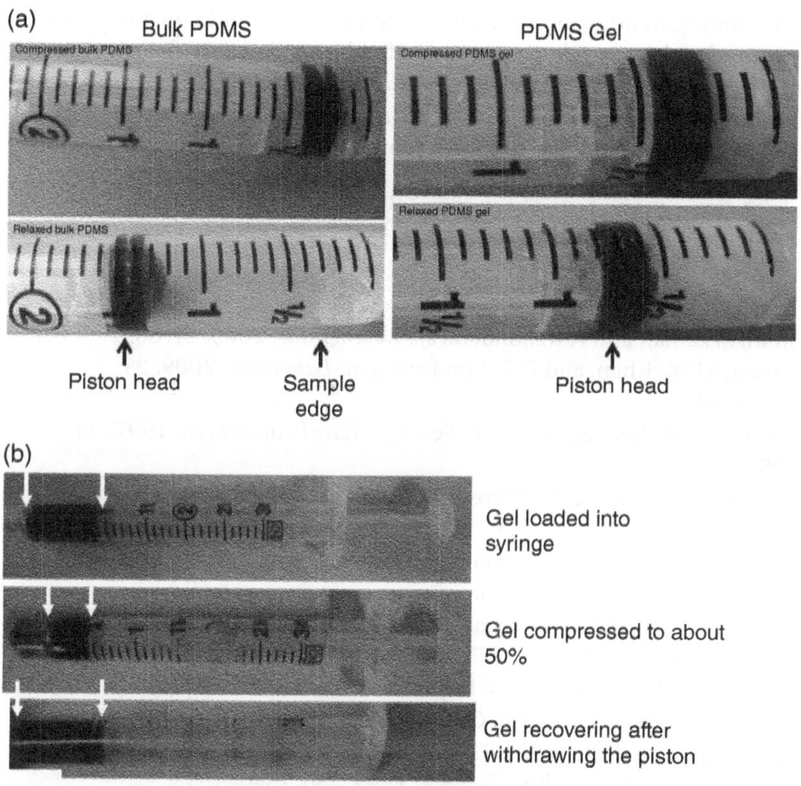

Figure 7.35 Simple compression/relaxation experiments to distinguish between a gel and a paste. (a) A comparison of bulk-PDMS and PDMS gel behavior. (b) Compression and recovery of PS-PTCDI-PS/*trans* decalin gel. (Source: Dahan and Sundararajan [83]. Reproduced with permission of Royal Society of Chemistry.)

The earlier discussion, as in the other chapters in the book, presented an overview of the gels related to polymers, oligomers, and their monomeric precursors. While the investigations on the gels caused by polymers such as PE and PS were aimed at understanding the phase separation/vitrification/gelation behavior of the polymer/solvent systems, those related to small-molecule gels (LMOG) channeled the efforts to *a priori* prediction of gelability of a gelator/solvent pair. As the materials presented earlier would show, there was a mutual convergence to integrate the studies on polymer/small-molecule complex, etc. This chapter was concerned only with organo- and polymer gels. A vast literature exists on hydrogels using synthetic as well as biopolymers, which are candidates for use in encapsulation and release of, for example, drug molecules. The literature on the organogels based on π-systems was reviewed by Ajayaghosh et al. [78, 88]. Most often, these molecules also bear groups to add hydrogen bonding in addition to π-interactions. One would consider the area of organo- and polymer gels as a mature field and that any future research could only be incremental. However, when we consider the advances in using, for example, block copolymers to design materials and applications in the nanoscale, the gel state of these materials could offer yet unexplored prospects and applications.

References

1 M. Moniruzzaman and P. R. Sundararajan, *Langmuir*, **2005**, *21*, 3802–3807.
2 S. Khanna, M. K. Khan, and P. R. Sundararajan, *Langmuir*, **2009**, *25*, 13183–13193.
3 J. van Esch, R. M. Kellogg, and B. L. Feringa, *Tetrahedron Lett.*, **1997**, *38*, 281–284.
4 K. Hanabusa, K. Shimura, K. Hirose, M. Yamada, and H. Shirai, *Chem. Lett.*, **1996**, *25*, 885–886.
5 J. van Esch, S. De Feyter, R. M. Kellogg, F. De Schryver, and B. L. Feringa, *Chem. Eur. J.*, **1997**, *3*, 1238–1243.
6 M. de Loos, J. van Esch, I. Stokroos, R. M. Kellogg, and B. L. Feringa, *J. Am. Chem. Soc.*, **1997**, *119*, 12675–12676.
7 J. van Esch, F. Schoonbeek, M. de Loos, H. Kooijman, A. L. Spek, R. M. Kellogg, and B. L. Feringa, *Chem. Eur. J.*, **1999**, *5*, 937–950.
8 C. Shi, Z. Huang, S. Kilic, J. Xu, R. M. Enick, E. J. Beckman, A. J. Carr, R. E. Melendez, and A. D. Hamilton, *Science*, **1999**, *286*, 1540–1543.
9 R. G. Weiss, *J. Am. Chem. Soc.*, **2014**, *136*, 7519–7530.
10 P. J. Flory, *Faraday Discuss. Chem. Soc.*, **1974**, *57*, 7–18.
11 K. Almdal, J. Dyre, S. Hvidt, and O. Kramer, *Polym. Gels Netw.*, **1993**, *1*, 5–17.
12 A. Keller, *Faraday Discuss.*, **1995**, *101*, 1–49.
13 R. M. Hikmet, S. Callister, and A. Keller, *Polymer*, **1988**, *29*, 1378–1388.
14 S. Wellinghoff, J. Shaw, and E. Baer, *Macromolecules*, **1979**, *12*, 932.
15 K-i. Murata and H. Tanaka, *Nat. Mater.*, **2012**, *11*, 436–443.
16 P. R. Sundararajan, *Polymer*, **2002**, *43*, 1691–1693.
17 G. Damaschun, *Kolloid-Z.*, **1962**, *180*, 65–67.
18 A. N. Falcâo, J. S. Pedersen, and K. Mortensen, *Macromolecules*, **1993**, *26*, 5350–5364.
19 T. Hajsz, I. Csetneki, G. Filipcsei, and M. Zrinyi, *Phys. Chem. Chem. Phys.*, **2006**, *8*, 977–984.
20 J. L. Braun, J. E. Mark, and B. E. Eichinger, *Macromolecules*, **2002**, *35*, 5273–5282.
21 E. Dahan and P. R. Sundararajan, *Langmuir*, **2013**, *29*, 8452–8458.
22 J. N. Lee, C. Park, and G. M. Whitesides, *Anal. Chem.*, **2003**, *75*, 6544–6554.
23 H. Jacobson and W. H. Stockmayer, *J. Chem. Phys.*, **1950**, *18*, 1600–1606.

24 P. J. Flory, V. Crescenzi, and J. E. Mark, *J. Am. Chem. Soc.*, **1964**, *86*, 146–152.
25 P. J. Flory and J. A. Symlyen, *J. Am. Chem. Soc.*, **1966**, *88*, 3209–3212.
26 P. J. Flory, *Statistical Mechanics of Chain Molecules*, **1969**, Interscience Publishers, New York (1989, republished by Hanser Publishers, New York).
27 M. Beltzung, C. Picot, P. Rempp, and J. Herz, *Macromolecules*, **1982**, *15*, 1594–1600.
28 G. L. Jadav, V. K. Aswal, and P. S. Singh, *J. Membr. Sci.*, **2011**, *378*, 194–202.
29 J. E. Mark, *J. Chem. Phys.*, **1968**, *49*, 1398–1402.
30 P. Vandeweerdt, H. Berghmans, and Y. Tervoort, *Macromolecules*, **1991**, *24*, 3547–3552.
31 K.-Y. Lin, D.-M. Wang, and J.-Y. Lai, *Macromolecules*, **2002**, *35*, 6697–6706.
32 M. Vaysse, M. K. Khan, and P. Sundararajan, *Langmuir*, **2009**, *25*, 7042–7049.
33 R. C. Domszy, R. Alamo, C. O. Edwards, and L. Mandelkern, *Macromolecules*, **1986**, *19*, 310–325.
34 E. K. M. Chan and L. Mandelkern, *Macromolecules*, **1992**, *25*, 5659–5664.
35 P. R. Sundararajan and T. A. Kavassalis, *J. Chem. Soc. Farad. Trans.*, **1995**, *91*, 2541–2549.
36 G. Ungar, J. Stejny, A. Keller, I. Bidd, and M. C. Whiting, *Science*, **1985**, *229*, 386–389.
37 G. Ungar and A. Keller, *Polymer*, **1986**, *27*, 1835–1844.
38 K. S. Lee and G. Wegner, *Makromol. Chem. Rapid Commun.*, **1985**, *6*, 203.
39 K. S. Lee, G. Wegner, and S. L. Hsu, *Polymer*, **1987**, *28*, 889.
40 W. M. Leung, R. St. John Manley, and A. R. Panaras, *Macromolecules*, **1985**, *18*, 746, 753, 760.
41 D. J. Abdallah and R. G. Weiss, *Langmuir*, **2000**, *16*, 352–355.
42 E. D. T. Atkins, D. H. Isaac, A. Keller, and K. Miyasaka, *J. Polym. Sci. Polym. Phys. Ed.*, **1977**, *15*, 211.
43 M. Girolamo, A. Keller, K. Miyasaka, and N. Overbergh, *J. Polym. Sci. Polym. Phys. Ed.*, **1976**, *14*, 39.
44 P. R. Sundararajan, *Macromolecules*, **1979**, *12*, 575–581.
45 P. R. Sundararajan, N. J. Tyrer, and T. L. Bluhm, *Macromolecules*, **1982**, *15*, 286–290.
46 P. R. Sundararajan and N. J. Tyrer, *Macromolecules*, **1982**, *15*, 1004–1012.
47 N. J. Tyrer, T. L. Bluhm, and P. R. Sundararajan, *Macromolecules*, **1984**, *17*, 2296–2303.
48 D. Y. Yoon, P. R. Sundararajan, and P. J. Flory, *Macromolecules*, **1975**, *8*, 776–783.
49 J.-M. Guenet, *Thermoreversible Gelation of Polymers and Biopolymers*, **1992**, Academic Press, London.
50 M. Klein and J. M. Guenet, *Macromolecules*, **1989**, *22*, 3716–3725.
51 V. Vittoria, R. Russo, and F. de Candia, *Polymer*, **1991**, *32*, 3371–3375.
52 G. Guerra, V. M. Vitagliano, C. De Rosa, V. Petraccone, and P. Corradini, *Macromolecules*, **1990**, *23*, 1539–1544.

53 G. Guerra, P. Musto, F. Karasz, and W. Macknight, *Makromol. Chem.*, **1990**, *191*, 2111–2119.

54 V. Vittoria, R. Russo, and F. de Candia, *Makromol. Chem. Macromol. Symp.*, **1990**, *39*, 317–332.

55 A. Grassi, P. Longo, and G. Guerra, *Makromol. Chem. Rapid Commun.*, **1989**, *10*, 687–690.

56 V. Vittoria, F. De Candia, P. Ianelli, and A. Immirzi, *Makromol. Chem. Rapid Commun.*, **1988**, *9*, 765–769.

57 A. Immirzi, F. De Candia, P. Ianelli, A. Zambelli, and V. Vittoria, *Makromol. Chem. Rapid Commun.*, **1988**, *9*, 761–764.

58 J. De Rudder, B. Bergé, and H. Berghmans, *Macromol. Chem. Phys.*, **2002**, *203*, 2083–2088.

59 C. Daniel, G. Guerra, and P. Musto, *Macromolecules*, **2002**, *35*, 2243–2251.

60 C. Daniel, D. Alfano, G. Guerra, and P. Musto, *Macromolecules*, **2003**, *36*, 1713–1716.

61 C. Daniel, D. Alfano, G. Guerra, and P. Musto, *Macromolecules*, **2003**, *36*, 5742–5750.

62 C. Daniel, A. Avallone, and G. Guerra, *Macromolecules*, **2006**, *39*, 7578–7582.

63 Y. Chatani, Y. Shimane, T. Inagaki, T. Ijitsu, T. Yukinari, and H. Shikuma, *Polymer*, **1993**, *34*, 1620.

64 C. De Rosa, P. Rizzo, O. Ruiz de Ballesteros, V. Petraccone, and G. Guerra, *Polymer*, **1999**, *40*, 2103–2110.

65 M. Kobayashi, *Macromol. Symp.*, **1997**, *114*, 1–12.

66 H. Itagaki, T. Tokami, and J. Mochizuki, *Polymer*, **2012**, *53*, 5304.

67 J. Mochizuki, T. Tokami, S. Kusuki, T. Sano, and H. Itagaki, *Macromol. Chem. Phys.*, **2013**, *214*, 1912–1920.

68 M. Suzuki, S. Owa, H. Shirai, and K. Hanabusa, *J. Polym. Sci. A Polym. Chem.*, **2006**, *44*, 3817–3824.

69 M. Suzuki, S. Owa, H. Shirai, and K. Hanabusa, *Macromol. Rapid Commun.*, **2005**, *26*, 803.

70 M. Suzuki, C. Setoguchi, H. Shirai, and K. Hanabusa, *Chem. Eur. J.*, **2007**, *13*, 8193–8200.

71 M. K. Khan, A. Chan, and P. R. Sundararajan, *Soft Matter*, **2012**, *8*, 10149–10157.

72 S. Khanna, M. Moniruzzaman, and P. R. Sundararajan, *J. Phys. Chem. B*, **2006**, *110*, 15251–15260.

73 W. C. van Bogart, P. E. Gibson, and S. L. Cooper, *J. Polym. Sci. Polym. Phys. Ed.*, **1983**, *21*, 65.

74 S. H. Wang, L. F. Silva, J. Kloss, M. Munaro, G. P. de Souza, M. A. Wada, J. G. C. Gomez, S. Zawadzki, and L. Akcelrud, *Macromol. Symp.*, **2003**, *197*, 255–264.

75 O. Jazkewitsch and H. Ritter, *Macromolecules*, **2011**, *44*, 375–382.

76 Z. Ge, J. Hu, F. Huang, and S. Liu, *Angew. Chem. Int. Ed.*, **2009**, *48*, 1798–1802.

77 A. Ajayaghosh and S. J. George, *J. Am. Chem. Soc.*, **2001**, *123*, 5148–5149.
78 A. Ajayaghosh and V. K. Praveen, *Acc. Chem. Res.*, **2007**, *40*, 644–656.
79 E. Dahan and P. R. Sundararajan, *Soft Matter*, **2014**, *10*, 5337–5349.
80 D. Yao, T. P. Bender, P. J. Gerroir, and P. R. Sundararajan, *Macromolecules*, **2005**, *38*, 6972–6978.
81 F. M. Menger and C. A. Littau, *J. Am. Chem. Soc.*, **1993**, *115*, 10083–10090.
82 F. M. Menger and J. S. Keiper, *Angew. Chem. Int. Ed.*, **2000**, *39*, 1906–1920.
83 E. Dahan and P. R. Sundararajan, *Eur. Polym. J.*, **2014**, *61*, 113–123.
84 E. Dahan and P. R. Sundararajan, *Eur. Polym. J.*, **2015**, *65*, 4–14.
85 S. Rakesh, C. P. Sakthidharan, M. Sarojadevi, and P. R. Sundararajan, *Eur. Polym. J.*, **2015**, *68*, 161–174.
86 J. H. van Esch and B. L. Feringa, *Angew. Chem. Int. Ed.*, **2000**, *39*, 2263–2266.
87 J.-M. Guenet, *Macomol. Symp.*, **2006**, *241*, 45–50.
88 S. S. Babu, V. K. Praveen, and A. Ajayaghosh, *Chem. Rev.*, **2014**, *114*, 1973–2129.

8

Small-Molecule Self-Assembly in Polymer Matrices

Polymers and small molecules are used together in a number of applications. The small molecule may be used to modify polymer properties, or the polymer could be a vehicle to just hold the functional small molecule. In the former category, the plasticization of polymers is well known. The small molecules, typically phthalates, are blended with the polymers to tailor the glass transition temperature (T_g) and enable processing at lower temperatures. For example, 15 wt% of diphenyl terephthalate (DPTP) in bisphenol-A polycarbonate [1] would reduce the T_g of the polymer by 50°C, from 150 to 100°C. The mechanical properties of the polymer would also be modified when it is plasticized. When an active layer in a device such as the photoreceptor uses a functional photoactive small molecule, it would usually be solution coated and it would also be required that it be dispersed molecularly without any aggregation. In such cases, the small molecule would be dissolved along with a polymer and coated to form a film for the device. The polymer then simply holds the small molecule as a dispersion. We will discuss the behavior of the small molecules in such films.

8.1 Phase Separation in Charge Transport Polymer Layers

With the organic photoreceptors that are used in photocopiers and printers, the charge transport molecule (CTM), usually of the tritolylamine (TTA) type shown in Figure 8.1, is dispersed in a polycarbonate matrix by solution coating. An unusually high loading of up to 50% by weight might be used. The small molecule should remain dispersed and any phase separation from the polymer would affect the functioning of the device. Similar to the case of plasticizers, the addition of the CTM would also reduce the T_g of the polymer matrix. With 50 wt% of TPD, the T_g of polycarbonate reduces [2] from 150 to 70°C. With TTA, which is a smaller molecule, the T_g reduces even more [3], from 150 to

Physical Aspects of Polymer Self-Assembly, First Edition. P. R. Sundararajan.

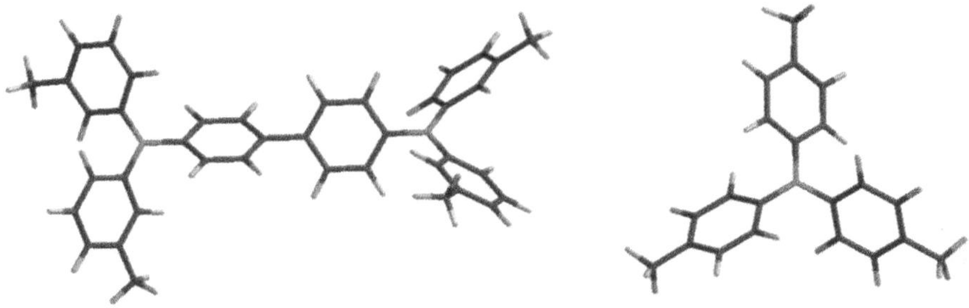

Figure 8.1 Molecular structures of hole transport molecules *N,N′*-diphenyl-*N,N′*-bis(3-methylphenyl)-[1,1′-biphenyl]-4,4′-diamine (TPD) and TTA (right).

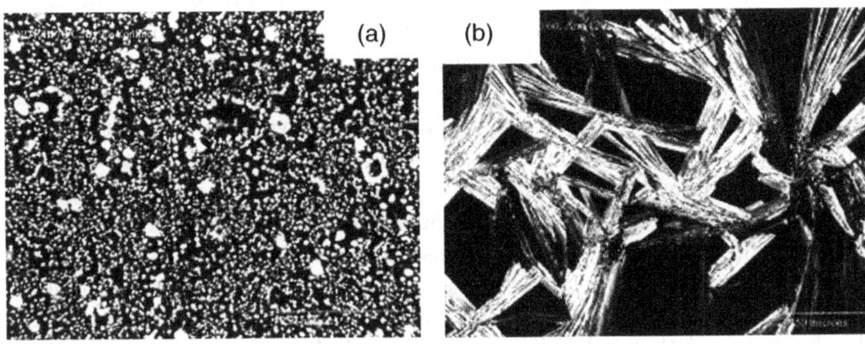

Figure 8.2 Optical micrographs of (a) DPTP/BPAPC (10/90) film annealed at 130°C for 2 h and (b) DPIP/BPAPC (40/60) film annealed at 120°C for 4 h. (Source: Tuteja et al. [1]. Reproduced with permission of Wiley.)

41°C. When these films are annealed at a temperature close to the original T_g of the polycarbonate, the CTM would diffuse to the surface of the film and crystallize. The extent of phase separation and crystallization would depend on the temperature of annealing and the type of small molecule. A few types of molecules such as the TTA would even sublime from the film [3]. Figure 8.2a and b shows the optical micrographs of phase separated crystals of DPTP and diphenyl isophthalate (DPIP), respectively, from the bisphenol-A polycarbonate (BPAPC) matrix upon annealing. In both cases, the films were sandwiched between two glass slides. In the absence of a cover slide, far fewer crystals were seen, due to the sublimation of the small molecule at the annealing temperature [1]. The phase separated morphology of the CTMs from the polymer matrix would be similar [3–6].

With the depletion of the CTM in the polymer due to phase separation, T_g of the composite would then recover to some extent. After annealing at 140°C for 2 h, and the resulting phase separation, the T_g's of polycarbonate/TPD and polycarbonate/TTA composites increased from 70 to 82°C, and from 41 to 127°C, respectively [2, 3]. The diffusion is much larger with the smaller size of TTA, compared to that of TPD.

8.2 Glass Transition and Diffusion of Small Molecules

Clearly, the aforementioned phase separation phenomenon involves diffusion of the small molecule in the polymer matrix. The absence of diffusion and aggregation is important for the stability of the aforementioned functional composites. It should be added that controlled movement of small molecules in polymer matrices is applicable to a few technologies. For example, in the case of liquid crystal (LC) displays, a liquid crystalline small molecule is dispersed as droplets in a polymer, and the orientation or randomness of the small molecule in the polymer is controlled by an electric field. Photorefractive polymers based on guest–host systems are prepared by incorporating a chromophore in a polymer with low T_g and birefringence is achieved by poling-induced chromophore orientation [7]. In these cases, the movement of the small molecule is controlled reversibly by an applied field.

Diffusion of small molecules through polymer matrices and the control of such movement are relevant for a number of other applications. Transport of molecules of liquids and gases/vapor through polymers is applicable to gas separation membranes, water purification systems, barrier layers for packaging materials, controlled drug delivery, etc. Such diffusion also affects the removal of residual solvent. Due to the commercial importance of the aforementioned applications, a number of techniques such as mass uptake, nuclear magnetic resonance (NMR), Fourier transform infrared (FTIR), and fluorescence spectroscopies have been used to study the diffusion of molecules [7–11], ranging from oxygen to short alkanes and other small molecules as well as fullerenes. Efforts to understand the factors that govern such diffusion in terms of the molecular size of the small molecule, effect of temperature, interaction with the polymer, liquid state versus glassy state, and developing models for interpreting experimental observations date back to the 1950s. Brandt [12] reported model calculations of activation energy of diffusion based on an activated state in which two polymer chains move and make room for the diffusing molecule. Such movement of the chains was modeled using an intermolecular term for the repulsion between segments of two neighboring chains and an intramolecular term to allow for the partial rotation of the chain units from their equilibrium conformation. The activation energy E_d is derived from Arrhenius-type equation

$$D = D_0 \exp\left(-\frac{E_d}{RT}\right) \tag{8.1}$$

in which D is the diffusion coefficient and D_0 is a prefactor. Systems obeying this relationship have been reported. However, a change in slope at the T_g has been noted. The Williams–Landel–Ferry (WLF) equation has also been applied for diffusion above the T_g of the polymer/composite to study the temperature dependence.

$$\log D(T) = \log D(T_g) + \frac{\xi C_{1g}(T - T_g)}{C_{2g} + (T - T_g)} \quad (8.2)$$

where $D(T_g)$ is the diffusion coefficient of the small molecule at the T_g of the polymer, ξ is the relative volume of the small molecule to that of the polymer, and C_1 and C_2 are the WLF parameters of the polymer. Even simple mass uptake measurements have been used to distinguish between Fickian and non-Fickian diffusion of small molecules in polymers [10, 13–15]. It was also found that the asymmetry of the molecular shape influences diffusion. The linear molecules such as n-alkanes showed higher diffusivities and lower E_d than spherical molecules of similar molar volume [16].

Since the penetrant molecule has to move between the polymer segments, models based on free volume were developed by Wilkens and Long [17] in 1957. Theoretical treatment of the dependence of small-molecule diffusion on free volume has been published [18]. The effect of free volume on penetrant diffusion has been studied by a number of authors. The Fox–Flory equation [19] for the fractional free volume is given by

$$v_f = K + (\alpha_R - \alpha_G)T \quad (8.3)$$

where α_R and α_G are the thermal expansion coefficients of the polymer in the rubbery and glassy states, respectively. Simha and Boyer [20] modified this equation such that at T_g,

$$(\alpha_R - \alpha_G)T_g = K_1 \quad (8.4)$$

Using the data on several polymers, Simha and Boyer [20] concluded that $K_1 = 0.113$, that is, the fractional free volume at T_g for any polymer is 11.3%. This is the reason for calling the T_g as the "iso-free-volume state." Techniques such as ortho-positronium (oPs) lifetimes [21–25] and fluorescence probe lifetimes [26, 27] have been used to measure the free volume in a number of polymers. Since the free volume would be expected to fluctuate with time with the segmental motion of the amorphous polymers, atomistic molecular simulations have been performed to calculate not only the fractional free volume but also its fluctuation. The methodology originally developed by Theodorou and Suter [28] for computationally building a model of the amorphous cell of polymers using periodic boundary conditions has been used in a number of publications [29–33] to study the free volume and its distribution in polymers.

Employing molecular dynamics simulations, this method includes a probe molecule in the amorphous cell and follows the probe movement with time to study its dynamics.

In the case of photoreceptor composite discussed earlier, when phase separation occurs due to thermal effects or mechanical abrasion, the small molecule would diffuse to the surface of the polymer film and crystallize. Plasticizers mentioned earlier can also phase separate and form crystals on the surface of the polymer. In most cases, the as-cast films would show no crystallization of the small molecules up to a certain concentration. Note that in these examples, although TTP, TPD, and DPTP are made up of aromatic segments, the molecules are nonplanar, and they do not "self-assemble." We will look at a few cases of dispersing self-assembling small molecules in polymer matrices.

In the aforementioned examples of TPD, TTA, and the terephthalate-type plasticizers, the interaction between these small molecules and a host polymer would involve mainly van der Waals forces, and no specific association might occur with the polymer. Since spontaneous association by specific intermolecular forces is a characteristic of a self-assembling molecule, the attraction between the self-assembling small molecules themselves would surpass the affinity if any to the polymer. However, the self-assembly would be affected by the presence of the polymer and the solvent used to cast the film. When a solution of a polymer and a (self-assembling) small molecule is cast as a film, initially, the polymer is swollen by the solvent. It is known that the T_g of a polymer will be depressed significantly in the presence of a solvent [34]. Jenckel and Heusch [35] reported that with about 25 wt% of chloroform, the T_g of polystyrene (PS) was reduced from 80 to 10°C, and with toluene, to −25°C. This gives rise to a large free volume for the small molecule to diffuse. In addition, the diffusion of the molecules in the polymer matrix leading to self-assembly would depend on the viscosity of the solvent and the rate of evaporation during drying of the film. As the solvent evaporates, the T_g of the polymer would recover. Khan et al. [2, 3] have shown that when the hole-transport molecule crystallizes out of the polymer, the T_g recovers significantly. Hence, the behavior of such a molecule in a polymer matrix would be different with time, and this would lead to morphologies that are different from the case of a plasticized polymer or a photoreceptor. We will discuss such a case with self-assembly promoted by hydrogen bond and by π-interactions.

8.3 Subsurface Self-Assembly of Small Molecules in Polymer Matrices

Consider the two molecules shown in Figures 3.4, 7.1, and 7.2. The carbamates and biscarbamates are model compounds for polyurethanes and have been considered for a number of applications, ranging from viscosity modification to ink jet printing. The biscarbamate contains two hydrogen bonding moieties, separated by a C_6 alkyl segment. Other spacers are known as well. The hydrogen bonding groups are flanked by alkyl chains, symmetrically varying from C_4 to C_{18}. With an increase in the length of the alkyl chain, the contribution from van der Waals interaction would exceed that from hydrogen bonding. The melting temperature increases from 91°C for the C_4 to 120°C for the C_{18} side chain. The melt viscosity is only a few centipoise and the material solidifies quickly from the melt, and this is one of the reasons for its consideration as ink vehicle for inkjet printing [36]. The half time ($t_{½}$) for completion of crystallization for these biscarbamates is 8 s for C_8 and 40 s for C_{12} alkyl side chains [37]. Interestingly, the rate of crystallization and spherulite growth showed a maximum for the C_8 side chain. The spherulite growth rates are 40, 120, 65, 30, and 30 µm/s with C_6, C_8, C_{12}, C_{16}, and C_{18} alkyl side chains, respectively [37, 38]. In contrast, the spherulite growth rate of TTA is much lower, with 1.2 µm/s. The rates increase from C_4 to C_8 in the case of biscarbamates, and a further increase in the alkyl chain length leads to a decrease in both crystallization and spherulite growth rates. This influences the morphology significantly when these biscarbamates are dispersed in a polymer matrix.

In the case of CTMs such as TPD and TTA, the as-cast films with BPAPC as the matrix (up to about 50 wt% concentration) would not show any aggregation or crystallization unless the composite film was annealed close to the T_g of BPAPC. However, in the case of self-assembling molecules, association occurs rapidly when the film is still wet and the solvent is present. It was found [39] that when a film was cast from a solution of biscarbamates with alkyl chain lengths of C_4–C_{18}, and BPAPC in chloroform, the small molecule self-assembled so fast that no diffusion to the surface took place and there was no molecular dispersion. The biscarbamates were not molecularly dispersed, as would be the case normally with the "non-self-assembling" system such as TPD/BPAPC. The self-assembled domains of biscarbamates were in the bulk of the film and not on the surface. This is seen in Figure 8.3. For C_4, C_6, and C_8 side chains, no clear crystals are seen as these are submerged in the polymer. Protrusions (shown by arrows) of growing crystals are seen for C_{12} and C_{16}, since the crystal growth rate (and self-assembly) is slower with longer alkyl chains. Hence, while most of the molecules are still below the surface, a few form needle-like crystals on the surface. The subsurface growth is even more discernible in Figure 8.3, as the concentration of C_4 was increased from 2 to 8 wt%. Patchy domains are seen, with no detectable crystals on the surface, indicative of assembly in the bulk of the film. As mentioned earlier, the mobility of the molecule depends on the matrix polymer. During the drying process soon after the film was cast, the self-assembly occurred due to the presence of the solvent as well as the depression of the glass transition of the polymer. In another experiment, the film was annealed for 2 or 30 min just above the melting temperature of the biscarbamates and cooled (taken off the hotplate).

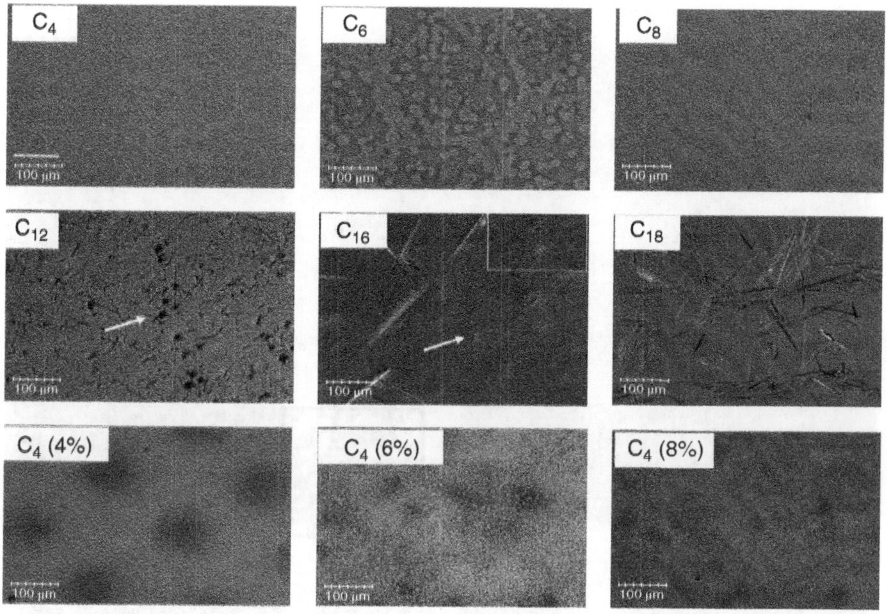

Figure 8.3 Optical micrographs (OMs) of biscarbamates with alkyl side chains C_4, C_6, C_8, C_{12}, C_{16}, and C_{18} (3 wt%) in polycarbonate films cast from chloroform. The OM of films with C4 biscarbamates with 4, 6, and 8 wt% concentrations are also shown. The scale bar is 100 μm. (Source: Islam and Sundararajan [39]. Reproduced with permission of NRC Research Press.)

This caused the diffusion of the melt of the biscarbamate in the polymer matrix, since the melt viscosity of these molecules is only a few centipoise. The T_g of the polycarbonate is 150°C, well above the melting temperatures of the biscarbamates. When the film was annealed for just 2 min and cooled, the self-assembly still occurred, within the polymer and not on the surface. However, after annealing for 30 min, which enabled extensive diffusion of the small molecules into the polymer matrix, there was no self-assembly upon cooling, and no crystallization was seen. This would imply that in the absence of the solvent, the polycarbonate matrix prevented the diffusion and aggregation/crystallization of the biscarbamate. Thus, although the hydrogen bond and van der Waals interactions in this case enable these molecules to self-assemble into domains in the polymer matrix, a molecularly dispersed, nonaggregated morphology can be achieved by the annealing step described earlier.

The monocarbamate shown in Figure 3.4 contains single hydrogen bonding unit, a C_{18} side chain on one side and the R varying from C_4 to C_{18} on the other side. Thus, it is asymmetric with respect to alkyl substitution, except when $R = C_{18}$. As the length of R increases from C_4 to C_{18}, the contribution from van der Waals interaction would exceed that from hydrogen bonding.

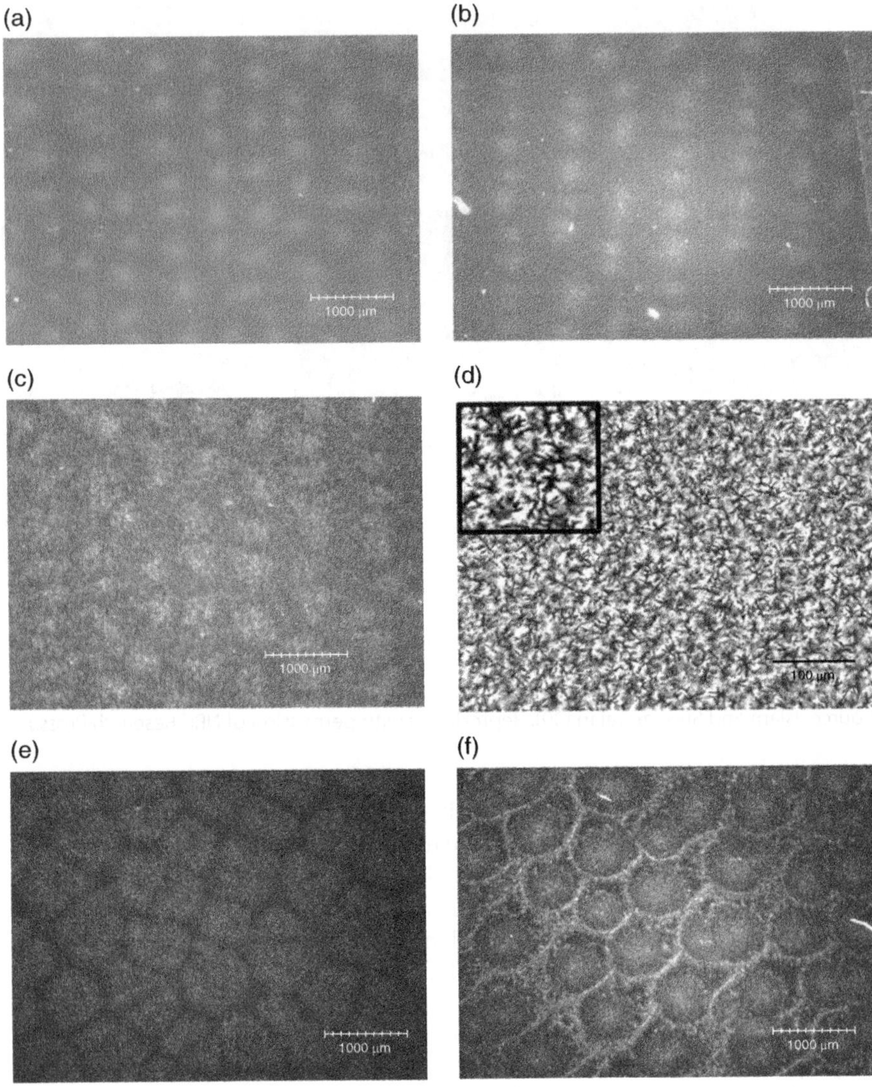

Figure 8.4 Optical micrographs of domains of monocarbamates with various alkyl side chain lengths in polycarbonate (a) 1% C_7, (b) 3% C_7, (c) 5% C_{12}, and (d) spherulitic structure inside the domains of (b); (e) 10% blend C_{12}/C_{16} (90/10) and (f) 15% blend. The scale bar is 100 μm for (d) and 1000 μm for others. (Source: Tuteja et al. [40]. Reproduced with permission of American Chemical Society.)

The rates of crystallization and spherulite growth are much faster than those of the biscarbamates. In this case too, the small molecule did not disperse molecularly in the matrix and the self-assembly occurred in the bulk of the film, but the morphology was very ordered [40]. Even with just 1 wt% of C_7

monocarbamate large spherical domains of about 340 μm appear in a rectangular lattice, and this persists with an increase in concentration (Figure 8.4). With a 3 wt% of C_7, the domain size increased to 375 μm. These sizes are much larger than the thickness of the film, which was only 25 μm. We still refer to them as "spherical" just as spherulites are referred to in thin films. These domains are not at the surface but in the bulk of the films. The origin of the uniform spacing between the domains is not known. Similar domains were seen with monocarbamates with longer alkyl chain lengths, as shown in Figure 8.4c. With 1 wt% concentration of the monocarbamate, the domain size increased from 340 for C_7 to 460 μm for C_{12}. A 90/10 blend of two monocarbamates (C_{12}/C_{16}) also behaved the same, with the domain size increasing from 390 to 675 μm with an increase in concentration of the blend from 1 to 15 wt%. Thus, the spontaneous assembly of the monocarbamates leads to such highly organized domains, and the size of the domains depends on the alkyl chain length and the concentration. The morphology seen here is similar to the organization of photonic crystals via self-assembly facilitated by van der Waals interactions between colloidal particles into opals and inverse opals.

The monocarbamates crystallize rapidly from solution or the melt. The domains seen in Figure 8.4 are not single spherulites. Note that the scale bar in Figure 8.4a–c, e, and f represents 1000 μm. Figure 8.4d shows the higher magnification image (scale bar is 100 μm), revealing the internal morphology of a domain. The morphology consists of densely packed spherulites, of about 50 μm. A still enlarged section of Figure 8.4d is shown in the inset, to show the spherulites. The domains then involve a three-level hierarchy of assembly. To verify this model, a few drops of the solution were placed on a microscope slide and observed *in situ* as the solvent evaporated. Very tiny crystals were seen initially, and these aggregated to form spherulites. These then associated into ordered domains, as illustrated in Figure 8.5. Thus, the colloidal sized

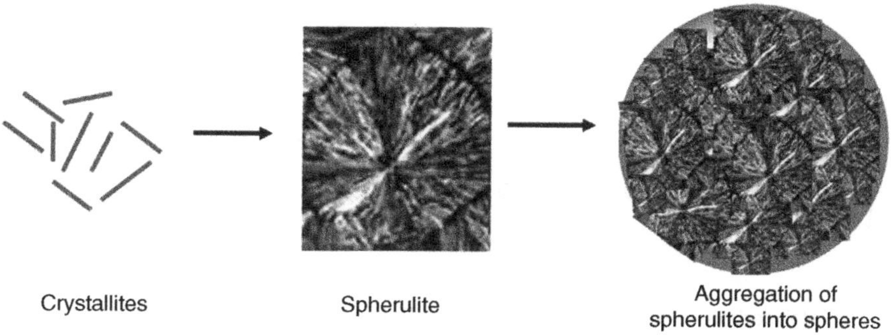

Figure 8.5 Illustration of hierarchical organization of small crystallites into spherulites and then to colloidal seize domains. (Source: Tuteja et al. [40]. Reproduced with permission of American Chemical Society.)

domains, which are crystalline and birefringent, are dispersed in the matrix of polycarbonate, which is amorphous.

8.4 Solvent Effect on Self-Assembly of Small Molecules in Polymer Matrices

The aforementioned studies with carbamates and biscarbamates used chloroform as the solvent to cast the films. Let us discuss a case [41] where the self-assembly is facilitated by π-interaction, and two different solvents and polymers were used, with the structures shown in Figure 8.6.

Films were cast from solutions of octabutoxy phthalocyanine (Pc) and polycarbonate (BPAPC) or poly(methyl methacrylate) (PMMA) in chloroform or 1,1,2,2-tetrachloroethane (TCE). Figure 8.7 shows the scanning electron microscope (SEM) image of the films of Pc/BPAPC cast from chloroform. The image consists of a smooth surface, does not show any features, and would tend to indicate that the small molecule was molecularly dispersed in the

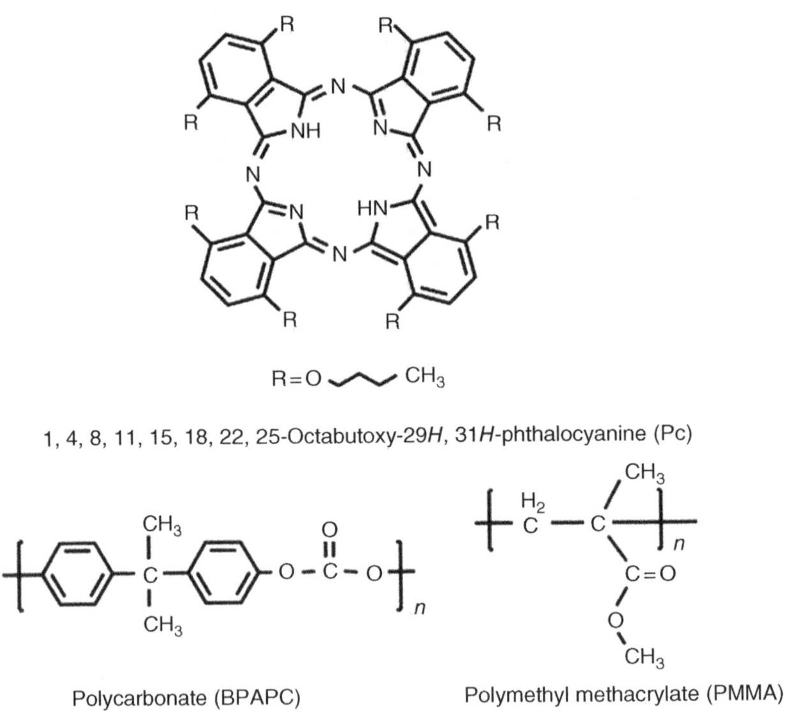

Figure 8.6 Molecular structures of octabutoxy phthalocyanines, bisphenol-A polycarbonate (BPAPC), and poly(methyl methacrylate) (PMMA).

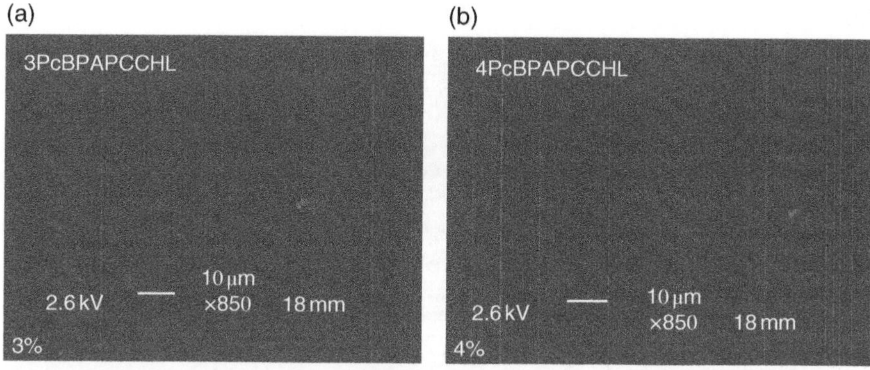

Figure 8.7 SEM images of Pc/BPAPC films with (a) 3 wt% and (b) 4 wt% of Pc cast from chloroform. (Source: Islam and Sundararajan [41]. Reproduced with permission of Wiley.)

polymer matrix. However, the optical microscopy (OM) images shown in Figure 8.8 indicate that self-assembly had occurred and that the domains are subsurface. With the concentration of Pc from 0.85 to 5 wt%, domains of a few microns are seen as bumps, below the surface. The droplets could be seen in the transmission OM, but the surface is smooth in the SEM image. Similar to the case of carbamates and biscarbamates discussed before, the self-assembly of Pc in the polycarbonate matrix in the presence of the solvent leads to such subsurface domains. With chloroform as the solvent, the same morphology was seen with PMMA as the matrix polymer, which has a lower T_g than the BPAPC. Similar was the morphology of films cast from methylene chloride solutions.

Figure 8.9 shows the OM and SEM of Pc/BPAPC films cast from TCE. Needle-like crystals are seen, with their length and population increasing with the concentration of Pc in the film. The percolation threshold was reached with 3 wt% of Pc, with the crystals overlapping with one another. The crystal growth in the film occurs during the solvent evaporation, and the *in situ* observation showed that the maximum length was reached in about 20 min. A feature seen in Figure 8.9e and f is that the crystals are tubular. They seem to wrap during formation. The same type of tubular needles was found with PMMA as the host. The authors found that when crystalline material was obtained from solutions of Pc alone in TCE, these were found to be tubular as well, with a diameter of about 30 μm and a wall thickness of 1–2 μm. Although substituted phthalocyanines tend not to be planar but be saddle-like [42, 43] and are considered tecton for tubular assembly, it is relevant to nanotube formation, but not microtubules. If the bent shape were sufficient to initiate tube formation, such tubular structure should have been observed with chloroform as the solvent as well, but was not.

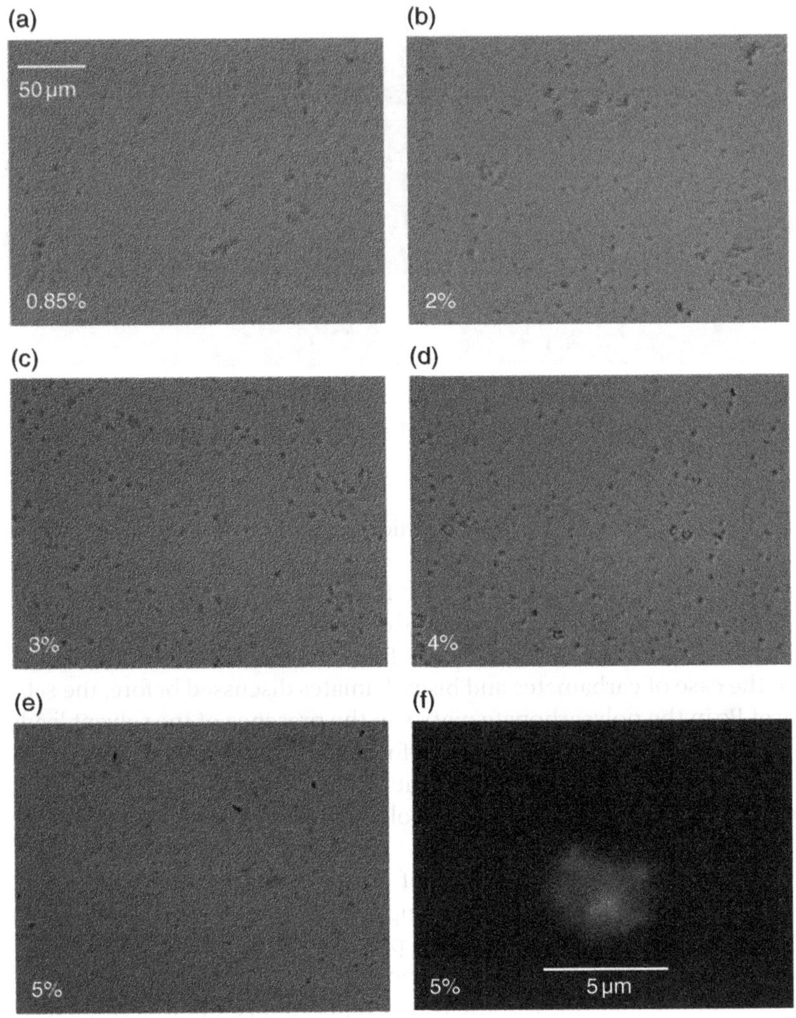

Figure 8.8 (a)–(e) OM images of Pc/BPAPC films (with 0.85–5 wt% of Pc) cast from chloroform. (f) Confocal microscope image of the Pc domain in the film. (Source: Islam and Sundararajan [41]. Reproduced with permission of Wiley.)

The difference in the morphology caused by the solvent was traced to the difference in the viscosity of these solutions and the diffusion of Pc. The viscosities of TCE and chloroform are 0.876 and 0.56 cp, respectively. The difference becomes more significant with the solutions. With 5 wt% of Pc in BPAPC, the solution viscosity with TCE is 1.35 cp, and it is 0.76 cp with chloroform. The solute concentration was 1 wt%. The higher viscosity and boiling

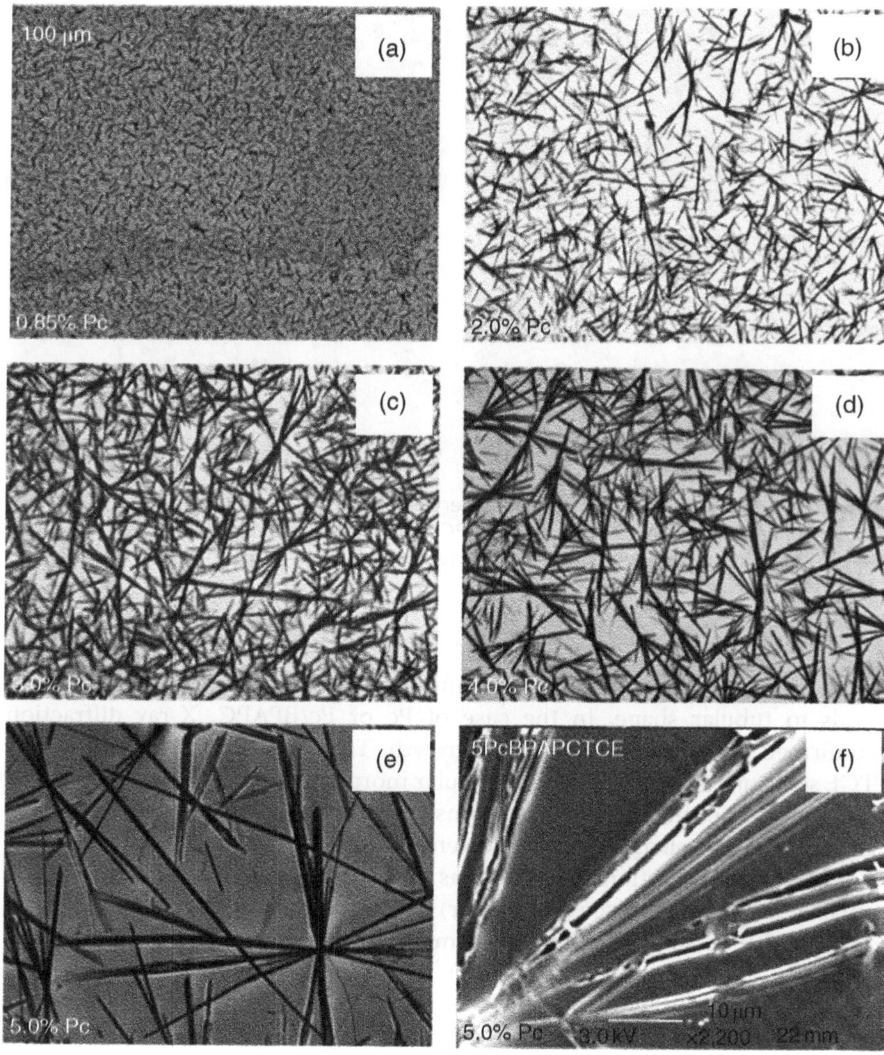

Figure 8.9 OM of Pc/BPAPC cast from TCE solutions with various concentrations (wt%) of Pc: (a) 0.85, (b) 2, (c) 3, (d) 4, and (e) 5. (f) SEM image of the film with 5% (wt) Pc in BPAPC. (Source: Islam and Sundararajan [41]. Reproduced with permission of Wiley.)

temperature of TCE leads to slow growth of Pc into tubular shape, irrespective of whether the solution contained only the Pc or the Pc and the polymer. When the crystal growth is limited by diffusion, such tubular morphology was seen in cases such as caffeine [44]. When the crystal growth is predominant in one direction, and if the diffusion is limited, the addition of molecules to the

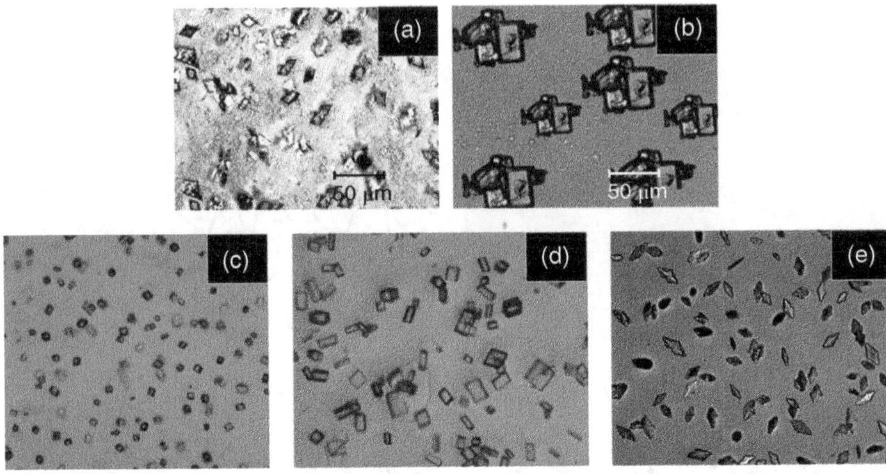

Figure 8.10 OM of solvent cast films of (a) perylene/PS (5/95 wt%), (b) PTCDI/PS (5/95), (c) perylene/BPAPC (2/98), (d) perylene/BPAPC (5/95), and (e) perylene/PMMA (2/98). (a)–(d) chloroform; (e) TCE. (Source: Dahan and Sundararajan [45]. Reproduced with permission of Elsevier.)

growing crystal face is slowed down, and a cavity forms within the crystal. This leads to tubular shape. In the case of Pc or Pc/BPAPC, X-ray diffraction confirmed predominant directional growth. Thus, the higher viscosity of the TCE solutions is the cause of such tubular morphology. The absorption spectra and the color of the solutions and films varied with the concentration of Pc in the polymer as well as the type of solvent.

The phase separation behavior was similar for perylene (which is much smaller than the Pc discussed earlier) in polycarbonate and PMMA (M. R. Islam and P. R. Sundararajan, Department of Chemistry, Carleton University, Ottawa, Canada, unpublished results). In this case again, the phase separated crystalline morphology was similar with polycarbonate and PMMA as the matrices, but different with the solvent used. With the slow evaporation of TCE, large crystals of about 100 μm were seen in the OM images whereas with chloroform, the crystals were only about 10 μm in size. The SEM images showed that the growth of crystals was subsurface due to the fast self-assembly of perylene in the matrix. Figure 8.10 shows the OM of films made with perylene or PTCDI in common polymers such as PS or polycarbonate (BPAPC). Even with very low concentrations of these molecules, profuse phase separation and crystallization occurred [45]. It was found [46] that the rate of crystal growth of perylene is faster than that of octabutoxy phthalocyanine. Ito et al. [47] used PMMA, perylene, and toluene to cast films and studied the effect of concentration on self-assembly. At low concentrations (<0.01 mol%) in the films, perylene was found to be in monomeric form, and with an increase in concentration, dimers and small aggregates developed into nanocrystals. They also followed the aggregation during solvent evaporation using dynamic fluorescence spectroscopy.

8.5 Polymer-Compatibilized Small-Molecule Assembly in Polymer Matrices

In the earlier examples of hydrogen-bond-mediated self-assembly of carbamates and biscarbamates and π–interaction facilitated self-assembly of phthalocyanines and perylenes, it was observed that molecular dispersion did not occur in polymer matrices even with very low concentrations of approximately 1 wt% because of the overwhelming association between these small molecules. The self-assembly and gelation of PTCDI substituted with oligomers of PEG (MJ–PTCDI, DJ–PTCDI), PDMS (PDI–PDMS, PDMS–PTCDI–PDMS) and polystyrene (PS–PTCDI–PS) were discussed in Chapters 4 and 7, respectively. These oligomers attached to PTCDI could act as "compatibilizers" when these molecules are dispersed in the corresponding polymers. The gelation of blends of these oligomer-substituted PTCDI and the corresponding polymers was described in Chapter 7. Dahan and Sundararajan [45] prepared solvent-cast films of PS–PTCDI–PS with PS with various solvents and the OM and SEM are shown in Figure 8.11. In contrast to the large crystals of PTCDI seen in Figure 8.10b, spherical domains of PS–PTCDI–PS with uniform size are seen in Figure 8.11 for the films cast from chlorobenzene or chloroform. The SEM image in Figure 8.11i further shows that these are nanometer-sized. In particular, Figure 8.11e shows densely packed spheres, which would indicate that percolation threshold which is necessary for charge hopping has been achieved. It was seen in Chapter 4 that the Gemini surfactant PS–PTCDI–PS forms vesicular morphology in solvent cast films. The same type of morphology was maintained when it was dispersed in the PS matrix. Thus, it is a supramolecular solution. It should also be added that the SEM of chloroform and chlorobenzene cast films (with 5 wt% PTCDI) did not show any features which would again mean that the domains were formed subsurface. Dense nanometer-sized crystals were seen only with 10 wt% of PTCDI (Figure 8.11i).

The subsurface self-assembly of molecules such as PTCDI might be an advantage for devices such as the photoreceptor. As mentioned earlier, CTMs such as the TPD are loaded to the extent of almost 50 wt% in the polycarbonate matrix for the charge transport layer (CTL) and is thus a metastable composite. For example, abrasion of the photoreceptor by brush-cleaning (during the copy/print cycles) could lead to a loss of the CTM. Overcoating the CTL with abrasion-resistant siloxane layers has been proposed to overcome the abrasion-induced depletion [48]. If, on the other hand, the CTM were to be one

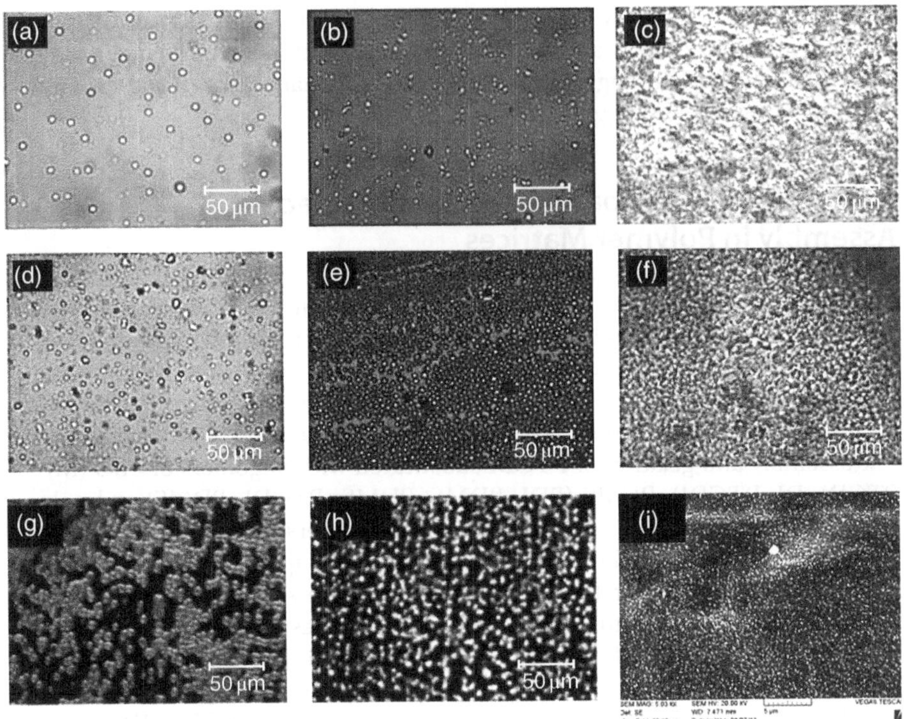

Figure 8.11 OM of PS/PS-PTCDI-PS films with (a, d) chlorobenzene, (b, e) chloroform, and (c, f) THF. The top row corresponds to 1 wt% PTCDI and the center row, 5 wt%. (g) and (h) are spin cast films from chloroform and THF, with 5 wt% PTCDI. (i) SEM of chloroform cast film (10 wt% PTCDI). (Source: Dahan and Sundararajan [45]. Reproduced with permission of Elsevier.)

with self-assembling properties and forms subsurface nanodomains, abrasion could be reduced or eliminated. We also discussed the use of rotaxanes for this purpose [49] in Chapter 6.

8.6 Polymerization-Induced Phase Separation and Reaction-Induced Phase Separation

Phase separation has been used advantageously in a number of applications. *Polymerization-induced* phase separation (PIPS) is a technique that was developed a couple of decades ago, and has been adapted to a few technologies. The kinetics of PIPS has been theorized by Chen and Chen [50], and a review was presented by Liu [51]. Complete miscibility of any two polymers is not common. Consider a case in which a polymer is dissolved in a monomeric solvent. By polymerizing the monomer, phase separation would occur as the polymerization proceeds. Since a monomer itself is a solvent, this process avoids the use of another solvent that has to be used for mixing the two polymers and had to be disposed of after the blend had been prepared. With blends of PS and polybutadiene (PB) or PS and polyisoprene, the polymers will phase separate and the rubber domains will usually be dispersed in the PS matrix. Similarly, in the case of rubber-toughened epoxy resins, the rubber forms spherical domains in the epoxy matrix. The manufacture of high impact PS (HIPS) uses this technique and certain specific processing conditions. Starting from a mixture of PB in styrene, peroxide radicals initiate the polymerization of styrene. As the polymerization proceeds, phase separation occurs. The initiator radicals also attack the PB chains and create a graft with PS. Upon completion of polymerization of styrene, the blend consists of PS matrix, with PB and the PB-*g*-PS domains. The final morphology depends on the stirring conditions, as described by Fischer and Hellmann [52]. Typical is a "salami" morphology with PS matrix with spherical domains consisting of PS enveloped by PB, with the PB-*g*-PS at the interface of the domains. Variations of the salami morphology as well as the synthetic procedures are known [53]. Zhang et al. [54] modified the spherical domains to disk-like domains by using a pressure-induced flow processing method.

Similar *reaction-induced* phase separation (RIPS) has been studied in the case of rubber-modified epoxy composites. For example, Yamanaka and Inoue [55] observed that a heterogeneous mixture of epoxy (diglycidyl ether of bisphenol A) and liquid rubber (carboxyl or amine-terminated butadiene-acrylonitrile) exhibited a single phase and upper critical solution temperature (UCST), upon heating to a curing temperature of 100°C. As the curing progressed, the UCST increased and phase separation occurred following spinodal decomposition. The system initially consisted of a co-continuous phase and with further curing, developed into discrete spherical domains of rubber in a matrix of epoxy resin. Kimura et al. [56] used photopolymerization to create phase separated morphologies. Methyl methacrylate monomer was photopolymerized in the presence of a PS derivative. As the molecular weight of the PMMA phase increases, phase separation occurs. During irradiation, the mixture would also shrink, since during the polymerization, the liquid monomer is converted to solid polymer. As the duration of irradiation increased, the phase separated structure changed from a co-continuous to salami to trilayer morphology.

8.7 PIPS for LC Displays

The PIPS has also been used in the design of LC displays using thermal or photo-polymerization methods. In a polymer dispersed LC (PDLC) display, a (nematic) liquid crystalline small molecule is dispersed in a polymer, by a process such as thermal quench. For example, the polymer and the small molecule are mixed together, sandwiched between two substrates, heated to form a homogeneous phase and quenched. The LC molecule would phase separate from the polymer and form the dispersed phase. This is called the "thermally induced phase separation (TIPS)." This method essentially takes advantage of the undesirable occurrence that was described earlier in the case of CTL of photoreceptors or plasticized polymer, in which phase separation of the small molecule occurs upon thermal annealing. Using another approach, a prepolymer and the LC small molecule could be mixed together, and the polymerization could be initiated. As the polymer grows, phase separation of the LC molecule would occur. For example, Kato et al. [57] mixed a commercially available ultraviolet (UV)-curable acrylate prepolymer and cyanobiphenyl derivatives (an LC small molecule) and sandwiched between two substrates with transparent electrodes. They then photopolymerized the acrylate prepolymer using UV light. The phase separated morphology of the TIPS and PIPS composites were found to be significantly different. These authors compared the performance of the two in multistacked polymer dispersed LC films for multipage display application and found the PIPS to be superior.

The same PIPS method was used to design a polymer-dispersed cholesteric LC system for color display applications. A prepolymer was mixed with cholesteric LC material, and photopolymerization was carried out, to obtain a polymer dispersed cholesteric LC device. Kim et al. [58] used a similar approach, but used thermal polymerization of an epoxy, in the presence of a nematic LC mixture. The curing was performed isothermally, at a temperature slightly above the nematic–isotropic transition temperature of the pure LC component, and time dependence of polymerization and phase separation was monitored using light scattering. While the PDLC consists of LC droplets in a polymer matrix, a polymer-stabilized cholesteric LC system is one in which a small amount of polymer is dispersed in a continuous cholesteric LC phase. The polymer is used to stabilize the cholesteric molecules in the on and off conditions [59]. As an example, Yang et al. [60] added monomers (2 wt%) based on acryloyl functionalized biphenyl derivatives along with a photo-initiator to the LC molecules. The monomer was then photopolymerized in the mixture. In another publication from this group [61], hydroxyl functionalized PMMA was used as the polymer component.

A further application of PIPS is the fabrication of holographic polymer-dispersed LCs (H-PDLCs) [62]. This technique again involves photopolymerization of a prepolymer or monomer, using an initiator, in the presence of an LC molecule, but the polymerization is controlled along a grating pattern. A review by Liu and Sun [63] described the advances in the fabrication, characterization, and applications of the H-PDLCs. The H-PDLCs have been applied in several technologies, such as flat panel displays, switchable gratings, and switchable microlens.

8.8 PIPS with Supramolecular Assembly

This is the case in which one of the components in the blend is a supramolecular polymer. It was mentioned in Chapter 4 that the groups of Sijbesma and Meijer [64] used a reactive hydrogen-bonding "synthon" with ureidopyrimidinone (UPy) and OH-telechelic polymers to synthesize supramolecularly chain-extended polymers. Using a similar strategy with UPy functionalized poly-THF and polycarbonate, they demonstrated the PIPS with supramolecular polymerization [65]. Acrylates and diacrylates in which the supramolecular polymer exhibited high solubility were chosen as the medium and as the blend component. The acrylates were then photopolymerized to create the PIPS blend. Various relative compositions of the acrylates and diacrylates with the poly-THF and polycarbonate were used. The mechanical property measurements showed that the supramolecular polymer behaved as a covalently linked high-molecular-weight polymer and not as a low-molecular-weight plasticizer. The morphology of the blend consisted of either macro- or microphase separated domains depending on the acrylate composition. The theory behind this phenomenon, in which there is a competition between self-assembly and phase separation, was presented by Dudowicz et al. [66].

8.9 PIPS for Porous Structures

In the earlier examples of PIPS, one component was dissolved or dispersed in a monomer and the latter was polymerized to obtain a phase separated blend. This method was also used to create porous polymers. Essentially, a template component with a specific structure/morphology is dispersed in a monomer, polymerization is carried out, and the template is extracted by washing or other means. The imprint of the template leaves pores in the polymer. The polymer is cross-linked prior to extraction of the template, to arrest the imprint. This is similar to the construction of nanoreactors in the case of block copolymers as described in Chapter 5. Such a template imprinting technique was used by Zhu et al. [67, 68] to synthesize a cross-linked, porous polymer resin with controlled pore sizes. An illustration of the method is shown in Figure 8.12.

Sodium bis(2-ethylhexyl) sulfosuccinate (known commonly as Aerosol OT or AOT) would form large surfactant aggregates by self-assembly, called "reverse micelles" in organic solvents, in the presence of water [70]. Beyond a water/surfactant molar ratio of 15, microemulsions would be formed, with droplets containing water molecules. Zhu et al. [67, 68] used this surfactant in a mixture of styrene (monomer), divinylbenzene (cross-linker, 20–50 vol%), and a free radical initiator. Upon addition of water (water/AOT molar ratio of 1–18%), reverse micelles formed. Styrene was then bulk-polymerized and

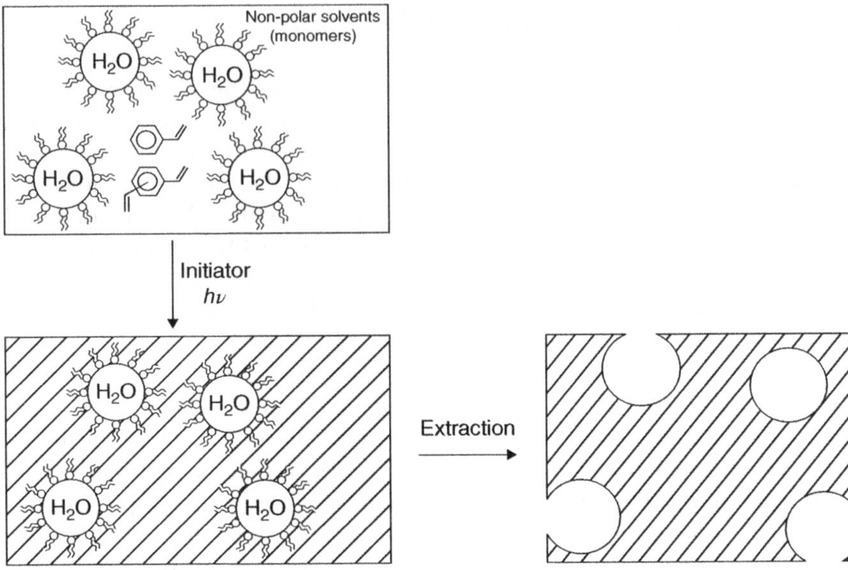

Figure 8.12 Illustration of the method used for making porous cross-linked resins using reverse micellar imprinting. (Source: Bleta et al. [69]. Reproduced with permission of American Chemical Society.)

cross-linked using UV radiation. The template was removed by successive washing with toluene, ethanol, and ether using a Soxhlet extractor. With an increase in the water/AOT molar ratio, the micelle size increased and, hence, the average pore diameter increased from about 10 to 20 nm.

A similar template method was used by Bleta et al. [69] for synthesizing mesoporous alumina with controlled pore size, by taking advantage of the micellar co-assembly of a triblock copolymer and methylated cyclodextrin. We discussed before in Chapter 6 that the triblock copolymer poly(ethylene oxide)–poly(propylene oxide)–poly(ethylene oxide) (PEO–PPO–PEO) forms pseudorotaxanes with cyclodextrins. Valero et al. [71] concluded that methylated β-cyclodextrins do not form pseudorotaxanes with PEO–PPO–PEO. However, Bleta et al. [69] found that the copolymer $PEO_{20}PPO_{70}PEO_{20}$ (Pluronic P123, $M_w = 5800$ g/mol) and a randomly methylated β-cyclodextrin (denoted by RAMEB) co-assembled in aqueous solutions. When varying amounts of RAMEB, up to 130 mg/ml, were added to the aqueous solution (7.8 wt%) of P123, the solutions remained transparent up to a concentration of 80 mg/ml, beyond which these became turbid suggesting the formation of large scattering objects. P123 by itself would form micellar structure in water. With the random substitution of the hydroxyl groups with methoxy group, the RAMEB has both hydrophilic and hydrophobic domains. This would influence

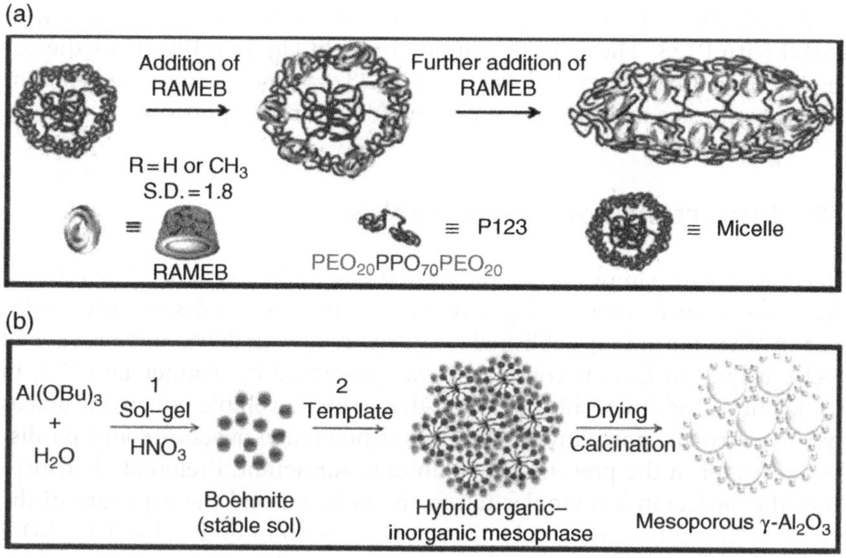

Figure 8.13 (a) Illustration of solubilization of RAMEB within the copolymer micelles. (b) Two-step nanoparticle procedure used for synthesizing mesoporous alumina. (Source: Bleta et al. [69]. Reproduced with permission of American Chemical Society.)

the interaction of RAMEB with the PEO and PPO segments of P123 and affect the micellar size. Dynamic light scattering (DLS) studies showed that for P123 as well as its mixtures with low RAMEB concentration of 5–30 mg/ml, the hydrodynamic radius (R_H) was about 9.1. This would mean that the cyclodextrin molecules are located in the corona of the micelles, interacting with the PEO segments via hydrogen bonds (Figure 8.13a). Upon increasing the concentration of RAMEB to 80 and 100 mg/ml, the R_H increased to 15.5 and 31 nm, respectively, indicating the swelling of the micelles by the cyclodextrin. This was interpreted by Bleta et al. [69] as due to the hydrophobic affinity between the OCH_3 groups of RAMEB and the PPO block of the copolymer, leading to the assembly of the cyclodextrins in the interfacial layer, as shown in Figure 8.13a. The location of the cyclodextrins in the interfacial layer also contributes to the change in the micellar curvature from spherical to ellipsoidal shape. Thus the micellar size can be tuned by adjusting the concentration of RAMEB.

As the first step in preparing mesoporous alumina, Bleta et al. synthesized Boehmite (γ-AlO(OH)) nanoparticles using a sol–gel method. The P123 and RAMEB were added to the sol containing these nanoparticles. Upon stirring and equilibration, the nanoparticles self-assemble around the P123/RAMEB assembly. After drying at 60°C for 48 h, the xerogels were calcinated at 500°C to remove the template and to convert Boehmite to γ-Al_2O_3. The pore size of mesoporous γ-Al_2O_3 without the micellar template was 5.6 nm, and it increased to 14.9 and 19.3 nm with 30 and 80 mg/ml concentrations, respectively, of RAMEB with P123. The steps are summarized in Figure 8.13b. Thus the co-assembly of methylated cyclodextrin and P123 was successfully used as templates to synthesize mesoporous alumina with controlled porosity.

8.10 Surfactant/Polymer Assembly

Surfactants are the simplest examples of self-assembling systems. The interactions between surfactants and polymers have been studied since the 1950s. Tam and Wyn-Jones [72] published a tutorial review in 2006, and a review of polyelectrolyte–surfactant complexes was presented by Thünemann [73] in 2002. Isemura and Imanishi [74] found that water-insoluble polymers such as poly(vinyl formal), poly(vinyl butyral), and poly(vinyl acetate) could be dissolved in water in the presence of an anionic surfactant. Preferential adsorption of the surfactant's nonpolar tail to the polymer and the exposure of the polar head group to water enable dissolution. Sodium dodecyl sulfate (SDS) (Figure 8.14) is the surfactant used in most of the studies. Based on the interaction between poly(vinyl acetate) and SDS, Botré et al. [75] concluded that micellar clusters are formed in the presence of the polymer, and results in a polymer–surfactant complex. The association of such micelles with polymers has since been investigated by a number of authors. Although the initial studies were concerned with nonpolar polymers, the majority of publications used polar, hydrophilic polymers. Since surfactants are used in a number of applications such as cosmetics and detergents, the polymer–surfactant complexation has been used to tailor the rheological and interfacial properties of these formulations. Surfactants, by themselves, form micelles beyond a certain concentration known as critical micelle concentration (*cmc*). When adding a surfactant to a polymer in solution, initially the surfactants remain as unimers, and as the concentration is increased, adhere to the polymer chain using it as a nucleating site (Figure 8.15a). Binding to the polymer and complexation begin (Figure 8.15b), and this is known as critical aggregation concentration (*cac*). These polymer-bound micelles are likened to a pearl necklace as illustrated in Figure 8.15c, as if the polymer chains wrap around the micelles. After the polymer chains become saturated with micelles, free micelles form in the solution.

Perhaps due to the polymer chain acting as a nucleating site, the *cac* is found to be smaller than cmc. It is rationalized that if *cmc* is smaller, the surfactant

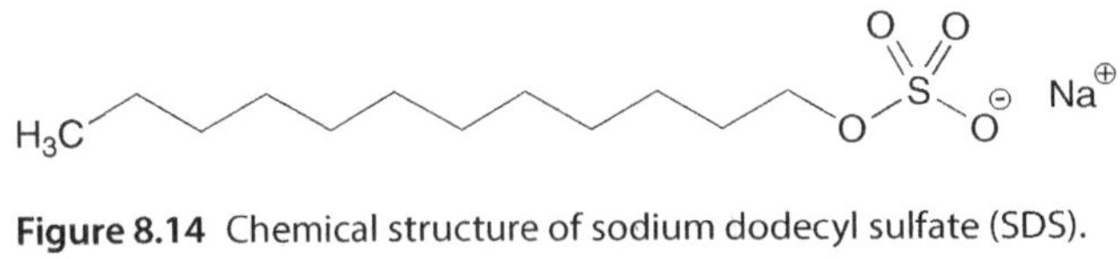

Figure 8.14 Chemical structure of sodium dodecyl sulfate (SDS).

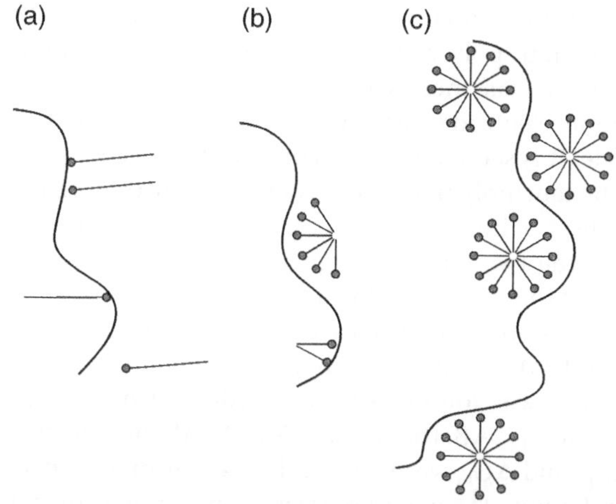

Figure 8.15 (a)–(c) Illustration of the onset of surfactant/polymer binding and micellization complex.

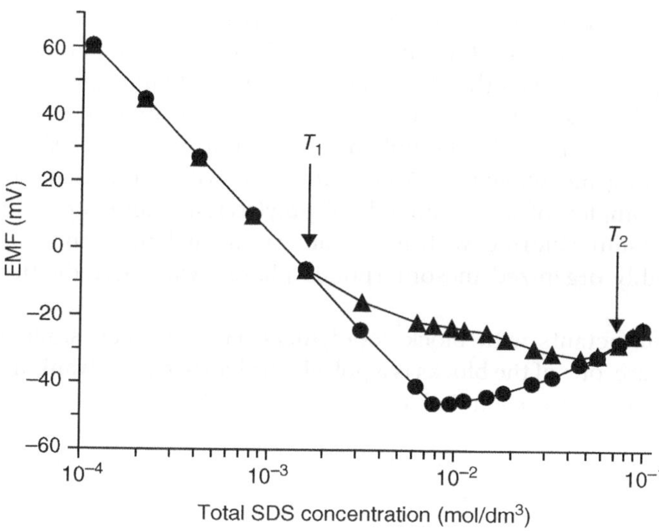

Figure 8.16 Variation of the emf of an SDS electrode (reference silver/Br^- electrode) with total SDS concentration with and without 1 w/v% PVP at 25°C: circle, pure SDS; triangle, SDS + 1 w/v% PVP. (Source: Li et al. [76]. Reproduced with permission of American Chemical Society.)

would prefer to form its own micelles than be associated with the polymer during the micellization. Methods using surfactant selective electrode and isothermal titration calorimetry were used in these studies, in addition to other techniques such as neutron scattering and electron microscopy. Figure 8.16

shows the variation of electromotive force (emf) with the concentration of the surfactant SDS with and without poly(vinyl pyrrolidone) [76]. The monomeric surfactant concentration was measured in this case using SDS-selective surfactant membrane electrode and the emf was measured with respect to a reference electrode. It is seen that as the concentration of SDS was increased, the curve for surfactant/polymer diverges at T_1 from that of the surfactant alone. This is the onset of binding (*cac*) and T_1 is less than the *cmc*. With further increase in SDS, the two curves merge again, as the saturation was reached at T_2 and the rest of the surfactant form free micelles.

Chu and Thomas [77] used fluorescent probes to study the interactions between poly(methacrylic acid) (PMA) and alkyltrimethylammonium bromide (C_nTAB, with $n = 6–16$). The *cac* was found to be 1–2 orders of magnitude lower than *cmc*. For example, for C_{12}TAB, the *cac* was 5×10^{-4} M and cmc was 1.5×10^{-2}. The calculated ΔG_{agg} and ΔG_m were −4.4 and −2.4 kcal/mol, respectively, indicating that the preference for complex formation is much higher than for micellization. The results were similar for C_nTAB with $n = 8–16$. The conformation of PMA in aqueous solutions was pH dependent. At low pH, it is a collapsed hydrophobic coil, and with an increase in pH, the chain becomes stretched out due to complete ionization of the carboxylic groups and repulsion between the anionic centers. Chu and Thomas [77] found that addition of C_nTAB to PMA at pH 8 collapses the chain. Their experiments indicated that PMA-C_{10}TAB form large structures, with about 100 C_{10}TAB molecules per chain (M_w of PMA was 1.1×10^4). The polymer–surfactant complexes show a variety of interesting morphologies. Antonietti et al. [78] found that the films cast from a complex of a sodium salt of poly(styrene sulfonate) and alkyltrimethylammonium chloride (with alkyl chain varying in length from C_{12} to C_{18}) showed highly organized mesomorphous phases, with periodicities of 2.9–4.1 nm.

Aggregation of surfactants with block copolymers has also been studied extensively. In this case, one of the blocks is a polyelectrolyte (e.g., a polyanion), and the other a nonionic, hydrophilic block. These are referred to as "block ionomer complexes" (BICs) by some authors. The cationic surfactant would form complex with the polyanion. Bronich et al. [79] used a block copolymer poly(ethylene oxide)-*b*-poly(sodium methacrylate) (PEO-*b*-PMANa) with cetylpyridinium bromide monohydrate (C_{16}PyBr) and dodecylpyridinium bromide (C_{12}PyBr). Depending on the ratio of concentration of the surfactant and the concentration of the ionic group on the polyanion ($Z = C_s/C_i$), spherical particles ranging in size from 40 to 100 nm were observed (see figures 5 and 6 of Bronich et al. [79]). An increase in the concentration of the surfactant results in progressive neutralization of the charge on the polyanion, and at $Z = 1$, the charges were neutralized completely. These complex particles remained in solution. A model was proposed that these are micellar aggregates, with the polyanion/surfactant cations complexed segments forming the core with a

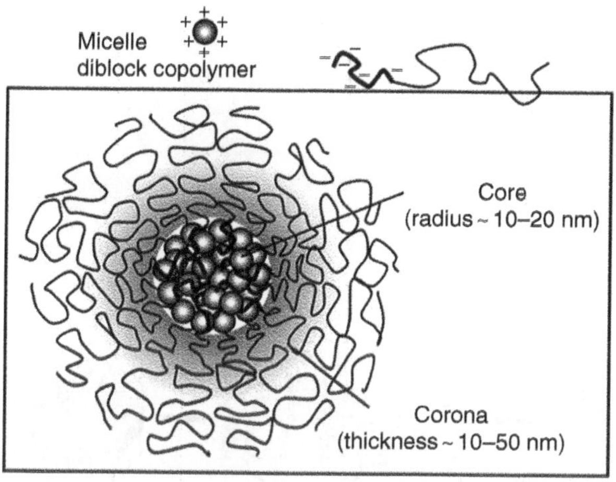

Figure 8.17 A schematic of a colloidal complex formed through the association of oppositely charged block copolymers and surfactants. A complex coacervation microphase of micelles connected by the polyelectrolyte blocks forms the core. The neutral segments form the corona. The radius of the core ranges from 10 to 20 nm and the corona thickness between 10 and 50 nm, depending on the copolymer molecular weight. (Source: Berret et al. [82]. Reproduced with permission of American Chemical Society.)

shell of PEO segments. Solomatin et al. [80] created BICs with various morphologies, using individual or mixtures of single- (HTAB), double- (DDDAB), and triple-tail (DODAB) surfactants with PEO-b-PMANa. Berret et al. [81] studied the complex formation between dodecyl trimethyl ammonium or sodium dodecyl sulfate and poly(sodium acrylate)-*b*-poly(acrylamide) using various scattering techniques and microscopy. They also conclude that the assembly is made up of core–shell particles, with an approximately 15 nm core and approximately 40 nm corona, as shown in Figure 8.17. Although this is similar to the core–shell organization of polymeric micelles of diblock copolymers, based on neutron scattering data, Berret et al. [81, 82] argued that the colloidal complexes of charged-neutral block copolymers and the surfactants bearing the opposite charge are nonequilibrium structures and are not the result of self-assembly mechanism as in the case of micelle-forming hydrophobic–hydrophobic diblock copolymers. This was based on the results that for the same solute concentration and polymer/surfactant ratio, the radius of the core varied from 12.8 to 23.5 nm, depending on the procedure used for solution preparation.

Using a double-hydrophilic block copolymer poly(ethylene glycol)-*b*-poly(sodium glutamate) (PEG113-*b*-PGlu50 and PEG113-*b*-PGlu100) with dodecyl trimethylammonium bromide, Han et al. [83] found that with PEG_{113}-*b*-$PGlu_{50}$ complexation with the surfactant leads to spherical or

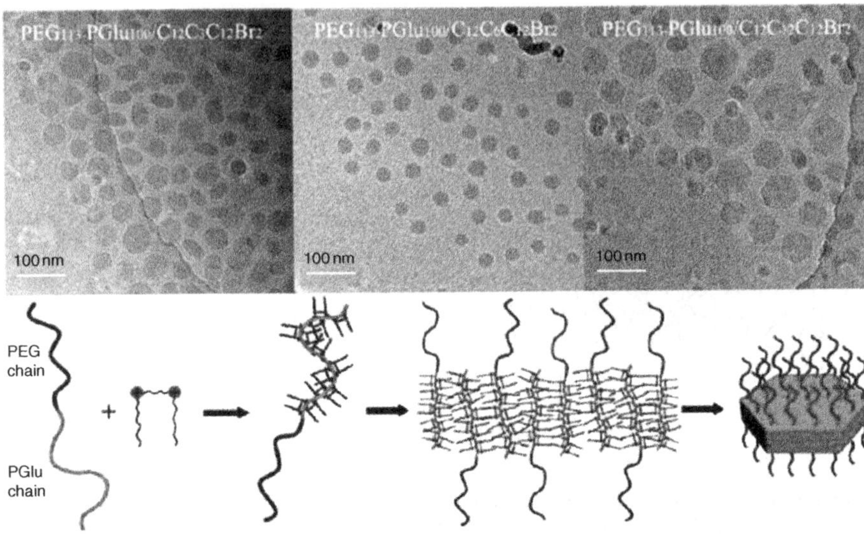

Figure 8.18 CryoTEM images of complexes of PEG_{113}-*b*-$PGlu_{100}$ with Gemini surfactants having $(CH_2)_3$, $(CH_2)_6$, and $(CH_2)_{12}$ spacers. The model for the assembly is also shown. (Source: Han et al. [84]. Reproduced with permission of American Chemical Society.)

rod-like morphology and with PEG_{113}-*b*-$PGlu_{100}$ a vesicular structure. While Solomatin et al. [80] created various complex morphologies with double- and triple-tail surfactants, Han et al. [84] also used cationic Gemini surfactants of the type $[C_{12}H_{25}(CH_3)_2N(CH_2)_xN(CH_3)_2C_{12}H_{25}]Br_2$, with $x = 3$, 6, and 12. The complexation of the Gemini surfactants with the Glu segment of PEG_{113}-*b*-$PGlu_{100}$ forms the hydrophobic part and the PEG, the hydrophilic section. Association of the hydrophobic segments leads to a sandwich type structure, with the PEG segments on the outer side of the sandwich (Figure 8.19). The TEM images in Figure 8.18 show the resulting morphology with $x = 3$, 6, and 12.

In the case of charged polymers, the complexation is with the oppositely charged head of the surfactant. However, the hydrophobic interactions could promote interactions with the apolar segment of polymers. It was observed that when the degree of neutralization of poly(acrylic acid) (PAA) was lower than 0.2, the alkyl segment of SDS was binding to the apolar segment of PAA. Such nonpolar association between the polymer and surfactant, driven by hydrophobic interactions, has also been studied with hydrophobically modified polymers [72].

These are either nonionic water-soluble polymers or polyelectrolytes, to which short hydrophobic side groups or end groups are attached. In the absence of a surfactant, depending on the concentration of these polymers

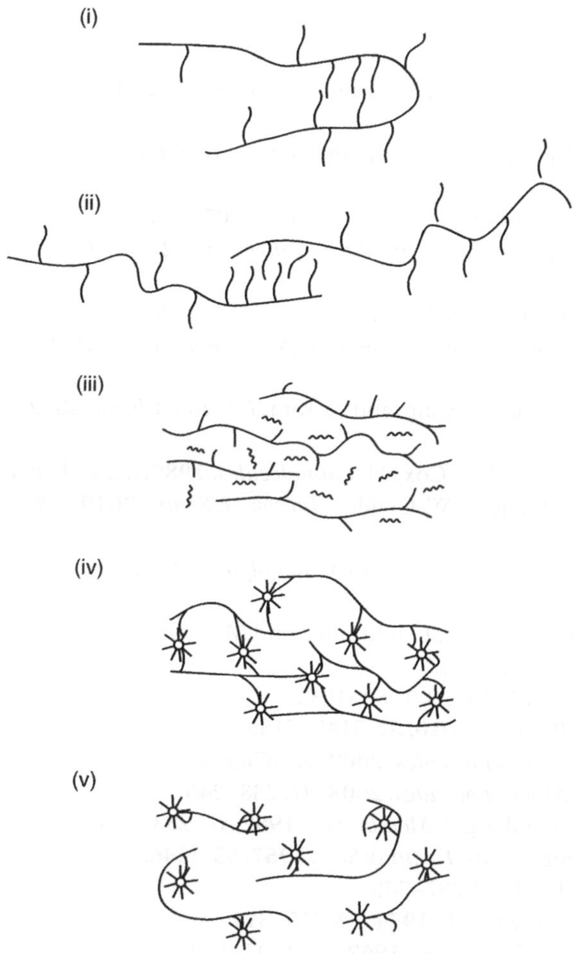

Figure 8.19 (i)–(v) Illustration of the types of structures that occur with the addition of SDS to a hydrophobically modified polymer. (Source: Tam and Wyn-Jones [72]. Reproduced with permission of Royal Society of Chemistry.)

in water, the hydrophobic parts could associate themselves intramolecularly (via, e.g., chain folding) or intermolecularly ((i) and (ii) in Figure 8.19). Such aggregation has been evidenced by shear viscosity measurements. As SDS is added, the alkyl chains of the surfactant associate themselves with the hydrophobic segment of the polymer, and the *cac* is reached. Further addition of SDS promotes intermolecular noncovalent cross-linking (Figure 8.19 (iv)), with the micelles providing the bridge. Beyond this stage, a breakdown of the network occurs. These are schematically shown in Figure 8.19.

References

1 B. Tuteja, F. Khan, and P. Sundararajan, *Macromol. Chem. Phys.*, **2005**, *206*, 354–363.
2 F. Khan, A.-M. Hor, and P. R. Sundararajan, *J. Phys. Chem. B*, **2004**, *108*, 117–126.
3 F. Khan, A.-M. Hor, and P. R. Sundararajan, *Synth. Met.*, **2005**, *150*, 199–211.
4 F. Khan, A.-M. Hor, and P. R. Sundararajan, *Pure Appl. Chem.*, **2004**, *76*, 1509–1520.
5 F. Khan and P. R. Sundararajan, *Org. Electron.*, **2006**, *7*, 410–422.
6 F. Khan, S. Khanna, A.-M. Hor, and P. R. Sundararajan, *Can. J. Chem.*, **2010**, *88*, 247–259.
7 H. Chun, I. K. Moon, D.-H. Shin, S. Song, and N. Kim, *J. Mater. Chem.*, **2002**, *12*, 858–862.
8 R. F. Storey, K. A. Mauritz, and B. D. Cox, *Macromolecules*, **1989**, *22*, 289–294.
9 F. Nilsson, M. S. Hedenqvist, and U. W. Gedde, *Macromol. Symp.*, **2010**, *298*, 108–115.
10 G. Bernardo, R. P. Choudhury, and H. W. Beckham, *Polymer*, **2012**, *53*, 976–983.
11 N. D. Treat, T. E. Mates, C. J. Hawker, E. J. Kramer, and M. L. Chabinyc, *Macromolecules*, **2013**, *46*, 1002–1007.
12 W. W. Brandt, *J. Phys. Chem.*, **1959**, *63*, 1080–1084.
13 J. Guo and T. A. Barbari, *Polymer*, **2010**, *51*, 5145–5150.
14 J. Guo and T. A. Barbari, *Macromolecules*, **2009**, *42*, 5700–5708.
15 J. Guo and T. A. Barbari, *Macromolecules*, **2008**, *41*, 238–245.
16 A. R. Berens and H. B. Hopfenberg, *J. Membr. Sci.*, **1982**, *10*, 283–303.
17 J. B. Wilkens and F. A. Long, *Trans. Faraday Soc.*, **1957**, *53*, 1146.
18 P. Meares, *Eur. Polym. J.*, **1993**, *29*, 237–243.
19 T. G. Fox and P. J. Flory, *J. Polym. Sci.*, **1954**, *14*, 315–319.
20 R. Simha and R. F. Boyer, *J. Chem. Phys.*, **1962**, *37*, 1003–1007.
21 G. M. Geise, C. M. Doherty, A. J. Hill, B. D. Freeman, and D. R. Paul, *J. Membr. Sci.*, **2014**, *453*, 425–434.
22 P. Winberg, M. Eldrup, and F. H. J. Maurer, *J. Chem. Phys.*, **2012**, *136*, 244902/1–244902/6.
23 H. Svajdlenkova, M. Iskrova, O. Sausa, G. Dlubek, J. Kristiak, and J. Bartos, *Macromol. Symp.*, **2011**, *305*, 108–115.
24 V. P. Shantarovich, *J. Polym. Sci. Polym. Phys.*, **2008**, *46*, 2485–2503.
25 H. Higuchi, Z. Yu, A. M. Jamieson, R. Simha, and J. D. McGervey, *J. Polym. Sci. Polym. Phys.*, **1995**, *33*, 2295–2305.
26 R. O. Loutfy, *Pure Appl. Chem.*, **1986**, *58*, 1239–1248.
27 N. Felorzabihi, J. C. Haley, G. R. Bardajee, and M. A. Winnik, *J. Polym. Sci. Polym. Phys.*, **2007**, *45*, 2333–2343.
28 D. N. Theodorou and U. W. Suter, *Macromolecules*, **1985**, *18*, 1467–1478.

29 H. Takeuchi, *J. Chem. Phys.*, **1990**, *93*, 4490–4491.
30 H. Takeuchi, R.-J. Roe, and J. E. Mark, *J. Chem. Phys.*, **1990**, *93*, 9042–9048.
31 M. L. Greenfield and D. N. Theodorou, *Macromolecules*, **1993**, *26*, 5461–5472.
32 A. A. Gray-Weale, R. H. Henchman, R. G. Gilbert, M. L. Greenfield, and D. N. Theodorou, *Macromolecules*, **1997**, *30*, 7296–7306.
33 Z.-W. Wang, P.-L. Wang, and C.-Y. Hu, *Packag. Technol. Sci.*, **2010**, *23*, 457–469.
34 D. J. Plazek and K. L. Nagai, in *Physical Properties of Polymers Handbook*, 2nd Ed., J. E. Mark, Ed., **2007**, Chapter 12, Springer.
35 E. Jenckel and R. Heusch, *Kolloid Z.*, **1953**, *130*, 19.
36 B. Goodbrand, D. Boils, P. R. Sundararajan, and R. Wong, Acoustic printing inks containing bis(carbamates) in vehicle, US Patent 6,414,051, **2002**.
37 S. Khanna, M. Moniruzzaman, and P. R. Sundararajan, *J. Phys. Chem. B*, **2006**, *110*, 15251–15260.
38 M. K. Khan and P. R. Sundararajan, *J. Phys. Chem. B*, **2008**, *112*, 4223–4232.
39 M. R. Islam and P. R. Sundararajan, *Can. J. Chem.*, **2008**, *86*, 600–607.
40 B. Tuteja, M. Moniruzzaman, and P. R. Sundararajan, *Langmuir*, **2007**, *23*, 4709–4711.
41 M. R. Islam and P. R. Sundararajan, *Chem. Eur. J.*, **2011**, *17*, 6098–6108.
42 T. Kojima, R. Harada, T. Nakanishi, K. Kaneko, and S. Fukuzumi, *Chem. Mater.*, **2007**, *19*, 51–58.
43 C. M. Drain, A. Varotto, and I. Radivojevic, *Chem. Rev.*, **2009**, *109*, 1630–1658.
44 M. D. Eddleston and W. Jones, *Cryst. Growth Des.*, **2010**, *10*, 365–370.
45 E. Dahan and P. R. Sundararajan, *Eur. Polym. J.*, **2015**, *65*, 4–14.
46 M. K. Khan and P. Sundararajan, *Chem. Eur. J.*, **2011**, *17*, 1184–1192.
47 F. Ito, Y. Kogasaka, and K. Yamamoto, *J. Phys. Chem. B*, **2013**, *117*, 3675–3681.
48 B. S. Ong, B. Keoshkerian, and D. K. Murti, Acoustic printing inks containing bis(carbamates) in vehicle, US Patent 4,835,081, **1989**.
49 P. R. Sundararajan, D. K. Murti, P. G. Odell, and G. K. Hamer, Photoreceptors including rotaxanes, US Patent 6,096,464, **2000**.
50 S.-H. Chen and W.-J. Chen, *Phys. A*, **1995**, *221*, 216–222.
51 Y. Liu, *J. Appl. Polym. Sci.*, **2013**, *127* (5), 3279–3292.
52 M. Fischer and G. P. Hellmann, *Macromolecules*, **1996**, *29*, 2498–2509.
53 F. Soriano-Corral, G. Morales, P. Acuña, E. Díaz-Barriga, B. Arellano, C. Vargas, and O. De la Paz, *Macromol. Symp.*, **2013**, *325–326*, 177–183.
54 S. Zhang, S. Zhu, X. Feng, K. Han, Q. Huan, J. Song, Y. Ma, and M. Yu, *RSC Adv.*, **2013**, *3*, 6879–6887.
55 K. Yamanaka and T. Inoue, *J. Mater. Sci.*, **1990**, *25*, 241–245.
56 N. Kimura, K. Kawazoe, H. Nakanishi, T. Norisuye, and Q. Tran-Cong-Miyata, *Soft Matter*, **2013**, *9*, 8428–8437.
57 K. Kato, K. Tanaka, S. Tsuru, and S. Sakai, *Jpn. J. Appl. Phys.*, **1993**, *32*, 4594–4599.
58 J. Y. Kim, C. H. Cho, P. Palffy-Muhoray, M. Mustafa, and T. Kyu, *Phys. Rev. Lett.*, **1993**, *71*, 2232–2234.

59 K. Kato, K. Tanaka, S. Tsuru, and S. Sakai, *Jpn. J. Appl. Phys.*, **1993**, *32*, 4600–4604.
60 D. K. Yang, L.-C. Chien, and J. W. Doane, *Appl. Phys. Lett.*, **1992**, *60*, 3102–3104.
61 J. L. West, R. B. Akins, J. Francl, and J. W. Doane, *Appl. Phys. Lett.*, **1993**, *63*, 1471–1473.
62 T. J. Bunning, L. V. Natarajan, V. P. Tondiglia, and R. L. Sutherland, *Annu. Rev. Mater. Sci.*, **2000**, *30*, 83–115.
63 Y. J. Liu and X. W. Sun, *Adv. Optoelectron.*, **2008**, 52p, doi: 10.1155/2008/684349.
64 B. J. B. Folmer, R. P. Sijbesma, R. M. Versteegen, J. A. J. van der Rijt, and E. W. Meijer, *Adv. Mater.*, **2000**, *12*, 874–878.
65 H. M. Keizer, R. P. Sijbesma, J. F. G. A. Jansen, G. Pasternack, and E. W. Meijer, *Macromolecules*, **2003**, *36*, 5602–5606.
66 J. Dudowicz, J. F. Douglas, and K. F. Freed, *J. Phys. Chem. B*, **2009**, *113*, 3920–3931.
67 X. X. Zhu, K. Banana, and R. Yen, *Macromolecules*, **1997**, *30*, 3031–3035.
68 X. X. Zhu, K. Banana, H. Y. Liu, M. Krause, and M. Yang, *Macromolecules*, **1999**, *32*, 277–281.
69 R. Bleta, C. Machut, B. Leger, E. Monflier, and A. Ponchel, *Macromolecules*, **2013**, *46*, 5672–5683.
70 M. P. Pileni, *J. Phys. Chem.*, **1993**, *97*, 6961–6973.
71 M. Valero, I. Grillo, and C. A. Dreiss, *J. Phys. Chem. B*, **2012**, *116*, 1273.
72 K. C. Tam and E. Wyn-Jones, *Chem. Soc. Rev.*, **2006**, *35*, 693–709.
73 A. F. Thünemann, *Progr. Polym. Sci.*, **2002**, *27*, 1473–1572.
74 T. Isemura and A. Imanishi, *J. Polym. Sci.*, **1958**, *33*, 337–352.
75 C. Botré, F. D. Martiis, and M. Solinas, *J. Phys. Chem.*, **1964**, *68*, 3624–3628.
76 Y. Li, R. Xu, D. M. Bloor, J. Penfold, J. F. Holzwarth, and E. Wyn-Jones, *Langmuir*, **2000**, *16*, 8677–8684.
77 D. Y. Chu and J. K. Thomas, *J. Am. Chem. Soc.*, **1986**, *108*, 6270–6276.
78 M. Antonietti, J. Conrad, and A. Thuenemann, *Macromolecules*, **1994**, *27*, 6007–6011.
79 T. K. Bronich, A. V. Kabanov, V. A. Kabanov, K. Yu, and A. Eisenberg, *Macromolecules*, **1997**, *30*, 3519–3525.
80 S. V. Solomatin, T. K. Bronich, A. Eisenberg, V. A. Kabanov, and A. V. Kabanov, *Langmuir*, **2007**, *23*, 2838–2842.
81 J. F. Berret, G. Cristobal, P. Hervé, J. Oberdisse, and I. Grillo, *Eur. Phys. J. E*, **2002**, *9*, 301–311.
82 J. F. Berret, B. Vigolo, R. Eng, P. Herve, I. Grillo, and L. Yang, *Macromolecules*, **2004**, *37*, 4922–4930.
83 Y. Han, L. Xia, L. Zhu, S. Zhang, Z. Li, and Y. Wang, *Langmuir*, **2012**, *28*, 15134–15140.
84 Y. Han, W. Wang, Y. Tang, S. Zhang, Z. Li, and Y. Wang, *Langmuir*, **2013**, *29*, 9316–9323.

Index

Printed and bound by CPI Group (UK) Ltd, Croydon, CR0 4YY
17/09/2026
14964286-0005